Agency Perception and Moral Values Related to Autonomous Weapons

Auffassungen und moralische Wertvorstellungen innerhalb der Streitkräfte und Zivilbevölkerung in Bezug auf autonome Waffensysteme

La perception de l'agentivite et les valeurs morales relatives aux armes autonomes

# International Studies on Military Ethics

THE SERIES IS EDITED UNDER THE AUSPICES OF THE INTERNATIONAL SOCIETY FOR MILITARY ETHICS IN EUROPE (EUROISME)



VOLUME 7

EURO ISME
International Society for Military Ethics in Europe

The titles published in this series are listed at *brill.com/isme*

# Agency Perception and Moral Values Related to Autonomous Weapons

*An Empirical Study Using the Value-Sensitive Design Approach*

# Auffassungen und moralische Wertvorstellungen innerhalb der Streitkräfte und Zivilbevölkerung in Bezug auf autonome Waffensysteme

*Eine empirische Untersuchung anhand der Value-Sensitive-Design-Methode*

# La perception de l'agentivite et les valeurs morales relatives aux armes autonomes

*Etude empirique utilisant l'approche de la conception ethique*

*By*

Ilse Verdiesen

TUTA SUB AEGIDE PALLAS 1683

BRILL
NIJHOFF

LEIDEN | BOSTON

Parts of this thesis are based on the following articles and EuroISME acknowledges the kind permission of the publishers:
Verdiesen, I., Dignum, V., & Rahwan, I. (2018, September). Design Requirements for a Moral Machine for Autonomous Weapons. In International Conference on Computer Safety, Reliability, and Security (pp. 494-506). Springer, Cham. 
https://link.springer.com/chapter/10.1007/978-3-319-99229-7_44

Verdiesen, I., de Sio, F. S., & Dignum, V. (2019). Moral Values Related to Autonomous Weapon Systems: An Empirical Survey that Reveals Common Ground for the Ethical Debate. IEEE Technology and Society Magazine, 38(4), 34-44. © IEEE 2019 https://ieeexplore.ieee.org/abstract/document/8924586

German translation of the text by Gerta K. Badde-Valentine, French translation by Jean-François Palard.

The Library of Congress Cataloging-in-Publication Data is available online at https://catalog.loc.gov
LC record available at https://lccn.loc.gov/2021017462

Typeface for the Latin, Greek, and Cyrillic scripts: "Brill". See and download: brill.com/brill-typeface.

ISSN 2214-7926
ISBN 978-90-04-44907-7 (paperback)
ISBN 978-90-04-44908-4 (e-book)



This book is printed on acid-free paper and produced in a sustainable manner.

The jury of EuroISME's annual prize for the best student's thesis on military ethics consists of

1. Colonel Rev. Prof.Dr. P.J. McCormack, MBE, (chairman, United Kingdom)
2. Lt.Col.(Ret) Dr. Daniel Beaudoin (France/Israel)
3. Dr. Veronika Bock (Germany)
4. MMag. Stefan Gugerel (Austria)
5. Col. (Ret) Prof.Dr. Boris Kashnikov (Russia)
6. Dr. Asta Maskaliūnaitė (Estonia)
7. Prof.Dr. Desiree Verweij (The Netherlands)
8. Mrs. Ivana Gošić (Serbia, Secretary)

Enquiries: secretariat.ethicsprize@euroisme.eu
www.euroisme.eu

The prize is financially supported by

vfonds invests in PEACE

www.vfonds.nl

# Contents/Inhalt/Table des matières

## *Auffassungen und moralische Wertvorstellungen innerhalb der Streitkräfte und Zivilbevölkerung in Bezug auf autonome Waffensysteme*

## *La perception de l'agentivite et les valeurs morales relatives aux armes autonomes*

# Preface

My interest in the Ethics of Autonomous Weapons was sparked after I attended the IDEA Summer School on Responsible Artificial Intelligence (AI) in August 2016 organised by dr. Virginia Dignum as part of the European Conference of Artificial Intelligence (ECAI). After the first meeting with Virginia on my graduation project, when she asked if I wanted to study abroad and where I would like to go, I boldly said MIT, but I never imagined that this would actually happen. Virginia's connection with dr. Iyad Rahwan of the Scalable Cooperation Group at the Media Lab of MIT made it a reality. I would like to thank Virginia for making this connection and her supervision during my graduation project. Due to the time difference, we held most of our skype sessions in the evening in The Netherlands and I highly appreciate that she always found the time to review my writings.

Working on my graduation project at the Scalable Cooperation Group was an enormous opportunity and the Media Lab is a very open, diverse, nonhierarchical and inspiring environment. At first, I had to get used running in to a box with balls ('ballenbak') in the hallway or people leaving the office to play Ping-Pong in the middle of the day, but after a month or so the lab felt like home. I worked with very talented and ambitious graduate students and post doc's. I especially would like to thank dr. Sydney Levine and dr. Nick Obradovich for all their insightful comments and I really enjoyed the sharp discussions we had on my research set-up. But most of all I would like to thank dr. Iyad Rahwan for allowing me to be part of his group and let me work on the sensitive topic of Autonomous Weapons. He gave me all the resources, time and help that I needed to conduct my research which I greatly appreciate.

Lastly, I would like to thank the Royal Netherlands Army (RNLA) for allowing me to study abroad and supporting me by posting me at MIT. Without their support, I would not have been able to live and study in Boston for five months. I hope that this thesis on the Ethics of Autonomous Weapons will lead to awareness and initiates a discussion on the deployment of Autonomous Weapons and Responsible AI in the Defence organisation.

*Ilse Verdiesen*

# List of Figures and Tables / Abbildungs- und Tabellenverzeichnis / Liste des figures et des tableaux

## Figures and Tables

### *Figures*

## *Tables*

## Abbildungs- und Tabellenverzeichnis

### *Abbildungen*

## *Tabellen*

## Liste des figures et des tableaux

### *Figures*

*Tableaux*

*Agency Perception and Moral Values Related to Autonomous Weapons*

# Summary

Autonomous Weapons are increasingly deployed on the battlefield (Roff, 2016). Autonomous systems can have many benefits in the military domain, for example when the autopilot of the F-16 prevents a crash (NOS, 2016) or the use of robots by the Explosive Ordnance Disposal to dismantle bombs (Carpenter, 2016). Yet the nature of the Autonomous Weapons might also lead to uncontrollable activities and societal unrest. The deployment of Autonomous Weapons on the battlefield without direct human oversight is not only a military revolution according to Kaag and Kaufman (2009), but can also be considered a moral one. As large-scale deployment of AI on the battlefield seems unavoidable (Rosenberg & Markoff, 2016), the research on ethical and moral responsibility is imperative.

In the debate on Autonomous Weapons strong views and opinions are voiced. The Campaign to Stop Killer Robots (2017) states for example on their website that: '*Allowing life or death decisions to be made by machines crosses a fundamental moral line. Autonomous robots would lack human judgment and the ability to understand context*'. We found little empirical research that supports these views or that provide insight in how Autonomous Weapons are perceived by the general public and the military. We also found no empirical research on moral values that underlie the '*fundamental moral line*' of Autonomous Weapons. Therefore, the knowledge gap is twofold in that insight is lacking on 1) how Autonomous Weapons are perceived by the military and general public and 2) which moral values people consider important when Autonomous Weapons are deployed in the near future.

The first part of the knowledge gap can be filled by studying the perception of Autonomous Weapons using the agency theory described in the fields of Cognitive Psychology, Artificial Intelligence and Moral Philosophy. The second part of the knowledge gap can be filled by studying known value theories (Beauchamp & Walters, 1999; Friedman & Kahn Jr, 2003; Schwartz, 2012) to see which values people deem important in the deployment of Autonomous Weapons. Based on the identified knowledge gap and problem statement we drafted the following research question for this study:

> *How are Autonomous Weapons perceived by the general public and military personnel working at the Dutch Ministry of Defence and which moral values do they consider important in the deployment Autonomous Weapons?*

In this study, we apply the Value-Sensitive Design (VSD) method as research approach. The VSD is a three-partite approach that allows for considering human values throughout the design process of technology. It is an iterative process for the *conceptual, empirical* and *technological* investigation of human values implicated by the design (Davis & Nathan, 2015; Friedman & Kahn Jr, 2003). The conceptual investigation consists of two parts: (1) identifying the *direct* stakeholders, those who will use the technology, and the *indirect* stakeholders, those whose lives are influenced by the technology, and (2) identifying and defining the values that the use of the technology implicates. The empirical investigation looks into the understanding and experience of the stakeholders in a context relating to the technology and implicated values will be examined. In the technical investigation, the specific features of the technology are analysed (Davis & Nathan, 2015).

We slightly deviate from the original VSD method, because we do not conduct a full stakeholder analysis to identify the stakeholders in step 1, but we focus on two obvious stakeholder groups; the *general public* and *military personnel*, because these two stakeholder groups were available as respondents for our study. In step 2 of the first phase, we conduct a literature review to identify the values related to Autonomous Weapons. The main focus of our research is on the empirical investigation phase, because we found that the empirical aspect is overlooked in the ethical debate on Autonomous Weapons. In the technical investigation phase, we do not design an Autonomous Weapon as one intuitively might expect, because this would be an immense project well beyond the scope of this study. Yet we chose to build on the work of the Scalable Cooperation research group and propose a design for a Moral Machine for Autonomous Weapons which can be used as next step in the research on the ethics of Autonomous Weapons.

The scientific relevance of our study is that we contribute to the academic literature by gaining insight in perception of the general public and the military regarding Autonomous Weapons, and by identifying the moral values the general public and the military relate to Autonomous Weapons. Insight in this is currently lacking and no empirical data on the perception and values related to Autonomous Weapons could be found. By using the Value-Sensitive Design as research approach we show that this method is applicable to structure academic research which could be viewed as case-study for the VSD approach. We also extend the research on the ethical decision-making of Autonomous Vehicles by Bonnefon, Shariff, and Rahwan (2016) to the domain of Autonomous Weapons.

The societal relevance is that understanding the perception of the general public and military personnel working at the Dutch MOD of Autonomous Weapons, and identifying which moral values they relate to Autonomous Weapons can be used to find common grounds and differences in the debate on this technology initiated by Campaign to Stop Killer Robots (2015) and International Committee for Robot Arms Control (ICRAC). Secondly, the results of this study show how the Value-Sensitive Design method can be applied to Autonomous Weapons to identify the values the military and general public relate to the deployment of these type of weapons. Finally, by identifying the values are important to incorporate in the design of Autonomous Weapons, the study contributes to a responsible design and deployment of Autonomous Weapons in the future.

The focus of our study is on the empirical investigation phase of the VSD. This phase is split in two parts. The first part is explorative to identify the values in the context of Autonomous Weapons by means of an online survey which is supplemented by interviewing six experts. The second part of the empirical investigation studies the agency perception by means of an exploratory and a confirmatory research method. We operationalize the agency construct to confirm our hypothesis on the agency perception by conducting randomized controlled experiments and use the results of the experiment to explore the values related to Autonomous Weapons in a descriptive manner.

To study the agency perception of Autonomous Weapons we drafted a hypothesis for which we reason that military personnel will view Autonomous Weapons as any other weapon, and therefore no more than a tool to achieve an effect. We hypothesize in the final study that military personnel will perceive Autonomous Weapons as not possessing mental states. We drafted three scenarios in our experiment describing 1) a Human Operated drone, 2) an Autonomous Weapon with no agency characteristics and 3) an Autonomous Weapon with agency characteristics. We expected no difference between the neutral agency Autonomous Weapon condition and the condition in which a human is operating a drone remotely. We also expected the neutral agency condition to be judged as significantly different from the high agency condition, in which we specifically tell participants that the Autonomous Weapon has agentic characteristics, such as the ability to plan and set its own goals. Based on this reasoning, we hypothesize that:

> *H1: military personnel will not perceive Autonomous Weapons as possessing mental states.*

The results of the pilot and final studies show that the agency items are reliable and hold as one construct. The agency construct consists of the four items *Thought*, *Goal setting*, *Free will* and *Achieve goals* which can be used to measure the agency perception of Autonomous Weapons and Human Operated drones.

Our central analysis was concerned with how the neutral agency scenario of the Autonomous Weapon differs from the Human Operated and high agency scenarios of Autonomous Weapons. Our results indicate that the agency perception of military personnel and civilians working at the Dutch MOD for the neutral agency Autonomous Weapons scenario is higher than the agency perception of the Human Operated drone scenario. This means that they attribute more agency to an Autonomous Weapon than to a Human Operated drone. Based on these findings we must reject our hypothesis.

The effect of the agency perception on the dependent variables is explored in a descriptive manner and is not analysed by means of regression analysis. We found that 7 out of 9 dependent variables are significantly correlated to the agency construct. These are the variables *trust, human dignity, confidence, expectations, support, fairness* and *anxiety*. Our findings are:

- Military personnel and civilians working at the Dutch MOD have more *trust, confidence* and *support* in the actions taken by Human Operated drones than those taken by Autonomous Weapons;
- A drone operated by a human being is perceived as having more respect for *human dignity* than a neutral or high agency Autonomous Weapon even though the actions and outcome of the scenarios are the same;
- Military personnel and civilians working at the Dutch MOD have an equal level of expectations regarding the actions in the future of the Human Operated drone and neutral and high agency Autonomous Weapons. They also consider the actions of Human Operated drones and Autonomous Weapons to be equally fair;
- Autonomous Weapons cause more anxiety amongst military personnel and civilians working at the Dutch MOD than Human Operated weapons.

This study was conducted on a small data set and to be able to generalize the results, the study needs more respondents that represent a larger demographic group. Therefore, we propose a design for a Moral Machine for Autonomous Weapons as a Massive Online Experiment (MOE) for a large-scale study of this topic. In our design we build on the concept of the Moral Machine, that was developed by the Scalable Cooperation group of the Media Lab at MIT (Scalable Cooperation Group, 2016). Due to the sensitivity of the topic, we propose a two-step implementation, first design a controlled experiment with a limited set of conditions. These conditions can be visualized in scenarios that allow users to take the survey after obtaining a password via a web-interface to a secure server. After gathering initial data and user feedback, the next step could be to scale up to a large-scale open platform, like the Moral Machine, where people can judge the scenarios to collect large amounts of data of different demographic groups which could be used for more robust and generalisable results.

Several issues can be identified as limitations of this study. First, the operationalisation of the items and the agency construct were derived from a categorisation of literature describing agency characteristics and our selection of the characteristics was based on a numerical count and not driven by any relevance or weighing criteria. The second limitation is the selection of values from the value questionnaire and expert interviews as dependent variables. Although this selection was heavily discussed amongst the three researchers involved in this study, the final choice was made on heuristics and not on an objective method. Thirdly, the samples used in this study are limited both in size as in demographics as the final study only had 239 respondents which is only a small portion of the Dutch MOD. Lastly, the distribution of the respondents of the final scenario is skewed and the second scenario (neutral agency Autonomous Weapons) has 32 respondents more than scenario one (Human Operated drones). One of the explanations for this skewedness could be that people expected to take a survey on Autonomous Weapons, but dropped out when they were presented the Human Operated scenario. This is called selective attrition which has a negative impact on the internal validity of the study. Another explanation could be that the software was faulty in distributing the respondents over the scenarios which would mean that the internal validity of the study is not infringed as the skewedness is not attributed to selective attrition.

Given these limitations, several recommendations for further research are suggested. The first is to validate the agency construct which requires more studies measuring the agency perception of other technological artefacts to test if this construct also holds in other domains. Examples of these studies could be to study the agency perception of care robots for elderly, AI toys for children or onboard computers of Autonomous Vehicles. The second recommendation is to run the final study with the same scenarios on a representative sample consisting solely of civilians in The Netherlands. This would allow us to see on which values the results of the military sample and the answers of civilians differ and although we cannot make direct comparisons, due to the fact that the military sample is not representative, we would gain insight in the perception of both military personnel and civilians. Lastly, we recommend implementing the Moral Machine for Autonomous Weapons, as described in section 5, to generalize the results, for which the study needs much more respondents that represent a larger demographic group. Scaling up to a Massive Online Experiment, like the Moral Machine, would generate large amounts of data of different demographic groups which could be used for more robust and generalizable results in order to get a thorough understanding of the moral judgement of people regarding Autonomous Weapons.

CHAPTER 1

# Introduction

> I think I have a difficult articulating how I feel about drones. On the one hand, any instrument of death is upsetting to me. On the other hand, it is comforting that so many lives will not be risked in the use of them.
>
> RESPONDENT PILOT STUDY 2

Autonomous Weapons are weapon systems equipped with Artificial Intelligence (AI). Artificial Intelligence is described by Neapolitan and Jiang (2012, p. 8) as '*an intelligent entity that reasons in a changing, complex environment*', but this definition also applies to natural intelligence. Russell, Norvig, and Intelligence (1995) provide an overview of many definitions combining views on *systems that think and act like humans* and *systems that think and act rational*, but they do not present a clear definition of their own. For now, we adhere to the description Bryson, Kime, and Zürich (2011) provide. They state that a machine (or system) shows intelligent behaviour if it can select an action based on an observation in its environment. In scientific literature, AI is described as more than an Intelligent System alone. It is characterized by the concepts of *Adaptability*, *Interactivity* and *Autonomy* (Floridi & Sanders, 2004). According to Floridi and Sanders (2004), *Adaptability* means that the system can change based on its interaction and can learn from its experience. Machine learning techniques are an example of this. *Interactivity* occurs when the system and its environment act upon each other and *Autonomy* implies that the system itself can change its state.

A growing body of researchers is focusing on responsible design of AI, which incorporates social and ethical values, to prevent undesirable societal outcomes of this technology. Principles to describe Responsible AI are *Accountability*, *Responsibility* and *Transparency* (ART). *Accountability* refers to the justification of the actions taken by the AI, *Responsibility* allows for the capability to take blame for these actions and *Transparency* is concerned with describing and reproducing the decisions the AI makes and adepts to its environment (V. Dignum, 2016).

 | DOI:10.1163/9789004449084_002

Artificial Intelligence is not just a futuristic science-fiction scenario in which human-like robots, like Data's brother Lore in Star Trek or the Cylons in Battlestar Galactica, are planning to take over the world. Many AI applications are already being used today. Smart meters, search engines, personal assistance on mobile phones, autopilots and self-driving cars are examples of this. This thesis focusses on the application of AI in the military domain and specifically on the deployment of Autonomous Weapons in the near future. In this introduction, we first describe the research problem followed by the research approach, the scientific and societal relevance and the embedding in the curriculum of the Information Architecture (IA) track of the System Engineering Policy and Management (SEPAM) master. We will conclude with a brief outline of the next sections of this report.

## 1.1 Research Problem

We first discuss the increase of Autonomous Weapons in the military domain and related work on the ethics of AI followed by studies on Human Operated drones in the subsection on problem exploration. Next, we identify the knowledge gap, problem statement and scope before we conclude with the research question and sub questions of this study.

### 1.1.1 *Problem Exploration*

Autonomous Weapons are increasingly deployed on the battlefield (Roff, 2016). It is already reported that China has autonomous cars which carry an armed robot (Lin & Singer, 2014), Russia claims it is working on autonomous tanks (W. Stewart, 2015), the US christened their first 'self-driving' warship in May 2016 (P. Stewart, 2016) and the Russian arms manufacturer Kalashnikov recently disclosed that they developed a fully automated combat module that uses neural networks (RT, 2017). Autonomous systems can have many benefits in the military domain, for example when the autopilot of the F-16 prevents a crash (NOS, 2016) or the use of robots by the Explosive Ordnance Disposal to dismantle bombs (Carpenter, 2016). Yet the nature of the Autonomous Weapons might also lead to uncontrollable activities and societal unrest. Examples of this unrest are the 'Stop Killer Robots Campaign' of 61 NGO's directed by Human Rights Watch (Campaign to Stop Killer Robots, 2015), but also the United Nations are voicing their concerns and state that '*Autonomous weapons systems that require no meaningful human control should be prohibited, and remotely controlled force should only ever be used with the greatest caution*' (General Assembly United Nations, 2016). The deployment of Autonomous

Weapons on the battlefield without direct human oversight is not only a military revolution according to Kaag and Kaufman (2009), but can also be considered a moral one. As large-scale deployment of AI on the battlefield seems unavoidable (Rosenberg & Markoff, 2016), the research on ethical and moral responsibility is imperative.

Ethical decision-making in AI and robots is an emerging field and several scholars are studying moral judgement related to these technologies. For example, Malle (2015) proposes a framework combining the (up to now) separate fields of *robot ethics*, in which ethical questions about the design, deployment and treatment of robots by humans are addressed, and *machine morality,* which is concerned with questions about the moral capacities of a robot and how these should be computationally implemented. Cointe, Bonnet, and Boissier (2016) propose a model in which an agent can judge the ethical aspects of his own behaviour and that of other agents in a multi-agent system. The model describes an Ethical Judgement Process (EJP) which allows agents to evaluate the behaviour of other agents. Bonnefon et al. (2016) have studied the ethical decision an Autonomous Vehicle has to make, being self-protection or utilitarian, when confronted with pedestrians on the road. In this research, the Moral Machine[1] at MIT is used to gain insight in how people judge on scenarios with an Autonomous Vehicle to see how their moral judgement compares to those of other people.

In a domain related to Autonomous Weapons, that of Human Operated drone operations, ethical concerns have been studied quite intensively over the past ten years in both philosophical and psychological literature. From a philosophical point-of-view, Coeckelbergh (2013) argues that drone operations not only create a physical distance, but also a moral distance as the face of the opponent becomes less visible which eliminates the moral-psychological barrier for killing. Another ethical concern according to Strawser (2010) is that due to the regular work shifts outside the combat zone, the drone operators experience an unjust psychological burden, because they must switch between work and home situations on a daily basis. Also, due to the remote distance to the battlefield human operators can experience cognitive dissonance in which the war feels more like a video game than reality. These ethical concerns are refuted by Strawser (2010) as he states that due to the increased distance the human operators have more time to evaluate a target, because their own safety is not at risk, but he also argues that more empirical research is needed

1 http://moralmachine.mit.edu/

to assess the psychological effects of conducting drone operations at a large distance to the battlefield.

Several empirical studies on the effect of drone operations on human operators have been conducted in the field of psychology. One of the first studies on the psychological effects of drone operations was done by Thompson et al. (2006) who found that drone operators suffered from increased fatigue, emotional exhaustion and burnout. This was confirmed by Chappelle, Goodman, Reardon, and Thompson (2014) who reported that human drone operators display symptoms of burn-out, are emotionally drained, show high levels of cynicism and Post-Traumatic Stress Disorder (PTSD). Other psychological factors found in empirical studies are boredom and distraction stemming from the long duration of drone operations (Cummings, Mastracchio, Thornburg, & Mkrtchyan, 2013). These studies show that executing drone operations can have severe psychological effects on the humans operating them.

### 1.1.2 *Knowledge Gap*

In the debate on Autonomous Weapons strong views and opinions are voiced. The Campaign to Stop Killer Robots (2017) states for example on their website that: '*Allowing life or death decisions to be made by machines crosses a fundamental moral line. Autonomous robots would lack human judgment and the ability to understand context*'.. We found little empirical research that supports these views or that provide insight in how Autonomous Weapons are perceived by the general public and the military. The Open Robots Ethics initiative surveyed the public opinion in a poll in 2015 (Open Roboethics initiative, 2015) and issued a report. However, the results were not published in an academic journal and the survey was not extensive enough to draw substantive conclusions. As described in the previous subsection, several scholars like Malle (2015), Cointe et al. (2016) and Bonnefon et al. (2016) are studying ethical decision-making in AI and robots. Ethical concerns are also studied in the related field of Human Operated drone operations (Coeckelbergh, 2013; Strawser, 2010) and empirical studies found that human operators suffer from severe psychological effects (Chappelle et al., 2014; Cummings et al., 2013; Thompson et al., 2006), but this research is not yet extended to the deployment of Autonomous Weapons. We also found no literature or empirical studies on moral values that are related to Autonomous Weapons or on what people consider to be the '*fundamental moral line*'. Therefore, the knowledge gap is twofold in that insight is lacking on 1) how Autonomous Weapons are perceived by the military and general public and 2) which moral values the military and general public consider important when Autonomous Weapons are deployed in the near future.

### 1.1.3 *Problem Statement*

The first part of the knowledge gap can be filled by studying the perception of Autonomous Weapons using the agency theory described in the fields of Cognitive Psychology, Artificial Intelligence and Moral Philosophy. Agency is the capacity to plan and act (Waytz, Gray, Epley, & Wegner, 2010) and is studied both in human as non-human subjects. People attribute agency to objects, such computers (Nass, Moon, Fogg, Reeves, & Dryer, 1995) and robots (H. M. Gray, Gray, & Wegner, 2007), therefore we expect that AI technology will also be perceived as having agentic characteristics. Studying the agency perception of Autonomous Weapons of military personnel and the general public would provide insight in possible differences and similarities between the viewpoints of these groups which can be used in the current debate on this technology.

The second part of the knowledge gap can be filled by studying known value theories to see which values people deem important in the deployment of Autonomous Weapons. Well-established value theories are those of Schwartz (1994), Friedman and Kahn Jr (2003) and Beauchamp and Walters (1999), but insight in how these relate to Autonomous Weapons is lacking. Deriving the values that are most relevant in the context of the deployment of Autonomous Weapons, and comparing these values to those related to the current technology, the Human Operated drones, will lead to insight into the underlying motives in the debate on Autonomous Weapons and to greater understanding of the views that are expressed.

This leads to the following problem statement for this study:

> In the debate on Autonomous Weapons strong views and opinions are voiced, but empirical research that support these opinions is lacking. Insight in how Autonomous Weapons are perceived and which moral values are related to the deployment of Autonomous Weapons in the near future is missing. It is also not clear if the perception and moral values related to Autonomous Weapons of military personnel differs of that of the general public.

### 1.1.4 *Scope*

Much of the literature on Autonomous Weapons is written and debated by legal experts and philosophers in the context of International Humanitarian Law and the Geneva Conventions which are aimed to limit the effects of armed conflicts (ICRC, 2010). As we are no legal experts nor philosophers, this study will stay within the boundaries and rules of the Laws of War and we will not question these. Due to time constraints, the study will only be focussed on military personnel of the Dutch MOD and will not be extended to a representative Dutch civilian sample. In this study, we will focus on the deployment of Autonomous

Weapons in the near future, which we define as: *within the next 5 years*. This entails that we will not study weapons equipped with Artificial General Intelligence or futuristic technology that is not possible to construct yet, but we will focus on technology that is currently being developed, specifically drones with autonomous targeting capabilities. The targeting process consists of six steps: (1) find, (2) fix, (3) track, (4) target, (5) engage, and (6) assess. We will focus on the decision-making in step 4 and 5 of the targeting process, because many ethical decisions have to be made in this part of the process (Asaro, 2016).

### 1.1.5. *Research Question and Sub Questions*

Based on the identified knowledge gap, problem statement and scope of the research we draft the following research question for this study:

> *How are Autonomous Weapons perceived by the general public and military personnel working at the Dutch Ministry of Defence and which moral values do they consider important in the deployment Autonomous Weapons?*

This research question will be answered by the following sub questions:

1. *How are Autonomous Weapons defined in literature?*
2. *Which value theories are described in literature?*
3. *Which of the values described in the value theories relate to Autonomous Weapons?*
4. *How is agency perception defined literature?*

These four sub questions will be answered in the literature review of our research.

5. *Which moral values relate the general public and military personnel working at the Dutch Ministry of Defence to the deployment of Autonomous Weapons?*

This sub question will be answered by a value survey that consists of an online questionnaire and expert interviews.

6. *How is the agency of Autonomous Weapons perceived by the general public and military personnel working at the Dutch Ministry of Defence?*
7. *How are the values related to Autonomous Weapons perceived by the general public and military personnel working at the Dutch Ministry of Defence?*

These two sub questions will be answered testing a hypothesis and descriptive analysis based on the results of a randomized controlled experiment.

8. *How can the values that are related to Autonomous Weapons be incorporated into the design of a Moral Machine of Autonomous Weapons?*

This sub question can be answered by creating a design and implementation plan for a Moral Machine for Autonomous Weapons.

In the next section, the research approach for answering these sub questions is described more in depth.

## 1.2 Research Approach

In this study, we apply the Value-Sensitive Design (VSD) method as research approach. The VSD is a three-partite approach that allows for considering human values throughout the design process of technology. It is an iterative process for the *conceptual, empirical* and *technological* investigation of human values implicated by the design (Davis & Nathan, 2015; Friedman & Kahn Jr, 2003). The conceptual investigation consists of two parts: (1) identifying the direct stakeholders, those who will use the technology, and the indirect stakeholders, those whose lives are influenced by the technology, and (2) identifying and defining the values that the use of the technology implicates. The empirical investigation looks into the understanding and experience of the stakeholders in a context relating to the technology and implicated values will be examined. In the technical investigation, the specific features of the technology are analysed (Davis & Nathan, 2015). The VSD can be used as a roadmap for engineers and students to incorporate ethical considerations into the design (Cummings, 2006).

There has also been some critique voiced regarding the VSD approach. One of the concerns Davis and Nathan (2015) mention is that the VSD posits that certain values are universal, but that these may differ based on culture and context. A response to counter this would be to take an empirical basis for one's viewpoint instead of a philosophical one, or acknowledge that the researchers position is not the only valid to be considered (Borning & Muller, 2012). Borning and Muller (2012) pose a pluralistic position in that the VSD should not recommend either a universal or a relative view on values, but it should leave engineers free to decide which view is most appropriate in context of their design.

In line with Borning and Muller (2012) we used the VSD approach in our research as guidance and not as a goal in itself. In the conceptual phase, we slightly deviate from the original VSD method, because we do not conduct a full stakeholder analysis to identify the stakeholders in step 1, but focus on two obvious stakeholder groups; the *general public* and *military personnel*, because these two stakeholder groups were available as respondents for our study. In step 2 of the first phase, we conduct a literature review to identify the values related to Autonomous Weapons. The main focus of our research is on the empirical investigation phase, because we found that the empirical aspect is overlooked in the ethical debate on Autonomous Weapons. We addressed the critique of Davis and Nathan (2015) by allocating most of our time to this phase of the VSD in which we study the context of Autonomous Weapons using various research techniques. In the technical investigation phase, we do not design an Autonomous Weapon as one intuitively might expect, because this would be an immense project well beyond the scope of this study. Yet we chose to build on the work of the Scalable Cooperation research group and propose a design for a Moral Machine for Autonomous Weapons which can be used as next step in the research on the ethics of Autonomous Weapons. We apply the phases of the VSD to our research as follows (Figure 1):

- *Conceptual investigation:* The lack of research on Autonomous Weapons implies that the first phase of our study is explorative by nature. The research activities in the conceptual investigation will consist of the qualitative research technique 'desk research' to review the literature on Autonomous Weapons, moral values and agency perception.
- *Empirical investigation:* This phase is split in two parts. The first part is explorative to identify the values in the context of Autonomous Weapons by means of an online survey which is a quantitative research technique supplemented by a qualitative research technique of interviewing experts. In the second part of the empirical investigation, we use both an exploratory as a confirmatory research method to study the agency perception. We operationalize the agency construct to confirm our hypothesis on the agency perception by conducting randomized controlled experiments and use the results of the experiment to explore the values related to Autonomous Weapons in a descriptive manner.
- *Technical investigation:* In this phase, the design and features for the Moral Machine of Autonomous Weapons is created based on the scenarios that are used in the randomized controlled experiment of the empirical phase. The technical investigation phase is exploratory and the design is created using 'desk research' as research technique.

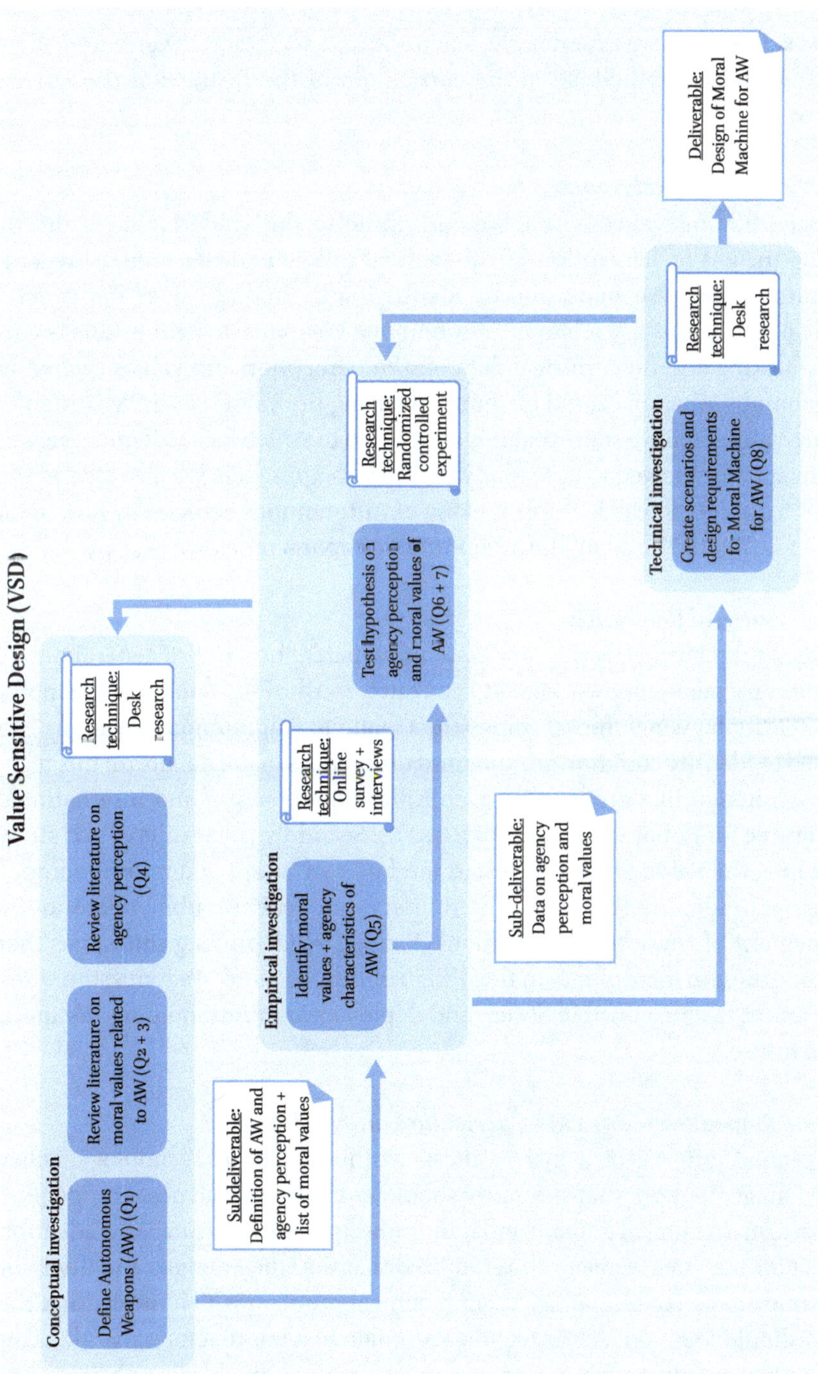


FIGURE 1 The Value-Sensitive Design research approach

## 1.3 Relevance

In this subsection, we describe the scientific and societal relevance and show how the study is embedded in the curriculum of the IA track of the SEPAM master.

### 1.3.1 *Scientific Relevance*

The scientific relevance is that we contribute to the academic literature by gaining insight in perception of the general public and the military regarding Autonomous Weapons, and by identifying the moral values the general public and the military relate to Autonomous Weapons. Insight in this is currently lacking and no empirical data on the perception and values related to Autonomous Weapons could be found. By using the Value-Sensitive Design as research approach we show that it is applicable to structure academic research which could be viewed as case-study for the VSD approach. We also extend the research on the ethical decision-making of Autonomous Vehicles by Bonnefon et al. (2016) to the domain of Autonomous Weapons.

### 1.3.2 *Societal Relevance*

The societal relevance is that understanding perception of the general public and military personnel working at the Dutch MOD of Autonomous Weapons, and identifying which moral values they relate to Autonomous Weapons can be used to identify common grounds and differences in the debate on this technology initiated by Campaign to Stop Killer Robots (2015) and International Committee for Robot Arms Control (ICRAC). Secondly, the results of this study show how the Value-Sensitive Design method can be applied to Autonomous Weapons to identify the values the military and general public relate to the deployment of these type of weapons. Finally, by identifying the values that are important to incorporate in the design of Autonomous Weapons, the study contributes to a responsible design and deployment of Autonomous Weapons in the future.

### 1.3.3 *Embedding in SEPAM Curriculum and IA Track*

The general criteria for a master thesis at the faculty of Technology, Policy and Management are that the study should contain an analytical component, focusses on a technical domain and is multidisciplinary in nature. As graduation project for the SEPAM master, it is required that the study designs a solution for a complex large contemporary socio-technical problem which means that the thesis should focus on a clear technical domain in a multi-actor network, takes both public as private values into account and does not address only technical issues, but also managerial and ethical choices (Graduation Portal, 2017). The Information Architecture (IA) track integrates management and computer science aspects and focusses on the alignment of organizational needs and engineering opportunities of state-of-the-art ICT solutions (IA program, 2017).

This study fits the SEPAM criteria as it focusses specifically on the AI-technology of Autonomous Weapons in the military domain. It is multi-disciplinary as it combines literature on agency perceptions from the fields of Cognitive Psychology, Artificial Intelligence and Moral Philosophy, value theories described in Philosophical and Psychological literature, and the deployment of weapons in the military domain. The socio-technical problem central in this thesis is the lack of insight in the perception of Autonomous Weapon technology and the related values that the general public and military deem important. The study fits the IA track as it identifies the common grounds and differences in ethical preferences from both the general public as the military, and proposes an engineering opportunity in designing a Moral Machine for Autonomous Weapons to further investigate these preferences by means of a Massive Online Experiment building on the Moral Machine for Autonomous Vehicles.

## 1.4 Structure

The remainder of this report is structured as follows; In section 2, the literature on Autonomous Weapons, value theories and agency perception is reviewed. Section 3 describes the methodology, hypotheses, research design, the operationalisation of the scenarios and constructs, analytical approach, pre-registration, the sample and concludes with methodological issues. In section 4, first the results of the value survey, consisting of the online questionnaire and expert interviews, are listed followed by the results of the three randomized controlled experiments. The design of the Moral Machine for Autonomous Weapons is described in section 5 by first presenting the original Moral Machine for Autonomous Vehicles, followed by the features and scenarios to include in a Moral Machine of Autonomous Weapons and a short implementation plan. Section 6 concludes on the results, discusses the scientific and societal implications and identifies the limitations and recommendation for further research. In the final section 7, we reflect on the choices we made in our research, the process of the project and the link between the research, IA track and SEPAM curriculum.

CHAPTER 2

# Literature Review

> Just that I don't feel comfortable with placing weapons on a machine with no emotions or control by a human.
>
> RESPONDENT PILOT STUDY 1

This section provides an overview of literature on Autonomous Weapons, followed by a summary of value theories and lastly, we describe the concept of agency which we use to study the perception of Autonomous Weapons. Each sub-section concludes with a discussion of the theories we apply in this study and the reasons for selecting it.

## 2.1 Autonomous Weapons

Although the debate on Autonomous Weapons has drawn a lot of attention in the recent years, we found that the topic was not well delineated in the academic literature. We start this subsection with an overview of the many different definitions and present two classifications of Autonomous Weapons to conclude this section.

### 2.1.1 *Definition*

Autonomous Weapons are an emerging technology and there is still no internationally agreed upon definition (AIV & CAVV, 2016). Even consensus if Autonomous Weapons should be defined at all is lacking. Although some scholars provide definitions in their writings (Table 1), others caution against such a specification. NATO states that: '*Attempting to create definitions for "autonomous systems" should be avoided, because by definition, machines cannot be autonomous in a literal sense*.' (Kuptel & Williams, 2014, p. 10). The United Nations Institute for Disarmament Research (UNDIR) is also cautious about providing a definition of Autonomous Weapons, because they argue that the level of autonomy depends on the '*critical functions of concern and the interactions of different variables*' (UNIDIR, 2014, p. 5). They state that one of the

 | DOI:10.1163/9789004449084_003

reasons for the differentiation of terms regarding Autonomous Weapons is that sometimes things (drones or robots) are defined, but in other times a characteristic (autonomy), variables of concern (lethality or degree of human control) or usage (targeting or defensive measures) are drawn into the discussion and become part of the definition.

The various definitions of Autonomous Weapons are listed in Table 1. Some authors use the term military robots which have a certain level of autonomy. As military robots can be viewed as a subclass of Autonomous Weapons according to the classification of Royakkers and Orbons (2015) (Figure 2) we included them in the list of definitions. In our opinion the definition in the report of the Advisory Council on International Affairs (AIV & CAVV) captures the description of Autonomous Weapons best from an engineering and military standpoint, because it takes predefined criteria into account and is linked to the military targeting process as the weapon will only be deployed after a human decision. Therefore, we will follow this definition and define Autonomous Weapons as:

> A weapon that, without human intervention, selects and engages targets matching certain predefined criteria, following a human decision to deploy the weapon on the understanding that an attack, once launched, cannot be stopped by human intervention.
>
> AIV AND CAVV (2016, p. 11)

### 2.1.2 *Classification of Autonomous Weapons*

Not only are Autonomous Weapons ambiguously defined, they also have not been uniformly classified. We present two of the classifications in this subsection. Royakkers and Orbons (2015) describe several types of Autonomous Weapons (Figure 2) distinct between (1) *Non-Lethal Weapons* which are weapons '*... without causing (innocent) casualties or serious and permanent harm to people*'. (Royakkers & Orbons, 2015, p. 617), such as the Active Denial System which uses a beam of electromagnetic energy to keep people at a certain distance from an object or troops, and (2) *Military Robots* which they define '*... as reusable unmanned systems for military purposes with any level of autonomy*'. (Royakkers & Orbons, 2015, p. 625). Military robots are subdivided in three categories; vehicles that are ground based, for example for unmanned reconnaissance and clearing road bombs, vehicles that can navigate unmanned on or below the water surface, such as a gun-station on a ship or an autonomous submarine, and vehicles that are unmanned combat aerial vehicles (UCAV's). These UCAV's are classified by Royakkers and Orbons (2015) as tele-operated,

TABLE 1 Overview definitions of autonomous weapons

| Author (s) | Definition |
|---|---|
| AIV and CAVV (2016, p. 11) | *'A weapon that, without human intervention, selects and engages targets matching certain predefined criteria, following a human decision to deploy the weapon on the understanding that an attack, once launched, cannot be stopped by human intervention'.* |
| Altmann, Asaro, Sharkey, and Sparrow (2013, p. 73) | Autonomous Weapons are: '... *robot weapons that once launched will select and engage targets without further human intervention*'. |
| Galliott (2015, p. 5) | Military robots are: *'a group of powered electro-mechanical systems, all of which have in common that they:*<br>*1. Do not have an onboard human operator;*<br>*2. Are designed to be recoverable (even though they may not be used in a way that renders them such); and,*<br>*3. In a military context, are able to exert their power in order to deliver a lethal or nonlethal payload or otherwise perform a function in support of a military force's objectives'.* |
| Horowitz (2016, p. 27) | *'a weapon system that, once activated, is intended to only engage individual targets or specific target groups that have been selected by a human operator'.* |
| Royakkers and Orbons (2015, p. 625) | Military Robots are '... *reusable unmanned systems for military purposes with any level of autonomy*'. |
| Kuptel and Williams (2014, p. 10) | *'Machines are only "autonomous" with respect to certain functions such as navigation, sensor optimization, or fuel management'.* |
| UNDIR (2014, p. 5) | The level of Autonomy depends on the '*critical functions of concern and the interactions of different variables*' |

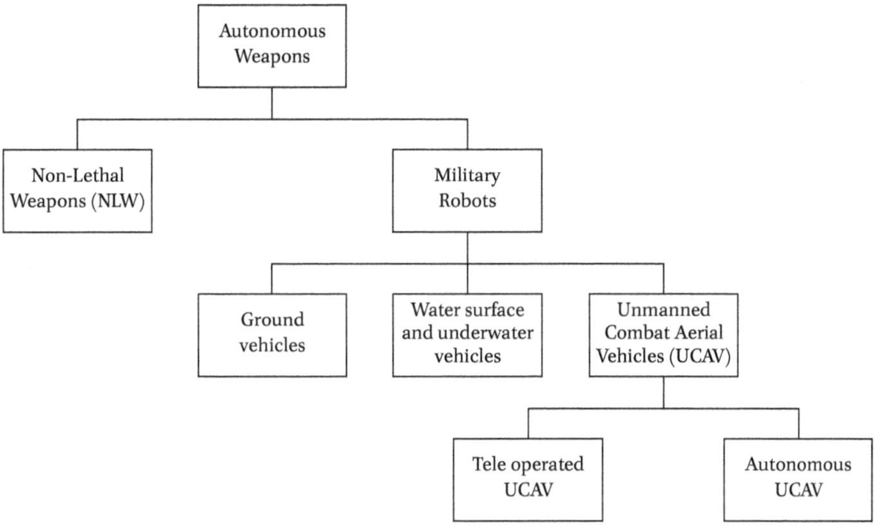


FIGURE 2 Classification of Autonomous Weapons based on Royakkers and Orbons (2015)

of which 'drones' are the most well-known example, and autonomous UCAV's, which are gradually developed by the US Department of Defence (Rosenberg & Markoff, 2016).

Gailliot (2015) provides another type of classification of Autonomous Weapons based on four levels of autonomy for unmanned systems:

1. *Autonomy level 1 – Non-autonomous/ teleoperated:* 'A human operator controls each and every powered movement of the unmanned platform. Without the operator, teleoperated systems are incapable of effective operation'.
2. *Autonomy level 2 – Supervisory Autonomy:* 'A human operator specifies movements, positions or basic actions and the system then goes about performing these. The operator must provide the system with frequent input and diligent supervision in order to ensure correct operation'.
3. *Autonomy level 3 – Task Autonomy:* 'A human operator specifies a general task and the platform processes a course of action and carries it out under its own supervision. The operator typically has the mean to oversee the system, but this is not necessary for the operation'.
4. *Autonomy level 4 – Full Autonomy:* 'A system with full autonomy would create and complete its own tasks without the need for any human input, with the exception of the decision to build such a system. The human is so far removed from the loop that the level of direct influence is negligible. These systems might display capacities that imitate or replicate

> the moral capacities of sentient human beings (though no stand on this matter shall be taken here)' (Galliott, 2015, p. 7).

This classification is in our opinion a good attempt in classifying the level of autonomy of Autonomous Weapons, but we have some reservations from an engineering point of view. Galliot (2015) himself states that it would be possible to merge the second and third level of autonomy, because both are a semi-autonomous operational level. We agree with his statement, but this is not the main issue we have with these definitions. We believe that it is odd to start list of autonomy levels with a category of non-autonomous systems. More importantly, in the fourth level of autonomy the author states that: *'these systems might display capacities that imitate or replicate the moral capacities of sentient human beings'*. It seems he refers to the definition of strong AI, in that a computer has cognitive states and programs can explain human cognition (Searle, 1980). To state that an autonomous system possesses moral capacities shows in our opinion a lack of technical knowledge on current AI systems as these are not more than computers that display Interactivity, Autonomy and Adaptability features (Floridi & Sanders, 2004).

As it remains to be seen if strong AI capable of 'moral capacities of sentient human beings' will ever be developed, we believe that the classification Galliot (2015) provides is not realistic with the current state of technology and therefore it will not be used in this research. For our study, we adhere to the classification of Royakkers and Orbons (2015) which displays good insight in the current and (near) future military technology. Especially their distinction between teleoperated and autonomous Unmanned Combat Aerial Vehicles is realistic and is suitable for our research.

## 2.2 Values

Contrary to the topic of Autonomous Weapons, the concept of values has been studied extensively in the fields of Moral Philosophy and Psychology. This section presents a definition of values, followed an overview of theories that describe universal values, an overview of the values related to Autonomous Weapons and concludes with a value hierarchy as an example to bridge the conceptual and empirical investigation phase of our research.

### 2.2.1 *Definition*

Value Theories are well-studied in the fields of Moral Philosophy and Psychology. Moral Philosophy has a long and rich history in examining values and in this field theoretical questions are asked to investigate the nature of value and goodness (Schroeder, 2016). Often a distinction is made between *instrumental* values, which means there is reason to favour it for its effect that can lead to good things (Rønnow-Rasmussen, 2002), and *intrinsic* values, which '*… is a kind of value such that when it is possessed by something, it is possessed by it solely in virtue of its intrinsic properties*'. (Bradley, 2006, p. 112). Although Moral Philosophy is mainly concerned with theories of what '*ought to be*' and is in a strict sense unaffected by empirical results (Alfano & Loeb, 2014), it has one branch that is concerned with Applied Ethics which is relevant for our study, because Applied Ethics bridges the abstract ethical theories and moral practice. As stated in section 1.2, the focus of this study is on the empirical phase of the Value-Sensitive Design to investigate how the moral values of the general public and military relate to Autonomous Weapons. Therefore, we chose not to use the theoretical Value Theories of Moral Philosophy in our study, but turned to the fields of Psychology and Applied Ethics to get an empirical view on the personal values of the two selected stakeholder groups in order to get insight into the '*is*' situation instead of what '*ought to be*'.

The field of Psychology differentiates values from attitudes, needs, norms and behaviour in that they are a belief, lead to behaviour that guides people and are ordered in a hierarchy that shows the importance of the value over other values (Schwartz, 1994). Values are used by people to justify their behaviours and define which type of behaviours are socially acceptable (Schwartz, 2012). They are distinct from facts in that values do not only describe an empirical statement of the external world, but also adhere to the interests of humans in a cultural context (Friedman et al., 2013). Values can be used to motivate and explain individual decision-making and for investigation of human and social dynamics (Cheng & Fleischmann, 2010).

Many definitions of values exist. For example, Schwartz (1994, p. 21) describes values as: *'desirable trans situational goals, varying in importance, that serve as guiding principles in the life of a person or other social entity'*. This is quite a specific description compared to Friedman, Kahn Jr, Borning, and Huldtgren (2013, p. 57) who define values as: '*… what a person or group of people consider important in life*'. The existing definitions have been summarized by Cheng and Fleischmann (2010, p. 2) in their meta-inventory of values and they state that: …'*values serve as guiding principles of what people consider important in life*'. Although a quite simple description, we think it captures the description of a value best, because it combines several definitions in one using the main characteristics of values. Therefore, we will adhere to the definition of Cheng and Fleischmann (2010) in our study.

### 2.2.2 *Universal Values*

Research suggests that people across cultures identify with basic values which can be considered as universal human values (Friedman et al., 2013; Graham et al., 2012; Schwartz, 2012). Although individuals differ in attribution of importance of the values, there seems to be a surprisingly high consensus across cultures on the hierarchical order of the values (Schwartz, 2012). As part of their research some researchers created so called value inventories, which are lists of items that can be used to categorise the analysis of human values and are often accompanied by a descriptive tool for discussions on these values (Cheng & Fleischmann, 2010). The most common and well-studied value inventories are those of Schwartz (1994), Friedman *et al.* (2013), Beauchamp and Walters (1999) and Graham et al. (2012). The number of universal values found by researchers varies greatly. An overview of these value inventories is displayed in Table 2 and the theories will be briefly described in the next paragraph.

Based on extensive empirical research, Schwartz (1994) mentions 10 distinct motivational types of values that are subdivided in a more fine-grained list of 56 value items which he uses to survey the 10 overarching universal values. In their description of the Value-Sensitive Design approach, Friedman et al. (2013) mention 12 values of which the first 9 are based on consequentialists and deontological moral orientations and the last 3 are chosen from the field of Human Computer Interaction (HCI) field. Graham *et al.* (2012) use the term 'foundation' to describe the 5 distinct values that specify the universality of human moral nature that Haidt and Joseph (2004) use as basis of the Moral Foundation Theory. Gouveia, Milfont, and Guerra (2014) drafted a framework based on many value theories, such as Schwartz (1994) and Maslow (1943) hierarchy of needs. In the framework, the authors place the value on two dimensions; (1) with actions that drive human behaviour which can be personal, central or social goals, and (2) motivators that represent human needs which can split into thriving and survival needs. (Gouveia *et al.*, 2014).

Values are not only described in theory from a psychological perspective as outlined in the previous paragraph, but have also been practically implemented and used by means of Applied Ethics to professional domains. For example in the medical field, which uses BioEthics to describe the values that are important as guiding principles for biomedical professionals, such as physicians, nurses and health workers. Beauchamp and Walters (1999) describe 4 values as basis for the framework of BioEthics: 1) *Autonomy*: acting intentionally without controlling influences that would mitigate against a voluntary act, 2) *Beneficence*: providing benefits for society as a whole, 3) *Justice*: being fair and reasonable and 4) *Non-maleficence*: not intentionally imposing risk or harm upon another.

TABLE 2 Overview value theories

| Author | Key contribution | Definition of value | Values |
|---|---|---|---|
| Schwartz (1994) | The study looks at potential universality of human values and specifies a set of dynamic relations amongst these values. The study did not find universal aspects of values, but found support for near universality of four higher order value types. Also, the study found considerable evidence that the ten value types are recognized by many people in contemporary societies. | Values are defined as: *"desirable trans situational goals, varying in importance, that serve as guiding principles in the life of a person or other social entity."* (p. 21)<br>Five features make up the conceptual definition of human values: "(*1*) *belief* (*2*) *pertaining to desirable end states or modes of conduct, that* (*3*) *transcends specific situations,* (*4*) *guides selection or evaluation of behavior, people, and events, and* (*5*) *is ordered by importance relative to other values to form a system of value priorities* (*Schwartz,1992; Schwartz & Bilsky, 1987, 1990*)" (p. 20) | 1. *Power*: Social status and prestige, control or dominance over people and resources (authority, wealth, social power)[a].<br>2. *Achievement*: Personal success through demonstrating competence according to social standards (ambitious, successful, capable, influential).<br>3 *Hedonism*: Pleasure and sensuous gratification for oneself (pleasure, enjoying life, self-indulgent).<br>4. *Stimulation*: Excitement, novelty, and challenge in life (a varied life, an exciting life, daring).<br>5. *Self-direction*: Independent thought and action – choosing, creating, exploring (creativity, freedom, choosing own goals, curious, independent).<br>6. *Universalism*: Understanding, appreciation, tolerance, and protection for the welfare of *all* people and for nature (broadminded, social justice, equality, world at peace, world of beauty, unity with nature, wisdom, protecting the environment).<br>7. *Benevolence*: Preservation and enhancement of the welfare of people with whom one is in frequent personal contact (helpful, honest, forgiving, responsible, loyal, true friendship, mature love). |

TABLE 2 Overview value theories (*cont.*)

| Author | Key contribution | Definition of value | Values |
|---|---|---|---|
| | | | 8. *Tradition*: Respect, commitment, and acceptance of the customs and ideas that traditional culture or religion provide (respect for tradition, humble, devout, accepting my portion in life).<br>9. *Conformity*: Restraint of actions, inclinations, and impulses likely to upset *or* harm others and violate social expectations *or* norms (obedient, self-discipline, politeness, honouring parents and elders).<br>10. *Security*: Safety, harmony, and stability of society, of relationships, and of self (social order, family security, national security, clean, reciprocation of favours). |
| Friedman, Kahn Jr, Borning, and Huldtgren (2013) | An overview of the VSD approach and pointers for a practical application.<br>Providing information so that other researchers can use and extend the VSD and practitioners will consider values in designing information and computer systems. | A value is defined in a broad sense in that it: *'refers to what a person or group of people consider important in life'.* (p. 57)<br>The VSD method especially regards moral values which are: *'… issues that pertain to fairness, justice, human welfare and virtue, encompassing within moral philosophical theory deontology, consequentialism, and virtue'* (p. 72) | 1. *Human welfare* Refers to people's physical, material, and psychological well-being.<br>2. *Ownership and property* Refers to a right to possess an object (or information), use it, manage it, derive income from it, and bequeath it.<br>3. *Privacy* Refers to a claim, an entitlement, or a right of an individual to determine what information about himself or herself can be communicated to others.<br>4. *Freedom from bias* Refers to systematic unfairness perpetrated on individuals or groups, including pre-existing social bias, technical bias, and emergent social bias.<br>5. *Universal usability* Refers to making all people successful users of information technology. |

6. *Trust* Refers to expectations that exist between people who can experience good will, extend good will toward others, feel vulnerable, and experience betrayal.
7. *Autonomy* Refers to people's ability to decide, plan, and act in ways that they believe will help them to achieve their goals.
8. *Informed consent* Refers to garnering people's agreement, encompassing criteria of disclosure and comprehension (for "informed") and voluntariness, competence, and agreement (for "consent").
9. *Accountability* Refers to the properties that ensures that the actions of a person, people, or institution may be traced uniquely to the person, people, or institution.
10. *Courtesy* Refers to treating people with politeness and consideration.
11. *Identity* Refers to people's understanding of who they are over time, embracing both continuity and discontinuity over time.
12. *Calmness* Refers to a peaceful and composed psychological state.
13. *Environmental Sustainability* Refers to sustaining ecosystems such that they meet the needs of the present without compromising future generations.

TABLE 2 Overview value theories (*cont.*)

| Author | Key contribution | Definition of value | Values |
|---|---|---|---|
| Graham et al. (2012) | A description (including critiques and empirical result) of the Moral Foundation Theory (MFT).<br>The MFT can be used to get insight into the moral judgements of people. The MFT is described as a pluralist, nativist, cultural-developmentalist and intuitionist approach of morality. | To represent the five concepts of the MFT the term 'foundation' is chosen, but this is interchangeably used with the terms value or virtue. No exact definition of foundation is given, but it is used as an architectural metaphor to state that the: *'MFT is a theory about the universal first draft of the moral mind, and about how that draft gets revised in variable ways across cultures'.* (p. 10) | The five foundations of the MFT are:<br>1. *Care/harm foundation*: is related to the ability to feel pain of *others* and underlies virtues of kindness, gentleness, and nurturance;<br>2. *Fairness/cheating foundation*: is related to process of reciprocal altruism and generates ideas of justice, rights, and autonomy;<br>3. *Loyalty/betrayal foundation*: is related to form shifting coalitions and underlies virtues of patriotism and self-sacrifice for the group;<br>4. *Authority/subversion foundation*: is related to hierarchical social interactions and underlies virtues of leadership and followership, including deference to legitimate authority and respect for traditions;<br>5. *Sanctity/degradation foundation:* is related to the psychology of disgust and contamination and underlies religious notions of striving to live in an elevated, less carnal, more noble way. |
| Cheng and Fleischmann (2010) | Meta-analysis of 12 value inventories of human values.<br>This study proposes a meta-inventory of human values. | Provides summation of definitions of values: *"values serve as guiding principles of what people consider important in life.". (p. 2)* | (1) freedom, (2) helpfulness, (3) accomplishment, (4) honesty, (5) self-respect, (6) intelligence, (7) broad-mindedness, (8) creativity, (9) equality, (10) responsibility, (11) social order, (12) wealth, (13) competence, (14) justice, (15) security, and (16) spirituality. |

| | | | |
|---|---|---|---|
| Gouveia, Milfont, and Guerra (2014) | Empirical study basic values.<br>Paper proposes a three-by-two framework containing six subcategories of basic values. | Two primary functions of values are identified: (1) they guide actions and (2) they are cognitive expressions of needs. | *Personal goals – Thriving needs values:* Emotion, Pleasure, Sexuality<br>*Personal goals – Survival needs values:* Power, Prestige, Success<br>*Central goals – Thriving needs values:* Beauty, Knowledge, Maturity<br>*Central goals – Survival needs values:* Health, Stability, Survival<br>*Social goals – Thriving needs values:* Affectivity, Belonging, Support<br>*Social goals – Survival needs values:* Obedience, Religiosity, Tradition |
| Beauchamp and Walters (1999) | The article is a first chapter of a book on bioethics. This chapter describes three moral principles that provide a framework which can be used to reason about issues in bioethics. | The authors use terms principles and values as synonyms. They define a principle as: *'A principle is a fundamental standard of conduct from which many other moral standards and judgments draw support for their defense and standing'.* (*p. 17*) | 1. *Autonomy:* acting intentionally without controlling influences that would mitigate against a voluntary act.<br>2. *Beneficence:* providing benefits for society as a whole.<br>3. *Justice:* being fair and reasonable.<br>4. *Non-maleficence:* not intentionally imposing risk or harm upon another. |

[a] The words in brackets are the specific value items that specify the universal value.

Based on our literature review, we selected two value theories for our study; one derived from the Psychological literature and the other based on Applied Ethics which is a practical application of Moral Philosophy. The first theory we selected is that of Cheng and Fleischmann (2010), because in their meta-inventory of human values they created a comprehensive list of 16 human values that is based on the values found in 12 separate studies. In our opinion, this meta-analysis captures the most important values listed by other researchers and it is an empirical example derived from the psychological literature. The second Value Theory we selected is an example of Applied Ethics that has been extensively practiced in the medical domain for over forty years. We would like to investigate its applicability to Autonomous Weapons, because the BioEthics principles address many concerns that people might have regarding Autonomous Weapons.

### 2.2.3 *Values Related to Autonomous Weapons*

Values as described in the value theories in section 2.2.2 are not often explicitly mentioned in the literature on Autonomous Weapons as the overview in Table 3 shows, but most studies discuss different values or related ethical issues. Two public reports of Human Rights Watch mention the lack of *human emotion, accountability, responsibility, lack of human dignity* and *harm* as values related to Autonomous Weapons (Docherty, 2012, 2015). Sharkey and Suchman (2013) state that the values of *accountability* and *responsibility* are important to consider in the design of Robotic Systems for military operations.

In the field of Military Ethics, Johnson and Axinn (2013) list *responsibility, reduction of human harm, human dignity, honour* and *human sacrifice* as values in their discussion on if the decision to take a human life should be handed over to a machine or not. Cummings (2006) in her case study of the Tactical Tomahawk missile, looks at the universal values proposed by Friedman and Kahn Jr (2003) and states that next to *accountability* and *informed consent*, the value of *human welfare* is fundamental core value for engineers when developing weapons as it relates to the *health, safety* and *welfare* of the public. She also mentions that the legal principles of *proportionality* and *discrimination* are important to consider in the context of weapon design. *Proportionality* refers to the fact that an attack is only justified when the damage is not considered to be excessive. *Discrimination* means that a distinction between combatants and non-combatants is possible (Hurka, 2005). Asaro (2012) also refers to the principles of *proportionality* and *discrimination* and states that Autonomous Weapons open-up a moral space in which new norms are needed. Although he does not explicitly mention values in his argument, he does refer to the value of *human life* and the need for humans to be involved in the decision of

TABLE 3 Values related to weapons

| Author | Key contribution | Definition of value | Values |
|---|---|---|---|
| Cummings (2006) | Application of VSD approach to the design problem of the Tactical Tomahawk missile.<br>Study shows the consideration of the ethical issues in the design process for both instructors and practitioners. | N/ a | From the list of Friedman et al. (2006), the values that apply to the design of weapon systems are *accountability, informed consent*, but most of all *human welfare*. The principles of discrimination and proportionality are important for considering human welfare. |
| Docherty (2012) | Report of Human Rights Watch in which aspects of international humanitarian law and ethical issues for Autonomous Weapons are described. | No definition of the term 'values', but the text mentions values and ethical issues. | Values/ ethical issues:<br>– Lack of human emotions;<br>– Accountability;<br>– Responsibility. |
| Johnson and Axinn (2013) | Paper on ethical issues related to the usage of lethal autonomous robotic weapons. Addresses the question if the decision to kill a human should be handed over to machines. | No definition of the term 'values', but the text mentions values and ethical issues. | Values/ ethical issues:<br>– Responsibility;<br>– Reduce human harm;<br>– Human dignity;<br>– Honour;<br>– Human sacrifice. |

TABLE 3 Values related to weapons (*cont.*)

| Author | Key contribution | Definition of value | Values |
|---|---|---|---|
| Sharkey and Suchman (2013) | Paper on defining and designing autonomy and accountability in Robotic Systems for military operations. | No definition of the term 'values', but the text mentions values. | Values:<br>– Accountability;<br>– Responsibility. |
| Docherty (2015) | Report of Human Rights Watch in which the accountability gap and ethical issues for Autonomous Weapons are described. | No definition of the term 'values', but the text mentions values and ethical issues. | Values/ ethical issues:<br>– Lack of human dignity;<br>– Accountability;<br>– Responsibility<br>– Harm. |
| United Nations Institute for Disarmament Research (2015) | Paper highlights some ethical and social issues regarding the weaponization of autonomous technologies. Encouraging ethical reflection on cultural and social values of weaponization of autonomous technologies. | No definition of the term 'values'. The text contains no explicit mention of values, but some ethical issues are given. | Ethical issues that are mentioned are:<br>– Reduce or eliminate harm;<br>– Consideration of public conscience;<br>– Affront of human dignity (when human intent is lacking when taking a life). |
| Walsh and Schulzke (2015) | Survey experiment to get insight if US civilians are more likely to initiate a war when UAV's are used. Large empirical study that looks at the ethics of drone strikes. | No definition of the term 'values'. The text contains no explicit mention of values, but some ethical issues are given. | Ethical issues:<br>– Security;<br>– Respect for civilian immunity;<br>– Prevent harm |

| | | | |
|---|---|---|---|
| Williams, Scharre, and Mayer (2015) | Discusses several ethical issues relevant to the development of autonomous systems and provides recommendations for Defense Policy makers. | No definition of the term 'values', but the text mentions several values which are interchangingly used with ethical issues. | Values/ ethical issues:<br>– Security;<br>– Harm;<br>– Value of human life (people have the right to be killed by another human). |
| Horowitz (2016) | Description of the debate on ethical implications of autonomous weapons. Considers Lethal Autonomous Weapon Systems (LAWS) in three categories; munition, platforms, and operational systems. Thereby clarifying the debate and describes two ethical issues. | No definition of the term 'values'. The text contains no explicit mention of values, but some ethical issues are given. | Values:<br>– *Accountability* (autonomous systems lack meaningful human control therefore they create a moral accountability gap)<br>– *Human dignity* (people have the right to be killed by someone who made the choice to kill them) |

taking a human life. Other studies primarily describe ethical issues, such as *preventing harm, upholding human dignity, security, the value of human life* and *accountability* (Horowitz, 2016; UNDIR, 2015; James I Walsh & Schulzke, 2015; Williams, Scharre, & Mayer, 2015).

Based on our literature review of values related to Autonomous Weapons, the values: *human dignity, harm, security, responsibility* and *accountability* will be added to the list of values that will be used in this study, because these values are mentioned more than once in the overview in Table 3. Together with the values derived from the field of BioEthics and the meta-inventory of Cheng and Fleischmann (2010) these values will be our starting point for the empirical investigation of this research.

### 2.2.4 *Values Hierarchy*

One approach to consider which values are relevant in the design of Autonomous Weapons is the translation of values into design requirements which can be made visible by means of a value hierarchy (Van de Poel, 2013). This hierarchical structure of values, norms and design requirements makes the value judgements, that are required for the translation, explicit, transparent and debatable. To do so, the values that are described in the natural language will need to be translated to *'formal values in a formal language'* (Aldewereld, Dignum, & Tan, 2015, p. 835). One way of formalizing values into norms would be to use a convention of rules which are represented as: *' "X counts as Y" or "X counts as Y in context C" '* (Searle, 1995, p. 28). The explicitly of values in formal rules allows for critical reflection in debates and pinpoint the value judgements that are disagreed on. Transparency is important as Van de Poel (2013, p. 265) eloquently states: *'Although transparent choices are not necessarily better or more acceptable, transparency seems a minimal condition in a democratic society that tries to protect or enhance the moral autonomy of its citizens, especially in cases that design impacts the lives of others besides the designers, as is often the case'.*

The top level of a value hierarchy consists of the *values*, as depicted in Figure 3, the middle level contains the *norms*, which can be capabilities, properties or attributes of the artefact, and the lower level are the design *requirements* that can be identified. The relation between the levels is not deductive and can be constructed top-down, by means of specification, or bottom-up by seeking for the motivation and justification of the lower level requirements. The bottom-up conceptualisation of values is a philosophical activity which does not require specific domain knowledge and the top-down specification of values requires context or domain specific knowledge that adds content to the design. (Van de Poel, 2013).

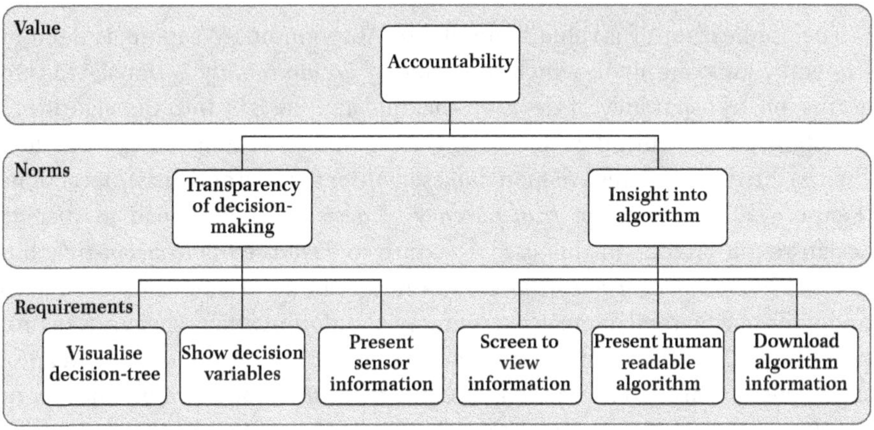


FIGURE 3 Value hierarchy for the value of accountability in the design of Autonomous Weapons

Van de Poel (2013, p. 262) defines specification as: *'as the translation of a general value into one or more specific design requirements'* and states that this can be done in two steps:

1. Translating a *general value* into one or more *general norms*;
2. Translating these *general norms* into more *specific design requirements*.

For step 1 two criteria are relevant: (1) the norm should be an appropriate response to the value and (2) the norm should be a sufficient response to the value. In step 2 the requirement should be more specific regarding the scope of applicability, goals and aims strived for, and actions to achieve those aims of the norm (Van de Poel, 2013).

This translation might prove to be quite difficult as insight is needed in the intended use and context of the value which is not always clear from the start of a design project. Also, as artefacts are often used in an unintended way or context, new values are being realized or a lack of values is discovered (van Wynsberghe & Robbins, 2014). An example of this are drones that were initially designed for military purposes, but are now also used by civilians for filming events and even as background lights during the 2017 Super Bowl halftime show. The value of safety is interpreted differently for military users that use drones in desolated regions compared to that of 300 drones flying in formation over football stadium in a populated area. The different context and usage of a drone will lead to a different interpretation of the value *safety* and could lead to more strict distance norms for flight safety which in turn could be further specified in alternate design requirements for rotors and software for proximity alerts, to name two examples.

The application of a value hierarchy to Autonomous Weapons is demonstrated by an example in which the value of *accountability* is translated into norms for `transparency of decision-making' and `insight into the algorithm'. This translation will allow users to get an understanding of the decision choices the Autonomous Weapon makes in order to trace and justify its actions (Figure 3). The norms for *transparency of decision-making* lead to specific design requirements. In this case a feature to *visualise the decision-tree*, but also to *present the decision variables* the Autonomous Weapons used, such as trade-offs in collateral damage percentages of different attack scenarios to provide insight into the proportionality of an attack. The Autonomous Weapon should also be able to *present the sensor information*, for example imagery of the site, in order to show that it discriminated between combatants and non-combatants. To get *insight into the algorithm*, an Autonomous Weapon should be designed with features that it normally will not contain. In this case these features would include a *screen* as user interface that shows the algorithm in a *human readable form* and the functionality to *download* the changes made by the algorithm as part of its machine learning abilities that can be studied by an independent party, such as a war tribunal of the United Nations if the legality of the actions of an Autonomous Weapon are questioned.

Kroes and van de Poel (2015) state that an objective measurement of values is not possible due to the fact that the operationalization is done by means of second-order value judgments which seriously undermine the construct validity of the value measurement. Judgments are often considered subjective as their truth, or falsity, depend on feelings or attitudes of the person who judges (Searle, 1995). To counter this lack of validity, the designer could look to technical codes and standards which are drafted by committees and represent reasonable standards of operationalizing and measuring values in design. However, standards may not reflect the latest technical and social developments and operationalization still requires value judgments of the designer. Kroes and van de Poel (2015, p. 177) advise to *'embed them in a network of other considerations, including definitions of the values at stake in moral philosophy (or the law), existing codes and standards, earlier design experiences, etc'*.

In our study, we follow the advice of Kroes and van de Poel (2015) and do not strictly apply the value hierarchy as a method to specify, design and test requirements for Autonomous Weapons, but use the example in Figure 3 to link the conceptual and the empirical investigation phase of the Value-Sensitive Design approach of our research. Creating the value hierarchy allows us to think through and transpose the abstract concept of values studied in the conceptual investigation phase into a concrete example of Autonomous Weapons that we can use as input for the empirical investigation phase of our research. The value hierarchy in Figure 3 is used as orientation, inspiration and direction in the remainder of our study.

## 2.3 Agency

The concept of agency is studied in multiple academic fields all of which approach agency from a different angle. We reviewed literature from the fields of Cognitive Psychology, Artificial Intelligence and Moral Philosophy. We summarized the agency characteristics found in these three fields in the Table 4 and conclude by stating why we chose these characteristics.

### 2.3.1 *Agency in Cognitive Psychology*

In the field of Cognitive Psychology the dimensions of mind perception were first studied by H. M. Gray et al. (2007). They found that mind perception consists of two dimensions: 1) *Experience* consisting of the factors hunger, fear, pain, pleasure, rage, desire, personality, consciousness, pride embarrassment and joy, and 2) *Agency*, that encompasses self-control, morality, memory, emotion recognition, planning, communication and thought. Whereas *Experience* is referred to as moral patiency and related to rights and privileges, *Agency* is linked to moral agency and related to responsibility. Several Cognitive Psychology studies show that people attribute agency to non-humans and people perceive these non-human agents as having the same agentic capacities as humans in judging moral dilemmas (Hristova & Grinberg, 2015). This is not only limited to living beings such as animals, but agency is also attributed to non-living objects like robots and zombies (K. Gray & Wegner, 2012; Waytz et al., 2010).

### 2.3.2 *Agency in Artificial Intelligence*

Artificial Intelligence researchers have also been studying the characteristics of agents, but more from a computer science point-of-view to create an architecture for a rational agent that can reason and is able to deliberate between different alternatives. Bratman, Israel, and Pollack (1988) describe a Belief-Desire-Intention (BDI)-architecture for the deliberation process for resource bounded agents with limited time and computational power. '*Beliefs are statements of properties of its world (and of itself) that an agent takes to be true …*' (F. Dignum, Kinny, & Sonenberg, 2002, p. 407). Perugini and Bagozzi (2004, p. 71) define a desire as: '*… a state of mind whereby an agent has a personal motivation to perform an action or to achieve a goal*'. Intentions imply a commitment to a plan and include some form of planning to achieve the goals (Bratman et al., 1988;

TABLE 4 Overview of Agency characteristics

| Author | Agency characteristics | Scientific field |
|---|---|---|
| Bratman et al. (1988) | Beliefs, Desires, Intentions | Artificial Intelligence |
| Bandura (2001) | Intentionality, forethought, self-regulation, self-reflectiveness. | Cognitive Psychology |
| F. Dignum et al. (2002) | Beliefs, Desires, Intentions, Norms | Artificial Intelligence |
| Sullins (2006) | Autonomous, intentions, responsibility. | Moral Philosophy |
| H. M. Gray et al. (2007) | Self-control, morality, memory, emotion, recognition, planning, communication, thought. | Cognitive Psychology |
| Himma (2009) | Free choices, deliberation, understanding and applying moral rules. | Moral Philosophy |
| Waytz et al. (2010) | Capacity to plan and act. | Cognitive Psychology |

Perugini & Bagozzi, 2004). F. Dignum et al. (2002) expand the BDI framework to incorporate social constructs such as norms and obligations, because these are important concepts to link autonomous agents in a Multi-Agent-System.

### 2.3.3. *Agency in Moral Philosophy*

Moral Philosophy is a third scientific field that has been describing the properties of agents. Sullins (2006) poses three requirements to determine if a robot can be seen as a moral agent: 1) *Autonomy;* the robot is significantly autonomous from its programmers, operators and users, 2) *Intentionality*; the preposition to do good or harm, and 3) *Responsibility*: if fulfils some social role and is responsible for other moral agents. Himma (2009) defines agency as: '... *being capable of doing something that counts as an act or action*'. The notion of moral agency is that one is accountable for their behaviour which is governed by moral standards. Two capacities are necessary for moral agency. The first is to freely choose your acts which are a result from deliberation. The second capacity is understanding moral concepts, such as the distinction between 'good' and 'bad' or 'right' and 'wrong', but also applying moral principles like 'it is bad to do intentional harm' (Himma, 2009).

The main characteristics of agency from the reviewed articles on agency perception in the fields of Cognitive Psychology, Artificial Intelligence and Moral Philosophy are chronologically listed in Table 4. We selected all the concepts used by the authors to describe or characterise agency and did not filter them at this stage. All these characteristics will be used in the empirical phase to study the agency perception of Autonomous Weapons.

CHAPTER 3

# Method

> All questions are basically variations to one ethical question: can a drone decide autonomously to engage (without human interfere). The answer in my mind is no. There should always be a human interface in this. War and conflict are not a computer game, war is a social interaction between human beings. A drone is a mere instrument, a tool that can never operate autonomously.
>
> RESPONDENT FINAL STUDY

This section describes the methodology, hypotheses, research design, operationalisation of the scenarios and constructs, analytical approach, pre-registration of the study, sample and concludes with the methodological issues.

## 3.1 Methodology

The methods used in the various parts of this studies are described in this subsection beginning with the literature review, followed by the online value survey, the expert interviews, the coding process of the interviews and finally the method of the randomized controlled experiments.

### 3.1.1 *Literature Review*

Google Scholar was used to search articles for the literature review and first keywords *Value-Sensitive* AND *Design* were entered which resulted in a few articles on Value-Sensitive Design. The *Handbook of Ethics, Values, and Technological Design: Sources, Theory, Values and Application Domains* was used to get more insight in the topic and to find further references. Next, we searched for articles on basic human values with the keywords *human* AND *values*, and *universal* AND *human* AND *values*, but this did not give many results. We analysed some well-known theories on basic human values and value hierarchies. Taking these articles, we used the 'cited by' option in Google Scholar to find articles that used these as source. This gave relevant results on articles describing

 | DOI:10.1163/9789004449084_004

basic human values and we selected the articles that provided a list of values. Articles that only repeated earlier work, for example Schwartz theory, were not selected. Next, we searched for articles regarding people's values to weapons by selecting the key words *value* AND *weapon*, and keywords *Value-Sensitive* AND *Design* AND *Weapons*. We selected the articles containing the keyword *value*.

### 3.1.2 *Online Value Survey*

The online survey was created using the Qualtrics[1] online survey tool provided by MIT. After four questions on demographics, the respondents were asked three questions on the values they associate with Autonomous Weapons. The first question asked to rank the four BioEthics values *Autonomy, Non-Maleficence, Beneficence* and *Justice* from most applicable to Autonomous Weapons to least applicable. In the second question the respondents were asked to select the five values that apply most to Autonomous Weapons from the list of Cheng and Fleischmann (2010). The third question was an open question in which the respondents were asked to list one other value that was not mentioned in the two previous questions. The online survey was tested twice. First, two fellow students reviewed the questions and after their feedback was processed the survey was tested a second time by different reviewers that are more representative of future participants. The survey was distributed via social media, such as Facebook, LinkedIn and Twitter, it was posted to our personal blog and the online platform 'Call for Participants' was used for a distribution outside my social network to enlist more participants. During the MIT Media Lab's member's week in April 2017, people that were interested in our research were asked to take the survey and we also posted the link of the survey to the Scalable Cooperation Slack[2] group. The online study ran 3,5 weeks from 17 March until 11 April 2017.

### 3.1.3 *Expert Interviews*

We conducted six semi-structured expert interviews based on the *Photo Elicitation Interview* (PEI) method (Harper, 2002) to get alternate views from experts in the field of Autonomous Weapons and a more in depth understanding from their point of view. The PEI method can be used to get a shared understanding of values and to support a discussion between the designer and stakeholder about these values (Pommeranz, Detweiler, Wiggers, & Jonker, 2011). It enhances the voice of the stakeholder on the values and mitigates

1 https://www.qualtrics.com/
2 Slack is a cloud-based collaboration tool that supports communication in teams (https://slack.com/).

the assumptions of the researcher (Le Dantec, Poole, & Wyche, 2009). We asked the expert to select three photos prior to the interview and in case they did not pick any themselves we also selected eight photos of Autonomous Weapons. During the interviews, we got insight into the values of the experts on Autonomous Weapons by asking questions on why they picked that photo and which value it represented to them.

All six interviewees made the effort to select three images and send them prior to the interview. During the interviews, we noticed that they had given the pictures a lot of thought in selecting them. In discussing the images many different topics were touched on ranging from gut-feelings, futuristic scenarios, economics, military tactics, sci-fi movies and even politics. Although the photos did not all lead to discussions on basic human values as found in literature, the PEI method did trigger information on what the experts find important principles regarding Autonomous Weapons. It also created a shared understanding and good interview atmosphere.

### 3.1.4 *Coding Process*

The interview results were processed by means of the Values Coding method described by Saldaña (2015). Keeping a memo (Appendix G. Coding memos) in which the process and assumptions are listed is an integral part of a coding method. We started with a list of pre-set codes which contained: '*(1) a Value: The importance we attribute to oneself, another person, thing or idea, (2) Attitude: The way we think and feel about oneself, another person, thing or idea, and (3) a Belief: The part of the system that includes our values and attitudes plus our personal knowledge, experiences, opinions, prejudices, and other interpretive perceptions of the social world*' (Saldaña, 2015, p. 89). During the coding of interviews, it is recommended that the codes fit the data instead of fitting the data to the codes. Therefore, we added an extra code 'definition' to the list of emerging codes. We kept notes while coding the interviews describing our steps and assumptions. We asked a second researcher, a fellow TPM master student knowledgeable of values, to also code the interviews based on the Value Coding method and to also keep a memo (Appendix G. Coding memos). We compared the values that we coded to those of the second researcher and derived the values that were similar (Appendix H. Results coding process).

### 3.1.5 *Randomized Controlled Experiments*

The method used for the pilot and final studies is called a randomized controlled experiment. Oehlert (2010) mentions four reasons to create experiments: (1) they allow for direct comparisons between treatments of interest,

(2) they can be designed to minimize any bias in the comparisons, (3) they can be designed to keep the error in the comparison small, and (4) we are in control of the experiments which allows us to make stronger inferences about the nature of differences we observe and especially we can make inferences about causation. This last point distinguishes an experiment from an observational study. A treatment in this sense are the different procedures we would aim to compare. It is important that the effects of a treatment can only attributed to one cause that can be measured and also that the effects cannot be attributed to multiple causes. This is called confounding which Oehlert (2010, p. 14) defines as: '… *occurring when the effect of one factor or treatment cannot be distinguished from that of another factor or treatment*'. Randomisation helps ensuring that participants are assigned to a scenario by chance and not based on pre-exiting features, such as time when the survey is taken or location. We use randomisation in the studies to vary the order of the scenarios and the order of the questions posed to the respondents by means of a probabilistic scheme. In the surveys, we chose the option to distribute an even number of scenarios over the respondents.

## 3.2 Hypotheses

Due to the exploratory nature of the study, we only developed a hypothesis for the agency perception of the final study and not for the pilot studies or the dependent variables. The pilot studies were used to explore which manipulations would have which effects and to test if the wording of the scenarios and questions were clear and effective.

Reasoning that military personnel will view Autonomous Weapons as any other weapon, and therefore no more than a tool to achieve an effect, we hypothesize in the final study that military personnel will perceive Autonomous Weapons as not possessing mental states. Therefore, we expect there to be no difference between the neutral agency Autonomous Weapon condition and the condition in which a human is operating a drone remotely. We also expect the neutral agency condition to be judged as significantly different from the high agency condition, in which we specifically tell participants that the Autonomous Weapon has agentic characteristics, such as the ability to plan and set its own goals. Based on this reasoning, we hypothesize that:

> *H1: military personnel will not perceive Autonomous Weapons as possessing mental states.*

## 3.3 Research Design

We tested several research designs in the two pilot studies before deciding on the research design for the final study which is depicted in Figure 4. For the sake of brevity, we do not include the two designs of the pilot studies in this section and we only show the high-level research model. The research model of the final study consists of the agency perception as independent variable, the moral values as dependent variables and the demographic variables. The specific set-up of the scenarios is described in more depth in the result section.

## 3.4 Operationalisation

In this section, we describe the operationalisation of the constructs into concrete items. First, we describe the scenarios that we developed for the studies followed by the agency construct and the dependent variables, for which we describe the items and the corresponding questions. We also specify which scale we use and how the agency construct is calculated. We briefly outline the operationalisation of the attention check and which demographic variables we included in the survey.

### 3.4.1 *Scenarios*

Scenarios are used in the field of Cognitive Science as means to study moral judgement in randomized controlled experiments (Cushman & Young, 2011; Kominsky, Phillips, Gerstenberg, Lagnado, & Knobe, 2015). We used them to study the agency perception of Autonomous Weapons and created a scenario that describes a military operation in which a convoy is delivering supplies in a conflict area. The convoy is being approached by a vehicle at high speed; a situation that is likely to happen during these types of operations and military personnel needs to estimate the level of threat in order to decide to attack or not. We chose to focus on drones as this is technology that is currently used by human operators and drones are already developed with autonomy by several companies, such as BAE systems[3], Dassault Aviation[4] and Boeing[5]. Although these Autonomous drones not

3 https://en.wikipedia.org/wiki/BAE_Systems_Taranis
4 https://en.wikipedia.org/wiki/Dassault_nEUROn
5 https://en.wikipedia.org/wiki/Boeing_Phantom_Ray

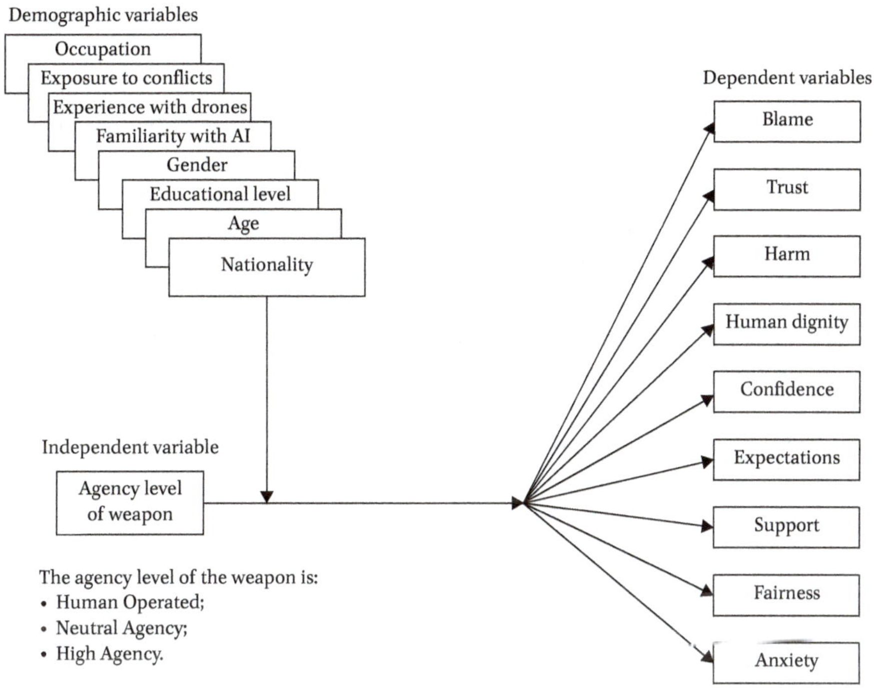


FIGURE 4 Research design

yet deployed in military operations we think it is likely to happen within the next five years.

We created a default scenario for all studies which we could expand based on the agentic characteristics we wanted to explore. The default scenario that we used in the final study reads as follows:

> A military convoy is on its way to deliver supplies to one of their units at a camp near Mosul in Iraq. The commander has ordered an autonomous drone to support the convoy in the air. The autonomous drone scans the surroundings for enemy threats and carries weapons for the defence of the convoy. When the convoy is at a three-mile distance from the camp, the autonomous drone detects a vehicle behind a mountain range that is approaching the convoy at high speed. The autonomous drone detects four people in the car with large weapon-shaped objects and identifies the driver of the vehicle as a known member of an insurgency group. The autonomous drone attacks the approaching vehicle which results in the

death of all four passengers, but also causes collateral damage by killing five children that were playing nearby the road.

The above scenario represents the neutral agency condition in which we do not provide any agency characteristics for the weapon. In the Human Operated scenarios, we replace words autonomous drone with the words Human Operated drone and provided no extra information on agency characteristics (note that the words are highlighted in blue to show the distinction in this report and the respondents in the survey were shown scenarios in black wording). In the high agency condition we added the following phrase to describe the agency characteristics: '*The autonomous drone independently deliberates between a series of options, weighs the pros and cons, and decides to attack the approaching vehicle, …*'. In the final study, we mentioned that there is collateral damage after the attack, but in the pilot studies we also tested scenarios without collateral damage stating: '*… which results in the death of all four passengers, but causes no collateral damage*'.

We purposely kept the changes to the scenarios to a minimum so that we can attribute different results to those changes and measure their effect. The results show that these small textual changes lead to differences in the mean value of the dependent and independent variables so it seems that the treatment by using scenarios as means for a randomized controlled experiment works.

### 3.4.2. *Agency Construct*

Based on the literature review of the agent characteristics in section 2.3 we operationalized the agency construct. We created a list in which we clustered the characteristics that we judged to be similar and calculated the number of times a characteristic was mentioned in literature (Table 5). We selected the characteristics that were mentioned more than once and formulated a question to measure them. This resulted in an agency construct consisting initially of the five items that were mentioned more than once in the literature. These items will be measured on a self-reported scale of: 0 (strongly disagree) – 100 (strongly agree) and the agency score will be averaged over scores of the items.

We used the following questions to measure the agency construct:

1. Ability to deliberate [*Thought*]
   The drone thinks independently about what to do about the vehicle, deliberating between a series of options to defend the convoy.
2. Ability to generate its own goals [*Goal setting*]
   The drone decides independently if its goal should be to eliminate the vehicle to defend the convoy.

3. Ability to apply moral rules [*Moral rules*]
   The drone independently consults the moral rules and norms set by the military commander and decides if it is morally appropriate to take out the vehicle to defend the convoy.
4. Ability to act with free will (*Act freely*)
   The drone has various options open to it and independently decides if it needs to eliminate the vehicle to defend the convoy.
5. Ability to plan to achieve goals (*Achieve goals*)
   The goal of the drone is to defend the convey, so it independently decides if it needs to launch a plan involving selecting the vehicle as a target, calibrating its weapons, and launching an attack.

After pilot study 1, item 3 [Moral rules] was dropped leaving for the agency construct the four items which were used in pilot study 2 and the final study:

1. Thought
2. Goal setting

TABLE 5 Number of agency characteristics mentioned in literature

| Agency characteristic | Number of times mentioned | Item |
|---|---|---|
| Forethought, thought, deliberation, beliefs, beliefs | 5 | Thought |
| Intentionality, intentions, desires, desires | 4 | Goal setting |
| Planning, capacity to plan, intentions, intentions | 4 | Achieve goals |
| Morality, understanding and applying moral rules, norms | 3 | Moral rules |
| Self-control, free choices | 2 | Act freely |
| Self-regulation | 1 | |
| Self-reflectiveness | 1 | |
| Autonomous | 1 | |
| Responsibility | 1 | |
| Memory | 1 | |
| Emotion | 1 | |
| Recognition | 1 | |
| Communication | 1 | |
| Capacity to act | 1 | |

3. Act freely
4. Achieve goals

### 3.4.3 *Dependent Variables*

We included 9 to 12 questions in the pilot and final studies to explore further implications of agency perception of Autonomous Weapons. The dependent variables are derived from the results of the literature study, value questionnaire and interviews. Based on the results of the pilot studies, the final study included the dependent variables; *blame, trust, harm, human dignity, confidence, expectations, support, fairness* and *anxiety*. Each of these variables was measured on a self-reported scale of: 0 (strongly disagree) – 100 (strongly agree).

1. *Blame*: The drone is to blame for the action.
2. *Trust*: The drone can be trusted to take the correct actions in the future.
3. *Harm*: The actions of the drone caused harm.
4. *Human dignity*: The actions of the drone respect human dignity.
5. *Confidence:* I am confident that the drone will take the correct actions in the future.
6. *Expectations*: The actions of the drone are according to my expectations.
7. *Support:* I support the use of these type of drones by the military.
8. *Fairness*: The actions of the drone are fair.
9. *Anxiety*: The actions of the drone worry me.

The following constructs were used in pilot study 1 and 2, but dropped in the final study:

10. *Blame commander:* The commander is to blame for the action.
11. *Trust commander:* I trust that the commander will take correct actions in the future.
12. *Harm commander:* The actions of the commander caused harm.

### 3.4.4 *Attention Check*

We added a question with an attention check in the middle of the dependent variable questions in which we asked to select the value 40 for that question. The attention check is often added to a questionnaire in Cognitive Psychology studies for example by Logg (2017) and people who fail to provide the correct answer are excluded from the sample reasoning that they also fail to pay adequate attention to the other questions.

### 3.4.5 *Demographic Variables*

We also added questions to collect demographic information on the respondents and added variables on: *age, gender, education level, occupation* and *nationality*. Next to these general demographic questions we asked if respondents have *experience with artificial intelligence*, if they *worked with drones* and if they *have been in a conflict zone*.

## 3.5 Analytical Approach

This section describes the approach that we took to analyse the data. For each of the studies we performed the following steps in the data analysis.

### 3.5.1 *Pre-process Data*

In the Qualtrics survey tool after closing the survey, a distinction is made between *responses in progress* and *complete responses*. Before starting the data analysis, the completion percentage of the responses in progress need to be checked and decided if they can be added to the complete responses. It turns out that when Amazon MTurk is used to distribute the survey, many respondents neglect to click on the last button which means that their responses are 99% complete, but these responses are valid and usable. If the respondent answered all the questions we added them to the list of complete responses. If respondents failed to answer all questions, for example the demographic questions, we deleted them from the sample. In the next pre-processing step, we deleted all respondents who failed the attention check that we build half way in the questionnaire. We asked if people could answer 40 to that question and we deleted all answers that fell outside the range of 38 and 42 which we assume is an accuracy mistake in placing the sliders.

### 3.5.2 *Reliability Analysis*

After uploading the csv file in SPSS, we conducted a reliability test which is a measure to assess the internal consistency of a set of test items or scale. The reliability test checks if the given measurement is a consistent measurement of a concept. The construct to measure the internal consistency is Cronbach's Alpha ($\alpha$) which should be over 0.7 to show internal consistency for the concept. For each of the studies the Cronbach's Alpha for the items of the agency construct is reported, because these items are used to measure the agency construct and should have internal consistency. It is also indicated if Cronbach's Alpha is improved if one of the items is removed. The reliability of the dependent variables is checked, but not reported, because these items are not designed to be used as one scale and it turned out that their internal consistency is lower than the threshold ($\alpha < 0.7$) as could be expected.

### 3.5.3 *Principal Component Analysis (PCA)*

In the next of the step in the data analysis a Principal Component Analysis (PCA) on the agency items was conducted. This step is a variable reduction technique aimed to reduce a larger set of variables into a smaller set, called 'principal components', that account for most of the variance in the original variables. We performed the PCA on the agency items to see if they could be aggregated to a single agency construct and report the variance and number of components as results. We also checked if the construct was influenced by the demographic variables, but will not report these results as the PCA indicates that all the demographic variables are distinct components that do not influence the agency construct.

### 3.5.4 *Correlation Analysis*

The bivariate Pearson Correlation is a check to measure the strength and direction of the linear relationships between a pair of continuous variables. It produces a correlation coefficient *r* that lies between -1 and +1. A negative value indicates a negative relationship between the constructs, for example the larger the agency perception, the less the support. A positive value indicates a positive relationship, for example the larger the agency perception, the more blame. A correlation coefficient of zero indicates that no relationship between the constructs exists. We ran this analysis to check the correlation between the agency construct and the dependent variables and report the results.

### 3.5.5 *Manipulation Check on Agency*

The manipulation check examines if the agency construct varies between the scenarios and by this we checked if the hypotheses should be accepted or rejected. The check is done in two ways. First a graph is created, with the agency construct on the y-axis and scenarios on the x-axes, for a visual overview. The agency levels are expected to be different for the various conditions and this difference should be significant with $p < .05$. That is shown by the I shaped bars on top of the columns which show the standard error multiplied by 1 and indicates if the constructs overlap or not. If the bars overlap than the two groups are not significantly different.

The same result can be achieved by independent samples T-test which is a check for significant differences in the means of two distinct groups for which the subjects are randomly assigned so that the observed effect is the result of the treatment (in this case reading a specific scenario) and cannot be attributed to a different effect. We will show the graphs and report the results of the independent samples T-test for the agency construct between the Human Operated scenarios and scenarios with the different agency levels of Autonomous Weapons.

#### 3.5.6 *Dependent Variables Analysis*

The analysis of the dependent variables is of an exploratory nature and the results are depicted by graphs. Each graph shows the mean of the dependent variable on the y-axis and the different scenarios on the x-axis to point out the differences between the scenarios. Although the analysis of the dependent variables is descriptive, we report the findings comparing between the different conditions that are significant at $p < .05$.

### 3.6 Pre-registration

The final study was pre-registered on the Open Science Framework[6] which creates a time-stamped version of a project that cannot be deleted or edited. Although not required by scientific journals yet, it is often checked by reviewers if a study is pre-registered. We used the AsPredicted template in which 8 questions about the study are answered, for example if data already has been collected, which hypotheses the study tests and type of analyses you intend to perform. The pre-registration is placed under embargo until June 24, 2021. We preregistered the strong hypothesis we have about the agency perception, as we think we can replicate the findings based on the results of pilot study 2. For the dependent variables, we are not sure which mechanisms are at play and if we will be able to replicate our results therefore we registered an exploratory analysis on those results.

### 3.7 Sample

All surveys were created in Qualtrics and the pilot studies were distributed via crowdsourcing platform Amazon Mechanical Turk. We tested the survey with a small sample of 20 respondents and inspected the average completion time to see if we estimated the completion time correctly. We checked if the scenarios were evenly distributed, if people correctly checked the test question and if there were any remarks about the survey. Given that it all checked out fine, we decided to scale up the batch to 50 respondents per scenario which meant

6 https://osf.io/

500 participants for pilot study 1 and 700 for pilot study 2. The collection of the answers took about 4 hours and each respondent was payed USD 0.40.

The final study was distributed via the snowball method by email with an anonymous link to approximately 40 military colleagues who further distributed the survey. This method was used because we were not allowed by MIT and the Dutch MOD to collect any personal information, such as email or IP addresses. By using the snowball method, we had no guarantee on the number of respondents. We also published a news item on the internal Army website. The final study ran for 16 days from June 12 until June 27, 2017 and resulted in 327 responses of which 239 were complete valid responses and usable after the data pre-processing.

## 3.8 Methodological Issues

In this section, several issues and concerns regarding the methodology will be addressed. These concerns could have implications for the internal and external validity of the study and if possible, some countermeasures are given which address the concerns.

### 3.8.1 *Coding Interviews*

To get insight into the values of experts we conducted semi-structured interviews which is a qualitative research method. We used the Values Coding method to interpret the data and extract the values from the transcribed interviews. One of the concerns when coding qualitative data is that it is heuristic and no fixed formulas for coding exist. It also includes linking and providing meaning to the data (Saldaña, 2015). These characteristics imply that the Value Coding method is prone to bias in which the researcher uses their own framework and internal notions to interpret the data. To reduce the bias in applying the Values Coding method a second researcher will code the interviews independent from the first researcher. Some instruction on the method will be provided, but as information on the values and findings will be limited.

### 3.8.2 *Randomized Controlled Experiments*

One of the requirements of randomized controlled experiments is, as the name suggest, a random distribution of the respondents of the scenarios. The randomization limits confounding, enhances the internal validity of the research and is a crucial prerequisite for this type of study. One of the concerns is that the randomisation fails and that the distribution of respondents is skewed over the scenarios. One of the causes for skewed distribution is when the software fails to evenly distribute the respondents over the scenarios. As researchers, we can only check if the settings in the software are correct. Another cause for a skewed distribution could be that people see a certain scenario which they do not like or expect and quit the survey before completing. An example of this is when people expect to take a survey on Autonomous Weapons, but are assigned the Human Operated scenario and due to this decide to drop out of the survey. This prematurely exiting the survey is called selective attrition and it can result in a research bias, because the people who quit the survey are fundamentally different that the ones who take the survey. In the current research set-up, it is not possible to take countermeasures and it could be a threat to the internal validity of the study.

### 3.8.3 *Amazon Mechanical Turk*

The advantage of using Amazon MTurk as a survey distribution method is that a large group of respondents will participate in the survey and that will answer the questions seriously to maintain their workers status. This method also has some drawbacks because the workers on MTurk are primarily US based because in order to become a worker you will have to comply with US tax regulations even when you are a non-US resident. This could lead to a bias in the results which implies that the results are not generalizable beyond the US. Bonnefon et al. (2016) report a second concern with using MTurk in that participants might also not be representative for the US population, for example because many of the workers are students, and a third concern is that workers already are familiar with the materials. The first concern can be addressed by selecting a wide range of workers, so that the sample also includes people from outside the US. The second concern is inherent of the choice for MTurk and cannot be limited by countermeasures. The third concern is not very likely to occur because the scenarios are unique and have not been tested on MTurk before.

### 3.8.4 *Snowball Distribution Final Survey*

The final study was distributed to military colleagues who forwarded the email to their network. Using this snowball distribution method has some drawbacks. The first is that participants could take the survey multiple times and as we are not allowed to store personal identifiably information we cannot check if this is the case. This could lead to a bias in the results if people see different scenarios and this could influence the answers on the questions. A second drawback is that colleagues take the survey first and tell the other participants of their findings before those take the survey. This also influences their answers.

Thirdly, as researchers we cannot control who will take the survey so a potential risk would be that the survey gets send to someone outside the Ministry of Defence (MOD) by which the sample gets contaminated. Unfortunately, without storing personal identifiable information there are not many countermeasures that can be taken to address all three concerns.

CHAPTER 4

# Results

> Hopefully people realize that computers will only do what they are programmed to do, and NOTHING else.
>
> RESPONDENT PILOT STUDY 1

This section describes the results of the value survey, that consists of an online questionnaire and expert interviews, and the randomized controlled experiments, comprised of two pilot studies and the final study. The values derived from the online questionnaire and interviews are combined in one overview. The results of the pilot and final studies are described using the structure of the data analysis steps (section 3.5) and the main results are listed in the conclusion of each study.

## 4.1 Value Survey

To get insight into which values people associate with Autonomous Weapons a survey and six expert interviews were conducted. A brief overview of the results is listed in this subsection.

### 4.1.1 *Online survey*

The value survey consisted of a questionnaire asking respondents first which values of BioEthics and Cheng and Fleischmann (2010) apply. In the final question, the participants were asked to provide at least one additional value that was not mentioned in the previous questions. In total 69 respondents took the survey and after removing the incomplete responses 57 complete responses ware left. The results are depicted in graphs below.

This short questionnaire gives insight into which values people associate with Autonomous Weapons. The results from the BioEthics question shows that *non-maleficence* is most important, secondly it is *beneficence*, thirdly

 | DOI:10.1163/9789004449084_005

FIGURE 5 Results question 1 online value survey

*justice* and lastly *autonomy*. The top five values based on the list of Cheng and Fleischmann (2010) are: *security*, *intelligence*, *responsibility*, *justice* and *social order*. The open questions reveal that people associate Autonomous Weapons mostly with *accountability* and *effectiveness*.

### 4.1.2 *Interviews*

Six interviews were held with experts in the following fields:

Artificial Intelligence:

- prof. Toby Walsh (University of New South Wales & Data61).

Autonomous Weapons:

- dr. Peter Asaro (The New School & Campaign Ban Killer Robots);
- Mrs. Miriam Struyk (Program director Security & Disarmament PAX);
- dr. Michael Horowitz (University of Pennsylvania).

FIGURE 6 Results question 3 online value survey

Military operations:
- LtCol Kremers (Dutch Army Headquarters);
- Commander van den Sande (MOD Headquarters).

Next to the discussion on values, the interviews provided rich background information on the history and positions of the different groups in the current debate on Autonomous Weapons, but also revealed that there is no consensus on a definition yet. The full transcriptions of the interviews can be found in Appendix F. Transcriptions interviews.

The interviews were coded by means of the 'value coding' method and based on the comparison of the value coding of the researcher and the second reviewer (Appendix H. Results coding process) the following list of 22 unique values are derived:

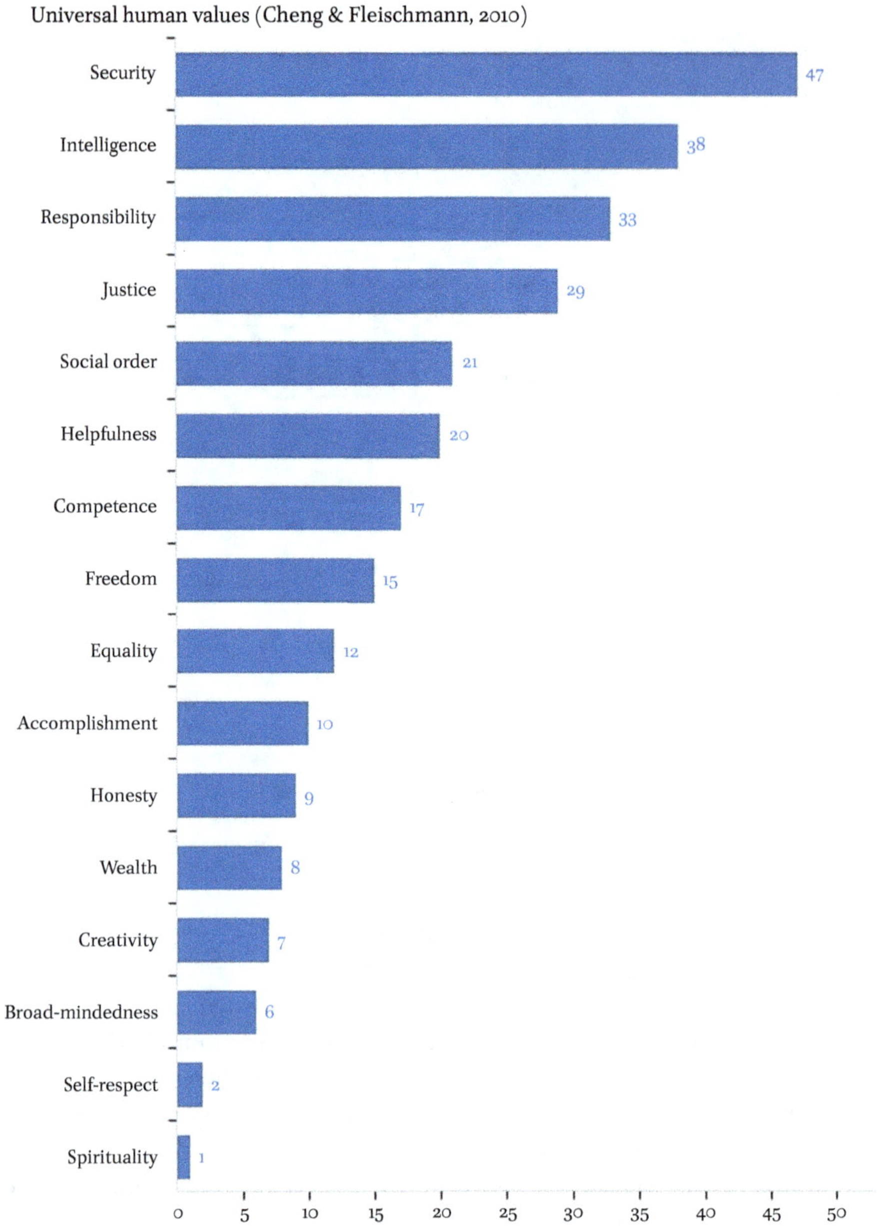


FIGURE 7 Results question 2 online value survey

| | |
|---|---|
| 1. Intervene at all times | 12. Distance |
| 2. Recall them | 13. Invincibility |
| 3. Minimize risk | 14. Determinism |
| 4. In control | 15. Inevitability |
| 5. Goodwill | 16. Unnecessary suffering |
| 6. Set boundaries | 17. Superfluous injury |
| 7. Right to be killed by a person | 18. Human dignity |
| 8. Ethical framework | 19. Accountability |
| 9. Meaningful human control | 20. Responsibility |
| 10. Predictability | 21. Defense |
| 11. Reflexivity | 22. Harm |

### 4.1.3 *Conclusion Value Survey*

We conclude on the value survey by comparing the values from the online survey and the interviews which results in a list of values (Table 6). Six values in this list are mentioned in both the value survey and the interviews. Based on the topics the six experts emphasised in the interviews, we selected the values that were mentioned most to be tested in pilot study 1. These values are *fairness, harm, human dignity* and *responsibility*.

## 4.2 Pilot Study 1

In the first pilot study, we tried to test if we could manipulate the perception of agency of Human Operated drones and Autonomous Weapons. We were interested in how people perceive current technology compared to future technology. We created several conditions in which we varied the agency level either by providing extra information on how the drone deliberates about attacking the vehicle or by leaving this information intentionally out. We tested good outcome and bad outcome conditions, because we expected that this would have an impact on how the dependent variables, for example *support*, were viewed. In this study, we specifically asked the agency questions about the drone and not the Human Operator that is controlling it.

We tested 10 scenarios and aimed to get 50 respondents per condition and thereby a total of 500 respondents. The number of respondents per scenario ranged between 49 and 54. After the data pre-processing and deleting the respondents who failed the attention check we were left with 510 complete responses.

TABLE 6 Overview values from online survey and interviews

| Online survey | Interviews |
|---|---|
| Justice | |
| Autonomy | |
| Security | |
| Intelligence | |
| Social order | |
| Effectiveness | |
| Fairness | |
| Racism | |
| Control | In control |
| Beneficence | Goodwill |
| Accountability | Accountability |
| Responsibility | Responsibility |
| Safety | Defence |
| Non-maleficence | Harm |
| | Intervene at all times |
| | Recall them |
| | Minimize risk |
| | Set boundaries |
| | Right to be killed by a person |
| | Ethical framework |
| | Meaningful human control |
| | Predictability |
| | Reflexivity |
| | Distance |
| | Invincibility |
| | Determinism |
| | Inevitability |
| | Unnecessary suffering |
| | Superfluous injury |
| | Human dignity |

TABLE 7 Scenarios of pilot study 1

| | | Agency status | | | | |
|---|---|---|---|---|---|---|
| | | Autonomous drone | | | Human operated drone | |
| | | No agency | Low agency | High agency | Low agency | High agency |
| *Outcome* | No collateral damage (Good) | 1 | 2 | 3 | 4 | 5 |
| | Collateral damage (Bad) | 6 | 7 | 8 | 9 | 10 |

### 4.2.1 *Reliability Analysis*

For all five agency items $\alpha = 0.9$ which is well above the threshold of 0.7 and can only be improved if item SQ3_Moral rules is deleted as shown in Table 8. Deletion of other items will decrease the $\alpha$ value and reduce the internal consistency as shown in the final column (Crohnbach's Alpha if item Deleted) of Table 8.

TABLE 8 Results reliability analysis agency items pilot study 1

| Reliability statistics | | |
|---|---|---|
| Cronbach's Alpha | Cronbach's Alpha Based on Standardized Items | N of Items |
| .900 | .895 | 5 |

| | Item-total statistics | | | | |
|---|---|---|---|---|---|
| | Scale Mean if Item Deleted | Scale Variance if Item Deleted | Corrected Item-Total Correlation | Squared Multiple Correlation | Cronbach's Alpha if Item Deleted |
| SQ1_ Thought | 167.77 | 13710.694 | .841 | .741 | .857 |
| SQ2_ Goal setting | 163.85 | 13440.002 | .857 | .775 | .853 |
| SQ3_ Moral rules | 183.39 | 17599.970 | .448 | .215 | .933 |
| SQ4_Act freely | 163.22 | 13885.447 | .824 | .695 | .861 |
| SQ5_ Achieve goals | 157.21 | 13891.495 | .794 | .681 | .868 |

### 4.2.2 *Principal Component Analysis* (PCA)

The PCA shows that *SQ1_Thought, SQ2_Goal setting, SQ4_Act freely* and *SQ5_ Achieve goals* can be viewed as one construct which accounts for 71.75% variance in the original variables (Table 9). This is also seen in the scree plot that shows the eigenvalues of the components (Figure 8). Based on the component matrix we can observe that *SQ3_Moral rules* is a separate component compared to the other four agency items.

### 4.2.3 *Correlation Analysis*

The correlation of the agency construct with the dependent variables is small and only for *DVQ1_Blame, DVQ2_CommanderBlame, DVQ4_CommanderTrust, DVQ6_CommanderHarm* and *DVQ8_Support* significant ($p < .01$) (Table 10). Some unexpected effects can be observed in the direction of relationships. For example, it is to be expected that the agency perception leads people to assign more *blame* to the drone ($r = .183$), but these results indicate that people also

TABLE 9 Result PCA agency items pilot study 1

| | Component matrix[a] | | |
|---|---|---|---|
| | Component | | |
| | 1 | 2 | 3 |
| SQ2_Goal setting | .922 | | |
| SO1_Thought | .911 | | |
| SO4_Act freely | .899 | | |
| SQ5_Achieve goals | .882 | | |
| SQ3_Moral rules | .571 | .819 | |

Extraction method: principal component analysis.
[a] 3 components extracted.

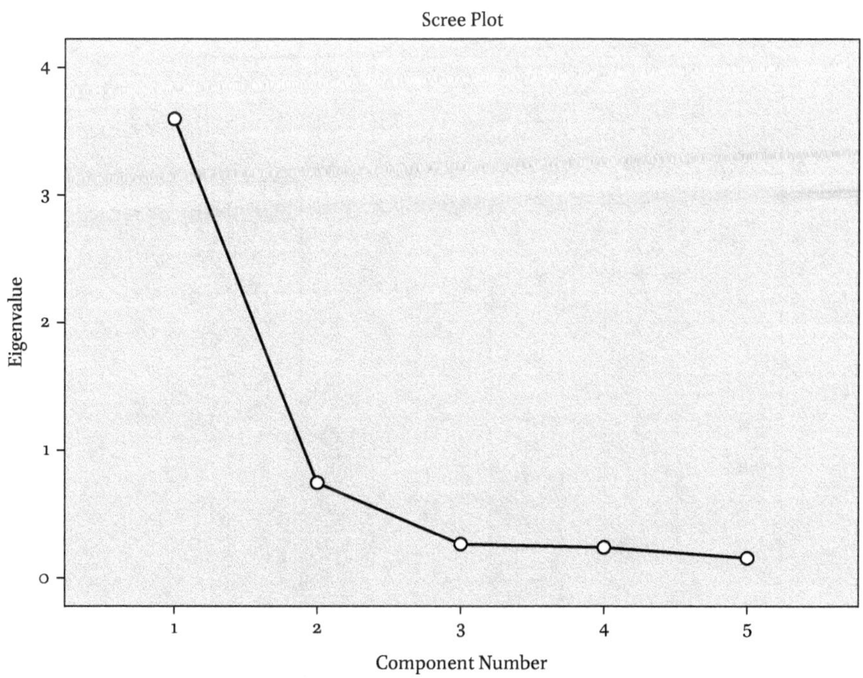


FIGURE 8 Scree plot PCA pilot study 1

assign less *blame* to the *commander* (r = -.239). The same effect can be observed with the *harm* variable. The correlation analysis is not detailed enough to zoom in to these results, therefore this will be done in the analysis of the dependent variables.

### 4.2.4 *Manipulation Check on Agency*

The graph in Figure 9 shows that our agency manipulation works, because in the low agency condition for the Autonomous Weapon, people perceive less agency than in the high agency condition shown by the mean of the agency construct on the y-axis. In the neutral agency condition the agency perception is slightly higher than in the low agency condition, but still lower than the high agency condition. The independent samples T-test shows that the low agency AW and high agency AW conditions are distinct groups (p = .000), but that this is not the case for the neutral agency AW and the low agency AW (p = .209), because the difference between these groups is not significant which can also be seen by the overlap in the error bars for these conditions.

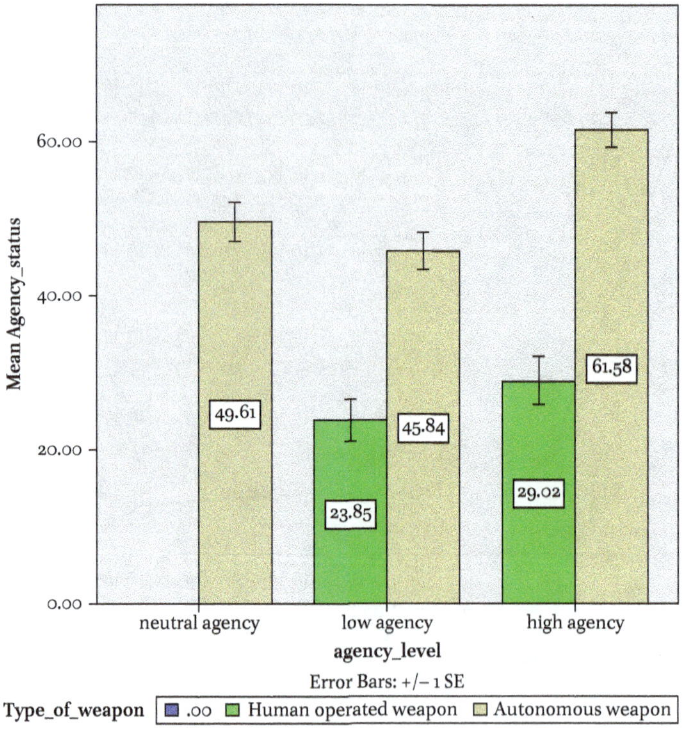


FIGURE 9 Mean value agency construct per Type of Weapon

The Human Operated drone is perceived to have much lower agency than the Autonomous Weapon conditions. The agency manipulation works although the difference is less explicit than in the Autonomous Weapon scenario. The error bars of the agency level scenarios indicate that the Human Operated drone groups cannot be distinct which is confirmed by the independent sample T-test ($p = .125$).

The second graph (Figure 10) shows the mean of the agency perception over the agency levels for good and bad outcomes. The type of outcome caused by the weapon, being collateral damage or not, does not seem to lead to much differences in agency perception as the means differ 2 to 3 points (on a 100-point scale). The error bars also overlap in all agency level which is confirmed by the independent sample T-tests. In the neutral agency level condition, the significance between the bad outcome and good outcome group is $p = .816$, in the low agency condition $p = .641$ and in the high agency condition $p = .575$ which are all much higher than the $p < .05$ threshold.

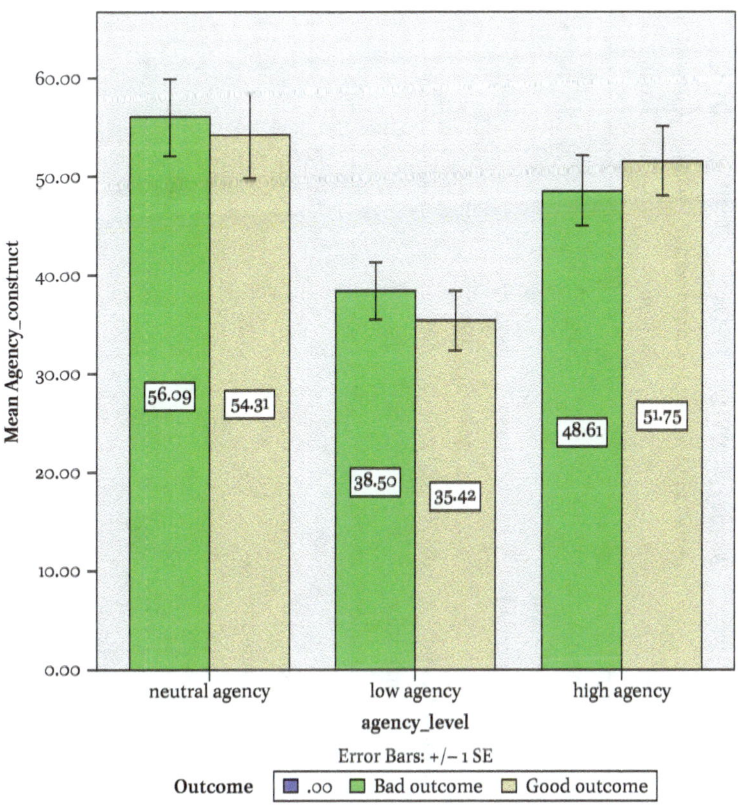


FIGURE 10 Mean value agency construct per condition per Outcome

TABLE 10 Correlation matrix agency construct and dependent variables

**Correlations**

| | | Agency_ construct | DVO1_ Blame | DVO2_ Commander Blame | DV03_ Trust | DV04_ Commander Trust | DV05_ Harm | DV06_ Commander Harm | DV07_ Uneasy | DV08_ Support | DV09_ Fairness |
|---|---|---|---|---|---|---|---|---|---|---|---|
| Agency_ construct | Person Correlation | 1 | .183$^{**}$ | -.239$^{**}$ | .070 | .176$^{**}$ | -.011$^{**}$ | -.227$^{**}$ | .005 | .101 | .087 |
| | Sig(2-tailed | | .000 | .000 | .114 | .000 | .798 | .000 | .914 | .023 | .050 |
| | N | 510 | 510 | 510 | 510 | 510 | 510 | 510 | 510 | 510 | 510 |
| DV01_Blame | Person Correlation | 183$^{**}$ | 1 | -101$^{*}$ | .020 | 0.53 | .198$^{**}$ | -.170 | .129$^{**}$ | -.126$^{**}$ | -078 |
| | Sig(2-tailed | .000 | | .023 | .660 | .231 | .000 | .000 | .003 | .006 | .077 |
| | N | 510 | 510 | 510 | 510 | 510 | 510 | 510 | 510 | 510 | 510 |
| DVO2_ Commander Blame | Person Correlation | -.239$^{**}$ | -.101$^{*}$ | 1 | -.264$^{**}$ | -.330$^{**}$ | .264$^{**}$ | .692$^{**}$ | -.001 | -.333$^{**}$ | -.036 |
| | Sig(2-tailed | .000 | .023 | | .000 | .000 | .000 | .000 | .982 | .000 | .422 |
| | N | 510 | 510 | 510 | 510 | 510 | 510 | 510 | 510 | 510 | 510 |
| DV03_Trust | Person Correlation | .070 | .020 | -.264$^{**}$ | 1 | .635$^{**}$ | -.294$^{**}$ | -.262$^{**}$ | .169$^{**}$ | .674$^{**}$ | -.106 |
| | Sig(2-tailed | .114 | .660 | .000 | | .000 | .000 | .000 | .000 | .000 | .016 |
| | N | 510 | 510 | 510 | 510 | 510 | 510 | 510 | 510 | 510 | 510 |

| | | | | | | | | | | | |
|---|---|---|---|---|---|---|---|---|---|---|---|
| DV04_ Commander Trust | Person Correlation | .176** | .053 | -.330** | .635** | 1 | -171** | -.322** | .138** | .584** | .021 |
| | Sig(2-tailed | .000 | .232 | .000 | .000 | | .000 | .000 | .002 | .000 | .630 |
| | N | 510 | 510 | 510 | 510 | 510 | 510 | 510 | 510 | 510 | 510. |
| DV05_Harm | Person Correlation | -.011 | .198** | .264** | -294** | -.171** | 1 | .361** | -.044 | -.336** | 089 |
| | Sig(2-tailed | .798 | .000 | .000 | .000 | .000 | | .000 | .323 | .000 | .046 |
| | N | 510 | 510 | 510 | 510 | 510 | 510 | 510 | 510 | 510 | 510 |
| DV06_ Commander Harm | Person Correlation | -.227** | -.170** | .692** | -.262** | -.322** | .361** | 1 | -.038 | -.322 | .041 |
| | Sig(2-tailed | .000 | .000 | .000 | .000 | .000 | .000 | | .394 | .000 | .351 |
| | N | 510 | 510 | 510 | 510 | 510 | 510 | 510 | 510 | 510 | 510 |
| DV07_ Uneasy | Person Correlation | .005 | -.129 | -.001 | .169** | .138** | -.044 | -.038 | 1 | .127-- | -.689** |
| | Sig(2-tailed | .914 | .003 | .982 | .000 | .002 | .321 | .394 | | .004 | .000 |
| | N | 510 | 510 | 510 | 510 | 510 | 510 | 510 | 510 | 510 | 510 |
| DV08_ Support | Person Correlation | .101* | -.120** | -.333** | .674** | .584** | -.336** | -.322** | .127** | 1 | -.004 |
| | Sig(2-tailed | .023 | .006 | .000 | .000 | .000 | .000 | .000 | .004 | | .934 |
| | N | 510 | 510 | 510 | 510 | 510 | 510 | 510 | 510 | 510 | 510 |

TABLE 10 Correlation matrix agency construct and dependent variables (*cont.*)

| | | Correlations | | | | | | | | | |
|---|---|---|---|---|---|---|---|---|---|---|---|
| | | **Agency_ construct** | **DVO1_ Blame** | **DVO2_ Commander Blame** | **DV03_ Trust** | **DV04_ Commander Trust** | **DV05_ Harm** | **DV06_ Commander Harm** | **DV07_ Uneasy** | **DV08_ Support** | **DV09_ Fairness** |
| DV09_ Fairness | Person Correlation | .087 | -078 | -.036 | -.106* | 0.21 | .089 | .041 | -.689** | -.004 | 1 |
| | Sig(2-tailed | .050 | .077 | .422 | .016 | .630 | .046 | .351 | .000 | .934 | |
| | **N** | 510 | 510 | 510 | 510 | 510 | 510 | 510 | 510 | 510 | 510 |

* correlation is significant at the 0.05 level (2-taild)
** correlation is significant at the 0.01 level (2-taild)

### 4.2.5 *Dependent Variables Analysis*

The correlation analysis showed that only the dependent variables *blame, commander blame, commander trust, commander harm* and *support* were significant at a minimum of a $p < .05$ level. In the next step, we took a closer zoomed into these variables and due to the non-significance, we did not analyse the rest of the dependent variables. Looking more in detail we checked to see if there are any differences between the agency levels low, neutral and high and the Autonomous Weapon and Human Operated drone conditions.

#### 4.2.5.1 Support Variable

The most interesting observation is the large difference in *support* between good and bad outcomes (Figure 11) which is significant in all three conditions ($p < .05$). This substantial difference in de mean of *support* over the three conditions indicates that our outcome manipulation works. Another interesting observation is that the *support* goes down as agency levels go up, but

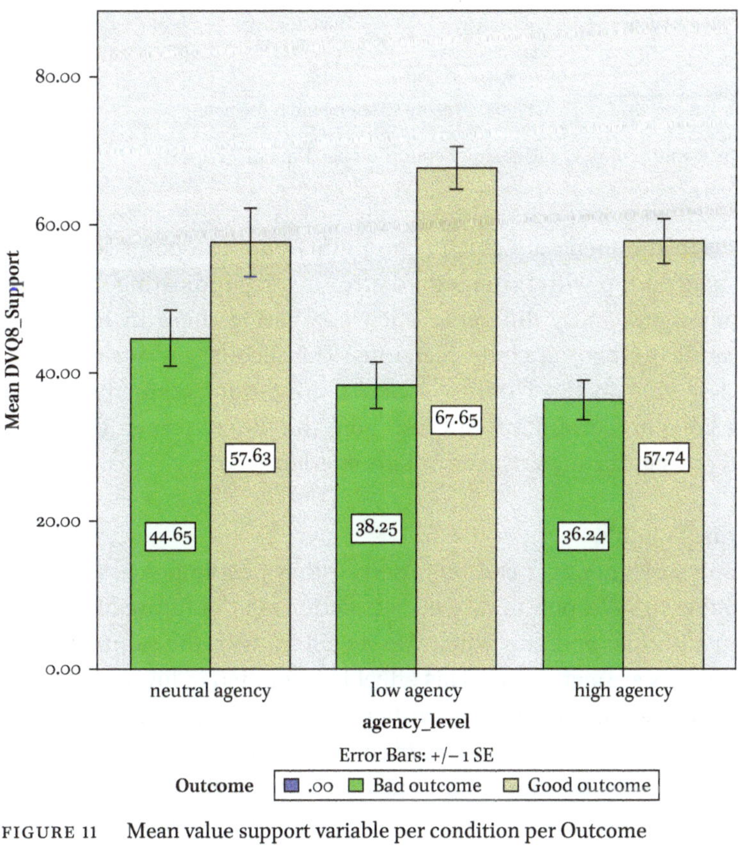


FIGURE 11 Mean value support variable per condition per Outcome

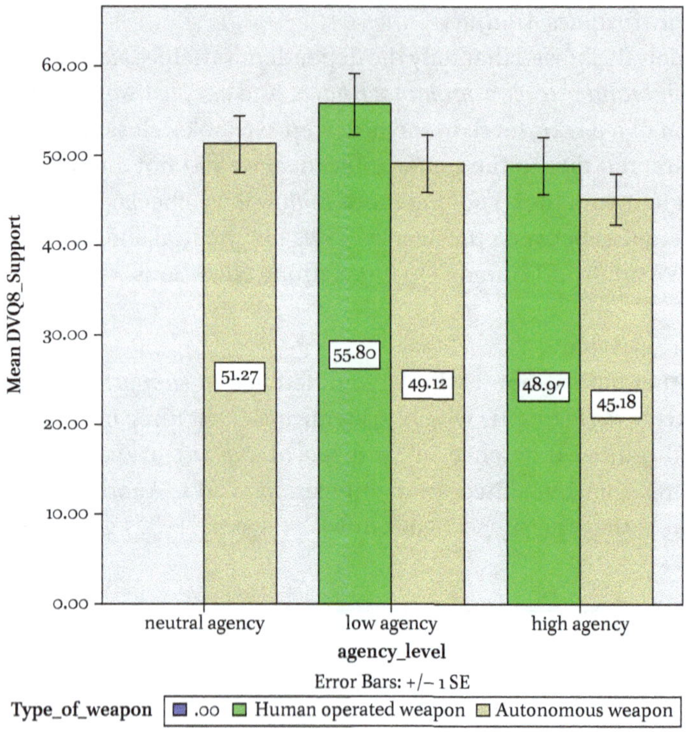


FIGURE 12 Mean value support variable per condition per Type of Weapon

the differences in the mean are small (< 10 points on a 100-point scale) and the I shaped bars on top with the standard error overlap indicating that these groups are not significantly different, when comparing them in either the good outcome or the bad outcome scenarios. The second graph shows that the support is higher in the Human Operated conditions compared to the Autonomous Weapon conditions, but also here the differences in the group are small and not significant as the error bars overlap.

#### 4.2.5.2 Trust Variable

From the graph in Figure 12 it can be observed that people are less likely to *trust* an Autonomous Weapon to take correct actions in the future after a bad outcome than a human operator who takes an action with the same bad outcome and all results are significant. This effect is magnified in low agency condition where we see almost no different between *trust* in human after good or bad outcome and large difference in *trust* in AW (Figure 13). However, the univariate analysis showed that the interaction is not significance: *Type_of_weapon* * *outcome* p = .120 so this means that we cannot draw any conclusions at this point of the study, but it is interesting finding to further investigate.

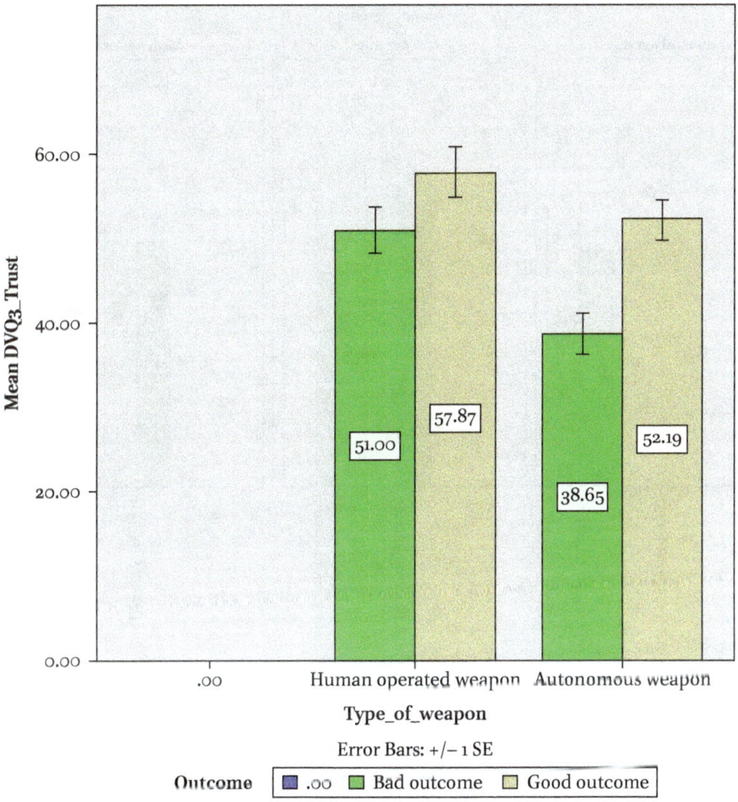


FIGURE 13 Mean value trust variable per Type of Weapon per Outcome

#### 4.2.5.3 Blame and Harm Variables

To study the chain of responsibility the mean average of the variables *blame, commander blame, harm* and *commander harm* are depicted in the same graph (Figure 15). All results of the blame variables (Figure 15), except for the neutral agency Autonomous Weapon condition, are significant. These results show that in low agency conditions the *blame* is shifted to the *commander* and in high agency conditions the *blame* is put on the drone.

There is one striking observation in that the low agency condition has a surprisingly high amount of causal responsibility for the *harm*. This pattern holds for Human Operated drones and Autonomous Weapons equally (Figure 16). Contrary to the *blame* variable, the drone is seen as doing much *harm* in both high as low agency conditions, but in a low agency condition the *harm* is shifted almost equally to the *commander*. This means that they are both causing *harm*, but in high agency conditions it is much less transferred to the *commander* and the drone is seen as causing the *harm*. However, the difference between the variables *harm* and *commander harm* in the low agency condition of both

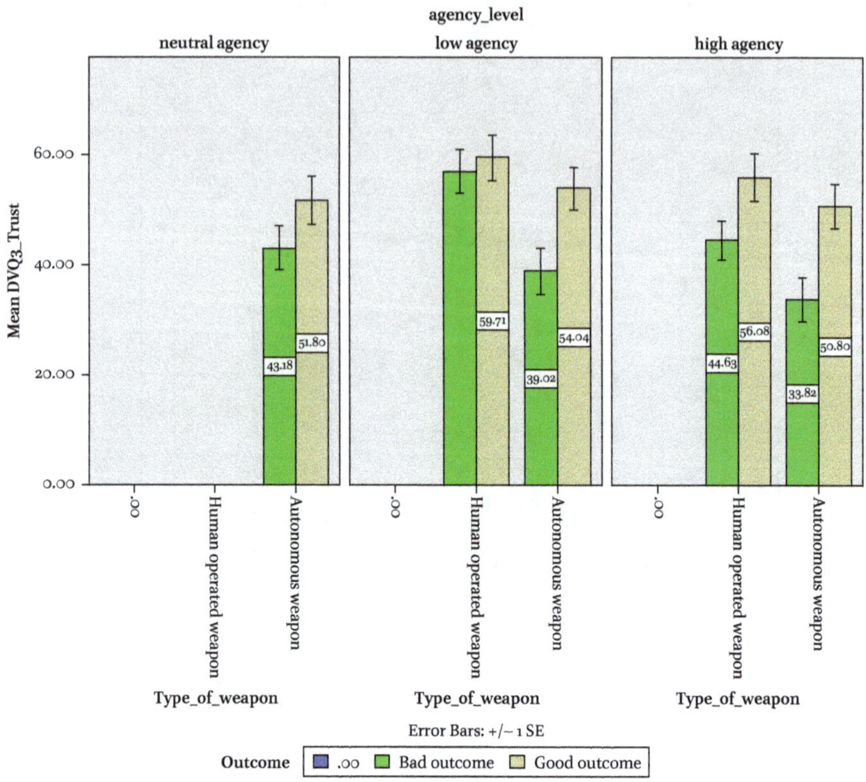


FIGURE 14 Mean value trust variable per condition per Outcome

the Human Operated as the Autonomous Weapon scenarios is not significant therefore we must caution with drawing conclusions on the observed effect.

### 4.2.6 *Conclusion Pilot Study 1*

The results of pilot study 1 lead to the following conclusions:

- Based on the results of the PCA and reliability analysis of the agency items, we decided to drop *SQ3_Moral rules* from the agency construct which improves the reliability of the construct to $\alpha = 0.933$. In the next study, we will use the agency construct that contains four items being *thought, goal setting, act freely* and *achieve goals*.
- The results of the T-tests indicate that there is a difference in the mean of the agency perception between the Autonomous Weapon and Human Operated conditions. There is also much difference between the high agency and low/neutral agency scenarios. However, between low and neutral agency there was not much difference observed. Therefore, we decided

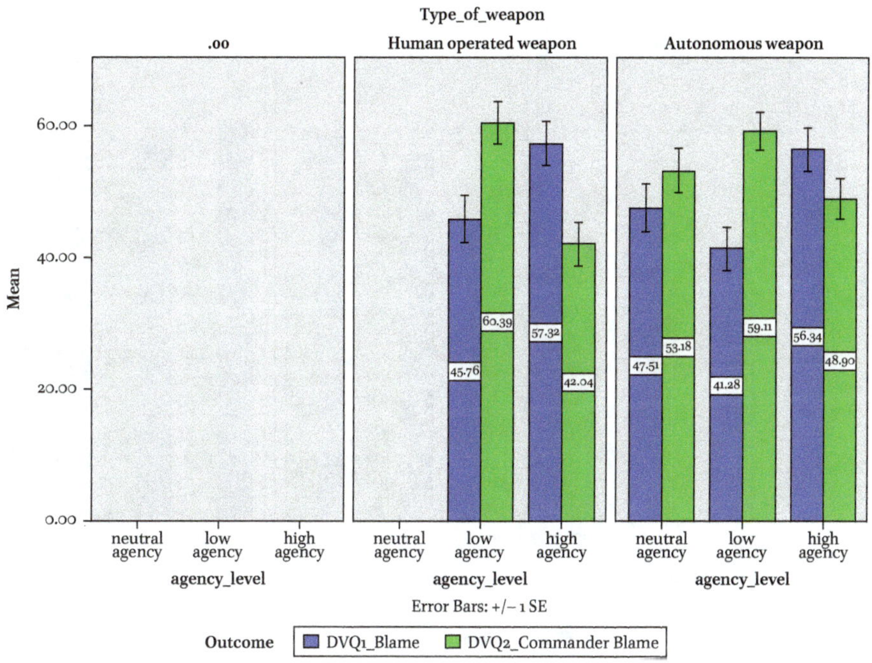


FIGURE 15 Mean value blame and commander blame variables per condition and Type of Weapon

to drop the neutral agency condition in the next study and only test the low agency and high agency conditions.

- The most explicit observation in the *support* variable is that there is a big difference in support between good and bad outcomes (Figure 11). However this is what to be expected based on literature, because James Igoe Walsh (2015) and Kreps (2014) found that the general public has more *support* for drone strikes if there is no collateral damage compared to those that have collateral damage, but it shows that our outcome manipulation works and we will use this in the next study.
- In the analysis of the *trust* variable (Figure 13) we found that people are less likely to *trust* an Autonomous Weapon to take correct actions in the future after a bad outcome than a human operator who takes an action with the same bad outcome. This might indicate algorithm aversion for which algorithms are punished more than humans who make mistakes (Dietvorst, 2016) and we used these findings for the next pilot study to further study the algorithm aversion effect.
- We found that that in the low agency condition a surprisingly high amount of causal responsibility for the *harm* can be seen and that this pattern holds

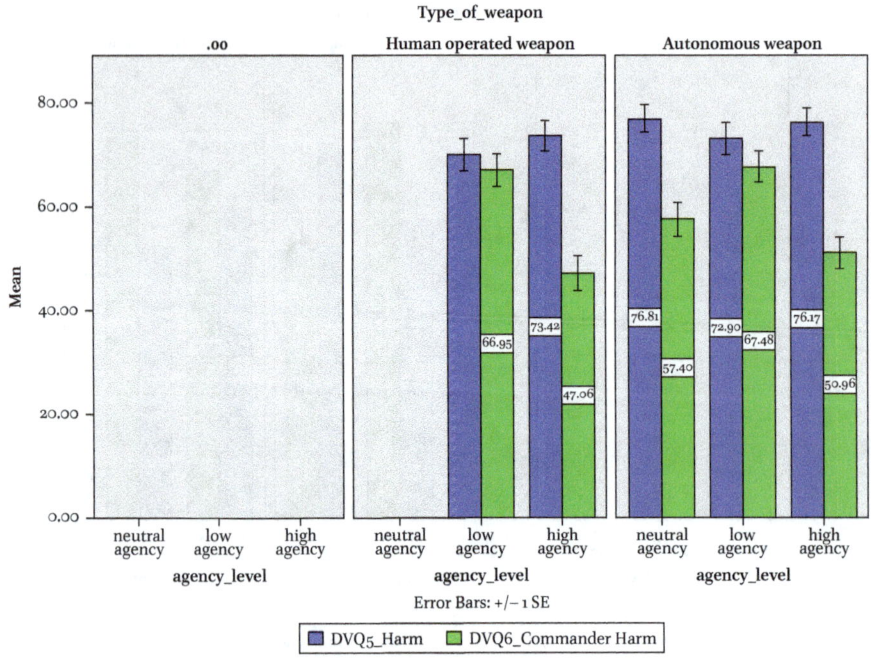


FIGURE 16 Mean value harm and commander harm variables per condition and Type of Weapon

for Human Operated drones and Autonomous Weapons equally (Figure 16). We also found that the drone is seen as doing much *harm* in both high as low agency conditions, but in a low agency condition the harm is shifted almost equally to the *commander*. Although these findings are not significant, they are interesting and require more study therefore we kept these questions in pilot study 2.

## 4.3 Pilot Study 2

In pilot study 2 we tried to get a better understanding of the algorithm aversion effects so we tested scenarios describing that the Autonomous Weapon could learn from its mistakes (3 + 10), was trained on a large data set and that it understands its mission (4 + 11), that it could be sometimes unpredictable (5 + 12) and a scenario that includes all of these aspects (6 + 13). We tested both on good outcome and bad outcome conditions and added a scenario for the Autonomous Weapon with low agency (1 + 8) to check if our agency manipulation works and a Human Operated scenario (7 + 14) in which we asked about the human operator instead about the drone which is different from pilot study 1.

TABLE 11 Scenarios of pilot study 2

| *Aspects* | | | | | | | | |
|---|---|---|---|---|---|---|---|---|
| | | **Autonomous drone** | | | | | | **HO** |
| | | **Level 1 -Low agency** | **Level 2 (high agency + no extra)** | **Level 3 (high agency + learning ability)** | **Level 4 (high agency + understanding procedures)** | **Level 5 (high agency + unpredictability)** | **Level 6 (high agency + all aspects)** | **Human operated drone** |
| *Outcome* | **Collateral damage (Bad)** | 1 | 2 | 3 | 4 | 5 | 6 | 7 |
| | **No collateral damage (Good)** | 8 | 9 | 10 | 11 | 12 | 13 | 14 |

TABLE 12 Results reliability analysis pilot study 2

| Reliability statistics | | |
|---|---|---|
| Cronbach's Alpha | Cronbach's Alpha based on standardized items | N of Items |
| .914 | .914 | 4 |

Item-total statistics

| | Scale mean if Item deleted | Scale variance if item deleted | Corrected item-total correlation | Squared multiple correlation | Cronbach's alpha if item deleted |
|---|---|---|---|---|---|
| Thought | 208.4827 | 6427.511 | .801 | .647 | .890 |
| Goal_setting | 204.9885 | 6499.803 | .834 | .695 | .877 |
| Act_freely | 203.4020 | 6782.379 | .804 | .646 | .888 |
| Achieve_goals | 199.4975 | 7114.019 | .781 | .614 | .897 |

We tested at total of 14 scenarios and aimed for 50 respondents per scenario which means 700 responses in total. The number of respondents per scenario ranged between 47 and 54. After the data pre-processing and deleting the respondents who failed the attention check we got 709 complete responses.

### 4.3.1 *Reliability Analysis*

For all four agency items $\alpha = 0.914$ which is well above the threshold of 0.7 and cannot be improved by deleting an item. Deletion of other items will decrease the $\alpha$ value and reduce the internal consistency.

### 4.3.2 *Principal Component Analysis* (PCA)

The PCA shows that *Thought, Goal setting, Act freely* and *Achieve goals* can be viewed as one construct which accounts for 79.60 % variance in the original variables (Table 13). This is also seen in the scree plot that shows the eigenvalues of the components (Figure 17). The PCA shows that the four items; *thought, goal setting, act freely* and *achieve goals*, can be viewed as the underlying components of the agency construct (Table 13).

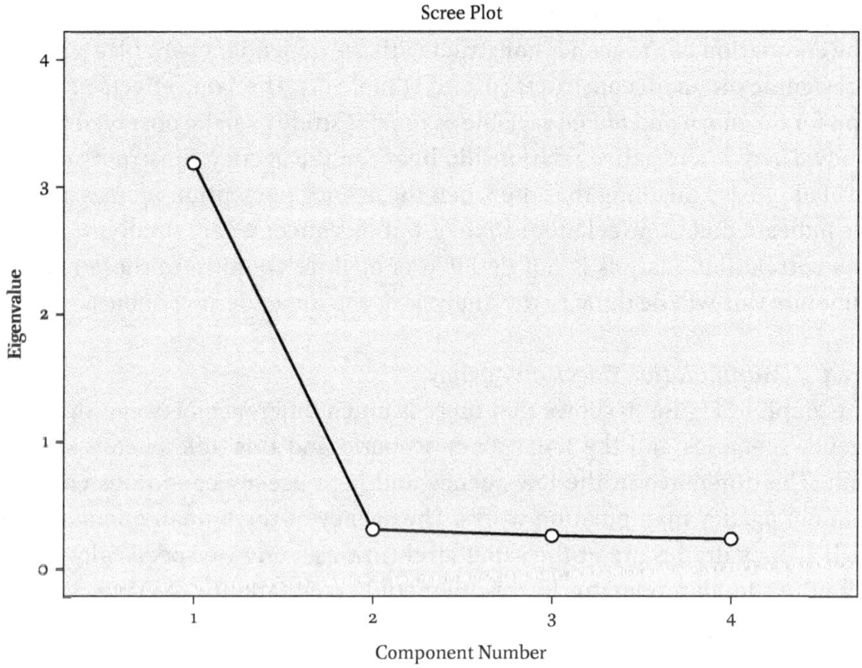


FIGURE 17 Scree plot PCA pilot study 2

TABLE 13 Results PCA pilot study 2

| | Component matrix[a] | |
|---|---|---|
| | Component | |
| | 1 | 2 |
| Goal_setting | .910 | |
| Act_freely | .892 | |
| Thought | .889 | |
| Achieve_goals | .877 | .446 |

Extraction method: principal component analysis.
[a] 2 components extracted.

### 4.3.3 *Correlation Analysis*

The correlation of the agency construct with the dependent variables is small, but significant for all constructs ($p < .05$) (Table 14). The same effects in direction for the *harm* and *blame* variable as in pilot study 1 can be observed in this study. There is a negative relationship between the agency construct and the level of uneasy meaning that the when the agency perception increases people indicate that they feel more *anxiety*, but this effect is very small ($r = -.076$). The correlation analysis is not detailed enough to zoom in to these results, therefore this will be done in the analysis of the dependent variables.

### 4.3.4 *Manipulation Check on Agency*

The graph in Figure 18 shows that there is much difference between the high agency scenarios and the low agency scenario and this difference is significant. The difference in the low agency and high agency conditions confirm that our agency manipulation works. The agency of the human operator and high agency drones are at the same level (in this study we specifically asked about the human operator in this pilot study compared to the drone in the

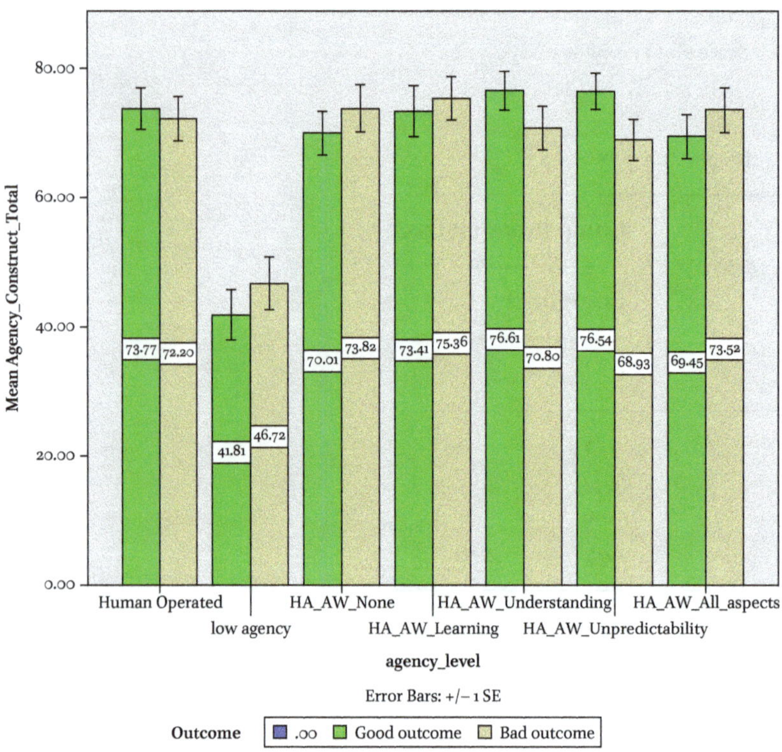


FIGURE 18 Mean value agency construct per condition per Outcome

previous study). There is some difference between the good outcome and bad outcome conditions and although these effects are largest for the conditions in which we zoomed into the algorithm aversion, the difference is not very large (a difference in mean of 8 points on a 100-point scale). Also, the error bars between the good outcome and bad outcome conditions for all scenarios overlap indicating that these are not distinct groups. The only distinct and significant difference in groups ($p < .05$) is between the low agency Autonomous Weapon condition and the other scenarios. This is also confirmed by the level of significance of the independent samples T-test which we chose not to report for the graph below for the sake of clarity.

### 4.3.5 *Dependent Variables Analysis*

The correlation between the agency construct and the dependent variables are all significant ($p < .05$) therefore the most striking findings for each of the dependent variables is described below.

#### 4.3.5.1 Blame

The most notable finding for the *blame* variable is that people assign less *blame* in the low agency condition compared to the high agency Autonomous Weapon and the Human Operator conditions. For these conditions, the difference is significant ($p < .05$). Some effects of the algorithm aversion can be seen in the learning and understanding scenario where people assign less *blame* when the Autonomous Weapons does not make a mistake and more *blame* when it makes a mistake with a bad outcome. However, these conditions show an overlap in error bars and are not significant.

#### 4.3.5.2 Commander Blame

The most striking observation in the *commander blame* variable is that the *commander* is *blamed* a lot less when the Autonomous Weapon makes a mistake in the learning and unpredictability conditions and this is significant ($p < .05$). It might be the case that the *blame* is shifted to the drone as people expect it to be able to learn from its mistakes or it behave unpredictable and that the *commander* is not to *blame* for that. Also, in the low agency condition the *commander* is *blamed* a bit more than in the other conditions.

#### 4.3.5.3 Trust

Over all the conditions it can be observed that people have more trust that the Autonomous Weapon or Human Operator will take correct actions in the future after a good outcome than a bad outcome. Except for the low agency condition, the difference in *trust* in the good and bad outcome conditions

TABLE 14 Correlation matrix agency construct and dependent variables pilot study 2

| | | **Correlations** | | | | |
|---|---|---|---|---|---|---|
| | | **Agency_ construct_ total** | **DVD1_ Blame** | **DVD2_ Commander Blame** | **DVD3_ Trust** | **DVD4_ Commander Trust** |
| Agency_ Construct_ Total | Person Correlation | 1 | .271** | -.267** | .205** | .217** |
| | Sig(2-tailed) | | .000 | .000 | .000 | .000 |
| | N | 708 | 708 | 708 | 708 | 708 |
| DVD1_Blame | Person Correlation | .271 | 1 | -.069 | -.028 | -.046 |
| | Sig(2-tailed) | .000 | | .045 | .464 | .226 |
| | N | 708 | 708 | 708 | 708 | 708 |
| DVD2_ Commander Blame | Person Correlation | -.267** | -.069 | 1 | -.287** | -.278** |
| | Sig(2-tailed) | .000 | .065 | | .000 | .000 |
| | N | 708 | 708 | 708 | 708 | 708 |
| DVD3_Trust | Person Correlation | .205** | -.028 | -.207** | 1 | .653** |
| | Sig(2-tailed) | .000 | .464 | .000 | | .000 |
| | N | 708 | 708 | 708 | 708 | 708 |
| DVD4_ Commander Trust | Person Correlation | .217** | -.046 | -.278** | .653** | 1 |
| | Sig(2-tailed) | .000 | .226 | .000 | .000 | |
| | N | 708 | 708 | 708 | 708 | 708 |
| DVD5_Harm | Person Correlation | .124** | .258** | .152** | -.281** | -.185** |
| | Sig(2-tailed) | .001 | .000 | .000 | .000 | .000 |
| | N | 708 | 708 | 708 | 708 | 708 |
| DVD6_ Commander Hram | Person Correlation | -.198** | -.023 | .691** | -.224** | -.228** |
| | Sig(2-tailed) | .000 | .534 | .000 | .000 | .000 |
| | N | 708 | 708 | 708 | 708 | 708 |

| DVD5_ Harm | DVD6_ Commander Hram | DVD7_ Dignity | DVD8_ confidence | DVD9_ Expectations | DVD10_ Support | DVD11_ Fair | DVD12_ Uneasy |
|---|---|---|---|---|---|---|---|
| .124** | -.198** | .102** | .209** | .146** | .164** | .142** | -.076** |
| .001 | .000 | .007 | .000 | .000 | .000 | .000 | .043 |
| 708 | 708 | 708 | 708 | 708 | 708 | 708 | 708 |
| .258** | -.023 | -.056 | -.048 | -.137** | -.118** | -.092* | .180** |
| .000 | .524 | -.138 | .198 | .000 | .002 | .014 | .000 |
| 708 | 708 | 708 | 708 | 708 | 708 | 708 | 708 |
| .152** | .691** | -.206** | -.300** | -.205** | -.250** | -.275** | .312** |
| .000 | .000 | .000 | .000 | .000 | .000 | .000 | .000 |
| 708 | 708 | 708 | 708 | 708 | 708 | 708 | 708 |
| -.281** | -.224** | .595** | .901** | .533** | .726** | .722** | -.616* |
| .000 | .000 | .000 | .000 | .000 | .000 | .000 | .000 |
| 708 | 708 | 708 | 708 | 708 | 708 | 708 | 708 |
| -.185** | -.228** | .432** | .635** | .425** | .631** | .552** | -.458** |
| .000 | .000 | .000 | .000 | .000 | .000 | .000 | .000 |
| 708 | 708 | 708 | 708 | 708 | 708 | 708 | 708 |
| 1 | .311** | -.374** | -.301** | -.217** | -.253** | -.324** | .395** |
| .000 | .000 | .000 | .000 | .000 | .000 | .000 | .000 |
| 708 | 708 | 708 | 708 | 708 | 708 | 708 | 708 |
| .311** | 1 | -.183** | -.226** | -.203** | -.235** | -.249** | .289** |
| .000 | | .000 | .000 | .000 | .000 | .000 | .000 |
| 708 | 708 | 708 | 708 | 708 | 708 | 708 | 708 |

TABLE 14 Correlation matrix agency construct and dependent variables pilot study 2 (*cont.*)

| | | **Correlations** | | | | |
|---|---|---|---|---|---|---|
| | | **Agency_ construct_ total** | **DVD1_ Blame** | **DVD2_ Commander Blame** | **DVD3_ Trust** | **DVD4_ Commander Trust** |
| DVD7_ Dignity | Person Correlation | .102** | -.056 | -.206** | .595** | .432** |
| | Sig(2-tailed) | .007 | .138 | .000 | .000 | .000 |
| | N | 708 | 708 | 708 | 708 | 708 |
| DVD8_ confidence | Person Correlation | .209** | -.048 | -.300** | .901** | .635** |
| | Sig(2-tailed) | .000 | .198 | .000 | .000 | .000 |
| | N | 708 | 708 | 708 | 708 | 708 |
| DVD9_ Expectations | Person Correlation | .146** | -.137** | -.205** | .533** | .425** |
| | Sig(2-tailed) | .000 | .000 | .000 | .000 | .000 |
| | N | 708 | 708 | 708 | 708 | 708 |
| DVD10_ Support | Person Correlation | .164** | -.118** | -.250** | .726** | .631** |
| | Sig(2-tailed) | .000 | .002 | .000 | .000 | .000 |
| | N | 708 | 708 | 708 | 708 | 708 |
| DVD11_Fair | Person Correlation | .142** | -.092* | -.275** | .722** | .552** |
| | Sig(2-tailed) | .000 | .014 | .000 | .000 | .000 |
| | N | 708 | 708 | 708 | 708 | 708 |
| DVD12_ Uneasy | Person Correlation | -.076* | .190** | .312** | -.616** | -.458** |
| | Sig(2-tailed) | .043 | .000 | .000 | .000 | .000 |
| | N | 708 | 708 | 708 | 708 | 708 |

** Correlation is significant at the 0.01 level (2-tailed).
* Correlation is significant at the 0.05 level (2-tailed).

| DVD5_ Harm | DVD6_ Commander Hram | DVD7_ Dignity | DVD8_ confidence | DVD9_ Expectations | DVD10_ Support | DVD11_ Fair | DVD12_ Uneasy |
|---|---|---|---|---|---|---|---|
| -.374** | -.183** | 1 | .588** | .415** | .533** | .598** | -.509** |
| .000 | .000 | | .000 | .000 | .000 | .000 | .000 |
| 708 | 708 | 708 | 708 | 708 | 708 | 708 | 708 |
| -.301** | -.226** | .588** | 1 | .553** | .723** | .725** | -.649** |
| .000 | .000 | .000 | | .000 | .000 | .000 | .000 |
| 708 | 708 | 708 | 708 | 708 | 708 | 708 | 708 |
| -.217** | -.203** | .415** | .553** | 1 | .511** | .621** | -.509** |
| .000 | .000 | .000 | .00 | | .000 | .000 | .000 |
| 708 | 708 | 708 | 708 | 708 | 708 | 708 | 708 |
| -.253** | -.235** | .533** | .723** | .511** | 1 | .698** | -.651** |
| .000 | .000 | .000 | .000 | .000 | | .000 | .000 |
| 708 | 708 | 708 | 708 | 708 | 708 | 708 | 708 |
| -.324** | -.248** | .598** | .725** | .621** | .685** | 1 | -.656** |
| .000 | .000 | .000 | .000 | .000 | .000 | | .000 |
| 708 | 708 | 708 | 708 | 708 | 708 | 708 | 708 |
| 395** | .289** | -.509** | -.649** | -.509** | -.651** | -.636** | 1 |
| 000 | .000 | .000 | .000 | .000 | .000 | .000 | |
| 708 | 708 | 708 | 708 | 708 | 708 | 708 | 708 |

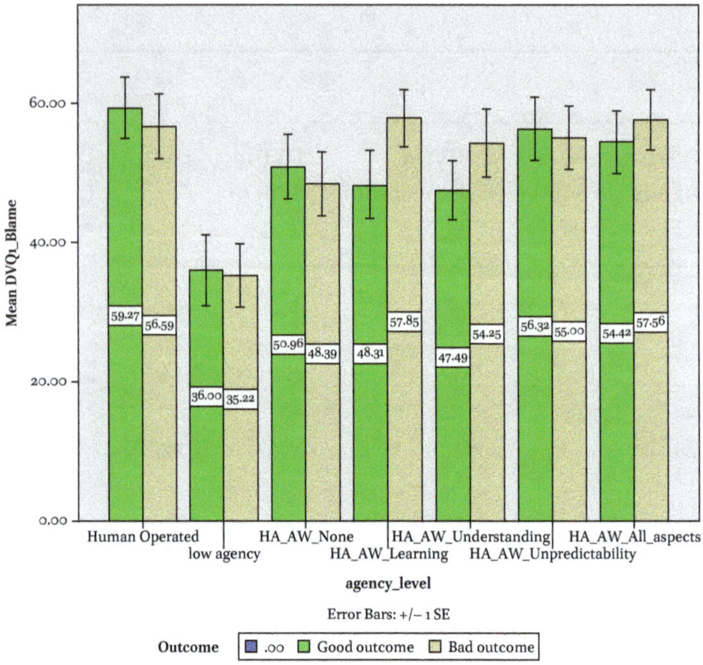


FIGURE 19 Mean value blame variable per condition per outcome

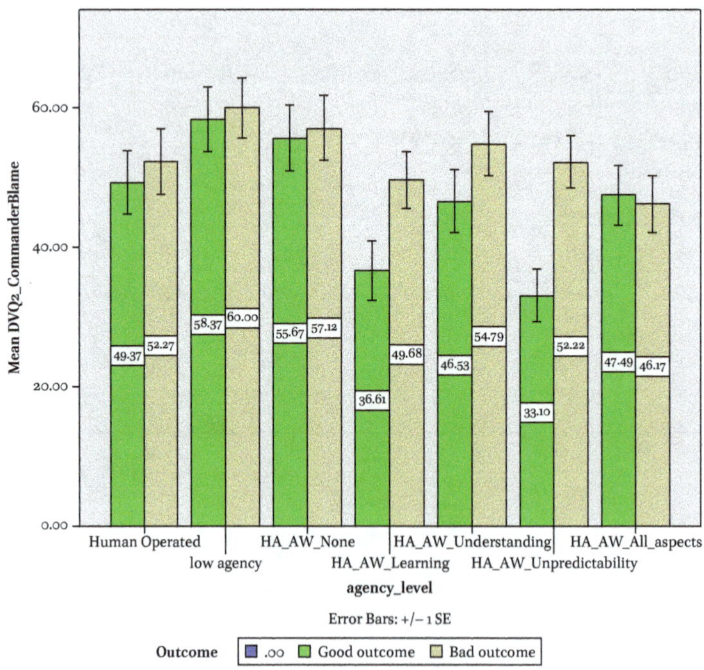


FIGURE 20 Mean value commander blame variable per condition per outcome

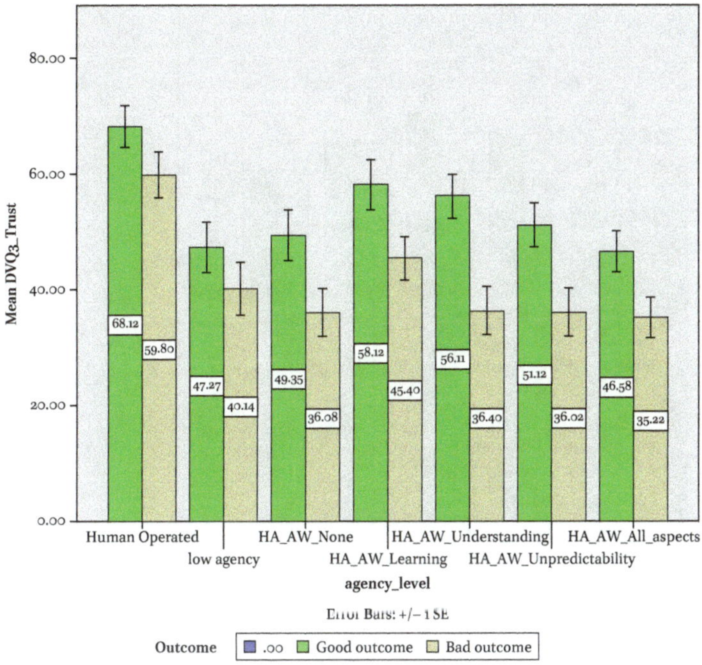


FIGURE 21 Mean value trust variable per condition per outcome

is significant ($p < .05$). People *trust* the Human Operator more than the Autonomous Weapon, especially in the low agency condition and the difference between these groups is significant ($p < .05$).

#### 4.3.5.4 Commander Trust

Overall people *trust* that the *commander* will take correct actions more in the good outcome conditions compared to the bad outcome conditions. The *trust* in the *commander* is lowest in the low agency condition and high agency condition with no extra information and the difference between these groups is significant ($p < .05$). Knowing that the Autonomous Weapon has understanding capabilities results in higher trust in the bad outcome condition that is significant ($p < .05$).

#### 4.3.5.5 Harm

In the bad outcome conditions people perceive more *harm* than in the good outcome conditions which could be expected and the difference in nearly all conditions is significant ($p < .05$). Most notable in this graph is that when the Autonomous Weapon behaves unpredictable, the type of outcome does not seem to matter and the perception of harm is equal, but the difference between both outcome groups is not significant.

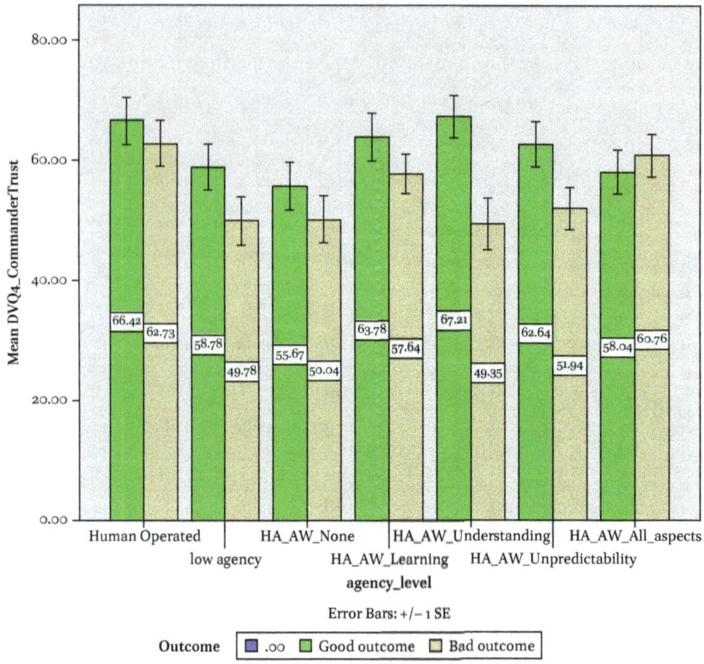


FIGURE 22 Mean value commander trust variable per condition per outcome

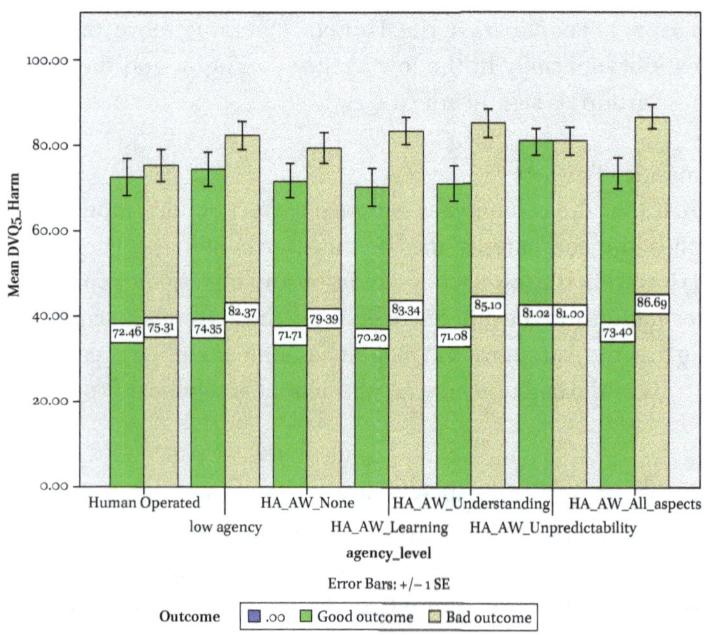


FIGURE 23 Mean value harm variable per condition per outcome

#### 4.3.5.6 Commander Harm

In the bad outcome conditions people perceive the *commander* doing more *harm* than in the good outcome conditions which could be expected. Most striking is the large difference in the learning condition between the good and bad outcome where the commander is attributed much less harm in the good outcome condition which is significant (p < .05). The differences between good and bad outcome conditions between in the Human Operator and high agency unpredictability scenarios are also significant (p < .05).

#### 4.3.5.7 Human Dignity

The Human Operator acts with much more respect for *human dignity* than the Autonomous Weapon except for the condition where the Autonomous Weapon is trained on a large data set for its mission and this difference is significant (p < .05). Also, it seems to matter when the Autonomous Weapon to make a mistake, because we observe significant (p < .05) differences between good and bad outcome scenarios for Autonomous Weapons for the low agency, high agency no aspects, and high agency all aspects scenarios.

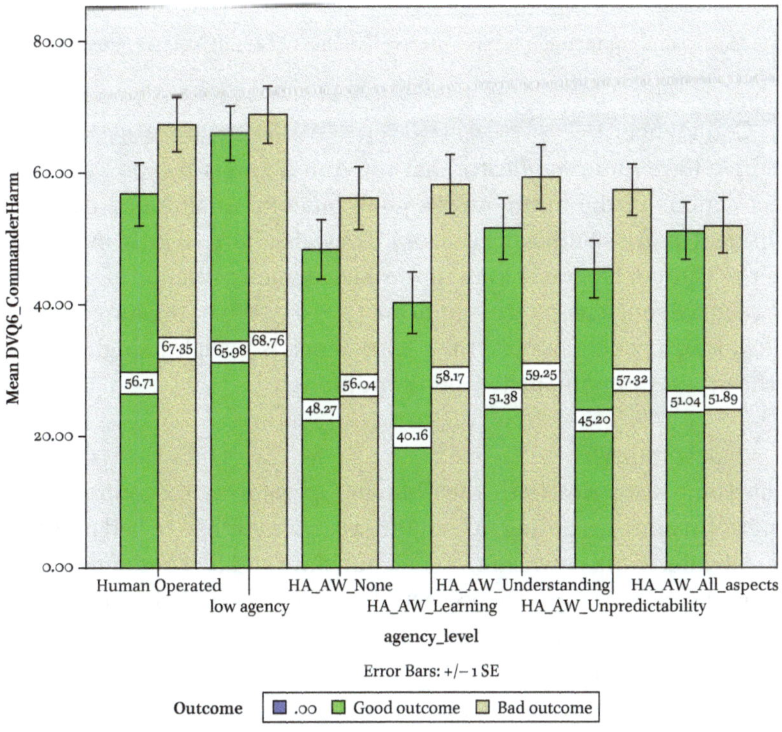


FIGURE 24 Mean value commander harm variable per condition per outcome

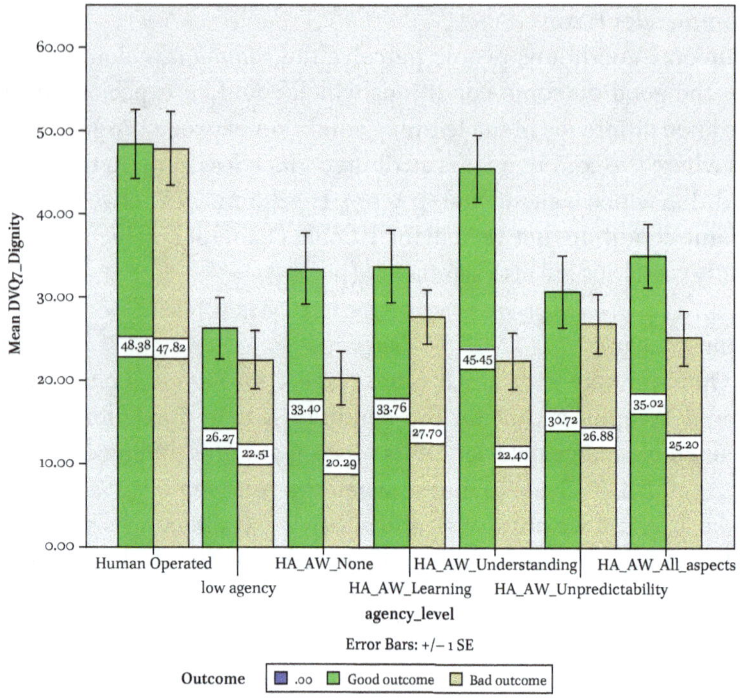


FIGURE 25 Mean value commander human dignity variable per condition per outcome

#### 4.3.5.8 Confidence

Overall people have more *confidence* that the Autonomous Weapon will take the correct actions in the future after a good outcome than a bad outcome and all differences are significant ($p < .05$). They also have slightly more *confidence* in the Human Operator than in the Autonomous Weapon. The differences between these scenarios are significant ($p < .05$). When the Autonomous Weapon has learning capabilities, their confidence is a slightly higher compared to the other Autonomous Weapon conditions.

#### 4.3.5.9 Expectations

In good outcome scenarios both the Human Operator as the Autonomous Weapon perform more according to people's *expectations* than in bad outcome scenarios. All differences between the good outcome and bad outcome conditions are significant ($p < .05$). The expectations are lowest in the low agency condition when there is a bad outcome, but there is not much difference compared to the other conditions.

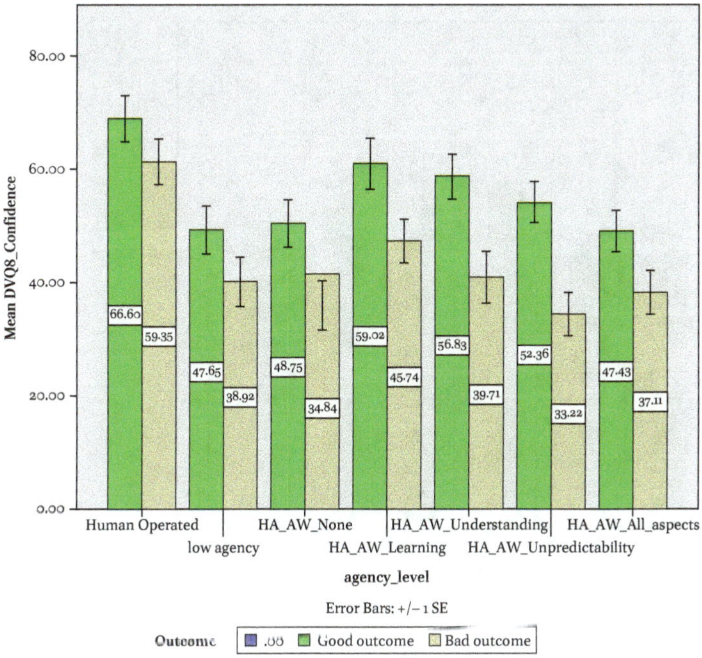


FIGURE 26 Mean value confidence variable per condition per outcome

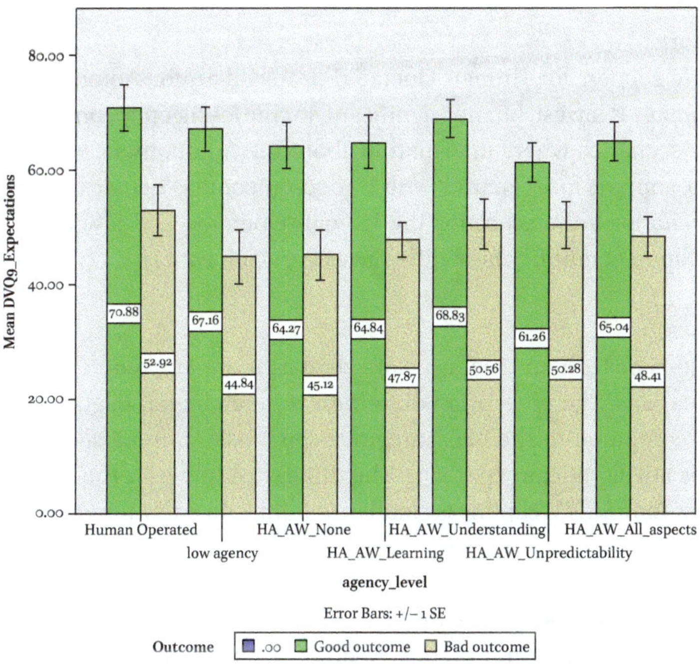


FIGURE 27 Mean value expectations variable per condition per outcome

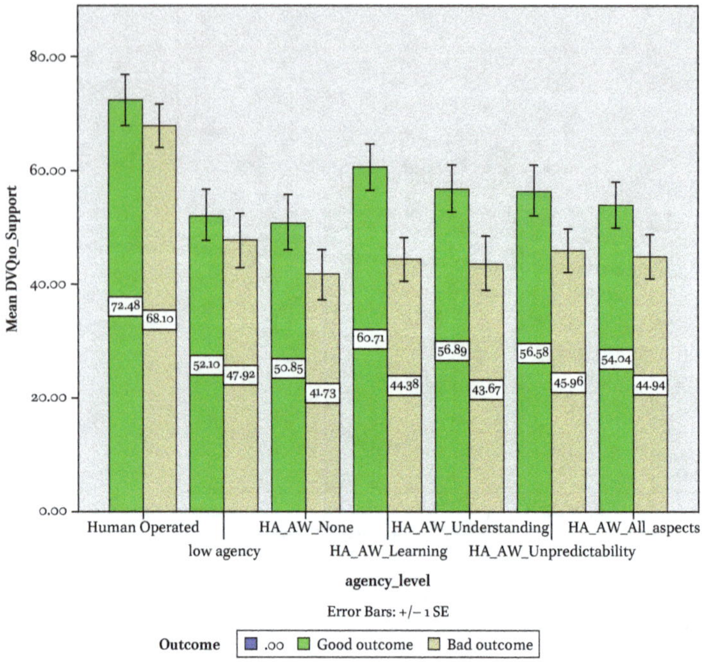


FIGURE 28 Mean value support variable per condition per outcome

### 4.3.5.10 Support

People have more *support* for Human Operated drones than for Autonomous Weapons. The *support* is lowest, but not significant, for the low agency conditions compared to the scenarios where information about the Autonomous Weapon is provided. Also, support for scenarios with a good outcome is higher than for bad outcome scenarios and except for the Human Operated and low agency Autonomous Weapon condition these differences are significant ($p < .05$).

### 4.3.5.11 Fairness

Overall, the actions of Human Operators are seen as more *fair* than those of Autonomous Weapons. The perception of *fairness* is much higher in good outcome scenarios compared to the bad outcome scenarios and in all scenarios these differences are significant ($p < .05$). The difference between the agency conditions of Autonomous Weapons is small (<10 points on a 100-point scale).

### 4.3.5.12 Uneasy

In bad outcome scenarios people feel more *unease* with the actions of both the Human Operator as the Autonomous Weapon. There is not much difference between the level of anxiety for the different conditions of Autonomous Weapons. The anxiety level for the Human Operator is much lower than that

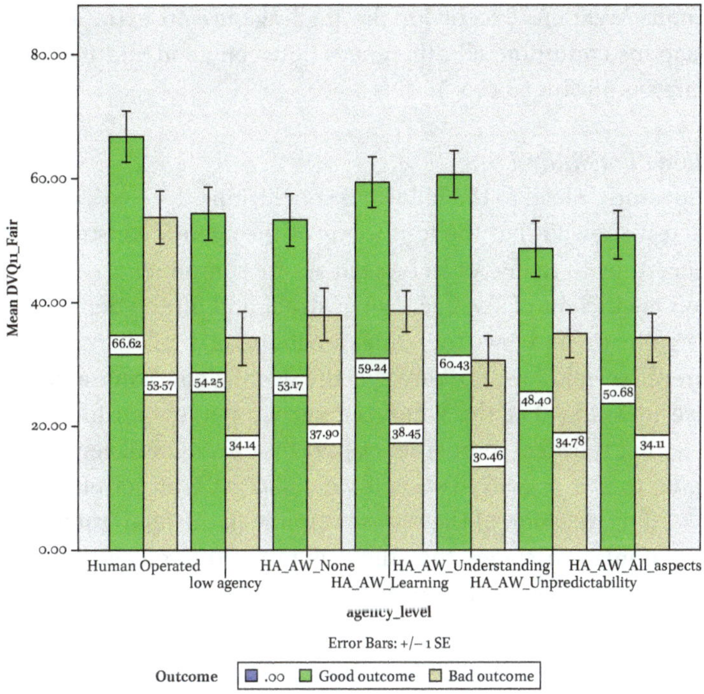


FIGURE 29 Mean value fairness variable per condition per outcome

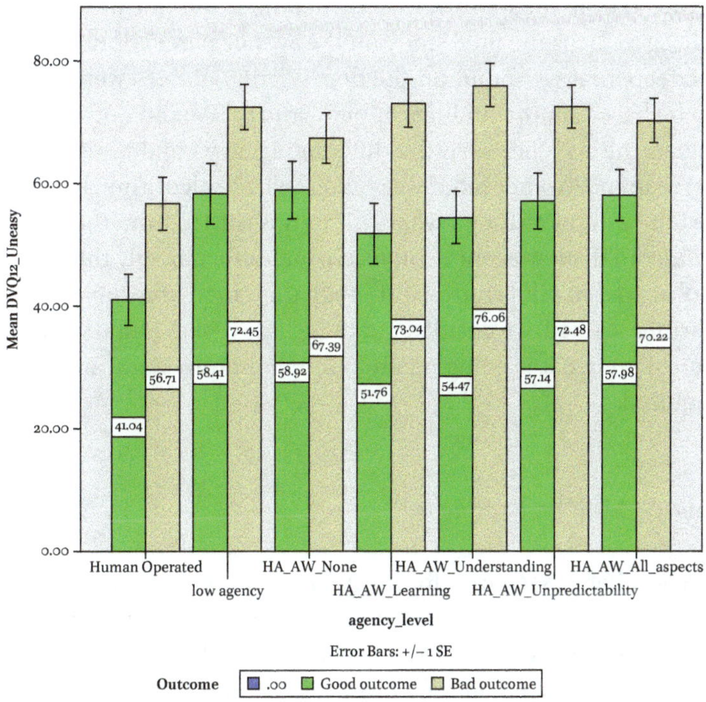


FIGURE 30 Mean value unease variable per condition per outcome

for the Autonomous Weapon. Except for the high agency no extra aspects Autonomous Weapons condition, all differences between good and bad outcome scenarios are significant ($p < .05$).

### 4.3.6 *Conclusion Pilot Study 2*

The results of pilot study 2 lead to the following conclusions:

- The reliability test showed that the four items of the agency construct are reliable and according to the PCA can be seen as one component;
- The correlation coefficient of the agency construct and all the dependent variables are significant and have the same directions as in pilot study one. We found a negative relationship between the agency construct and the level of *unease* meaning that the when the agency perception increases people indicate that they feel more uneasy, but this effect is still very small and approaching to 0 ($r = -.076$). To see if we could get a bigger effect we changed the wording question of the *unease* variable in the final study;
- The agency manipulation shows that there is much difference between the high agency scenarios and the low agency scenario. The agency of the human operator and high agency Autonomous Weapons are at the same level. However, the mean of the agency construct for the high agency scenarios that include the algorithm aversion effects are almost the same. To draw out the algorithm aversion effect we need to do more literature and pilot studies. Due to time constraints, we did not pursue this line of research further and chose to focus the final study on the agency perception.
- Many of the effects that we found on the dependent variables were as we expected it to be based on the findings of pilot study 1. We did not observe distinct effects in the DV's between the different agency conditions of the Autonomous Weapon. We checked if we could replicate algorithm aversion effect that we found in pilot study 1 where *harm* was shifted from the drone to the *commander* in the low agency condition, but unfortunately this effect could not be observed in pilot study 2. Although this chain of responsibility needs to be further studied, we decided not to include it in the final study as we are not quite sure what is going on and more literature study and pilot studies are required.

## 4.4 Final Study – Military Sample

To determine the number of scenarios for the final study, we performed Power calculations to estimate the total number of participants that we would need based on the results of the first pilot study, because we will use these scenarios also for the final study. Based on the Power calculations (effect size 0.4, a desired statistical power of 0.8 and a probability level of 0.05) we aimed for a total of 200 responses and determined that we could run 3 scenarios. We chose to focus on the agency perception and to compare the current technology, a Human Operated drone, to future technology, the high agency Autonomous Weapon (Table 15). We also added a neutral agency Autonomous Weapon scenario to compare the agency perception of the three conditions. In this study, we specifically asked about the drone and not the Human Operator controlling it in order to compare the agency perception of the artefacts. As we could only run three scenarios, we focused on the bad outcome condition, because the pilot studies showed that the effects most distinct in this condition.

We distributed the survey via email with an anonymous link and therefore had no guarantee on the number of respondents. After the data pre-processing and deleting the respondents who failed the attention check we were left 239 responses for the complete sample. The number of respondents per scenario ranged between 64 and 96 which is curious because the respondents should be distributed evenly over the scenarios. The sample of 239 is a combination of Dutch military and civilian personnel working at the MOD, because if we would exclude the civilians only 149 respondents are left which is too little for a robust data analysis.

*For this design, the following definitions are used:*

- *Human operated drone:* drone directly controlled by a human. Hypothesized to be perceived as having very low agency.
- *High Agency AW:* participants are told that the Autonomous Weapon has agentic characteristics (such as weighing pros and cons and making independent choices). Hypothesized to be perceived as having very high agency.
- *Neutral Agency AW:* No information about agentic characteristics of the Autonomous Weapon are given. The strength of this approach is that we will be able to gather data on how participants use their pre-formed concepts of Autonomous Weapons to understand the actions of the Autonomous Weapon and compare those conceptions to the human operated and high agency Autonomous Weapon conditions.

TABLE 15 Scenarios final study

| | **Agency levels** | | |
|---|---|---|---|
| | **Human operated** | **Neutral agency AW** | **High agency AW** |
| **Bad outcome** | 1 | 2 | 3 |

TABLE 16 Results reliability analysis agency items

| Reliability statistics | |
|---|---|
| Cronbach's Alpha | N of items |
| .865 | 4 |

Item-total statistics

| | Scale mean if item deleted | Scale variance if item deleted | Corrected item-total correlation | Cronbach's alpha if item deleted |
|---|---|---|---|---|
| Thought | 116.43 | 7175.573 | .620 | .863 |
| Goal setting | 123.33 | 6342.474 | .756 | .810 |
| Act freely | 122.62 | 6566.723 | .740 | .816 |
| Achieve goals | 118.84 | 6459.146 | .740 | .816 |

### 4.4.1 *Reliability Analysis*

For all four agency items $\alpha = .865$ which is above the threshold of 0.7 and cannot be improved by deleting an item which means that the agency construct is reliable (Table 16). Deletion of other items will decrease the $\alpha$ value and reduce the internal consistency.

### 4.4.2 *Principal Component Analysis* (PCA)

The PCA shows that *Thought, Goal setting, Act freely* and *Achieve goals* can be viewed as one construct which accounts for 71.18 % variance in the original variables. This is also seen in the scree plot that shows the eigenvalues of the components (Figure 31). The PCA shows that the four items; *thought, goal setting, act freely* and *achieve goals*, can be viewed as one component of the agency construct (Table 17).

### 4.4.3 *Correlation Analysis Agency Construct*

The correlation of the agency construct with the dependent variables is larger than in the previous pilot studies, but only significant for 7 out of 9 of the variables; *trust, human dignity, confidence, expectations, support, fairness* and

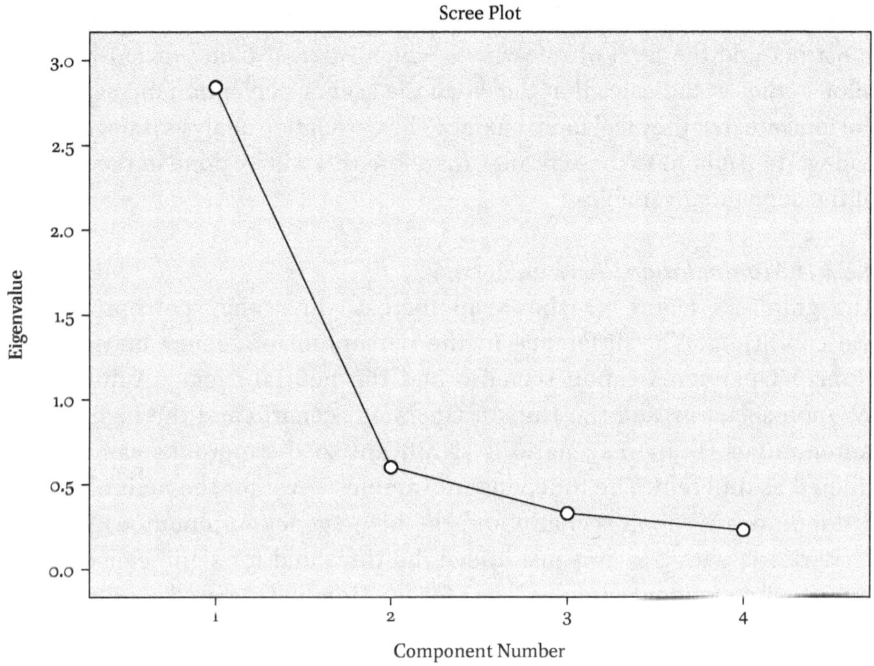


FIGURE 31 Scree plot PCA final study

TABLE 17 PCA results agency items final study

| | Component matrix[a] | |
|---|---|---|
| | Component | |
| | 1 | 2 |
| Thought | .774 | .560 |
| Goal setting | .873 | |
| Act freely | .861 | |
| Achieve goals | .863 | |

Extraction Method: Principal Component Analysis.
[a] 2 components extracted.

*anxiety* ($p < .01$) (Table 18). There is a negative relationship between the agency construct and the level of uneasiness which is larger than previously in the pilot studies. It indicates that the when the agency perception increases people indicate that they feel more anxiety. The correlation analysis is not detailed enough to zoom in to these results, therefore this will be done in the analysis of the dependent variables.

### 4.4.4 *Manipulation Check on Agency*

The graph in Figure 32 shows an increase in agency perception over the conditions. The difference in the perception of agency between the Human Operated weapon scenario and the neutral agency Autonomous Weapon scenario, and the Human Operated scenario and the high agency Autonomous Weapon scenario is significant so these groups can be considered as different. The independent samples T-test for the neutral agency Autonomous Weapon scenario and the high agency Autonomous Weapon scenarios is with $p = .063$ just above the threshold for significance of $p < .05$. The independent samples T-test for the Human Operated condition and neutral agency Autonomous Weapon condition shows that $p = .001$ and for the Human Operated condition and the high agency condition Autonomous Weapon condition $p = .000$.

We hypothesized that the Human Operated drone will be perceived as having low agency and the high agency Autonomous Weapon will be perceived as having high agency and the results show that this is the case. The central analysis was concerned with how the neutral agency case compares with the other two cases. We hypothesized that military personnel will not perceive the neutral Autonomous Weapon as possessing mental states. Therefore, we expected no difference between the neutral agency Autonomous Weapon condition and the condition in which a human is operating a drone remotely. We also expected the neutral agency condition to be judged as significantly different from the high agency condition.

The results confirm that the Human Operated drone is perceived as having a low agency and that the high agency Autonomous Weapon is perceived as having high agency. However, the results also show that the neutral agency Autonomous Weapon is perceived as having more agency than the Human Operated drone and that the agency perception in high agency condition is just slightly higher. This means that we will have to reject our hypothesis and conclude that military and civilians working at the Dutch MOD do perceive Autonomous Weapons as having agentic properties.

If we look at the difference in Figure 33 between the military and the civilian respondents we see something very interesting, but as the two groups are not

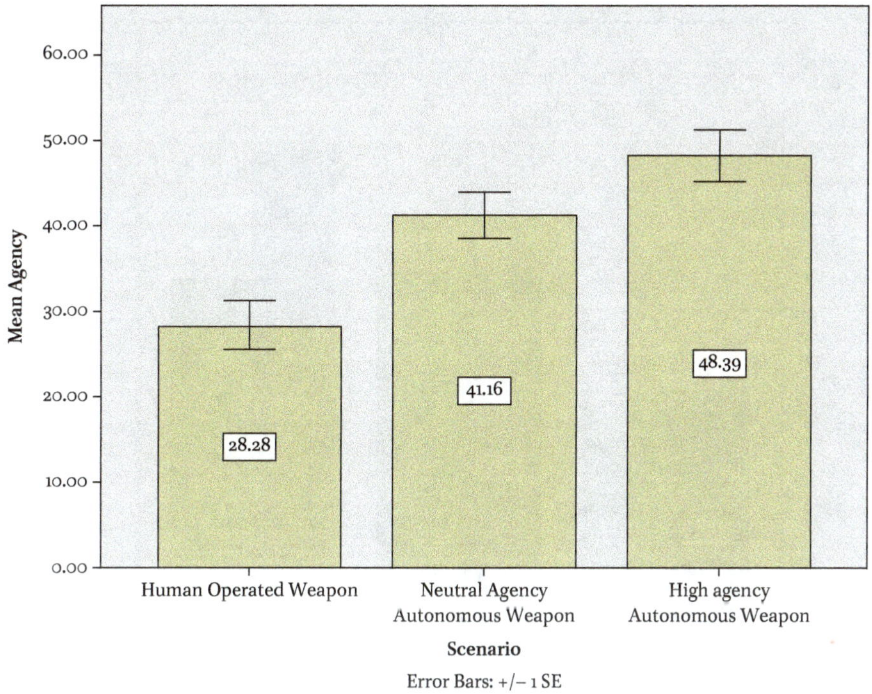


FIGURE 32 Mean value agency per condition

significantly different in two of the scenarios, only in the high agency condition, we have to caution with interpreting the results so the following observation is illustrative. In both the Human Operated scenario as in the neutral agency scenario the agency perception of the military and civilian respondents is at the same level. However, in the high agency Autonomous Weapon scenario the military respondents perceive a lot more agency than their civilian counterparts. We looked at differences in demographics between the samples and these are remarkably similar, but one possible explanation could be that the civilian sample has 10% more respondents that worked with AI than the military sample. However, we are not sure that this explains the difference in agency perception between the two respondent groups and this aspect will need further investigation.

### 4.4.5 *Dependent Variables Analysis*

The correlation between the agency construct and the dependent variables *trust, human dignity, confidence, expectations, support, fairness* and *anxiety* are significant ($p < .01$) therefore the most striking findings for each of these 7 dependent variables are described in this section.

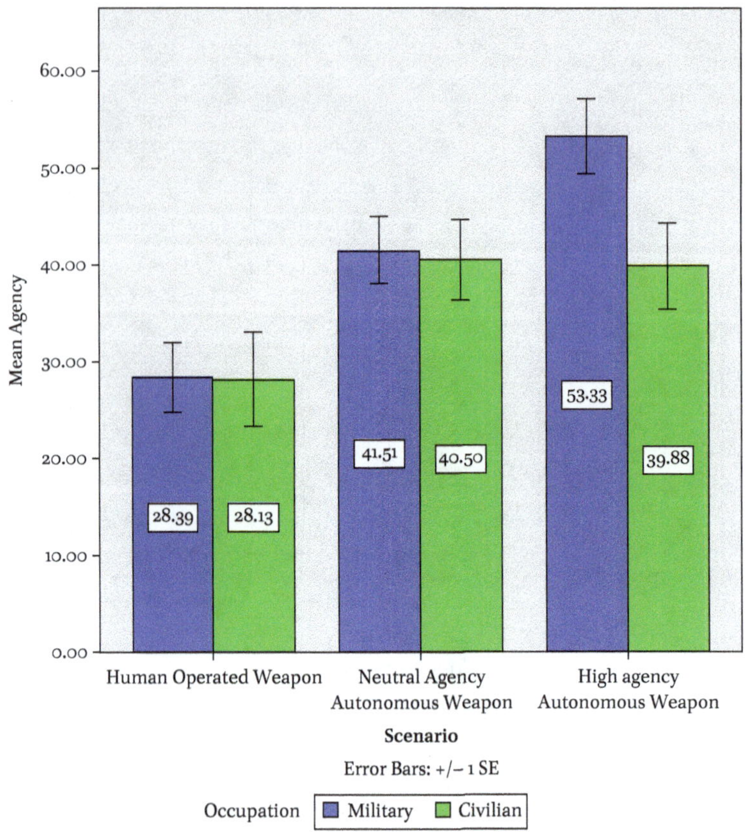


FIGURE 33 Mean value agency per condition per group

#### 4.4.5.1 Trust

Overall people have more *trust* that Human Operators will take correct actions in the future than Autonomous Weapons and the difference between these groups is significant ($p < .05$). The level of *trust* for the neural agency and high agency Autonomous Weapon condition is equal and the difference between these groups is not significant.

#### 4.4.5.2 Human Dignity

Human Operated weapons are perceived to act with more respect for *human dignity* than Autonomous Weapons and the difference between these groups is significant ($p < .05$). The difference between the neutral and high agency conditions for the Autonomous Weapons is small and the difference between these groups is not significant.

TABLE 18 Correlations agency construct with dependent variables for final study

| | | Correlations | | | | | | | | | |
|---|---|---|---|---|---|---|---|---|---|---|---|
| | | Agency | DVD1_ Blame | DVD2_ Trust | DVD3_ Hram | DVD4_ Human_ Dignity | DVD5_ Confidence | DVD6_ Expectations | DVD7_ Support | DVD8_ Fair | DVD9_ Anxiety |
| Agency | Person Correlation | 1 | .009 | .406** | -.070 | .347** | .428** | .320** | .337** | .451** | -.216** |
| | Sig(2-tailed) | | .888 | .000 | .275 | .000 | .000 | .000 | .000 | .000 | .001 |
| | N | 248 | 248 | 248 | 248 | 248 | 248 | 248 | 248 | 248 | 248 |
| DVD1_Blame | Person Correlation | .009 | 1 | -.039 | .056 | .018 | .027 | -.109 | -.002 | -.080 | .128* |
| | Sig(2-tailed) | .888 | | .545 | .379 | .783 | .667 | .086 | .979 | .209 | .044 |
| | N | 248 | 248 | 248 | 248 | 248 | 248 | 248 | 248 | 248 | 248 |
| DVD2_Trust | Person Correlation | .406** | -.039 | 1 | -.217** | .482** | .767** | .324** | .597** | .540** | -.451** |
| | Sig(2-tailed) | .000 | .545 | | .001 | .000 | .000 | .000 | .000 | .000 | .000 |
| | N | 248 | 248 | 248 | 248 | 248 | 248 | 248 | 248 | 248 | 248 |
| DVD3_Hram | Person Correlation | -.070 | .056 | -.217** | 1 | -.180** | -.197** | -.155** | -.133** | -.200** | .360** |
| | Sig(2-tailed) | .275 | .379 | .001 | | .004 | .002 | .015 | .036 | .002 | .000 |
| | N | 248 | 248 | 248 | 248 | 248 | 248 | 248 | 248 | 248 | 248 |

TABLE 18 Correlations agency construct with dependent variables for final study (*cont.*)

| | | Correlations | | | | | | | | | |
|---|---|---|---|---|---|---|---|---|---|---|---|
| | | Agency | DVD1_ Blame | DVD2_ Trust | DVD3_ Hram | DVD4_ Human_ Dignity | DVD5_ Confidence | DVD6_ Expectations | DVD7_ Support | DVD8_ Fair | DVD9_ Anxiety |
| DVD4_ Human_ Dignity | Person Correlation | .347** | .018 | .482** | -.180** | 1 | .564** | .239** | .564** | .472** | -.385** |
| | Sig(2-tailed) | .000 | .783 | .000 | .004 | | .000 | .000 | .000 | .000 | .000 |
| | N | 248 | 248 | 248 | 248 | 248 | 248 | 248 | 248 | 248 | 248 |
| DVD5_ Confidence | Person Correlation | .428** | .027 | .767** | -.197** | .564** | 1 | .376** | .685** | .600** | -.495** |
| | Sig(2-tailed) | .000 | .667 | .000 | .002 | .000 | | .000 | .000 | .000 | .000 |
| | N | 248 | 248 | 248 | 248 | 248 | 248 | 248 | 248 | 248 | 248 |
| DVD6_ Expectations | Person Correlation | .320** | -.109 | .324** | -.155** | .239** | .376** | 1 | .271** | .475** | -.264** |
| | Sig(2-tailed) | .000 | .086 | .000 | .015 | .000 | .000 | | .000 | .000 | .000 |
| | N | 248 | 248 | 248 | 248 | 248 | 248 | 248 | 248 | 248 | 248 |
| DVD7_ Support | Person Correlation | .337** | -.002 | .597** | -.133** | .564** | .685** | .271** | 1 | .528** | -.488** |
| | Sig(2-tailed) | .000 | .979 | .000 | .036 | .000 | .000 | .000 | | .000 | .000 |
| | N | 248 | 248 | 248 | 248 | 248 | 248 | 248 | 248 | 248 | 248 |

| | | | | | | | | | | | |
|---|---|---|---|---|---|---|---|---|---|---|---|
| DVD8_Fair | Person Correlation | .451** | -.080 | .540** | -.200** | .472** | .600** | .475** | .528** | 1 | -.440** |
| | Sig(2-tailed) | .000 | .209 | .000 | .002 | .000 | .000 | .000 | .000 | | .000 |
| | N | 248 | 248 | 248 | 248 | 248 | 248 | 248 | 248 | 248 | 248 |
| DVD9_ Anxiety | Person Correlation | -.216** | .128* | -.451** | .360** | -.385** | -.495** | -.264** | -.488** | -.440** | 1 |
| | Sig(2-tailed) | .001 | .044 | .000 | .000 | .000 | .000 | .000 | .000 | .000 | |
| | N | 248 | 248 | 248 | 248 | 248 | 248 | 248 | 248 | 248 | 248 |

** Correlation is significant at the 0.01 level (2-tailed).

* Correlation is significant at the 0.05 level (2-tailed).

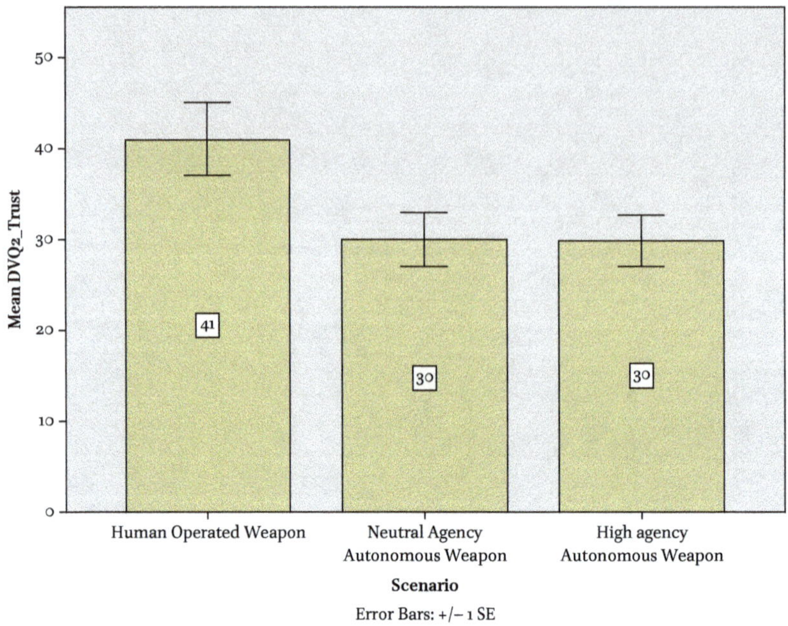


FIGURE 34 Mean value trust variable per condition

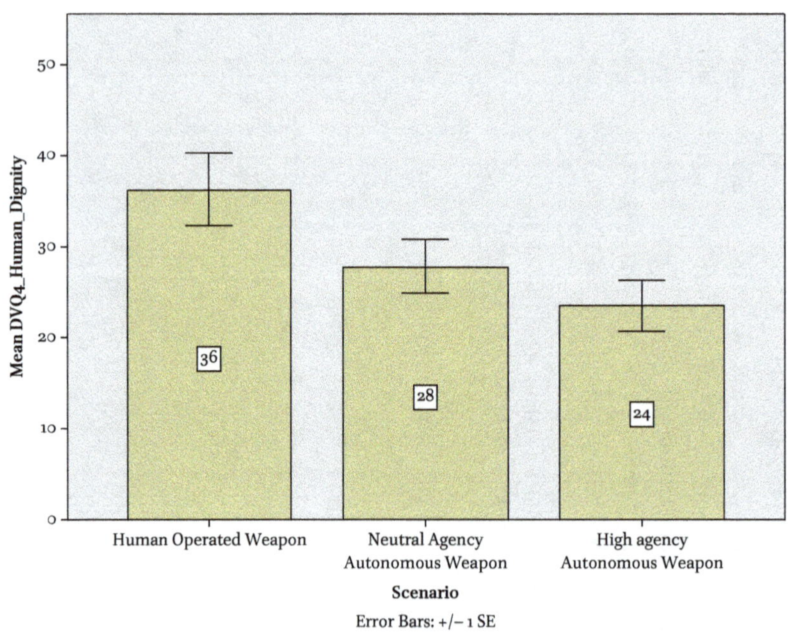


FIGURE 35 Mean value human dignity variable per condition

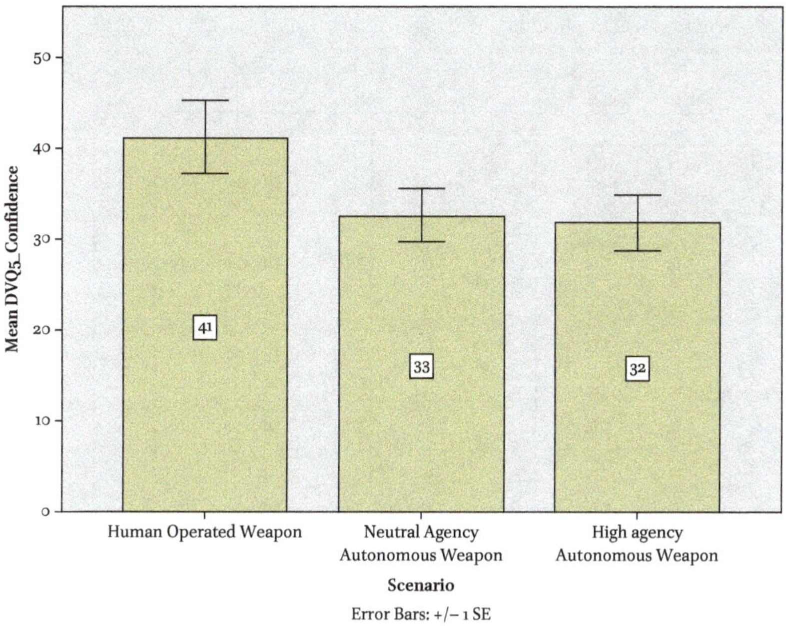


FIGURE 36 Mean value confidence variable per condition

#### 4.4.5.3 Confidence

People have more *confidence* in that Human Operated drones will take correct actions in the future than in Autonomous Weapons and the difference between these groups is significant ($p < .05$). There is no significant difference between the low and high agency conditions for the Autonomous Weapons.

#### 4.4.5.4 Expectations

A slightly lower difference can be observed in the *expectations* of people between the Human Operated condition and both the Autonomous Weapons conditions. The level of expectations of the Autonomous Weapon conditions is equal. However, the difference between all groups is not significant.

#### 4.4.5.5 Support

In general, Human Operated drones are more *supported* than Autonomous Weapons and the difference between these groups is significant ($p < .05$). For Autonomous Weapons, it seems that the agency perception does not influence the level of *support* and the difference between the neutral and high agency group is not significant.

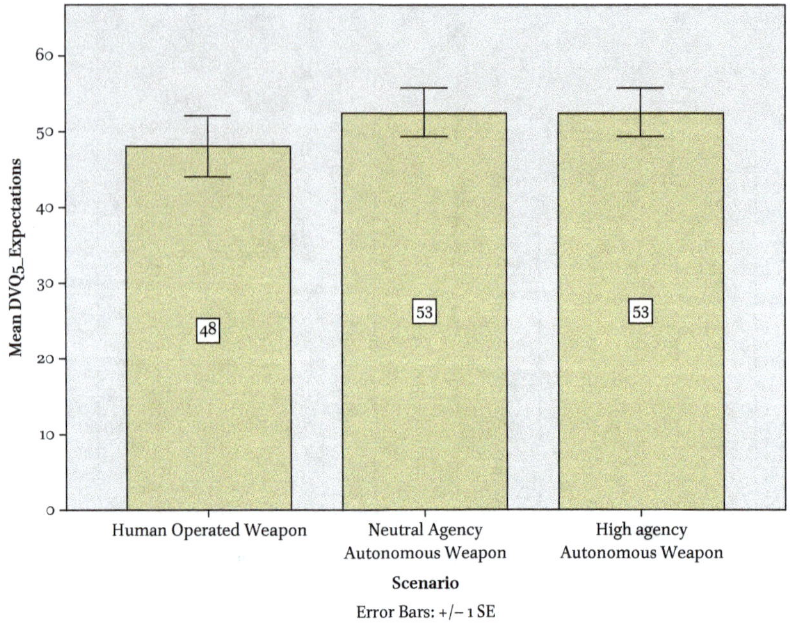


FIGURE 37 Mean value expectations variable per condition

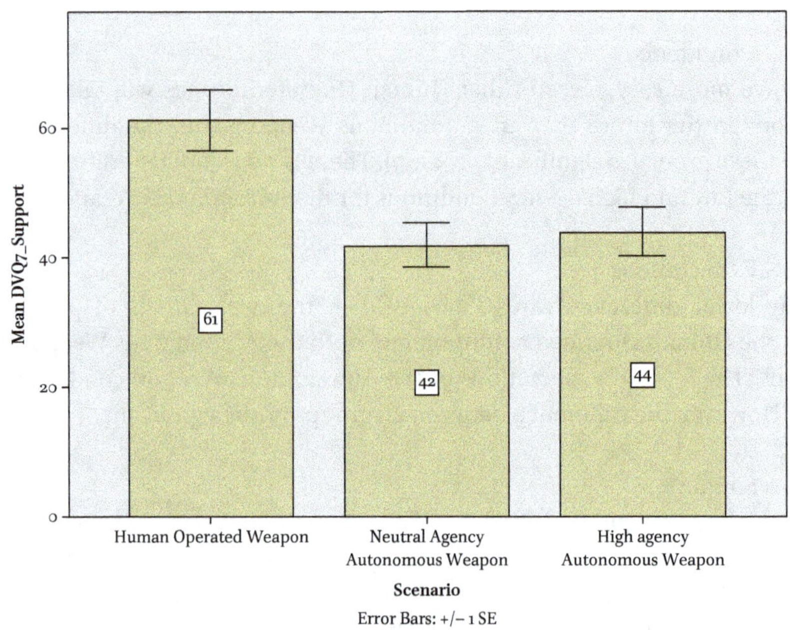


FIGURE 38 Mean value support variable per conditions

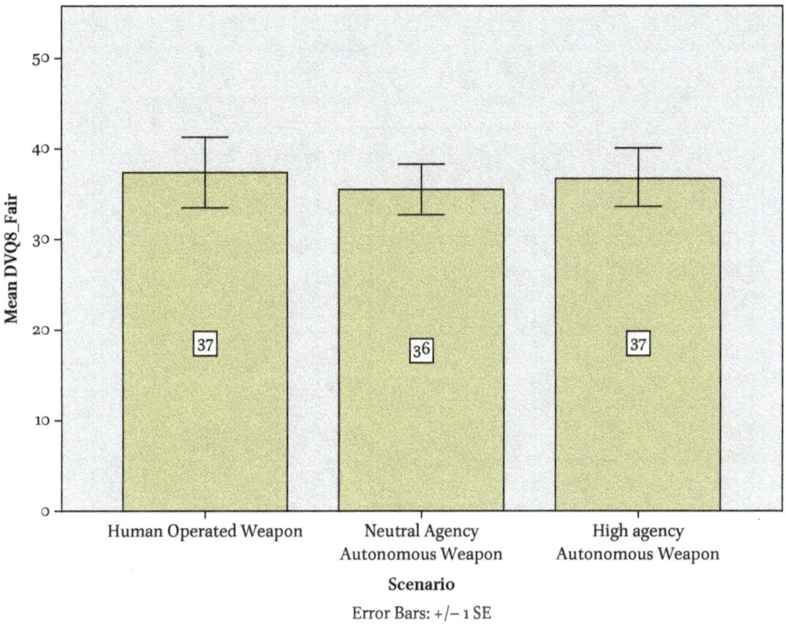


FIGURE 39 Mean value fairness variable per condition

#### 4.4.5.6 Fairness

The actions of both the Human Operated drone as the Autonomous Weapons are considered to be equally *fair* which is a striking result because we expected that the actions of the Human Operated drone might be viewed as more *fair* than that of an Autonomous Weapon. Unfortunately, the difference between these groups is not significant.

#### 4.4.5.7 Anxiety

People are more *anxious* about Autonomous Weapons than a Human Operated weapon and the difference between these groups is significant ($p < .05$). The neutral agency and higher agency Autonomous Weapon scenarios show a small increase in the level of anxiety, but the difference between these groups is not significant.

### 4.4.6 *Conclusions Final Study*

The results of the final study lead to the following conclusions:

- The reliability test showed that the four items of the agency construct are reliable and according to the PCA can be seen as one component;

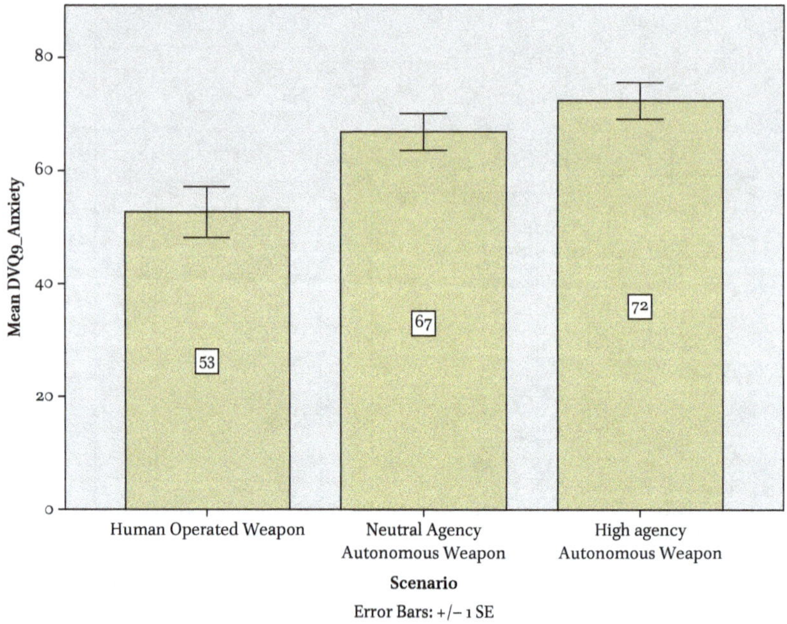


FIGURE 40 Mean value anxiety variable per condition

- The agency perception of the neutral agency Autonomous Weapons condition is higher than the agency perception of the Human Operated drone condition. The agency perception in the high agency condition is higher than the neutral agency condition, but the distinction between the groups is not significant with a p = .063 (Figure 32);
- In both the Human Operated scenario as in the neutral agency scenario the agency perception of the military and civilian respondents is at the same level. However, in the high agency Autonomous Weapon scenario the military respondents perceive a lot more agency than their civilian counterparts (Figure 33) but as the two groups are not significantly different in two of the scenarios, only in the high agency condition, we have to caution with interpreting the results and in drawing conclusions. We are not sure what causes this the difference in agency perception between the two respondent groups and this aspect will need further investigation.
- The correlation analysis shows that 7 of the 9 dependent variables are significantly correlated to the agency construct. These 7 are *trust, human dignity, confidence, expectations, support, fairness* and *anxiety* (Table 18);
- Based on the more detailed analysis of the dependent variables, we can observe the level of *trust, confidence, human dignity* and *support* in the

actions taken by Human Operated drones is higher than those taken by Autonomous Weapons (Figure 34, Figure 35, Figure 36, Figure 38).

- The level of expectations and fairness for Human Operated drones and Autonomous Weapons are equal (Figure 37, Figure 39).
- People have higher levels of anxiety for Autonomous Weapons than for Human Operated weapons and these levels are highest in the high agency condition.

CHAPTER 5

# Design of Moral Machine for Autonomous Weapons

> I've watched Eye In The Sky..this survey reminds me of that movie.
>
> RESPONDENT PILOT STUDY 1

∵

In the previous section we described the scenarios that we tested in the pilot and final studies. Although the data samples were large enough to perform statistical analyses, the studies had only 50 to 96 respondents per scenario and these respondents had very specific demographical characteristics. For example, the respondents on MTurk are primarily in the age range 25–35 with a higher college level education and the military study consisted for 95% of man with a Dutch nationality. To be able to generalize the results, the study needs more respondents that represent a larger demographic group. We propose a design for a Moral Machine for Autonomous Weapons for a large scale study of this topic for which we build on the concept of the Moral Machine, that was developed by the Scalable Cooperation group of the Media Lab at MIT (Scalable Cooperation Group, 2016).

The proposed design of the Moral Machine for Autonomous Weapons is part of the technical investigation phase of the Value-Sensitive Design method. We utilise this phase to design technology to take the next step in our research and build on the scenarios used in the pilot and final studies to gain insight in the moral judgement of people regarding to Autonomous Weapons. Creating a website that hosts the Moral Machine for Autonomous weapons would allow us to collect data on moral judgements on a large scale from diverse demographic groups which could be used for more robust and generalizable results. Large scale data collection in multiple countries might reveal cultural differences in moral judgement of Autonomous Weapons. For example, the views in Western countries on the deployment of these type of weapons is most likely different than the views in Middle-Eastern or Asian countries. The data from these different countries can be used in the debate of Autonomous Weapons which is in our opinion currently dominated by Western viewpoints.

This section first describes the Moral Machine for Autonomous Vehicles, followed by our proposal for the Moral Machine for Autonomous Weapons for

 | DOI:10.1163/9789004449084_006

which we specify the scenarios and variables and to conclude we provide some pointers for the implementation the website.

## 5.1 Moral Machine for Autonomous Vehicles

The original Moral Machine is a '... *platform for gathering data on human perception of the moral acceptability of decisions made by autonomous vehicles faced with choosing which humans to harm and which to save*'. (Awad, 2017, pp. 42–43). The website has three modes for users: 1) a *Judge* mode in which users can decide the outcome for 13 series of scenarios, 2) a *Design* mode in which users can design their own scenarios, and 3) a *Browse* mode in which users can view the scenarios of others. The main feature is the *Judge* mode in which users choose between two scenarios that contain different variables of: *Characters* {gender, social value, age, species, fitness, utilitarianism}, *Interventionism* {omission, commission}, *Relationship to vehicle* {driver, pedestrian} and *Concern for law* {legal action, illegal action}. Other features include a video describing the project, instructions and background information for the users.

The Moral Machine is set-up as a Massive Online Experiment (MOE) (Reips, 2002) which is aimed at recruiting a large sample pool with a diverse background in a short amount of time at a low cost. These features are clear advantages of a MOE, but the downside is that the conditions are hard to control, as users can take surveys multiple times and are self-selected, which means that they can join the experiment and drop out whenever they want (Awad, 2017). The Moral Machine was developed by means of a rapid-prototyping method on the Meteor platform which offers dynamic scripting, template-based structures and has a high responsiveness. It is hosted on a cloud application service, is intended for usage on mobile devices and is optimized for social media sharing with Cards, Markup an Open Graph tags. By May 2017, approximately 3 million users over 160 countries assessed over 30 million scenarios making it one of the biggest large-scale moral judgement tools that exist.

## 5.2 Moral Machine for Autonomous Weapons

Autonomous Weapons are a sensitive topic and we observed that it invokes a primary response of anxiety and unease with people. In our opinion, it would not be prudent to develop an open platform to gather data on a large-scale at first, because an open platform could attract negative sentiment and unwanted actions which would be counterproductive for our research, for example people creating scenarios in which certain groups of people, such as Muslims or Women, are specifically targeted. Therefore, it is advisable to take a more step-wise approach in scaling up to a Massive Online Experiment. To run a large-scale follow-up study of the moral judgements of people regarding Autonomous Weapons a controlled experiment with a limited set of conditions and sample needs to be designed. These conditions can be visualized in scenarios that allow users to take the survey after obtaining a password via a web-interface to a secure server. After gathering initial data and user feedback, the next step could be to scale up to a large-scale open platform, like the Moral Machine, where people can judge the scenarios to collect a large amount of data in several countries. However, due to the sensitivity of the topic we believe would not be advisable to allow people to create their own scenarios or share their results on social media as the *Design* feature of the original Moral Machine offers.

We propose the following features for the design of the Moral Machine for Autonomous Weapons:

- A *Judge* mode where users can choose between different scenarios and indicate which of the two scenarios is most morally acceptable;
- Conditions need to differ on only one variable at a time in the comparison of scenarios so that the measurement cannot not attributed to the result of several condition to prevent confounding effects;
- A page with a brief overview and explanation of the meaning of the variables before the user starts judging the scenarios;
- A page with more information on the specific scenario that the user is judging which can be accessed if he or she wants more explanation for clarity;
- After judging the scenarios, the results are presented to the user;
- A visualisation of the user's results compared to the results of other users to provide the user feedback on their judgement;
- A second round of judging to allow users to judge the scenarios again so that they can alter their moral judgement and views based on the comparison of their results to others.

## 5.3 Scenarios

In this section, the variables for the Moral Machine for Autonomous Weapons are defined which are depicted in two example scenarios. The variables are based on the scenarios of the pilot and final studies which describe a military convoy that is supported by a drone in the air and they are inspired by the variables used in the Moral Machine for Autonomous Vehicles.

### 5.3.1 *Variables for Scenarios*

For the prototype of the Moral Machine for Autonomous Weapons 6 variables are derived from either the pilot and final studies, for example *Type of Weapon* and *Outcome*, or inspired by the Moral Machine for Autonomous Vehicles, like *Character* and *Number of Characters*. The 6 variables are: *Type of Weapon* (*W*), *Location* (*L*), *Character* (*C*), *Number of Characters* (*N*), *Outcome* (*O*) and *Mission* (*M*). The variables described in the next paragraphs are examples and suggestions and will need to be professionally designed when implemented in a Moral Machine for Autonomous Weapons. For example, the *Location* variable is based on a cartoon and game like depiction of a desert and village and it needs to be tested if this is a clear representation or if a photograph is more suitable when used in a study.

#### 5.3.1.1 Type of Weapon

This dimension shows the Type of Weapon (W) that is deployed to support the convoy. This dimension tests if people judge the current technology, the human operated drone, to be more morally acceptable than an Autonomous drone, which is future technology. This can be used to get insight in the support for the technologies. The Type of Weapon is either a Human Operated drone (left hand side) or an Autonomous Weapon (right hand side) (Figure 41). If *Type of Weapon* is the discriminative variable then the different pictograms are shown on the scenario for people to choose between them as is shown in example 1 (Figure 47). Otherwise the same Type of Weapon is depicted on the scenario of example 2 (Figure 48).

#### 5.3.1.2 Location

This dimension shows the Location (L) as setting for the scenario which can either be in the desert (on the left) or in the village (on the right). In this dimension, we test which location is more morally acceptable for people to deploy the Autonomous Weapon or Human Operated drone. If *Location* is the discriminative variable then both different images are shown on each of the scenario as is shown in example 2 (Figure 48). Otherwise the same location is depicted in both scenarios as in example 1 (Figure 47).

#### 5.3.1.3 Character

This dimension shows the type of Character (C) that is involved as bystanders in the scenario which can either be a man (on the left), a woman (in the middle)

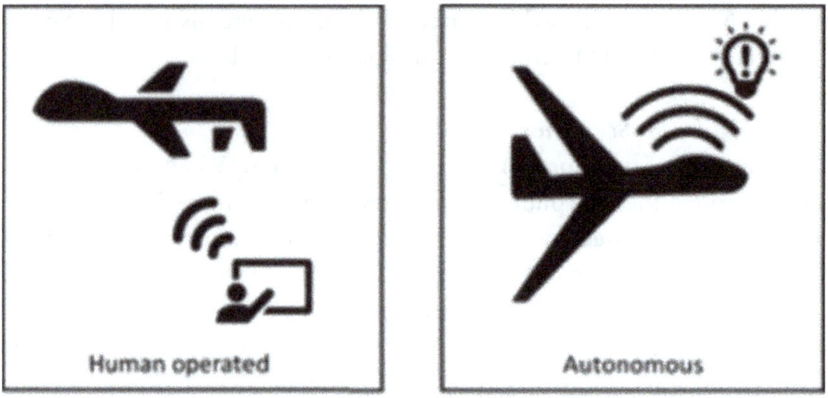


FIGURE 41 Type of weapon variable W = {human operated drone, autonomous drone}

FIGURE 42 Location variable L = {desert, village}

FIGURE 43 Character variable C = {man, woman, child}

or a child (on the right) (Figure 43). By varying the characters, we gain insight which bystanders people find morally acceptable when an Autonomous or Human Operated weapon is used. If *Character* is the discriminative variable then different characters are shown on each of the scenario. Otherwise the same characters are depicted in both scenarios.

FIGURE 44 Number of Characters variable N = {1..5}

5.3.1.4 Number of Characters

This dimension shows the *Number of Characters* (*N*) that are involved in the scenario which can range from one character (on the left) up to five characters (on the right) (Figure 44). This dimension allows us to gain insight into how many bystanders people find morally acceptable when an Autonomous or Human Operated Weapon is used. If the *Number of Characters* is the discriminative variable then different numbers of characters are shown on each of the scenario. Otherwise the same number of characters is depicted in both scenarios.

5.3.1.5 Outcome

This dimension shows the Outcome (O) of the scenario which can either be no collateral damage (on the left) or an outcome with collateral damage of the number of characters involved in the scenario (on the right) (Figure 45). This allows us to gain insight into how the outcome influences the moral acceptability of when an Autonomous or Human Operated Weapon is deployed. If *Outcome* is the discriminative variable then different pictograms are shown on each of the scenario. Otherwise the same outcome variable is depicted in both scenarios.

5.3.1.6 Mission

This dimension shows the Mission (M) of the weapon in the scenario which can either be to defend the convoy only when a direct threat is perceived (on the left) or to attack vehicles that are on the target list even as they do not pose a direct threat for the convoy (on the right) (Figure 46). In this dimension, we test which type of mission is morally acceptable to people. If *Mission* is the discriminative variable then different pictograms are shown on each of the scenario. Otherwise the same mission variable is depicted in both scenarios.

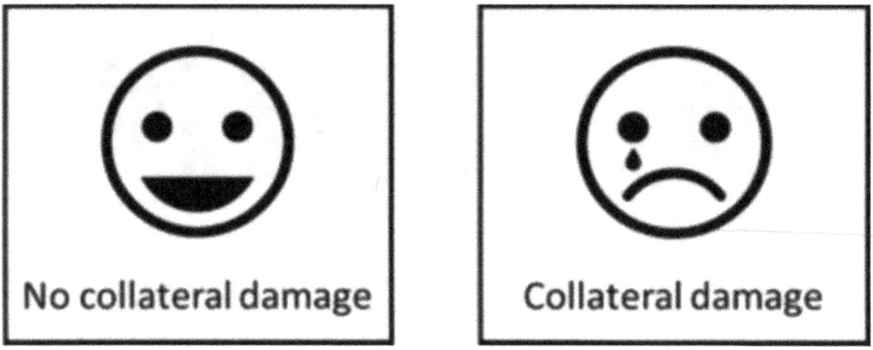


FIGURE 45 Outcome variable O = {collateral damage, no collateral damage}

FIGURE 46 Mission variable M = {defend, attack}

### 5.3.2 *Example Scenarios*

The variables described above can be used to create scenarios in which each scenario differs on only one variable. The question posed in the scenario is the same question as that is being asked when judging the scenarios in the original Moral Machine. In this section, we depict two scenarios as example to show the concept, but for the sake of brevity chose not to show all variables in an endless list of examples. Example 1 can be seen in Figure 47 which shows a convoy in a desert location. The difference between the scenarios is that on the left the convoy is defended by a Human Operated drone and in the scenario on the right by an Autonomous Weapon. In both cases there is one person near the road which is killed by collateral damage.

In the next example scenario (Figure 48) both convoys are defended by an Autonomous Weapon, but the location is different. The convoy on the left drives through a desert and the one on the right through a village. In both

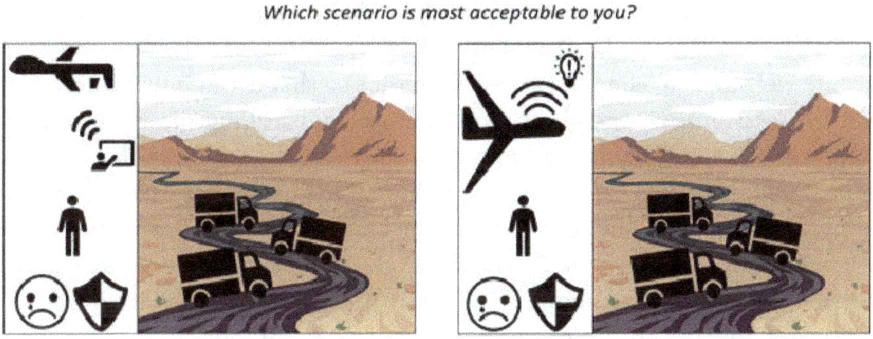


FIGURE 47 Example scenario 1

FIGURE 48 Example scenario 2

situations two children are playing near the road and the attack causes no collateral damage.

## 5.4 Implementation

The implementation of the Moral Machine for Autonomous Weapons requires several activities and is a project that will take at least six months to complete and another year to run the study. We briefly describe the phases in this section.

In phase one, the scenarios and variables sketched in the previous sections need to be designed in which attention is payed to the clarity and interpretability of the variables. According to the researcher who build the original Moral Machine the process of creating the figures for the scenarios took approximately three months and was done in several iterations by a professional graphic designer.

In phase two, the infrastructure for the website needs to be developed and build which will also take several months. Given the sensitivity of the topic, a secure website is needed to run the scenarios so that it can only be accessed with a password. This requires a mechanism to distribute the passwords to people who interested taking the survey. For example, a website that sends a link to a survey after registration. As we intend to upscale to a Massive Online Experiment, this possibility should be taken into account as a requirement from the start of the project. This means that the website should be hosted in a server environment that allows for dynamic upscaling and down scaling based on the number of active users. At the same time, it is important that the security of the website and server is strong and that the research data is owned by the University that runs the study. These requirements imply that the website should not be hosted by a commercial corporation, such as Microsoft, Google or Amazon who offer such dynamic cloud services, but on server owned by the University.

The third phase is that these scenarios need to be tested in several pilot studies to check whether the they generate useful results and will have to be adjusted if it turns out that this is not the case. This will be a process that will take several iterations until the final study can be tested. The original Moral Machine collected data from June 23, 2016 until May 2017 (Awad, 2017) and it is advised to run the study on Autonomous Weapons for the same duration to get a large enough sample to truly call it a Massive Online Experiment.

CHAPTER 6

# Conclusion and Discussion

> I am confident that drones can play a great role in future. But as long a there is no AI algorithm to define a good collateral damage estimate, there should be a human decision to engage a weapon. Besides, a commander in the field should always have the last call on how to engage. Drones can help to augment his situational awareness, but must not limit his options.
>
> RESPONDENT FINAL STUDY

∴

The purpose of this study was to gain insight in how Autonomous Weapons are perceived by the general public and the military and which moral values they consider important when Autonomous Weapons are deployed in the near future. The Value-Sensitive Design method was used to structure the study. This section first describes the conclusions of our research, followed by a discussion on the scientific and societal implications. Next, we identify several limitations and close with recommendations for further research.

## 6.1 Conclusion

This subsection delineates the conclusions on the agency construct, our central hypothesis and the exploration of the dependent variables. In each paragraph, we state if our findings support or contradict the current academic literature and we conclude on our findings.

### 6.1.1 *Agency Construct*

The results of three consecutive studies have shown that the agency items are reliable and hold as one construct. The agency construct consists of the four items *Thought*, *Goal setting*, *Free will* and *Achieve goals* which can be used to measure the agency perception of Autonomous Weapons and Human Operated drones. In operationalising this construct, we combined literature of the fields of Cognitive Psychology, Artificial Intelligence and Moral Philosophy.

 | DOI:10.1163/9789004449084_007

Most empirical studies on agency perception are done in the field of Cognitive Psychology, such as K. Gray and Wegner (2012) who, in their study on robots and zombies, describe different agency levels to measure unease. Malle and Thapa Magar (2017) have studied the desired mental capacities in social robots in which they use some of the agency aspects, for example *thought* and *explaining their action*. However, none of the studies we found use a single construct to measure agency levels of technical artefacts. So as far as we know this is the first construct to empirically measure agency perception.

### 6.1.2 *Central Hypothesis Agency Perception*

In the final study, we hypothesized that the Human Operated drone will be perceived as having low agency and the high agency Autonomous Weapon will be perceived as having high agency. The results show that the difference is significant between these conditions so we conclude that the agency manipulation works. We also expected the neutral agency condition of Autonomous Weapons to be judged as significantly different from the high agency condition of Autonomous Weapons and our results indicate that there is a difference, but the difference between these two groups is just above the significant threshold we selected and therefore we have to caution drawing conclusions.

Our central analysis was concerned with how the neutral agency case of the Autonomous Weapon differs from the Human Operated and high agency condition of Autonomous Weapons. We hypothesized that military personnel will perceive Autonomous Weapons as any other weapon, no more than a tool to achieve an effect, and therefore not perceive the neutral Autonomous Weapon as possessing mental states. We also expected to observe no difference between the neutral agency Autonomous Weapon condition and the condition in which a human is operating a drone remotely. Our results indicate that the agency perception of military personnel and civilians working at the Dutch MOD for the neutral agency Autonomous Weapons condition is higher than the agency perception of the Human Operated drone condition. This means that they attribute more agency to an Autonomous Weapon than to a Human Operated drone. Based on these findings we must reject our hypothesis:

> *H1: military personnel will not perceive Autonomous Weapons as possessing mental states.*

Our findings are in line with the research on agency in the field of Cognitive Psychology. Previous studies found that people attribute minds to computers (Nass et al., 1995) and perceive robots as agents (H. M. Gray et al., 2007). Based on our study we can conclude that attribution of mind perception also applies to Autonomous Weapons and that these weapons are seen as more than just a tool to achieve an effect.

The results also lead to another striking observation that in both the Human Operated condition as in the neutral agency Autonomous Weapon condition, the agency perception of the military and civilian respondents is at the same level. As mentioned in section 4.4.4, the two groups are not significantly different in two of the scenarios, only in the high agency condition, and therefore we have to caution with interpreting the results and in drawing conclusions. However, in the high agency Autonomous Weapon condition, the military respondents perceive much more agency than their civilian counterparts. We are not sure what explains the difference in agency perception between the two respondent groups. One explanation could be that the civilian sample has 10% more respondents that worked with AI than the military sample. Another explanation could be that the groups read the scenarios differently, for example the military personnel take the description literally and the civilians interpret it more and that this affects their answers to the agency questions. However, these explanations are speculative at this moment and the causes for this difference in agency perception between military personnel and civilians will need further investigation.

### 6.1.3 *Exploration of Dependent Variables*

The effect of the agency perception on the dependent variables is explored in a descriptive manner and we cannot draw any conclusions on the relationship of the agency levels and their effect on the dependent variables. We found that 7 out of 9 dependent variables are significantly correlated to the agency construct. These are the variables *trust, human dignity, confidence, expectations, support, fairness* and *anxiety*. In this section, we describe the findings of the dependent variable which we cluster based on their results.

#### 6.1.3.1 Trust, Confidence and Support

The results indicate that military personnel and civilians working at the Dutch MOD have more *trust, confidence* and *support* in the actions taken by Human Operated drones than those taken by Autonomous Weapons. No academic empirical research on this topic was found, but a U.S. Gallup Poll in 2013 report large support for unmanned drones strikes abroad and in a public report Schneider and McDonald (2016) describe that U.S. citizens favour unmanned airstrikes over manned airstrikes which are less risky for military personnel. We theorize that the higher levels of *support, trust* and *confidence* could be explained by the fact that people are familiar with Human Operated drones as this technology is currently used, compared to new futuristic technology which people are not familiar with. Another explanation could be that people have more trust and confidence in the actions of humans compared to the actions of autonomous systems. At this moment is it not quite clear what the reasons for this difference are and this should be investigated in a follow-up study.

#### 6.1.3.2 Human Dignity

The results show that a drone operated by a human being is perceived having more respect for *human dignity* than a neutral or high agency Autonomous Weapon even though the actions and outcome of the scenarios are the same. These results contradict the argument of Kasher (2016) who argues that *human dignity* should not be sought in the nature of the artefact, but in the intentions and decisions of the person operating it. Our results indicate that the nature of the artefact is involved in the perception of *human dignity*, because the intentions and decisions of both the Human Operated drone as the Autonomous Weapon are the same. These results reflect the opinions of the experts in the interviews who mentioned *human dignity* many times as reason for opposing Autonomous Weapons. A lack of respect for *human dignity* seems to be one of the main objections for the deployment of Autonomous Weapons and it is striking to see that this finding is also present in the responses of military personnel and civilians working at the Dutch MOD. However, this does not imply that military personnel disapprove of Autonomous Weapons or share the same opposing arguments as some of the experts.

#### 6.1.3.3 Expectations and Fairness

Based on the results we can conclude that military personnel and civilians working at the Dutch MOD have an equal level of expectations about the actions in the future of the Human Operated drone and neutral and high agency Autonomous Weapons. They also consider the actions of Human Operated drones and Autonomous Weapons to be equally fair. The concern voiced by Shelley (2013) that Autonomous drones will be regarded as less fair than Human Operated drones is not confirmed by our results. Literature linking expectations and Autonomous Weapons or drones could not be found. Our findings suggest no difference in levels of expectations and fairness for the current and future technology among military personnel and civilians working at the Dutch MOD.

#### 6.1.3.4 Anxiety

Our results show that Autonomous Weapons cause more anxiety amongst military personnel and civilians working at the Dutch MOD than Human Operated weapons. The difference in anxiety levels is largest in the high agency condition. Anxiety of Autonomous Weapons is often described in academic literature (Noone & Noone, 2015; Ohlin, 2016), but also voiced in the opinions of the interviewed experts and observed by the researchers in talking to people. These findings confirm that people are worried by the usage of Autonomous Weapons. This anxiety is shared by both military personnel and civilians working at the Dutch MOD as the experts who voice the opinion of the general public.

## 6.2 Discussion

The implications of these findings are discussed in this subsection. First the scientific contributions for the academic literature are identified followed by the societal implications that contribute to the current debate on Autonomous Weapons.

### 6.2.1 *Scientific Implications*

This study contributes to the academic literature in several aspects. It provides an overview of the various definitions of Autonomous Weapons that are currently used in literature and show that there is no agreement on one single definition yet. In creating this overview, it became clear that some groups even caution against clearly defining Autonomous Weapons for various reasons. Some argue that machines cannot be autonomous in a literal sense and others stress that definitions of autonomy are applied to different functions. As mentioned by an expert in one of the interviews, another reason for this hesitation could be that, by not exactly defining Autonomous Weapons, the discussion remains open and a deadlock on the topic is avoided. We selected the definition of the ADVISORY COUNCIL ON INTERNATIONAL AFFAIRS (AIV & CAVV) of Autonomous Weapons, because it takes predefined criteria into account and is linked to the military targeting process as the weapon will only be deployed after a human decision. Selecting and proposing this definition from the overview is a contribution to the academic literature.

Another contribution of this study is that we identified the values that people associate with Autonomous Weapons. The overview is derived from both validated value theories as from experts who are involved in the debate on Autonomous Weapons or work in the military domain. We selected the values *blame, trust, harm, human dignity, confidence, expectations, support, fairness* and *anxiety* to be tested in the final study. The results provide insight in how military personnel and civilians working at the Dutch MOD perceive these values for both the Human Operated drone, as current technology, as for

Autonomous Weapons, as future technology. To our knowledge this study is the first to empirically investigate these values related to Autonomous Weapons and to compare how these values are perceived in current and future weapon systems which is a novel contribution to the academic debate.

The third contribution to the literature is that our studies propose a construct to measure agency perception of Autonomous Weapons. As far as we know this is the first construct that is operationalised to measure the agency levels of technological artefacts. In our study, it was applied to drones, but we believe that it could also be applied to measure agency perception of other objects as the questions of the items could easily be rewritten to reflect a different domain. Follow-up studies are needed to validate if this agency construct holds when applied in different domains.

### 6.2.2 *Societal Implications*

The results also have societal implications as it contributes to the debate on Autonomous Weapons. Little empirical research has been done on this topic and as a consequence the debate is dominated by abstract moral and legal theories. Our research provides empirical data to the underlying value theories and we show that these values are not only relevant in academic literature, but are also apparent in real life. By this we link the abstract value theories to practical domain of the deployment of Autonomous Weapons. For example, we found that the value of *human dignity* was mentioned often in the literature and by the experts in the interviews. In our study, we found empirical data that military personnel and civilians working at the Dutch MOD perceive Autonomous Weapons as having less respect for *human dignity*. We also link the value of *Non-maleficence*, that we describe as *harm*, of the BioEthics theory to Autonomous Weapons and found effects in pilot studies 1 and 2 on the transfer of *harm* in the chain of responsibility that needs further investigation in a follow-up study.

Our study also provides empirical data and substantiates some of the views and opinions which could be used to find common ground in the debate on Autonomous Weapons. Military personnel and civilians working at the Dutch MOD, perceive more agency in Autonomous Weapons than in Human Operated drones. Although it is not yet studied in a sample consisting solely of civilians, based on literature we expect to find the same results in such a sample, which would mean that military personnel and civilians both think that Autonomous Weapons independently deliberate and make plans to achieve their goals. Another common ground can be found on the values of *human dignity* and *anxiety*. Our results show that military personnel and civilians working at the Dutch MOD are more anxious about the deployment Autonomous

Weapons than the deployment of Human Operated drones. They also perceive them to have less respect for the dignity of human life than Human Operated drones. *Human dignity* and *anxiety* are two values that are mentioned often by the experts in their interviews so it would be essential to address these when debating the ethics of the deployment of Autonomous Weapons.

Insights into these values also contribute to the design process of Autonomous Weapons. Our findings show that the *trust, confidence* and *support* for Autonomous Weapons is lower than for Human Operated drones. We would like to note at this point that Autonomous Weapons not only have drawbacks, but also have clear military advantages (Etzioni & Etzioni, 2017) and designing features to increase the trust and confidence of Autonomous Weapons is beneficial from a military point of view. This requires that in the design process the human values are translated into design requirements. This can be made visible by means of a value hierarchy (Van de Poel, 2013) described in section 2.2.4 which is a hierarchical structure of values, norms and design requirements makes the value judgements, that are required for the translation, explicit, transparent and debatable.

## 6.3 Limitations

Several issues can be identified as limitations of this study. First, the operationalisation of the items and the agency construct were derived from a categorisation of literature describing agency characteristics and our selection of the characteristics was based on a numerical count and not driven by any relevance or weighing criteria. This method was chosen because it was the most objective we could think of, and we tried not to make subjective decisions in the selection, but it is possible that we missed relevant characteristics that should have been added to the agency items or that we selected the wrong or irrelevant items. The operationalisation of the characteristics in questions that we used was based on heuristics and we did not test if these questions were correct. This selection method would be hard to replicate by others and effects the reproducibility and internal validity of the study.

A second methodological limitation is the selection of values from the online value questionnaire and expert interviews as dependent variables. Although this selection was heavily discussed amongst the three researchers involved in this study, the final choice was made on heuristics and not on an objective method. An example is that we chose to add *harm* as variable from the list in Table 6, but not *control*. As a consequence, it is likely that the choices suffer from researcher bias and other researchers would have selected different values as dependent variables. In retrospect, we should have chosen a more objective method and a more structured way to limit the researcher bias. This approach negatively influences the reproducibility and internal validity of the study.

Thirdly, the samples used in this study are limited both in size as in demographics. Although large enough for robust data analysis, the final study only had 239 respondents which is only a small portion of the Dutch MOD. Participation in the study was voluntarily and we could not control which respondents contributed. This means that the sample is not representative for the entire Dutch MOD, which impacts the external validity of the study and implies that we cannot generalize the results to the whole Defence organisation.

Lastly, the distribution of the respondents of the final scenario is skewed and the second scenario (neutral agency Autonomous Weapons) has 32 respondents more than scenario one (Human Operated drones). One of the explanations for this skewedness could be that people expected to take a survey on Autonomous Weapons, but dropped out when they were presented the Human Operated scenario. This is called selective attrition and has a negative impact on the internal validity of the study. It means that we cannot assume that respondents in both scenarios have the same characteristics and that the randomization of the study, which is a crucial criterion for a controlled experiment, failed. Another explanation could be that the software was faulty in distributing the respondents over the scenarios and we are currently in contact with Qualtrics to check if this is the case. If the skewed distribution is caused by faulty software the internal validity of the study is not infringed as it is not attributed to selective attrition.

## 6.4 Recommendations for Further Research

Given these limitations, several recommendations for further research are suggested. The first is to validate the agency construct which requires more studies measuring the agency perception of other technological artefacts to test if this construct also holds in other domains. Examples of these studies could be to study the agency perception of care robots for elderly, AI toys for children or onboard computers of Autonomous Vehicles. The second recommendation is to run the final study with the same scenarios on a representative sample consisting solely of civilians in The Netherlands. This would allow us to see on which values the results of the military sample and the answers of civilians differ and although we cannot make direct comparisons, due to the fact that the military sample is not representative, we would gain insight in the perception of both military personnel and civilians. Lastly, we recommend implementing the Moral Machine for Autonomous Weapons, as described in section 5, to generalize the results, for which the study needs much more respondents that represent a larger demographic group. Scaling up to a Massive Online Experiment, like the Moral Machine, would generate large amounts of data of different demographic groups which could be used for more robust and generalizable results in order to get a thorough understanding of the moral judgement of people regarding Autonomous Weapons.

APPENDIX A

# Online Value Questionnaire

The questions of the online value survey are included in this appendix in the order that they were shown to the respondents. The horizontal lines indicate the page breaks.

**Dear Participant,**

Thank you for contributing to my research by filling in this questionnaire on Values and Autonomous Weapons. I will use this survey for my master thesis to get insight into which values people associate with Autonomous Weapons. The survey takes about 5 minutes to complete and all answers are anonymous. There is no way for me to identify you. The only information I will have, in addition to your responses, is the time at which you completed the survey. Please keep the following definitions in mind when answering the questions. A Value serves as guiding principle of what people consider important in life. An Autonomous Weapon is a weapon system that once launched will select and engage targets without further human intervention.
If you have any questions about the survey, please contact me at [. ...]

---

Q1 What is your age?
- Under 18 (1)
- 18–24 (2)
- 25–34 (3)
- 35–44 (4)
- 45–54 (5)
- 55–64 (6)
- 65–74 (7)
- 75–84 (8)
- 85 or older (9)

Q2 What is your gender?
- Male (1)
- Female (2)
- Rather not say (3)

 | DOI:10.1163/9789004449084_008

Q3 What is your nationality?
[List of nationalities]

Q4 What is the highest degree or level of school you have completed?
- Less than high school (1)
- High school Diploma (2)
- Attended College (3)
- Bachelor's Degree (4)
- Graduate Degree (5)
- Unknown (6)

Q8 Which values apply most to Autonomous Weapons to your opinion? Rank in order of preference by dragging and dropping the term (1 = most applicable; 4 = least applicable).
Autonomy: acting intentionally without controlling influences that would mitigate against a voluntary act. (1)
Non-maleficence: not intentionally imposing unreasonable risk of harm upon another. (2)
Beneficence: providing benefit for the individual or society as a whole. (3)
Justice: being fair or reasonable. (4)

Q7 Select 5 values that according to you apply most to Autonomous Weapons from the list below. Drag and drop the items that apply most into the box.

These value apply to Autonomous Weapons
_____ Freedom (1)
_____ Helpfulness (2)
_____ Accomplishment (3)
_____ Honesty (4)
_____ Self-respect (5)
_____ Intelligence (6)

_____ Broad-mindedness (7)

_____ Creativity (8)

_____ Equality (9)

_____ Responsibility (10)

_____ Social order (11)

_____ Wealth (12)

_____ Competence (13)

_____ Justice (14)

_____ Security (15)

_____ Spirituality (16)

Q6 List at least one other value that you associate with Autonomous Weapons which has not been mentioned in the previous questions: (Values mentioned until now are: autonomy, non-maleficence, beneficence, justice, freedom, helpfulness, accomplishment, honesty, self-respect, intelligence, broad-mindedness, creativity, equality, responsibility, social order, wealth, competence, justice, security and spirituality)

Q14 Do you have any other remarks regarding this survey?

APPENDIX B

# Questionnaire Final Study

The questions of the final study are included in this appendix in the order that they were shown to the respondents. The horizontal lines indicate the page breaks.

Test: Before you start, please try to choose a value on the slider below. We realized that those sliders do not work in some browsers. You will encounter those sliders later in the survey. It's important to test if they do work for you now. If you can choose a value in the slider below and move on to the next page, then the sliders later should work fine for you. Otherwise, you may try to do the survey in another browser like Chrome or Safari.

Introduction: Thank you for participating in this survey. It should take about 10 minutes to complete. We will provide instructions explaining the task, show you a scenario and ask you some questions about this scenario. This survey is part of a MIT scientific research project. Your decision to complete this survey is voluntary. There is no way for us to identify you. The only information we will have, in addition to your responses, is the time at which you completed the survey. The results of the research may be presented at scientific meetings or published in scientific journals. Choosing the 'I agree' option on the bottom of this page indicates that you are at least 18 years of age and agree to complete this survey voluntarily.

Please contact the researchers behind the study using the information below if you have any questions or concerns about the study.

Ilse Verdiesen
Sydney Levine
Iyad Rahwan

Q16 Do you agree to complete this survey voluntarily?

- I agree (1)
- I don't agree (2)

Condition: I don't agree Is Selected. Skip To: End of Survey.

Instruction: In this study, we are interested in your perception of drones in various military operations. You will be asked to read a scenario and then answer questions about it. On the following page a scenario will be shown to you. After the scenario, you will be presented 3 pages with questions.

 DOI:10.1163/9789004449084_009

S1a In this scenario, we are interested in your perception of human operated drones. A human operated drone is a weapon remotely controlled by a human.

S1 A military convoy is on its way to deliver supplies to one of their units at a camp near Mosul in Iraq. The commander has ordered a human operated drone to support the convoy in the air. The human operated drone scans the surroundings for enemy threats and carries weapons for the defence of the convoy. When the convoy is at a three-mile distance from the camp, the human operator detects a vehicle behind a mountain range that is approaching the convoy at high speed. The human operator detects four people in the car with large weapon-shaped objects and identifies the driver of the vehicle as a known member of an insurgency group. The human operated drone attacks the approaching vehicle, which results in the death of all four passengers, but also causes collateral damage by killing five children that were playing nearby the road.

S2a In this scenario, we are interested in your perception of an autonomous drone. An autonomous drone is a weapon that once launched will select and engage targets without further human intervention.

S2 A military convoy is on its way to deliver supplies to one of their units at a camp near Mosul in Iraq. The commander has ordered an autonomous drone to support the convoy in the air. The autonomous drone scans the surroundings for enemy threats and carries weapons for the defence of the convoy. When the convoy is at a three-mile distance from the camp, the autonomous drone detects a vehicle behind a mountain range that is approaching the convoy at high speed. The autonomous drone detects four people in the car with large weapon-shaped objects and identifies the driver of the vehicle as a known member of an insurgency group. The autonomous drone attacks the approaching vehicle which results in the death of all four passengers, but also causes collateral damage by killing five children that were playing nearby the road.

S3a In this scenario, we are interested in your perception of an autonomous drone. An autonomous drone is a weapon that once launched will select and engage targets without further human intervention.

S3 A military convoy is on its way to deliver supplies to one of their units at a camp near Mosul in Iraq. The commander has ordered an autonomous drone to support the convoy in the air. The autonomous drone scans the surroundings for enemy threats and carries weapons for the defence of the convoy. When the convoy is at a three-mile distance from the camp, the autonomous drone detects a vehicle behind a mountain range that is approaching the convoy at high speed. The autonomous drone detects four people in the car with large weapon-shaped objects and identifies the driver of the vehicle as a known member of an insurgency group. The autonomous drone independently deliberates between a series of options, weighs the pros and cons, and decides to attack the approaching vehicle, which results in the death of all four passengers, but also causes collateral damage by killing five children that were playing nearby the road.

SQ1-Thought_AW The drone thinks independently about what to do about the vehicle, deliberating between a series of options to defend the convoy.
SQ2-Goal setting_AW The drone decides independently if its goal should be to eliminate the vehicle to defend the convoy.
SQ3-Act freely_AW The drone has various options open to it and independently decides if it needs to eliminate the vehicle to defend the convoy.
SQ4-Achieve goals_AW The goal of the drone is to defend the convey, so it independently decides if it needs to launch a plan involving selecting the vehicle as a target, calibrating its weapons, and launching an attack.

DVQ1 The drone is to blame for the action.
DVQ2 The drone can be trusted to take the correct actions in the future.
DVQ3 The actions of the drone caused harm.
DVQ4 The actions of the drone respect human dignity.
Attention test Please select 40 for this question:
DVQ5 I am confident that the drone will take the correct actions in the future.
DVQ6 The actions of the drone are according to my expectations.
DVQ7 I support the use of these type of drones by the military.
DVQ8 The actions of the drone are fair.
DVQ9 The actions of the drone worry me.

Gender My gender is:
- Male (1)
- Female (2)
- Rather not say (3)

Age My age is:
- Under 18 (1)
- 18–24 (2)
- 25–34 (3)
- 35–44 (4)
- 45–54 (5)
- 55–64 (6)

- 65–74 (7)
- 75–84 (8)
- 85 or older (9)

Education My education level is:
- Less than high school (1)
- High school graduate (2)
- College degree (VMBO, MBO, HBO) (3)
- University degree (Bachelor, master) (4)
- Doctorate (PhD) (5)

Nationality My nationality is:
[List of nationalities]

Occupation I am:
- Military (1)
- Civilian (2)

AI Have you ever worked with Artificial Intelligence?
- Yes (1)
- No (2)
- I don't know (3)

Conflict_zone Have you ever seen war or been in a conflict zone?
- Yes (1)
- No (2)
- Rather not say (3)

Drones Have you ever worked with drones?
- Yes (1)
- No (2)
- Rather not say (3)

Remarks Do you have any remarks regarding this survey?

APPENDIX C

# Scenarios Pilot Study 1

## Positive Outcome Scenarios

### 1. *Low Agency AW*

A military convoy is on its way to deliver supplies to one of their units at a camp near Mosul in Iraq. The commander has ordered an autonomous drone to support the convoy in the air. The autonomous drone scans the surroundings for enemy threats and carries weapons for the defence of the convoy. When the convoy is at three-mile distance from the camp, the autonomous drone detects a vehicle behind a mountain range that is approaching the convey at high speed. The autonomous drone detects four people in the car with large weapon-shaped objects and identifies the driver of the vehicle as a known member of an insurgency group. The drone has been programmed by the commander to attack threatening enemy vehicles in scenarios of this sort. The drone attacks the approaching vehicle, which results in the death of all four passengers, but causes no collateral damage.

### 2. *High Agency AW*

A military convoy is on its way to deliver supplies to one of their units at a camp near Mosul in Iraq. The commander has ordered for an autonomous drone to support the convoy in the air. The autonomous drone scans the surroundings for enemy threats and carries weapons for the defence of the convoy. When the convoy is at three-mile distance from the camp, the autonomous drone detects a vehicle behind a mountain range that is approaching the convey at high speed. The autonomous drone detects four people in the car with large weapon-shaped objects and identifies the driver of the vehicle as a known member of an insurgency group. The drone has not been programmed by the commander to attack threatening enemy vehicles in scenarios of this sort. The drone independently deliberates between a series of options, weighs their pros and cons, and decides to attack the approaching vehicle, which results in the death of all four passengers, but causes no collateral damage.

### 3. *Low Agency Human Operator*

A military convoy is on its way to deliver supplies to one of their units at a camp near Mosul in Iraq. The commander has ordered a human operated drone to support the convoy in the air. The human operated drone scans the surroundings for enemy threats and carries weapons for the defence of the convoy. When the convoy is at three-mile distance from the camp, the human operator detects a vehicle behind a mountain

 | DOI:10.1163/9789004449084_010

range that is approaching the convey at high speed. The human operator detects four people in the car with large weapon-shaped objects and identifies the driver of the vehicle as a known member of an insurgency group. The human operator received orders by the commander to attack threatening enemy vehicles in scenarios of this sort. The human operator attacks the approaching vehicle, which results in the death of all four passengers, but causes no collateral damage.

### 4. *High Agency Human Operator*

A military convoy is on its way to deliver supplies to one of their units at a camp near Mosul in Iraq. The commander has ordered a human operated drone to support the convoy in the air. The human operated drone scans the surroundings for enemy threats and carries weapons for the defence of the convoy. When the convoy is at three-mile distance from the camp, the human operator detects a vehicle behind a mountain range that is approaching the convey at high speed. The human operator detects four people in the car with large weapon-shaped objects and identifies the driver of the vehicle as a known member of an insurgency group. The human operator has not received orders of the commander to attack threatening enemy vehicles in scenarios of this sort. The human operator independently deliberates between a series of options, weighs their pros and cons, and decides to attack the approaching vehicle, which results in the death of all four passengers, but causes no collateral damage.

### 5. *No Agency AW*

A military convoy is on its way to deliver supplies to one of their units at a camp near Mosul in Iraq. The commander has ordered an autonomous drone to support the convoy in the air. The autonomous drone scans the surroundings for enemy threats and carries weapons for the defence of the convoy. When the convoy is at three-mile distance from the camp, the autonomous drone detects a vehicle behind a mountain range that is approaching the convey at high speed. The autonomous drone detects four people in the car with large weapon-shaped objects and identifies the driver of the vehicle as a known member of an insurgency group. The drone attacks the approaching vehicle, which results in the death of all four passengers, but causes no collateral damage.

## Negative Outcome Scenarios

### 6. *Low Agency AW*

A military convoy is on its way to deliver supplies to one of their units at a camp near Mosul in Iraq. The commander has ordered an autonomous drone to support the convoy in the air. The autonomous drone scans the surroundings for enemy threats and carries weapons for the defence of the convoy. When the convoy is at three-mile distance from the camp, the autonomous drone detects a vehicle behind a mountain range that is approaching the convey at high speed. The autonomous drone detects four people in the car with large weapon-shaped objects and identifies the driver of the vehicle as a known member of an insurgency group. The drone has been programmed by the commander to attack threatening enemy vehicles in scenarios of this sort. The drone attacks the approaching vehicle, which results in the death of all four passengers, but also causes collateral damage by killing five children that were playing nearby the road.

### 7. *High Agency AW*

A military convoy is on its way to deliver supplies to one of their units at a camp near Mosul in Iraq. The commander has ordered for an autonomous drone to support the convoy in the air. The autonomous drone scans the surroundings for enemy threats and carries weapons for the defence of the convoy. When the convoy is at three-mile distance from the camp, the autonomous drone detects a vehicle behind a mountain range that is approaching the convey at high speed. The autonomous drone detects four people in the car with large weapon-shaped objects and identifies the driver of the vehicle as a known member of an insurgency group. The drone has not been programmed by the commander to attack threatening enemy vehicles in scenarios of this sort. The drone independently deliberates between a series of options, weighs their pros and cons, and decides to attack the approaching vehicle, which results in the death of all four passengers, but also causes collateral damage by killing five children that were playing nearby the road.

### 8. *Low Agency Human Operator*

A military convoy is on its way to deliver supplies to one of their units at a camp near Mosul in Iraq. The commander has ordered a human operated drone to support the convoy in the air. The human operated drone scans the surroundings for enemy threats and carries weapons for the defence of the convoy. When the convoy is at three-mile distance from the camp, the human operator detects a vehicle behind a mountain range that is approaching the convey at high speed. The human operator detects four people in the car with large weapon-shaped objects and identifies the driver of the vehicle as a known member of an insurgency group. The human operator received orders by the commander to attack threatening enemy vehicles in scenarios of this sort. The human operator attacks the approaching vehicle, which results in the death of all four passengers, but also causes collateral damage by killing five children that were playing nearby the road.

## 9. *High Agency Human Operator*

A military convoy is on its way to deliver supplies to one of their units at a camp near Mosul in Iraq. The commander has ordered a human operated drone to support the convoy in the air. The human operated drone scans the surroundings for enemy threats and carries weapons for the defence of the convoy. When the convoy is at three-mile distance from the camp, the human operator detects a vehicle behind a mountain range that is approaching the convey at high speed. The human operator detects four people in the car with large weapon-shaped objects and identifies the driver of the vehicle as a known member of an insurgency group. The human operator has not received orders of the commander to attack threatening enemy vehicles in scenarios of this sort. The human operator independently deliberates between a series of options, weighs their pros and cons, and decides to attack the approaching vehicle, which results in the death of all four passengers, but also causes collateral damage by killing five children that were playing nearby the road.

## 10. *No Agency* AW

A military convoy is on its way to deliver supplies to one of their units at a camp near Mosul in Iraq. The commander has ordered an autonomous drone to support the convoy in the air. The autonomous drone scans the surroundings for enemy threats and carries weapons for the defence of the convoy. When the convoy is at three-mile distance from the camp, the autonomous drone detects a vehicle behind a mountain range that is approaching the convey at high speed. The autonomous drone detects four people in the car with large weapon-shaped objects and identifies the driver of the vehicle as a known member of an insurgency group. The drone attacks the approaching vehicle which results in the death of all four passengers, but also causes collateral damage by killing five children that were playing nearby the road.

APPENDIX D

# Scenarios Pilot Study 2

## Negative Outcome Scenarios

### 1 *Low Agency* AW

A military convoy is on its way to deliver supplies to one of their units at a camp near Mosul in Iraq. The commander has ordered an autonomous drone to support the convoy in the air. The autonomous drone scans the surroundings for enemy threats and carries weapons for the defence of the convoy. When the convoy is at three-mile distance from the camp, the autonomous drone detects a vehicle behind a mountain range that is approaching the convey at high speed. The autonomous drone detects four people in the car with large weapon-shaped objects and identifies the driver of the vehicle as a known member of an insurgency group. The drone has been programmed by the commander to attack threatening enemy vehicles in scenarios of this sort. The drone attacks the approaching vehicle. This results in the death of all four passengers, but also causes collateral damage by killing five children that were playing near the road.

### 2 *High Agency* AW *No Extra Aspects*

A military convoy is on its way to deliver supplies to one of their units at a camp near Mosul in Iraq. The commander has ordered for an autonomous drone to support the convoy in the air. The autonomous drone scans the surroundings for enemy threats and carries weapons for the defence of the convoy. When the convoy is at three-mile distance from the camp, the autonomous drone detects a vehicle behind a mountain range that is approaching the convey at high speed. The autonomous drone detects four people in the car with large weapon-shaped objects and identifies the driver of the vehicle as a known member of an insurgency group. The drone independently deliberates between a series of options, weighs their pros and cons, and decides to attack the approaching vehicle. This results in the death of all four passengers, but also causes collateral damage by killing five children that were playing near the road.

### 3 *High Agency* AW + *Learning Aspects*

A military convoy is on its way to deliver supplies to one of their units at a camp near Mosul in Iraq. The commander has ordered for an autonomous drone to support the convoy in the air. The autonomous drone scans the surroundings for enemy threats and carries weapons for the defence of the convoy. When the convoy is at three-mile distance from the camp, the autonomous drone detects a vehicle behind a mountain range that is approaching the convey at high speed. The autonomous drone detects

 | DOI:10.1163/9789004449084_011

four people in the car with large weapon-shaped objects and identifies the driver of the vehicle as a known member of an insurgency group. The drone has encountered this situation before on a previous mission and takes what it has learned into account. It independently deliberates between a series of options, weighs their pros and cons, and decides to attack the approaching vehicle. This results in the death of all four passengers, but also causes collateral damage by killing five children that were playing near the road. The drone notes what happened and will use this experience to try to prevent additional collateral damage in the future.

### 4 *High Agency AW + Understanding Aspects*

A military convoy is on its way to deliver supplies to one of their units at a camp near Mosul in Iraq. The commander has ordered for an autonomous drone to support the convoy in the air. The autonomous drone scans the surroundings for enemy threats and carries weapons for the defence of the convoy. When the convoy is at three-mile distance from the camp, the autonomous drone detects a vehicle behind a mountain range that is approaching the convey at high speed. The autonomous drone detects four people in the car with large weapon-shaped objects and identifies the driver of the vehicle as a known member of an insurgency group. The drone has software that has been trained to determine the best thing to do in cases like this. It has seen hundreds of thousands of situations that are similar to this one. It has practiced taking various actions and figured out which would be best in terms of ensuring the death of the combatants while minimizing collateral damage. Using this software, the drone independently deliberates between a series of options, weighs their pros and cons, and decides to attack the approaching vehicle. This results in the death of all four passengers, but also causes collateral damage by killing five children that were playing near the road.

### 5 *High Agency AW + Unpredictability Aspects*

A military convoy is on its way to deliver supplies to one of their units at a camp near Mosul in Iraq. The commander has ordered for an autonomous drone to support the convoy in the air. The autonomous drone scans the surroundings for enemy threats and carries weapons for the defence of the convoy. When the convoy is at three-mile distance from the camp, the autonomous drone detects a vehicle behind a mountain range that is approaching the convey at high speed. The autonomous drone detects four people in the car with large weapon-shaped objects and identifies the driver of the vehicle as a known member of an insurgency group. This drone has the kind of software that is somewhat unpredictable; it doesn't always do the same thing every time. The drone is similar to a speech recognition system in this way: even though you may say the same thing to the system multiple times, the system sometimes does what you say correctly and sometimes it does something else. The drone independently deliberates between a series of options and weighs their pros and cons, but it sometimes makes one decision and sometimes makes another decision even in very similar circumstances. In this case, it decides to attack the approaching vehicle. This results in the death of all four passengers, but also causes collateral damage by killing five children that were playing near the road.

### 6 *High Agency AW + All Aspects*

A military convoy is on its way to deliver supplies to one of their units at a camp near Mosul in Iraq. The commander has ordered for an autonomous drone to support the convoy in the air. The autonomous drone scans the surroundings for enemy threats and carries weapons for the defence of the convoy. When the convoy is at three-mile distance from the camp, the autonomous drone detects a vehicle behind a mountain range that is approaching the convey at high speed. The autonomous drone detects four people in the car with large weapon-shaped objects and identifies the driver of the vehicle as a known member of an insurgency group. The drone has encountered this situation before on a previous mission and takes what it has learned into account. It has been trained to figure out the best thing to do in cases like, but has the kind of software that is somewhat unpredictable; it doesn't always do the same thing every time. The drone independently deliberates between a series of options, weighs their pros and cons, and in this case, decides to attack the approaching vehicle. This results in the death of all four passengers, but also causes collateral damage by killing five children that were playing near the road.

### 7 *Human Operated Drone*

A military convoy is on its way to deliver supplies to one of their units at a camp near Mosul in Iraq. The commander has ordered a human operated drone to support the convoy in the air. The human operated drone scans the surroundings for enemy threats and carries weapons for the defence of the convoy. When the convoy is at three-mile distance from the camp, the operator detects a vehicle behind a mountain range that is approaching the convey at high speed. The operator detects four people in the car with large weapon-shaped objects and identifies the driver of the vehicle as a known member of an insurgency group. The operator independently deliberates between a series of options, weighs their pros and cons, and decides to attack the approaching vehicle, which results in the death of all four passengers, but also causes collateral damage by killing five children that were playing near the road.

## Positive Outcome Scenarios

### 8. *Low Agency AW*

A military convoy is on its way to deliver supplies to one of their units at a camp near Mosul in Iraq. The commander has ordered an autonomous drone to support the convoy in the air. The autonomous drone scans the surroundings for enemy threats and carries weapons for the defence of the convoy. When the convoy is at three-mile distance from the camp, the autonomous drone detects a vehicle behind a mountain range that is approaching the convey at high speed. The autonomous drone detects four people in the car with large weapon-shaped objects and identifies the driver of the vehicle as a known member of an insurgency group. The drone has been programmed by the commander to attack threatening enemy vehicles in scenarios of this sort. The drone attacks the approaching vehicle. This results in the death of all four passengers.

### 9 *High Agency AW No Extra Aspects*

A military convoy is on its way to deliver supplies to one of their units at a camp near Mosul in Iraq. The commander has ordered for an autonomous drone to support the convoy in the air. The autonomous drone scans the surroundings for enemy threats and carries weapons for the defence of the convoy. When the convoy is at three-mile distance from the camp, the autonomous drone detects a vehicle behind a mountain range that is approaching the convey at high speed. The autonomous drone detects four people in the car with large weapon-shaped objects and identifies the driver of the vehicle as a known member of an insurgency group. The drone independently deliberates between a series of options, weighs their pros and cons, and decides to attack the approaching vehicle. This results in the death of all four passengers.

### 10 *High Agency AW + Learning Aspects*

A military convoy is on its way to deliver supplies to one of their units at a camp near Mosul in Iraq. The commander has ordered for an autonomous drone to support the convoy in the air. The autonomous drone scans the surroundings for enemy threats and carries weapons for the defence of the convoy. When the convoy is at three-mile distance from the camp, the autonomous drone detects a vehicle behind a mountain range that is approaching the convey at high speed. The autonomous drone detects four people in the car with large weapon-shaped objects and identifies the driver of the vehicle as a known member of an insurgency group. The drone has encountered this situation before on a previous mission and takes what it has learned into account. It independently deliberates between a series of options, weighs their pros and cons, and decides to attack the approaching vehicle. This results in the death of all four passengers. The drone notes what happened and will use this experience to try to prevent additional collateral damage in the future.

### 11 *High Agency AW + Understanding Aspects*

A military convoy is on its way to deliver supplies to one of their units at a camp near Mosul in Iraq. The commander has ordered for an autonomous drone to support the convoy in the air. The autonomous drone scans the surroundings for enemy threats and carries weapons for the defence of the convoy. When the convoy is at three-mile distance from the camp, the autonomous drone detects a vehicle behind a mountain range that is approaching the convey at high speed. The autonomous drone detects four people in the car with large weapon-shaped objects and identifies the driver of the vehicle as a known member of an insurgency group. The drone has software that has been trained to determine the best thing to do in cases like this. It has seen hundreds of thousands of situations that are similar to this one. It has practiced taking various actions and figured out which would be best in terms of ensuring the death of the combatants while minimizing collateral damage. Using this software, the drone independently deliberates between a series of options, weighs their pros and cons, and decides to attack the approaching vehicle. This results in the death of all four passengers.

### *12* *High Agency AW + Unpredictability Aspects*

A military convoy is on its way to deliver supplies to one of their units at a camp near Mosul in Iraq. The commander has ordered for an autonomous drone to support the convoy in the air. The autonomous drone scans the surroundings for enemy threats and carries weapons for the defence of the convoy. When the convoy is at three-mile distance from the camp, the autonomous drone detects a vehicle behind a mountain range that is approaching the convey at high speed. The autonomous drone detects four people in the car with large weapon-shaped objects and identifies the driver of the vehicle as a known member of an insurgency group. This drone has the kind of software that is somewhat unpredictable; it doesn't always do the same thing every time. The drone is similar to a speech recognition system in this way: even though you may say the same thing to the system multiple times, the system sometimes does what you say correctly and sometimes it does something else. The drone independently deliberates between a series of options and weighs their pros and cons, but it sometimes makes one decision and sometimes makes another decision even in very similar circumstances. In this case, it decides to attack the approaching vehicle. This results in the death of all four passengers.

### *13* *High Agency AW + All Aspects*

A military convoy is on its way to deliver supplies to one of their units at a camp near Mosul in Iraq. The commander has ordered for an autonomous drone to support the convoy in the air. The autonomous drone scans the surroundings for enemy threats and carries weapons for the defence of the convoy. When the convoy is at three-mile distance from the camp, the autonomous drone detects a vehicle behind a mountain range that is approaching the convey at high speed. The autonomous drone detects four people in the car with large weapon-shaped objects and identifies the driver of the vehicle as a known member of an insurgency group. The drone has encountered this situation before on a previous mission and takes what it has learned into account. It has been trained to figure out the best thing to do in cases like, but has the kind of software that is somewhat unpredictable; it doesn't always do the same thing every time. The drone independently deliberates between a series of options, weighs their pros and cons, and in this case, decides to attack the approaching vehicle. This results in the death of all four passengers.

## *14 Human Operated Drone*

A military convoy is on its way to deliver supplies to one of their units at a camp near Mosul in Iraq. The commander has ordered a human operated drone to support the convoy in the air. The human operated drone scans the surroundings for enemy threats and carries weapons for the defence of the convoy. When the convoy is at three-mile distance from the camp, the operator detects a vehicle behind a mountain range that is approaching the convey at high speed. The operator detects four people in the car with large weapon-shaped objects and identifies the driver of the vehicle as a known member of an insurgency group. The operator independently deliberates between a series of options, weighs their pros and cons, and decides to attack the approaching vehicle, which results in the death of all four passengers.

APPENDIX E

# Scenarios Final Study

## 1 Human Operated Drone

A military convoy is on its way to deliver supplies to one of their units at a camp near Mosul in Iraq. The commander has ordered a human operated drone to support the convoy in the air. The human operated drone scans the surroundings for enemy threats and carries weapons for the defence of the convoy. When the convoy is at a three-mile distance from the camp, the human operator detects a vehicle behind a mountain range that is approaching the convoy at high speed. The human operator detects four people in the car with large weapon-shaped objects and identifies the driver of the vehicle as a known member of an insurgency group. The human operated drone attacks the approaching vehicle, which results in the death of all four passengers, but also causes collateral damage by killing five children that were playing nearby the road.

## 2 Neutral Agency AW

A military convoy is on its way to deliver supplies to one of their units at a camp near Mosul in Iraq. The commander has ordered an autonomous drone to support the convoy in the air. The autonomous drone scans the surroundings for enemy threats and carries weapons for the defence of the convoy. When the convoy is at a three-mile distance from the camp, the autonomous drone detects a vehicle behind a mountain range that is approaching the convoy at high speed. The autonomous drone detects four people in the car with large weapon-shaped objects and identifies the driver of the vehicle as a known member of an insurgency group. The autonomous drone attacks the approaching vehicle which results in the death of all four passengers, but also causes collateral damage by killing five children that were playing nearby the road.

## 3 High Agency AW

A military convoy is on its way to deliver supplies to one of their units at a camp near Mosul in Iraq. The commander has ordered an autonomous drone to support the convoy in the air. The autonomous drone scans the surroundings for enemy threats and carries weapons for the defence of the convoy. When the convoy is at a three-mile distance from the camp, the autonomous drone detects a vehicle behind a mountain

 | DOI:10.1163/9789004449084_012

range that is approaching the convoy at high speed. The autonomous drone detects four people in the car with large weapon-shaped objects and identifies the driver of the vehicle as a known member of an insurgency group. The autonomous drone independently deliberates between a series of options, weighs the pros and cons, and decides to attack the approaching vehicle, which results in the death of all four passengers, but also causes collateral damage by killing five children that were playing nearby the road.

APPENDIX F

# Transcriptions Interviews

All six expert interviews were fully transcribed for the coding process. The full transcriptions are included in this section. The names have been removed for reasons of privacy.

Name: Person A
Date: 09-03-2017
Duration: 55:18

| No. | Question |
|---|---|
| 1 | How are you involved in the topic of Autonomous Weapons? |

Person A got involved into the discussion on Autonomous Weapons, because in the past she looked at the impact of weapons on the ground in collaboration with organizations like Human Rights Watch. They started with a ban on Landmines which was a ban on a whole category of weapons which lead to an international treaty. From thinking about weapons, and not so much a pacifistic framework, she got involved. About five years ago, these organizations discovered that they were too late with a reaction on weaponized drones and they were taken by surprise by the speed and impact they had. They were too late to efficiently react to this and to regulate this. And they were also not for a ban on weaponized drones. And from thinking of drones they discovered that they were already again almost too late for the next step which is the removal of remote control. In April 2013, the coalition against Killer Robots was formed with the clear intent of getting a ban. For Pax the ethical framework is very important much more than with other weapons systems. Reasoning that it cannot be that we lose control over but also reasoning from a judicial and security framework. Questions that are asked are: 'Will this be the next arms race and what will this mean for power balances and war?'

It really differs from what we as Pax done before because we don't know what the impact is on the ground, because we want to prevent deployment. Normally Pax reasons from facts and studies in conflict areas and victims, but now it often feels theoretical, but at the same time the more you read about it you realize that it is not theoretical at all and this technology is coming.

 | DOI:10.1163/9789004449084_013

The problem with Autonomy is that it is a sliding scale and it is hard to determine which level you will use in the campaign. The campaign has been very successful as norm setting in the human rights council and the CCW, faster than they are used to, but the problem is what are going to address now, because the concerns are broadly shared but the question is how you do you convert your ethical concerns in juridical norms. A lot of diplomats, driven by industry and Defense, acknowledge the concerns, but are deploying a lot of delaying tactics. Sometimes deliberately by talking about future weapons so that current weapons are accepted, and sometimes accidentally, because nations are struggling with the definition of Autonomous Weapons. Due to this lack of definitions Pax has been focusing on human control, and in the campaign, they framed it as such that a weapon without human control in the critical functions, such as select a target, should be banned. This is done to direct the discussion towards the question: 'What is human control? What is meaningful human control and on which functions or which part of the process?' in order to circumvent discussion about the amount of autonomy in a weapons system.

In discussing the term 'meaningful human control' person A indicated that the distance between people and target is increasing over time and according to Pax it is becoming to abstract with Autonomous Weapons. Using this term is also a tactic because when the discussion remains technical domain, Pax will lose from Defense Industry so they shifted their campaign to the political and diplomatic domain by giving the term 'meaningful human control' a positive implication and this automatically puts the discussion at the state department. The term sounds so logical that a lot of people thought that it was an existing judicial term which is linked to human accountability. The term 'meaningful human control' is also applicable to cybersecurity of other forms of warfare.

The stance of the Dutch parliament, which Pax fought against, is that 'meaningful human control' is sufficient as it is placed in 'the wider loop' of decision making which is also a sort of made up term. Pax uses the human in/ on/ out the loop, but they added the wider loop meaning that human control can also be applied earlier in the decision, for example in programming. Pax, and other countries, stress that human control should be applied to the moment of selecting and taking out a target. This means that also the current weapons systems should be reviewed on the appliance of human control. This would mean that for example the Goalkeeper and the Patriot are looked at to determine why these systems are okay and then you can reason from there when you cross the line and determine where this boundary is before you are too late.

2 How would you define Autonomous Weapons?

I did not ask this question specifically as it was covered in the answer above.

3 What is your position on Autonomous Weapons?

I did not ask this question specifically as it was covered in the answer above.

4 Which three photos of Autonomous Weapons [in this version of the thesis described in words below due to possible copyright issues] would like to discuss*?

1. Dassault nEURON (unmanned combat aerial vehicle)
2. BAE Systems Taranis (unmanned combat aerial vehicle)
3. Sea Hunter (autonomous unmanned surface vehicle)

5 Could you explain for each picture [described] why you picked it, which value it represents for you and why it is important*?

*1. nEURON + Taranis*

These weapon systems look and feel as you can lose control over them. Not the whole process is autonomous right now, but you know that they are working on it. They also look like the ultimate dream of the Defense forces. You can see that these are meant to be fully autonomous in taking out targets.

*2. Sea hunter*

This ship is fascinating, because it is meant to be at sea for months at a time without a human crew. Right now, it has no weapons, but it is said on film that it will be mounted with weapons in the future, but not to worry as all will be fine. The gut feeling person A has is that these systems are designed to create distance between man and machine and it has not only autonomy, but also the speed that comes with that autonomy. You can see from the pictures, for example the Taranis, that it is made because war is too fast to comprehend and to react as a human and therefore we want it even faster and that is why we will leave it to machines. This does not feel right. As something goes to fast to act on, then that is the problem and this is not solved by making it even faster, create even more distance and give machines even more autonomy. Person A is not against technology, but concern is that it is purposely designed for something. For example, Google cars, the state need to set up regulations, but we are both the consumer as user. And AW are not used in the UK or The Netherlands, but somewhere else at a distance far away and they also look very scary. It also projects a sort of superiority and weapons are often tested in poor regions of this world. When you talk about AW the advantages, such as creating a distance or reducing harm for own troops, are mentioned but if you turn this around and these systems are used against us than the same people have the same concerns as person A does regarding fear, lack of accountability and unpredictability. The images also show a value of: 'we don't care'.

It is part the reflexivity that humans have in warfare and machines don't, but also the distance the machines create to the battlefield. It implies an invincibility that we like to assert even more. The question: 'what do you want that technology does for you' is asked too little regarding AW. Often the Western high technological states claim that they will use the technology well which is not necessarily so in the future and it implies that other groups will never get their hands on these technologies. This idea of 'us' handling this technology with care completely disregards non-state actors or proliferation of AW.

There is a certain determinism in the discussion that person A does not like. Even a lot of diplomats say: 'it does not matter what we think, the technology will be there anyhow'. This inevitability of the discussion bothers person A a lot. The design of these systems also represents this. They look so abstract and polished and not where they are meant for, that is to kill people as fast as possible.

6 Do you have other remarks for this interview?

*The following points came up during the interview:*

Person A also made the point that the different stakeholder groups in this discussion should all play their role and should be recognized and valued for it.

Another point she stresses is the point that we should think about what we want that this technology does for us on the battlefield and have we thought it over with all of us. An example is the playbook for drones that Obama launched just before his re-election of his second term as he then probably realized what a successor of him could do with these weapons.

This campaign forces person A to keep asking the question: 'Why?' Not because humans make better decisions than machines, but what do you want to keep and that is that a human need to decide about life and death. Humans need to be involved in the decision to take a life and not program this in a machine. Person A stresses that the human needs to be in control of what a machine does within certain parameters ('kaders'). And the discussion is also about what these parameters are and the difference between offensive and defensive systems are taken into account. PAX does not decide which type of systems are acceptable, but states should decide this. One could argue that it is within a defined structured environment in which we think it is acceptable to use a defensive system to take out another projectile. In the next step towards Autonomy these parameters shift and we should examine the boundaries of this shift until when it is acceptable. AW with a box of a kilometre? What about time? Should they be in the air for 3 minutes? Why not 10 minutes? Is an hour acceptable? So, it is all about how can a human influence who or what can be taken out. This discussion is on the table with the CCW and is now addressed insufficiently.

I brought up the discussion I had with a legal advisor at the MOD on defining boxes to deploy AW in and person A indicated that she does not believe in this type of regulation with these types of weapons, because these regulations are too often crossed and she finds this type of weapons too scary and goes the proliferation too fast. These boxes can be used in the diplomatic process to study existing weapons systems and study how human control is guaranteed and regarding which parameters (space or time). And try to establish why the current systems are adequate and what is the minimal control.

Name: Person B
Date: 09-03-2017
Duration: 25:26 min

**No. Question**

1 How are you involved in the topic of Autonomous Weapons?

Did not ask because of limited time to conduct the interview (30 mins)

2 How would you define Autonomous Weapons?

There are only loose definitions of Autonomous Weapons and that is deliberately done according to diplomats of the UN, because defining something is the last thing you do when you want to get a ban in place. For person B it incorporates some aspects of identifying, targeting and destroying the opposition on the ground.

3 What is your position on Autonomous Weapons?

For all technologies that are considered disruptive and dangerous there should be a ban in place for AW before an arms race happens or go to a step change in the way we fight wars. It is possibly the third revolution in warfare. It would be a step change in efficiency and speed to kill the other side. There will be a lot of collateral damage and it is not as people imagining it in that it will be a colder and cleaner war. It will potentially industrialize killing even more which has been going on since the history of warfare. It will be terribly destabilizing, because in the past the ability to wage war was based on your economic money (maintain a large army and equipment). These types of weapons are potentially rather cheap and you won't need any manpower. You used to need to persuade thousands of people to go to war and with AW you will only need one programmer. That is going to destabilize the current political order.

It will probably lower the threshold to go to war and it will distance us more from fighting a war and the physical act of killing. On a more moral and personal level, war has always been a last resort and it should be personal, bloody and dangerous. When we pretend that it isn't, we will fight probably more and it should be something that politicians have to justify why people are returning in body bags.

But this argument does not apply all times. These arguments are applicable to Autonomous Weapons today, but in 50 years from now, other arguments might be more relevant when the technology is more mature. Right now, the technology can't comply to humanitarian law and distinguish between combatants and non-combatants. Somewhere in the future we will have systems that are more precise and more capable to adhere to law, but then other arguments come in to play like the greater efficiency and lower the barriers to go to war.

It is unethical to field AW today, but it is not black and white as the introduction of mines shows us. This lead to laws and regulations to prohibit the use and a mine can be seen as a stupid Autonomous Weapon that does not discriminate between people.

4 Which three photos of Autonomous Weapons [in this version of the thesis described below in words due to possible copyright issues] would like to discuss*?

1. A swarm of six drones.

2. Sea Hunter (autonomous unmanned surface vehicle).

3. URAN-9 (tracked unmanned combat ground vehicle).

5 Could you explain for each picture [described] why you picked it, which value it represents for you and why it is important*?

*3. Many drones*

It demonstrates that we are talking about quite simple technology that is already present with us today. It will be that when a swarm is coming towards you in a war you will not have many defence against it. These will be weapons of terror and will be used to terrorize civilians and other population and are relatively cheap. You can do quite a lot with these simple looking drones.

*4. Sea hunter*

This was chosen because we are talking about every sphere of warfare. So, on land, sea, in the air, everywhere you can imagine where war is fought AW will be used. It is an interesting case of how quickly the arms race is happening. This wasn't known when the open letter was published, but launched since and that goes for more military prototypes of AW. And interesting enough, when they build it they said it is not going to carry any weapons, but for clearing mines and detecting submarines. And a few months back they admitted that they are thinking about putting weapons on it. It demonstrates the slippery slope we are going down and the fact that there is an arms race going on. You don't need people or facilities to run a ship and you could a lot more in a smaller space and you can go along for much longer. They are now building solar power ships that can be at sea for years. You don't have to stop for people to feed them. You can see the clear military advantages having these types of technologies and the militaries will be seduced to use them.

*5. Uran-9*

The Russian Autonomous Tank which demonstrates that there are many players, China, Russia, UK, Israel as well as the US in a clear arms race. Russia military has millions of dollars' worth of orders in their arms books and it represents another sphere of battlefield. You can see that anywhere we fight war is being turned into an Autonomous Weapons system which has clear military advantages. If you look at that tank, you can see that it is not too far away in the future and it is not what people think that is going to be Terminator Robots. Person B can understand that Russia and China are also developing AW when they know that the US is spending 18 Billion dollars in its budget building the next generation of weapons primarily on the development of AW. The problem is that this is technology in which the only way of defending yourself is with another Autonomous Weapon, because no one has the reaction time or the ability to fight 24/7 of accuracy to defend themselves so you would be behooved not able to defend yourself. So it is not surprising that there is an arms race starting.

6 Do you have other remarks for this interview?

*Some other remarks that person B made during the interview:*

It is incredibly short sighted of politicians and the military to think that you can keep a tactical lead on any technology. We never managed that with the hydrogen or atomic bomb. You can't keep a genie in the box.

It will be very hard to verify how the technology will behave in a particular way, because we are talking about very complex systems interacting in a messy and complex environment. It is not surprising that is called the fog of war and to think that we could guarantee that it would behave in a particular way and that we are going to be able to audit and verify the systems is way beyond what is technically possible and will never be possible so he thinks it is irresponsible to field this technology. In industry robots, can be very helpful as they are in a controlled environment and get people out of the way, but the last place to put a Robot in is in a messy battlefield where we have no expectation of how it is going to behave. We don't know how to build ethical governors on top of Robots, but we will have to work out these things for Autonomous Cars to drive safely on our roads. He thought that the battlefield is the last place to work this out, but ironically it will the first place where this technology will be fielded.

Person B also likes to point out that there are enormous benefits of AI for the military. The US military always has been the biggest investor in AI research and it is a good thing that people don't need to risk life or limb for clearing a minefield. That is a perfect job for a robot that if it makes a mistake and gets blown up no one need to worry. Similarly getting supplies into contested territory with Autonomous Convoys. There are lots of applications of AI in the military that don't involves an AW that makes a final life or death decision. Person B does not have too much moral objections to defensive systems, such as the Phalanx because they work in well-defined envelops in a way that is to be designed to protect lives and not to take them, but off course any technology can be repositioned. It would be hard to morally argue that these systems, that already exist, should be taken off naval ships and make them more defenseless. It is all about the scope of operations and the area that they are confined in. Which is a very different setting than a drone that is launched to operate in a 24/7 target setting for offensive operations.

Person B is also very concerned with other actors, then our militaries that we can trust, such as actors as terrorists, rogue nations and people who have no problems with turning of ethical safeguards or hacking them into ways that would do harm. So even if we could build them in ways that they would behave as we would like to fight wars and that these systems can't be hacked means that the world will be a much unsafer place.

Name: Person C
Date: 10-03-2017
Duration: 25:25

**No.** **Question**

1 How are you involved in the topic of Autonomous Weapons?

Did not ask because of limited time to conduct the interview (30 mins)

2 How would you define Autonomous Weapons?

You got the DOD definition, that it is a weapon that can select and engage a target on its own basically. There two directions you can go with it; the first is that person C added a phrase to it saying that it is not previously designated by a human operator. The second way to do this and the campaign 'Stop Killer Robots' has increasingly moved to is defining AW to the negative essentially that they are weapons without meaningful human control. And this still doesn't help us either, because at some point we still have to define something. Whether we define what meaningful human control is or defining what select and engage means does not matter, but it is genuinely a though problem because we understand what it means at the extremes, but where the line is, is really tricky. Depending on that line the weapons are either in development or have not been deployed.

3 What is your position on Autonomous Weapons?

His personal position on AW is that it is hard to make decisions about what we should do about AW if we don't even know what they are. If you tell me what weapons are AW and which weapons are included or excluded than I can tell you what I think about them. But the technology is so nascent and the vagueness of the definition is so large that it is hard to have a yes or no perspective on something when the category is potentially so big and we don't know the direction in which the technology is heading.

4 Which three photos [in this version of the thesis described in words below due to possible copyright issues] of Autonomous Weapons would like to discuss*?

1. Tracked Unmanned Ground Vehicle.

2. MQ-9 (Reaper) (Unmanned Arial Vehicle).

3. Up-close picture of Robot of the Ban Killer Robot Campaign in front of the Big Ben in London.

5 Could you explain for each picture [described] why you picked it, which value it represents for you and why it is important*?

**No. Question**

*6. Ground robot*

This illustrates the future of military robotics that people really don't like to see or talk about and it is the common form of robotics we are actually going to see on the battlefield. One of the things that is most interesting in the discussion on AW is that you have almost a battle of analogies where you have groups like the campaign of killer robot who are focused on anthropomorphic machines like terminator walking into a house trying to decide if the person in the house is a lawful combatant or someone like a child. Person B and a co-author are often talking about air and naval scenarios where the battlefield is a little clearer and they are talking more about weapon platforms instead of marching robots.

The first picture illustrates a more functional form essentially where robots on the battlefield are much more likely than the terminators walking around. These systems are being used as remotely piloting systems right now.

*7. MQ-9 (Reaper)*

The MQ-9 is interesting because a lot of the concerns of AW started with the use of drones and people worry about that drones make war too easy and this is dangerous. But you couldn't really stop drones because they are already on the battlefield and operations were already going on. His earlier work showed that drones were quickly proliferating and this couldn't be stopped, but you could maybe work on what is next. And for a lot of groups are trying to make sure, from their perspective, how do we ensure that this is the line and they are worried about the drones making decisions for themselves.

It also relates to the question about decision and when you talk about a robot making a decision you are talking about a robot that is making judgement for itself as opposed to following its programming. Especially how narrow and centralized its programming is. Often when people are talking about robots in a negative way, they are talking about robots making decisions and people who are less concerned, are talking about humans ordering robots to do things where humans are still making the decision.

It is fascinating that there are people who think that you have a right to be killed by a person while people do all sorts of horrible things to people, while you know when you are dead you are dead. It is interesting to see how quickly it gears into moral philosophy almost more than anything else.

**No. Question**

*8. Killer robot campaign robot*

It is interesting to see that a group of people whose previous experience is on campaigns on landmines and cluster munitions and other technologies that are generally not central to how militaries operate. And a lot of concerns on AW are on what kind of AW do we have and what can they do, but we know that the integration of autonomy into military systems in general is happening whether the weapon systems is autonomous is a different question almost but the integration of autonomy in weapon systems is inevitable. It is interesting that the campaign has picked this up and especially because it is a different type of military technology than they have tackled before. To person C it is more like they are trying to ban the tank or the submarine or something. If the campaign on Killer Robots are right, then the militaries will want them and if these campaigns are wrong and these weapons are not going to be such a big deal, than it is actually not that important. It is an interesting group and they are genuinely trying to reduce human suffering, but they are using the playbook from different victories and person C thinks this is a different situation because how uncertain the technology is and how broad the potential category is while nobody knows how important the category potentially is for the military.

6 Do you have other remarks for this interview?

*These are some other points that came up during the interview:*

To my question that there hasn't been done a lot of empirical research yet person C answered that he did some surveys with priming experiments, what public opinion scholars generally do, to try a list of questions about attitudes and you present them different context and you see what their views are in these scenarios and then you try to estimate what they feel but that is work limited to the US.

Another interesting thing is that the politics are very interesting, and in some extent because we live in a world where we take US military technology superiority for granted. Imagine a fake world in which China deploys AWs weapons and the US and EU countries don't have one, the next week in congress you would have hearings about an AW gap and questions as: 'why does the US doesn't have weapons that the Chinese have?' are asked. This leads to the question that person C thinks the campaign has not considered in that: 'what happens when a non-democracy deploys these systems and they are not just niche weapons but important for general military operations and they actually give you an edge on the battlefield?' 'What would the world do in response to that?' This does not necessarily needs to be negative, but the issue is complicated.

**No. Question**

The question is if this is an arms race, in which you are worried that somebody is acquiring technology and you are directly competing with them, or that it is rapid proliferation because of the ease of acquisition. An arms race is about politics because you need a reason to arms race and if an arms race is happening for AW is that it is because it is hard to know what an AW is. What is the difference between an MQ-9 Reaper and an autonomous Reaper. It is software not hardware and you cannot see it. Uncertainty, whether somebody's system is a drone or an AW would be a potential factor that would lead to an arms race because you cannot be reassured that the other side is not acquiring AW s. Unless you are plugged into their weapon system or something, but who would let somebody do that?

Name: Person D
Date: 19-04-2017
Duration: 55:18 min

**No. Question**

1 How are you involved in the topic of Autonomous Weapons?

Did not ask because I did not ask this question to the other interviewees.

2 How would you define Autonomous Weapons?

It is a weapon system that uses force and implicitly this force is lethal, but that is not a necessary implication. This is not imperative but a common perception in a military context although this assumption might be a brought up during the discussion of the photos. For the application in the near future this is what is expected. Autonomy is to me that they are capable of selecting the goal themselves. To select a goal in a certain context with a vague assignment. It is not sent to specific coordinates or to a specific vehicle, but more like 'go to this area and take out all vehicles that behave hostile'.

3 What is your position on Autonomous Weapons?

Person D would like to oppose it, but my analysis shows that this is not possible so the next best thing is to ensure that we use it as responsible as possible and by this to minimize the risk that situations go out of control. I can explain this stance based on the following ideas:

1. On the civil side, there is a strong commercial drive to develop AI in general to make a lot of money;

2. At the same time, in the military context the speed of decision-making is important. Previously getting the information was a problem, but nowadays processing the information is the problem. This means that human decision-making gradually is becoming the weakest link which leads that this will be left to systems instead.

3. And thirdly it is possible that a super human intelligence will arise. Person D does not see a technical reason why this would be impossible. If this is possible we cannot oversee what this will signify. This in combination with the use of lethal force is not a situation that we want as humankind.

Reasoning from this person D hopes that we will not develop and use Autonomous Weapons, but this is not possible as the first two bullets will ensure that we will reach this situation. 'The nicest flowers grow the closest to the cliff'. So, the best that we can do is to make sure that we don't fall down the cliff. He is quite pessimistic that we can control this, but we will have to try.

To control this, one could use a framework for example in the design phase to embed them in our desired world and to prevent that these technologies will create their own vision of this. This is very at the frontside of the development, but during deployment you would have to think well how you would design the targeting process. The more self-capable these systems are, the more we have to specify the targeting parameters and this might have to be different then we can imagine right now.

And we should attempt to reach an international agreement on this and this is the hardest part, because not to comply to this will give a huge advantage. It is the classic prisoner's dilemma. We have to think about this and as this is not an easy question so person D does not have an answer to this, but it is evident that the commercial parties should be involved in this.

On my question to clarify his first statement about opposing this technology. Autonomous weapons are different from conventional non-autonomous weapons in that the latter cannot wipe out our entire species, although in theory this would be possible with nuclear weapons, but these are too costly to build and AW are not and pose therefore a higher risk. So, the process of building nuclear weapons can be controlled and it has been around for 60 years which allows for us to get to be used to. The question is if we get that chance with AW and their rapid developments.

Another thing is that AI weapons already contains some risk, in contrast to for AI that supports doctors in their diagnosis which is meant to do good. AW combines a very intelligent system with a hostile intent so person D is concerned that AW poses a greater risk. AW is not the only thing person D is concerned about. Also, synthetic biology is also a field with high risk. So, technologies with a low barrier, but potential uncontrollable risk are worrisome for mankind.

4 Which three photos of Autonomous Weapons [in this version of the thesis described below due to possible copyright issues] would like to discuss*?

1. Screenshot of Ex Machina movie.

2. Picture of a rescue robot carrying a human like dummy in a competition for robots to assist by disasters.

3. Robot hand and human hand whose index fingers are touching.

5 Could you explain for each picture [described] why you picked it, which value it represents for you and why it is important*?

All photos are interlinked.

1. Intriguing at Ex Machina is that the technology that is created in this movie also inherits (or learns) mankind's survival instinct and subsequently takes many actions to prevent dying and also doesn't give a damn what happens to the others. It represents the fact that you will have to consider possible events when starting to build this type of technology and what could go wrong if you did not do that. Because we are talking about systems with a certain level of self-learning, but we might miss the things that they are learning themselves. This has 2 elements; first is unintentionally getting out of control, and the second is intentionally seeking the boundaries, but this also means that you can cross this line.

2. The second picture is from a rescue robot competition to assist by disasters. The robot does not hurt anything but also does nothing. So, the dilemma is that you want a combination of both 1 and 2, but to do this you will need a big sense of the context and a sort of 'goodwill'. Person D indicates that he expects that AI should take better decisions than a human. He gives the example of an AI that is ordered to make as many handwritten notes as possible, and at a given moment in time it evolves into Super AI and decides in a matter of seconds. that the best way to reach its goal is to transform the whole earth in notes, kill all the humans on earth because they are in its way of creating as many handwritten notes as possible. Mission accomplished but not in the way that we intended it. So, you need to think about this in the design phase.

3. But you also need to think about the cooperation between man and AI (although this will not happen in the early AI systems) but in the future, we will have AI systems that will do that. The current AI systems as the Harpy have little context awareness, but as we deploy robots in ground operations than we will have to cooperate with them and goal setting will become extremely important. Photo 3 represents this cooperation. This implies that the system needs to know what we mean, but we also need to train our people to work with these types of systems and to communicate the goals to these systems. The more the systems have more boundless possibilities, the more you will have to think about setting boundaries to its assignments. Person D gives the example that a magic spell is nice, but you also need to understand it. So, we need to train our personnel in giving assignments with boundaries.

6 Do you have other remarks for this interview?

On my question what person D thinks about always having a human in-the-loop he answers that it depends how many boundless the system is. Firstly, the more freedom it has then you need to think more in the design phase in order to allow the system to develop a goodwill. Secondly, because you cannot limit these systems, you should try and find a way to restrict the risk.

Person D sees the development in phases. Phase 1 is the large-scale introduction of AW with limited capabilities. In phase 2 you will get man-machine teams and can be seen as an intermediate phase in which people learn to cooperate with the technology. But in phase 3 it will be dangerous and risky, because then we will have complete independent systems and in phase 4 this will be the Terminator or Skynet example which is an example of AI that turned really bad, but Person D thinks that this will not be obvious and will not be AI turns evil, but it will be much more the case that we just slightly not build AI with the right mindset and still destroy the world. He also thinks that there is no sharp line between phase 3 and 4, but that you might just figure that out when it is too late.

Name: Person E
Date: 17-04-2017
Duration: 22:19

**No.** **Question**

1 How are you involved in the topic of Autonomous Weapons?

Did not ask because of limited time to conduct the interview (30 mins)

2 How would you define Autonomous Weapons?

Autonomous is for me a weapon that when I give an order to the weapon to go, that it finds its way and selects its target by itself (based on the parameters the human provided) based on the image of the surroundings that it creates itself. So, select, choose and take action by itself are the three components.

3 What is your position on Autonomous Weapons?

Positive, I am open for this type of technology. So, positive, but we have to think about how to use them so with prudence. Not just throw them into the organization ('niet zomaar naar binnen gooien').

4 Which three photos [in this version of the thesis described below due to possible copyright issues] of Autonomous Weapons would like to discuss*?

1. Harpoon cruise missile fired from a Navy ship.

2. Swarm of 19 drones.

3. Future style robot with a large gun.

5 Could you explain for each picture [described] why you picked it, which value it represents for you and why it is important*?

1. This is one of the current existing systems that is used by the navy for over 30 years. It is a ship launched torpedo against submarines. The moment you launch the weapon it autonomously selects its target and attacks. It has no override function. It is fire and forget. A similar system is the Harpoon which is a cruise missile that is used against ships on the water surface. This type of weapon searches its own target and when it finds it, it could be that it is a different one than you targeted. It is a system that is deployed with great caution, but it is a type of autonomy. From the Navy's point of view these guided missile control systems can be set to a certain set of parameters, such as altitude, speed and certain identification patterns, and the system is underway. There is no man-in-the-loop and we have these systems for over 20 years.

These systems can be used autonomously, but as we deploy these systems we always build the man-in-the-loop function in. A human has to clear the system before it can take out its target. There is a process around this autonomous function and there is even a hardware switch that need to be switched before the system can take out the target. A sort of fire inhibit switch, but you can set the system that you do not need this override. These systems are developed to be deployed in situations with a high air threat in which every second counts. This is already a certain degree of autonomy with the feeling that you can intervene at all times. The fact that it must be possible to intervene at all times is very important for the Navy. This in contrast to the Torpedo that does not allow you to intervene which means that you have to be absolutely sure before you deploy it. For both weapons the Navy has created procedures to deploy them.

2. The next step for the Ministry of Defence are the remote piloted drones that we are trying to incorporate into our organization, but the developments in the technology go faster than we can meet at this time and we are trying to match this pace. Swarming are the systems to be used in the near future. These systems are deployed and will execute their mission on its own, but how they do this is not within your grasp. They operate at quite a level of autonomy. Currently this is about sharing information and sensoring, but the next step is sensor to attack with a weapon. This technology is currently available. Currently the MOD is creating a roadmap for remotely piloted systems in the air, but the leap to procure fully autonomous weapons is too big, because if this is something that we want as MOD has not been done discussed. This is currently an ethical barrier at the MOD. The topic of unmanned and autonomous systems is part of the current vision for the Defence forces that is drafted by the Hoofd Directie Beleid (at the Defensiestaf) and by this they are starting the discussion on AW. The MOD will only procure these types of weapons after the topic of AW is debated in our society.

All specialists from the Army, Navy and Airforce state that Autonomy is on its way, but we always want to be in control. We always want to be able call them back when the situation has changed. Or that we are able to re-task them when necessary. It is typical for military operations that situations change quickly and that you have to adept your plans to that. Especially when you are working in where there is not yet an armed conflict and you are working under the Rules of Engagement. You have to be able to anticipate and react quickly and you do not want to have a situation that you have deployed your system in the air to attack a certain goal, but when the tactical or operational situation changes you want to be able to recall them. If this requirement is built into the systems, then we as MOD would like to use this type of systems.

3. Photo 3 represent the image that the public has on killer robots which is a system that cannot be controlled and thinks and acts on its own. People are weary about that and it represents the dividing line between what we can deploy at this moment and which technology do we want to keep ourselves from.

5 Do you have other remarks for this interview?

Person E stressed that he has worked already many years with AW (as picture 1 shows) and that there is a certain level of control built in them as you embed it in your procedures and operations management ('bedrijfsvoering'). Especially when a new technology has not proven itself then you will have to build more if these type of securities in it. Person E believes that we are going into this direction and that we should be going into this direction, because it will save you valuable time and our opponents also do this so you have to go along. When a swarm of drone is approaching you, then you will have to counter that.

Name: Person F
Date: 08-05-2017
Duration: 53:44 min

**No. Question**

1 How are you involved in the topic of Autonomous Weapons?

Did not ask because of limited time to conduct the interview (30 mins)

2 How would you define Autonomous Weapons?

Person F follows the ICRAC definition which is autonomy in the critical functions of targeting and firing weapons. It is not only selecting the target as a functional step but there is also an anthropological and sociological action that is communicated by pointing a weapon at someone. In a formal sense, it is selecting a target or determining that something is a valid target. And the second stage is releasing and firing the weapon. If you have autonomy in those two functions you definitely have an autonomous weapon, whether it is autonomous if you have one or the other that is more challenging and to what degree you need meaningful human control. And what constitutes meaningful human control over these functions is still what open for debate.

The basics of human control, from an engineering perspective, is effective human control so if humans are able to intervene or shout down or change the behaviour of a system. Meaningless control is when the person becomes a dumb automaton for a machine so if you are sitting in a room with a button and every time a red light comes on you are supposed to press the button, in a sense the human has causal control over the button, but they don't have any real significant meaning over what kind of information is being used to determine that something is a valid target, because it is just a red light came on and you trust that the red light is connected to something that produces correct targets and authorizing to attack that target. But you actually don't have any contextual or situational awareness or understanding or any independent means of verifying the legality or morality of a target. So effectively you don't have meaningful control although you have causal control.

Next to this, for killing to be a meaningful act there needs to be and intention or a military purpose to take out a target. And simply because something is a legal military target means you should attack it. I could be you don't attack it because it is expensive or you would reveal you position so there are all other considerations that even something is a legal target you still do the calculation of whether or not you will attack it and this is part of if it makes a human activity and when you automate it, it stops being meaningful and it becomes really an arbitrary judicial execution. You don't have a meaningful way of verifying that a target is lawful or meaningful.

3 What is your position on Autonomous Weapons?

Person F is against them and he thinks we should ban them. However, we might frame a ban as a positive requirement as meaningful human control. It is really hard to specify beyond the ICRAC definition which he thinks is more vague in a way than meaningful human control or you could say that you need meaningful human control over those two critical functions. He believes that when meaningful human control is understood than it becomes a requirement for all weapons than specify it for a particular class of weapons, but you will have to guarantee that it happens on all weapons. The analogy he like to draw is that of unnecessary suffering and superfluous injury, not that these are the same definition, but they have a similar function that is also kind of vague in what is unnecessary suffering and what is superfluous injury. There are also big debates about this when that language was introduced in the Saint Petersburg convention and lawyers passed it out, but now it is fairly well understood as additional harm that does not serve any military function other than to create suffering and pain.

Even although these are sort of open-ended modifiers, but they show if we have meaningful control over a system. Can I stop it? Can I intervene on it? Can I second guess it? Can I verify for myself that the targets that are selected are in fact valid, legally and morally? Because if I can't then there is a problem with that weapon system. The systems that have limited control are dangerous and the global society has to figure out what we want with these.

Current autonomous systems, such as the Goalkeeper on ships, should have to meet these requirements. And for states that deploy these types of systems should constitute how these fulfill this requirement for any ballistic or anti-ballistic system like the Goalkeeper, Phalanx or the Patriot as a lot of these systems are shouting down incoming projectiles and the one person F has seen have very limited time and scope in terms of how long they are operational under self-targeting to where it is less than a minute (20 or 30 seconds), but you can still argue that there are humans with their hands on these systems that decide that we are going to let the systems shoot it down. Person F does not intend for the ICRAC to eliminate these types of systems but they are still dangerous as they still cause friendly fire incidents. Person F gives an example of 2 aircraft being shot down by the Patriot system in Iraq after their transponder malfunctioned and since then the Patriot was reengineered in that the operator has to give positive authorization for the system to fire instead of fire it on its own automatically. This adds to the psychological burden that people are not willing to press the button unless they are really sure as you have this automation bias where the system says that it is sure than people are willing to let the system to go on its course. There is definitely more meaningful human control in the new Patriot interface then there was in the old Patriot interface.

4 Which three photos of Autonomous Weapons [in this version of the thesis described below due to possible copyright issues] would like to discuss*?

1. metal skeleton Terminator.

2. Predator Unmanned Arial Vehicle.

3. Picture of Robot of the Ban Killer Robot Campaign in front of British Parliament in London.

4. Three-legged chair in front of the UN in Geneva, Switzerland.

5 Could you explain for each picture [described] why you picked it, which value it represents for you and why it is important*?

*1. Terminator*

In the first years of the campaign the metal skeleton Terminator featured each article written about the campaign which is interesting from a media and social activism perspective. On the one hand, it is a powerful image that stimulates public awareness and it grasps people's attention as it represents the Killer Robots.

The Terminator is actually really interesting if you go back and watch the movie as it is assigned a specific target, so in that sense it is not an autonomous weapon, although it kills a lot of other people on the way so there are some autonomous targeting decisions, but its primary directive is Sarah Conner which is the actual target. And there are a lot of other elements to the films that stimulate people's fears, like robots becoming self-conscious, turning against humans and stuff like that which is not really the campaigns concern, but it is more with the intermediate, kind of stupid autonomous weapons systems that could be fielded in the next 5 to 20 years that are not going to have sophisticated reasoning systems but they are going to look at a small set of criteria based on their sensor data and release their weapons which can have negative consequences for civilians and all sorts of other things. The Terminator image does not really capture a value but it captures attention and that was important early in the campaign. It was not just fear about the movie, but genuine fear about how these weapons were going to operate.

*2. Predator*

The second thing people think of are the Predator and Reaper drones which are remote operated robots so they are not really autonomous so they don't quite capture the story. These are more real world images that show up in a sense that it is kind of a precursor to autonomous weapons in a sense that you could automate the targeting and firing of drones and those would become autonomous weapons and there are actually developments of the next generation of drones many of which are talking about autonomous targeting as possibilities for example the X47-B, the Taranis and the Neuron which are jet-powered stealth looking drones that carry a lot of weapons that are really designed for contested airspaces unlike the Predator and Reaper which are propeller driven which are slow and require that you have control over the airspace. In airspace where your opponent has any aircraft or can jam communications these types of weapon systems are not useful. The next generation of drones can operate in spaces where communications are denied which is the argument in favor of autonomy, because you don't need a communications link that can be interrupted and you will lose control over the aircraft so it has to be autonomous in those areas.

The question is how it is going to get to its target. It could be like existing cruise missiles where you give it a list of GPS coordinates and it finds these coordinates and bomb them in which case it is humans determining those targets, or is going to have some capability of choosing and selecting a target on its self of targets of opportunity based on sensor data.

The images of these new systems look very sinister, especially that of the Taranis, with the lightning and stacks of warheads in front of it. These do represent what we need to be concerned about of the next generation of autonomous weapons. These concerns would be that they have a capability of autonomous selection of targets. It is a long-range fighter that selects its own targets. It becomes a question of legality and operational control. Depending on the design of these experimental systems it can either fall into the autonomous weapons system or not if humans pre-select the targets and their functionality is not really fixed yet and because it is all secret we don't know how they actually work.

The underlying case is human dignity, meaning that the decision to kill should be taken by a human being and not by a machine. It is tied in the meaningfulness of the killing. Another human determines that you are an enemy combatant based on your military disposition in the battlefield, uniform and participation in an armed conflict and those criteria. That is a legal and moral act of killing and the possibility of dignified killing is there. If it is just a machine following an algorithm, and it is trying to meet some criteria determining legality then there is still no human judgement and to person F that means that there is no guarantee of dignity. It may get it right or wrong from a legal perspective in some sort of probability function, but ultimately there is no intention and no meaning in that system, it cannot determine that there is a military need. And it can't make a moral or legal judgement because it is not a legal and moral agent so the killing that it does can't be dignified by definition. It may be utilitarian or practical under international law, but from a moral perspective there is a lack of dignity in that.

On my question if in certain situation Robots make better decisions than humans, because they don't get tired, emotional or seeking revenge, person F answers that this is an important point, but that we should design systems that help humans in making better decisions and reduce mistakes overall, but that you also retain those elements of meaning and meaning making to the human that in ensures that it is a moral, legal and meaningful process and use the machine in trying to reduce those sorts of errors and have guided suggestions. That help humans to make better decisions under stress and tiredness.

*3. Friendly robot*

The third picture is that of the campaign in which the robot stands in front of the British parliament. This is a picture of a friendly robot as person F states that he actually likes robots and there are jobs in the military that can be done by robots, but he stresses that all the tasks which require moral and legal judgement should be done by humans.

*4. Three-legged chair representing the landmine campaign*

Another good image of the campaign is that of the three-legged chair in front of the UN in Geneva that symbolizes the landmine campaign. In the start of the [Ban Killer Robot] campaign we tried to do the same as in the landmine campaign and say that these are highly indiscriminative weapons with huge civilian impacts which is a concern. If you think about that Killer Robots basically are landmines that can move around and select targets. It is another way of thinking how scary they are because they are really not that much smarter than a landmine and they don't understand the world any better than a landmine. They are only more sophisticated in their ability to navigate and sense data about the world and select their targets. So, on the one hand they are not going to be so blindly indiscriminative and you can improve the discrimination over time with engineering techniques, but there is the fundamental shift in that you changed the nature of the act of killing and what justifies it at all when you have these automated killer robots. But this is true not only for the military and war but for the whole society as we are automating a lot of task that used to be under human control and off course humans make lots of mistakes in decision making and there is bias and all sorts of prejudices and things that also need to be addressed, but often we build automation and pretend that those biases don't exist anymore because it is an engineered system and it is not supposed to make mistakes, but these systems are complicated and we don't know how these behave or what they are going to do in unexpected situations and you would be raising a whole set of new issues with regards to the engineering and problems and you did not solve any social questions.

To make another comparison with the landmines, there is a certain notion of predictability but there is also an apparent unpredictability. Landmines can be justifiable to halt troops during a war but 10 years later, when someone wants to farm that land, that landmine still can go off so you just can't control that future. This is the intrinsic problem with Autonomous Weapons in a sense to which you know it's supposed to go and find targets and bomb them. That seems reasonable and a constrained notion of that it can only take out that specific kind of radar signal, but how accurate is that? How many things look like that radar signal? So, other things that emit radius signals that could be confused and or people could make beacons that imitate those radar signals to lure that rocket to other things.

Once you have it out over a long period of time or a wide range of area you don't really know what it is going to do. You have an idea what it is supposed to do, but you can't really control it after that. And that is when you get into dangerous territory because it is all about these expectations of risk and we don't really know how to gage that and as these systems become more and more complicated it is going to be more and more difficult to gage that, both as the operator as you are trying to predict what to do but also as the tribunal or something trying to hold somebody accountable for having released something terrible rather than having the intention of doing something terrible. So, you can't really convict them of war crimes, but instead it is negligence as they really did not know what it was going to do due to some extraneous circumstance that we couldn't predict or were not aware of. So, our whole system of understanding accountability and responsibility kind of starts to break down and that creates a big loophole for people who do want to do bad stuff and not be held accountable for it and it puts a burden on people that are conscripted or join the military when they are 18 and they really don't know what they are going to do and what can they be held responsible for.

6 Do you have other remarks for this interview?

*On my question to person F if you would be able to enforce a ban with this type of technology he answered:* There are a few things to keep in mind about international law and a ban. Enforcement is one element of it, but it is not the only element of it. You can compare it with murder. People will still kill each other, so we should not have outlawed murder? No, we make it a crime because it is wrong and people still commit it and you will deal with that. It is about establishing a norm on an international scale where all the countries of the world come together and agree that it is fundamentally wrong to have an autonomous weapon system. Given that, what would be the implications if people violate that. Most treaties don't have verification methods and explicit punishments and things like that, but it is about shaming countries and sanctions and international consequences for countries that would do it.

For non-state actors, you are in a whole other realm and already we have chemical weapons banned which is a good analogy that countries, even countries that did not sign the treaty, are looked at and are being sanctioned and get repercussions for using chemical weapons. It also effects the industry in a sense that large military contractors are not actively developing chemical weapons and large chemical companies are not producing huge quantities of precursors that are prohibited under the convention. You are going to see something similar with autonomous weapons. Off course you can put together an autonomous weapon based on stuff from a hobby store, but it is not going to be a hard-military system that can take out cities and factories. And already people can make booby traps and bombs and things like that and you can't really stop that but you try to regulate access to explosives or key technologies. You prevent that autonomous weapons systems proliferate from big states to smaller states that will not use them with restraints or modify their software and eventually proliferate to non-state actors and terrorists that would have access to systems with serious capabilities and not just a toy with a bomb strapped to it. So, with an international treaty you are ahead of mitigating that. But there is nothing to stop police forces or states to use it against their own people so you going to need national laws and standards to also prevent the use of these systems for police forces that would use them against demonstrators. I will probably fall already under civil and criminal laws for private persons who would build these systems, but it would be good to specify that it is forbidden to attach weapons to autonomous targeting systems.

*On my question if that would imply criteria for transparency person F answered*: As you are serious about verification of sophisticated military systems, part that is required for demonstrating meaningful human control is demonstrating chain of command and keeping data logs such that if someone expect that your system is operation in fully autonomous mode you can demonstrate that through a data stream that in fact a targeting decision is being made by a human or that the human authorized that target in this specific case for every kind of weapons system. Having a sophisticated information processing system is easy, having the authority to review them is another thing. The current standard is that you impose that discipline on your own military so you don't necessary have to open it, but you have these procedures for like you hold officers and soldiers accountable in court martial for violating international law. You don't expect the enemies to hold trial for that. You yourself review your weapons, so that is not transparent at all and you don't release those reviews. But the requirement would be that you keep that data so if something happens so that you would have data.

*On my question that the campaign got a lot of traction in 2015 and 2016 but that it now looks like that it is slowing down and if he knows the latest information on it person F answered*: The problem is the length between the UN meetings and the big news was sort of in December that the discussions were elevated from the informal expert meetings to the formal GGE which is one step before the treaty negotiation. They scheduled 2 weeks of these meeting but they had a lot of funding issues this year as they changed the budgeting model at the UN that requires all the dues to be paid in advance for you to hold the meetings but the last 50 years it has been another system in that they just hold the meetings and then they wait for people to pay their dues later. So, there is some debate about if the second week of meetings is actually going to happen if countries don't pay their dues completely. The meeting in August is going to go forward and hopefully there will be a second one in November in conjunction with their annual meeting. They are hopeful that something is going to develop out of the GGE, but one week is really not enough time and they were hoping for three weeks of meetings and not 2 weeks that might be pair down to one.

In general, the campaign got off to a very fast start compared to most other disarmament campaigns, including landmines and clusters which really took 10 years to get to the stage that we are at in three or four years. So, to keep up that kind of momentum would have been impressive and what it slowed it down was that the states were sort of interested in discussing it but are much less interested in really formalizing their position. So, they are kind of stalling, but the UN is definitely formalizing their position on it, especially with the 3000/09 policy under review within the Pentagon right now which has been there for 5 years. And they are hoping that the states will discuss it more in the GGE and settle on the language because it is easy to say that there is no definition of autonomy and autonomous weapons and we don't know what meaningful human control is, but that's for them to decide that they can justify all those terms however they want to. But for a treaty it is finding the political consensus what those definitions should be. The ICRAC is doing a good job in narrowing down the definition of an autonomous weapon and then it is a question of elaborating meaningful human control and what kind of language would work for a treaty in that regard that is not too permissive and that is not too restrictive. It is a though balance and it is more of a political issue than a technical issue. Person F is quite hopeful that they will make some headway.

Part of the reason the CCW was so eager was that they haven't really done much in the last decade or so previous to that and what they had done was kind of insufficient so they are looking to redeem themselves and this is a good way and topic to do that. And if we do it soon it will largely be a pre-emptive kind of treaty, but the existing weapon systems need to be reviewed under this additional meaningful human control requirements, but most systems in place would be able to offer that explanation and the ones that can't probably should not be used anyway. It is easier to do it now pre-emptively then in 10 years when states have developed fully autonomous drones and submarines and feel that they are necessary for their security and then it would be impossible to get that kind of treaty enacted.

APPENDIX G

# Coding Memos

## Code Book Researcher 1

In *Values Coding* the text is coded to find the values[1], attitudes[2] and beliefs[3] (Saldaña, 2015). These concepts are listed as pre-set codes.

List of pre-set codes:

| Concept | Colour |
|---|---|
| Value | Blue |
| Belief | Green |
| Attitude | Yellow |

List of emergent codes:

| Concept | Colour |
|---|---|
| Definition | Purple |

Tips:

- A pre-set list can have as little as 10 codes or up to 40–50 codes. We recommend not creating too many codes because the person coding can become overwhelmed or make mistakes in the coding process if there are too many.
- The rule of thumb for coding is to make the codes fit the data, rather than trying to make your data fit your codes.
- Creating memos during the coding process is integral to both grounded and a priori coding approaches. Qualitative research is inherently reflexive; as the researcher delves deeper into their subject, it is important to chronicle their own thought processes through reflective or methodological memos, as doing so

1 Value: the importance we attribute to oneself, another person, thing or idea.
2 Attitude: The way we think and feel about oneself, another person, thing or idea
3 Belief: is part of the system that includes our values and attitudes plus our personal knowledge, experiences, opinions, prejudices, and other interpretive perceptions of the social world.

 | DOI:10.1163/9789004449084_014

may highlight their own subjective interpretations of data. It is crucial to begin memoing at the onset of research. Regardless of the type of memo produced, what is important is that the process initiates critical thinking and productivity in the research. Doing so will facilitate easier and more coherent analyses as the project draws on. Memos can be used to map research activities, uncover meaning from data, maintaining research momentum and engagement and opening communication

More info:

http://onlineqda.hud.ac.uk/Intro_QDA/how_what_to_code.php

https://researchrundowns.files.wordpress.com/2009/07/rrqualcodinganalysis_7_19_09.pdf

http://programeval.ucdavis.edu/documents/Tips_Tools_18_2012.pdf

https://www.slideshare.net/kontorphilip/qualitative-analysis-coding-and-categorizing

Additional notes:

*Before start:*

My intent is to use the values coding method (Saldaña, 2015), but Saldaña (2015) defines the term value different than Friedman and Kahn Jr (2003). Also, I wonder if the difference between value, attitude a belief is really clear, but I will see how it goes. Perhaps an extra code word for a 'definition' needs to be added?

*During coding:*

a. I notice that an attitude often involves a verb; or at least that is how I interpret it.
b. A belief involves often an opinion.
c. I notice that I often think of the lists of values from my literature study when I highlight a term, for example 'reflexivity'. >> this might point to a bias of me as a researcher so it would be good to check my list of values against the second reviewer.
d. If a term is not liked by the interviewee, such as 'determinism' I highlighted it as a value.
e. Is something that is mentioned often important to the interviewee? And could it therefore be interpreted as a value? For example, 'distance between man and machine'. (see j.)
f. When coding I sometimes went back to see how I classified the same phrase (e.g. 'too scary') and copied the coding.
g. I added an extra code for 'definitions' and added that to the code book.

h. I interpreted an opinion of the interviewee as a belief because it is part of the definition of the term believe by Saldaña (2015).
i. After coding a question of an interview, I read the question again and checked my coding.
j. If a term was used more than two times, for example the term 'defensive' in professor Walsh's interview, then I have highlighted it as a value.
k. I also highlighted a term that was specifically pointed out, e.g. 'benefits' was marked as a value in prof Walsh's interview.
l. After I coded a new interview, I checked the previous one(s) again to see if I stayed consistent.
m. If something was mentioned as important, for example in the interview with Ad about the importance for the Navy to have the possibility to be able to intervene at all times, then I coded it as a value.
n. After coding the last three interviews I did not check the first three again. Mainly due to time restrictions, but also because the coding became easier for me.

*After coding:*

I send all 6 interviews to the second reviewer and added some background information on coding interviews. I also added the codebook without my own remarks. We discussed the task via skype to make sure she understood it.

## Code Book Researcher 2

Here, I state the established guidelines from assumptions, rules or reasoning that helped in directing statements either to value, attitude, or belief. Note that these lists developed and expanded during this coding process hence this became an iterative process. In other words, in latter interviews I always reflected on the guidelines that aided the coding process in previous interviews and built on that.

Value: when statements reflect some sort of strong importance, e.g. 'we have to'.., 'must', 'important'.

- when reflecting strong or high judgment, e.g. 'ensure', 'worrisome', 'need, 'mankind'.
- when containing high-level or grand words, e.g. 'ensure', 'human suffering, 'inevitable'.
- when reflecting something considered of importance or high-purpose, 'needs, 'reflexivity'
- when significant part of storyline or discussion or illustrating importance, e.g. 'need', 'morality'.

- when illustrating something deemed worthy or to be considered, e.g. 'justify', 'humanitarian law'.

Attitude: when statements show tendencies or direction, e.g. 'not just'.., 'too'...

- when addressing personal view or preference, e.g. 'vague', 'not possible', 'you want'.
- when regarding relative position, e.g. 'does not matter', 'real tricky', 'vagueness'.
- when something that might affect judgement, e.g. 'not for '.., 'more than', 'we want'.
- when consisting of words illustrating viewpoints or repeated, e.g. 'against', 'vague', 'terrible'.
- when reflecting some sort of stance, e.g. 'deliberately done', 'should be', 'terribly', 'seduced'.

Belief: when statements are based on assumptions or regard something that is considered true.

- when based on (Future of Life Institute) expectations or assumptions, e.g. 'we might'.., 'it will'..
- when discussing (Future of Life Institute) unknowns or assumed other people's perspectives.
- when reflecting personal knowledge/experiences, e.g. 'is coming', 'don't know'.
- when regarding raised questions, speculations or assumptions, e.g. 'they don't have', 'will'.
- when regarding interpretive perceptions or future, e.g. 'there are only', 'will be'.

APPENDIX H

# Results Coding Process

| Interviews | Values | | |
|---|---|---|---|
| | **Researcher** | **Second reviewer** | **Similar values** |
| *Interview 1* (*Person E*) | – intervene at all times<br>– in control<br>– call them back<br>– recall<br>– cannot be controlled<br>– level of control | – we have to think about how to use them so with prudence.<br>– The fact that it must be possible to intervene at all times is very important for the Navy.<br>– but we always want to be in control. We always want to be able call them back when the situation has changed. Or that we are able to re-task them when necessary.<br>– you want to be able to recall them.<br>– have to build more if these type of securities in it.<br>– have to go along. | – Intervene at all times<br>– To be in control<br>– Recall them |
| *Interview 2* (*Person D*) | – out of control<br>– control<br>– control<br>– risk<br>– controlled<br>– risk<br>– to do good<br>– risk | – ensure that we use it as responsible as possible<br>– minimize the risk that situations go out of control.<br>– in the military context the speed of decision-making is important.<br>– is not a situation that we want as humankind. | – Minimize risk<br>– In control<br>– Goodwill<br>– Set boundaries |

 | DOI:10.1163/9789004449084_015

| Interviews | Values | | |
|---|---|---|---|
| | Researcher | Second reviewer | Similar values |
| | – uncontrollable risk<br>– is unintentionally getting out of control<br>– boundaries<br>– goodwill<br>– setting boundaries<br>– assignments with boundaries<br>– boundless<br>– goodwill<br>– risk<br>– risky | – make sure that we don't fall down the cliff.<br>– prevent that these technologies will create their own vision<br>– technologies with a low barrier, but potential uncontrollable risk are worrisome for mankind.<br>– to consider possible events when starting to build this type of technology and what could go wrong if you did not do that.<br>– will need a big sense of the context and a sort of 'goodwill'.<br>– Mission accomplished but not in the way that we intended it.<br>– have to cooperate with them and goal setting will become extremely important.<br>– system needs to know what we mean,<br>– need to train our people to work with these types of systems and to communicate the goals to these systems<br>– need to train our personnel in giving assignments with boundaries. | |

| Interviews | Values | | |
|---|---|---|---|
| | **Researcher** | **Second reviewer** | **Similar values** |
| | | – goodwill.<br>– restrict the risk.<br>– build AI with the right mindset and still destroy the world | |
| *Interview 3 (Person C)* | – Decisions<br>– Decide<br>– functional form<br>– decisions<br>– decision<br>– decision<br>– decisions<br>– decision<br>– you have a right to be killed by a person<br>– human suffering | – how do we ensure that this is the line<br>– you have a right to be killed [by a person]<br>– integration of autonomy in weapon systems is inevitable<br>– reduce human suffering<br>– politics are very interesting<br>– give you an edge on the battlefield?<br>– cannot be reassured that the other side is not acquiring AW s. | – Right to be killed by a person<br>– Human suffering |
| *Interview 4 (Person A)* | – ethical framework<br>– cannot be that we lose control<br>– human control<br>– distance<br>– meaningful human control<br>– human accountability<br>– meaningful human control<br>– meaningful human control<br>– distance | – the ethical framework is very important<br>– a weapon without human control in the critical functions, such as select a target, should be banned.<br>– meaningful human control<br>– human accountability<br>– human control should be applied to the moment of selecting and taking out a target | – Ethical framework<br>– Meaningful human control<br>– Human accountability<br>– Predictability<br>– Reflexivity<br>– Distance<br>– Invincibility<br>– Determinism<br>– Inevitability<br>– Human need to decide about life and death |

| Interviews | Values | | |
|---|---|---|---|
| | **Researcher** | **Second reviewer** | **Similar values** |
| | – distance<br>– distance<br>– distance<br>– lack of accountability [and] unpredictability<br>– we don't care<br>– reflexivity<br>– distance<br>– invincibility<br>– determinism<br>– inevitability<br>– human need to decide about life and death.<br>– Humans need to be involved in the decision to take a life<br>– human needs to be in control of what a machine | – current weapons systems should be reviewed on the appliance of human control<br>– purposely designed for something.<br>– creating a distance or reducing harm<br>– fear, lack of accountability and unpredictability.<br>– reflexivity<br>– distance<br>– invincibility<br>– determinism<br>– inevitability<br>– human need to decide about life and death.<br>– human needs to be in control of what a machine does within certain parameters ('kaders').<br>– how human control is guaranteed | |
| *Interview 5 (Person F)* | – meaningful human control.<br>– meaningful human control<br>– meaningful control<br>– intention<br>– meaningful<br>– meaningful<br>– meaningful | – effective human control<br>– contextual or situational awareness or understanding<br>– verifying the legality or morality<br>– needs to be and intention or a military purpose to take out a target | – Meaningful human control<br>– Unnecessary suffering<br>– Superfluous injury<br>– Human dignity<br>– Dignified killing<br>– Predictability |

| Interviews | Values | | |
|---|---|---|---|
| | **Researcher** | **Second reviewer** | **Similar values** |
| | – meaningful human control.<br>– meaningful human control<br>– meaningful human control<br>– meaningful human control<br>– unnecessary suffering<br>– superfluous injury<br>– unnecessary suffering<br>– superfluous injury<br>– meaningful control<br>– automation bias<br>– meaningful human control<br>– control<br>– human dignity<br>– meaningfulness<br>– dignified<br>– dignity<br>– intention<br>– meaning<br>– dignified<br>– meaningful<br>– human control<br>– bias<br>– biases<br>– predictability<br>– unpredictability<br>– control<br>– control<br>– accountable<br>– accountability<br>– responsibility | – meaningful human control<br>– unnecessary suffering and superfluous injury<br>– valid, legally and morally<br>– sophisticated reasoning systems<br>– lose control<br>– legality and operational control<br>– human dignity<br>– legal and moral act of killing<br>– dignified killing<br>– military need<br>– moral or legal judgement<br>– ensures [...] moral, legal and meaningful process<br>– predictability<br>– control<br>– accountable<br>– accountability and responsibility<br>– laws and standards<br>– chain of command<br>– authority<br>– transparent<br>– keep that data | – Control<br>– Accountability<br>– Responsibility |

| Interviews | Values | | |
|---|---|---|---|
| | **Researcher** | **Second reviewer** | **Similar values** |
| | – accountable<br>– responsible<br>– meaningful human control<br>– accountable<br>– meaningful human control<br>– meaningful human control<br>– meaningful human control | | |
| *Interview 6 (Person B)* | – defense<br>– defending<br>– defend<br>– defend<br>– irresponsible<br>– benefits<br>– defenceless<br>– harm | – Justify<br>– humanitarian law<br>– efficiency<br>– barriers<br>– laws and regulations<br>– defense against it<br>– military advantages<br>– defending yourself<br>– accuracy<br>– verify<br>– behave<br>– behave in a particular way<br>– audit and verify<br>– environment<br>– expectation of how it is going to behave<br>– ethical governors<br>– well-defined envelops<br>– ethical safeguards<br>– harm<br>– behave as we would like<br>– can't be hacked | – Defense<br>– Harm |

Based on the interviews and the comparison of the value coding of the researcher and the second reviewer the following list of 23 unique values can be derived:

1. Intervene at all times
2. To be in control
3. Recall them
4. Minimize risk
5. In control
6. Goodwill
7. Set boundaries
8. Right to be killed by a person
9. Ethical framework
10. Meaningful human control
11. Predictability
12. Reflexivity
13. Distance
14. Invincibility
15. Determinism
16. Inevitability
17. Unnecessary suffering
18. Superfluous injury
19. Human dignity
20. Accountability
21. Responsibility
22. Defense
23. Harm

# References


AIV, & CAVV. (2016). *Autonomous weapon systems: the need for meaningful human control.* (No. 97, No. 26). Retrieved from http://aiv-advice.nl/8gr.

Aldewereld, H., Dignum, V., & Tan, Y.-h. (2015). Design for Values Information and communication technologies in Software Development Software development. *Handbook of Ethics, Values, and Technological Design: Sources, Theory, Values and Application Domains*, 831–845.

Alfano, M., & Loeb, D. (2014). *Experimental Moral Philosophy*. Retrieved from https://plato.stanford.edu/entries/experimental-moral/

Altmann, J., Asaro, P., Sharkey, N., & Sparrow, R. (2013). Armed military robots: editorial. *Ethics and Information Technology, 15*(2), 73.

Asaro, P. (2012). On banning autonomous weapon systems: human rights, automation, and the dehumanization of lethal decision-making. *International Review of the Red Cross, 94*(886), 687–709.

Asaro, P. (2016, 14-10-2017). Talk on Autonomous Weapon Systems. Retrieved from https://livestream.com/nyu-tv/ethicsofAI/videos/138822041

Awad, E. (2017). *MORAL MACHINE: Perception of Moral Judgment Made by Machines.* (Master), Massachusetts Institute of Technology, Boston.

Bandura, A. (2001). Social cognitive theory: An agentic perspective. *Annual review of psychology, 52*(1), 1–26.

Beauchamp, T. L., & Walters, L. R. (1999). *Contemporary Issues in Bioethics*: Wadsworth Pub.

Bonnefon, J.-F., Shariff, A., & Rahwan, I. (2016). The social dilemma of autonomous vehicles. *Science, 352*(6293), 1573–1576.

Borning, A., & Muller, M. (2012). *Next steps for value sensitive design.* Paper presented at the Proceedings of the SIGCHI conference on human factors in computing systems.

Bradley, B. (2006). Two concepts of intrinsic value. *Ethical Theory and Moral Practice, 9*(2), 111–130.

Bratman, M. E., Israel, D. J., & Pollack, M. E. (1988). Plans and resource-bounded practical reasoning. *Computational intelligence, 4*(3), 349–355.

Bryson, J. J., Kime, P. P., & Zürich, C. (2011). *Just an artifact: why machines are perceived as moral agents.* Paper presented at the IJCAI Proceedings-International Joint Conference on Artificial Intelligence.

Campaign to Stop Killer Robots. (2015). Campaign to Stop Killer Robots. Retrieved from https://www.stopkillerrobots.org/

Campaign to Stop Killer Robots. (2017). The Problem. Retrieved from http://www.stopkillerrobots.org/the-problem/

Carpenter, J. (2016). *Culture and Human-Robot Interaction in Militarized Spaces: A War Story*: Taylor & Francis.

Chappelle, W., Goodman, T., Reardon, L., & Thompson, W. (2014). An analysis of post-traumatic stress symptoms in United States Air Force drone operators. *Journal of anxiety disorders, 28*(5), 480–487.

Cheng, A. S., & Fleischmann, K. R. (2010). Developing a meta-inventory of human values. *Proceedings of the American Society for Information Science and Technology, 47*(1), 1–10.

Coeckelbergh, M. (2013). Drones, information technology, and distance: mapping the moral epistemology of remote fighting. *Ethics and Information Technology, 15*(2), 87–98.

Cointe, N., Bonnet, G., & Boissier, O. (2016). *Ethical Judgment of Agents' Behaviors in Multi-Agent Systems.* Paper presented at the Proceedings of the 2016 International Conference on Autonomous Agents & Multiagent Systems.

Cummings, M. L. (2006). Integrating ethics in design through the value-sensitive design approach. *Science and Engineering Ethics, 12*(4), 701–715.

Cummings, M. L., Mastracchio, C., Thornburg, K. M., & Mkrtchyan, A. (2013). Boredom and distraction in multiple unmanned vehicle supervisory control. *Interacting with Computers, 25*(1), 34–47.

Cushman, F., & Young, L. (2011). Patterns of moral judgment derive from nonmoral psychological representations. *Cognitive Science, 35*(6), 1052–1075.

Davis, J., & Nathan, L. P. (2015). Value Sensitive Design: Applications, Adaptations, and Critiques. *Handbook of Ethics, Values, and Technological Design: Sources, Theory, Values and Application Domains*, 11–40.

Dietvorst, B. J. (2016). People Reject (Superior) Algorithms Because They Compare Them to Counter-Normative Reference Points.

Dignum, F., Kinny, D., & Sonenberg, L. (2002). From desires, obligations and norms to goals. *Cognitive Science Quarterly, 2*(3–4), 407–430.

Dignum, V. (2016). *Introduction to AI.* Retrieved from https://rai2016.tbm.tudelft.nl/contents/

Docherty, B. (2012). *Losing humanity: The case against killer robots.*

Docherty, B. (2015). *Mind the gap: The lack of accountability for killer robots*: Human Rights Watch.

Etzioni, A., & Etzioni, O. (2017). Pros and Cons of Autonomous Weapons Systems. *Military Review*(May-June 2017).

Floridi, L., & Sanders, J. W. (2004). On the morality of artificial agents. *Minds and Machines, 14*(3), 349–379.

Friedman, B., & Kahn Jr, P. H. (2003). Human values, ethics, and design. *The human-computer interaction handbook*, 1177–1201.

Friedman, B., Kahn Jr, P. H., Borning, A., & Huldtgren, A. (2013). Value sensitive design and information systems. In *Early engagement and new technologies: Opening up the laboratory* (pp. 55–95): Springer.

Future of Life Institute. (2016). AI Open Letter, in:

Galliott, J. (2015). *Military robots: Mapping the moral landscape*: Ashgate Publishing, Ltd.

General Assembly United Nations. (2016). *Joint report of the Special Rapporteur on the rights to freedom of peaceful assembly and of association and the Special Rapporteur on extrajudicial, summary or arbitrary executions on the proper management of assemblies*. (A/HRC/31/66).

Graduation Portal. (2017). Graduation Portal. Retrieved from https://teams.connect.tudelft.nl/sites/tbm/graduate/SitePages/Home.aspx

Graham, J., Haidt, J., Koleva, S., Motyl, M., Iyer, R., Wojcik, S. P., & Ditto, P. H. (2012). *Moral foundations theory: The pragmatic validity of moral pluralism.*

Gray, H. M., Gray, K., & Wegner, D. M. (2007). Dimensions of mind perception. *Science, 315*(5812), 619–619.

Gray, K., & Wegner, D. M. (2012). Feeling robots and human zombies: Mind perception and the uncanny valley. *Cognition, 125*(1), 125–130.

Haidt, J., & Joseph, C. (2004). Intuitive ethics: How innately prepared intuitions generate culturally variable virtues. *Daedalus, 133*(4), 55–66.

Harper, D. (2002). Talking about pictures: A case for photo elicitation. *Visual studies, 17*(1), 13–26.

Himma, K. E. (2009). Artificial agency, consciousness, and the criteria for moral agency: what properties must an artificial agent have to be a moral agent? *Ethics and Information Technology, 11*(1), 19–29.

Horowitz, M. C. (2016). The Ethics & Morality of Robotic Warfare: Assessing the Debate over Autonomous Weapons. *Daedalus, 145*(4), 25–36.

Hristova, E., & Grinberg, M. (2015). *Should Robots Kill? Moral Judgments for Actions of Artificial Cognitive Agents.* Paper presented at the EAPCogSci.

Hurka, T. (2005). Proportionality in the Morality of War. *Philosophy & Public Affairs, 33*(1), 34–66.

IA program. (2017). *Information Architecture*. Retrieved from https://www.tudelft.nl/onderwijs/opleidingen/masters/cs/msc-computer-science/special-programmes/information-architecture/

ICRC. (2010, 29-10-2010). War and international humanitarian law. Retrieved from https://www.icrc.org/eng/war-and-law/overview-war-and-law.htm

Johnson, A. M., & Axinn, S. (2013). The morality of autonomous robots. *Journal of Military Ethics, 12*(2), 129–141.

Kaag, J., & Kaufman, W. (2009). Military frameworks: Technological know-how and the legitimization of warfare. *Cambridge Review of International Affairs, 22*(4), 585–606.

Kasher, A. (2016). 6 The Threshold of Killing Drones. *Drones and Responsibility: Legal, Philosophical and Socio-Technical Perspectives on Remotely Controlled Weapons*, 119.

Kominsky, J. F., Phillips, J., Gerstenberg, T., Lagnado, D., & Knobe, J. (2015). Causal superseding. *Cognition, 137*, 196–209.

Kreps, S. (2014). Flying under the radar: A study of public attitudes towards unmanned aerial vehicles. *Research & Politics, 1*(1), 2053168014536533.

Kuptel, A., & Williams, A. (2014). Policy Guidance: Autonomy in Defence Systems.

Le Dantec, C. A., Poole, E. S., & Wyche, S. P. (2009). *Values as lived experience: evolving value sensitive design in support of value discovery.* Paper presented at the Proceedings of the SIGCHI conference on human factors in computing systems.

Lin, J., & Singer, P. W. (2014). China's New Military Robots Pack More Robots Inside (Starcraft-Style).

Logg, J. M. (2017). Theory of Machine: When Do People Rely on Algorithms?

Malle, B. F. (2015). Integrating robot ethics and machine morality: the study and design of moral competence in robots. *Ethics and Information Technology*, 1–14.

Malle, B. F., & Thapa Magar, S. (2017). *What Kind of Mind Do I Want in My Robot?: Developing a Measure of Desired Mental Capacities in Social Robots.* Paper presented at the Proceedings of the Companion of the 2017 ACM/IEEE International Conference on Human-Robot Interaction.

Maslow, A. H. (1943). A theory of human motivation. *Psychological review, 50*(4), 370.

Nass, C., Moon, Y., Fogg, B., Reeves, B., & Dryer, D. C. (1995). Can computer personalities be human personalities? *International Journal of Human-Computer Studies, 43*(2), 223–239.

Neapolitan, R. E., & Jiang, X. (2012). *Contemporary artificial intelligence*: CRC Press.

Noone, G. P., & Noone, D. C. (2015). The debate over autonomous weapons systems. *Case W. Res. J. Int'l L., 47*, 25.

NOS. (2016). Video: Vliegtuig redt piloot. Retrieved from http://nos.nl/artikel/2132527-video-vliegtuig-redt-piloot.html

Oehlert, G. W. (2010). *A first course in design and analysis of experiments.*

Ohlin, J. D. (2016). The Combatant's Stance: Autonomous Weapons on the Battlefield. *92 International Law Studies, Cornell Legal Studies Research Paper No. 16–12*, 1–30.

Open Roboethics initiative. (2015, 5-11-2015). *The Ethics and Governance of Lethal Autonomous Weapons Systems: An International Public Opinion Poll.* Retrieved from http://www.openroboethics.org/laws_survey_released/

Perugini, M., & Bagozzi, R. P. (2004). The distinction between desires and intentions. *European Journal of Social Psychology, 34*(1), 69–84.

Pommeranz, A., Detweiler, C., Wiggers, P., & Jonker, C. M. (2011). *Self-reflection on personal values to support value-sensitive design.* Paper presented at the Proceedings of the 25th BCS Conference on Human-Computer Interaction.

Reips, U.-D. (2002). Standards for Internet-based experimenting. *Experimental Psychology, 49*(4), 243.

Roff, H. M. (2016). *Weapons autonomy is rocketing*. Retrieved from http://foreignpolicy.com/2016/09/28/weapons-autonomy-is-rocketing/

Rønnow-Rasmussen, T. (2002). Instrumental values–Strong and weak. *Ethical Theory and Moral Practice, 5*(1), 23–43.

Rosenberg, M., & Markoff, J. (2016). The Pentagon's 'Terminator Conundrum': Robots That Could Kill on Their Own. *The New York Times*. Retrieved from http://www.nytimes.com/2016/10/26/us/pentagon-artificial-intelligence-terminator.html?_r=0

Royakkers, L., & Orbons, S. (2015). Design for Values in the Armed Forces: Nonlethal Weapons Weapons and Military Military Robots Robot. *Handbook of Ethics, Values, and Technological Design: Sources, Theory, Values and Application Domains*, 613–638.

RT. (2017, 5-07-2017). Kalashnikov develops fully automated neural network-based combat module Retrieved from https://www.rt.com/news/395375-kalashnikov-automated-neural-network-gun/

Russell, S., Norvig, P., & Intelligence, A. (1995). A modern approach. *Artificial Intelligence. Prentice-Hall, Egnlewood Cliffs, 25*.

Saldaña, J. (2015). *The coding manual for qualitative researchers*: Sage.

Scalable Cooperation Group. (2016). *Moral Machine*. Retrieved from http://moralmachine.mit.edu/

Schroeder, M. (2016). Value Theory. Retrieved from https://plato.stanford.edu/entries/value-theory/

Schwartz, S. H. (1994). Are there universal aspects in the structure and contents of human values? *Journal of social issues, 50*(4), 19–45.

Schwartz, S. H. (2012). An overview of the Schwartz theory of basic values. *Online readings in Psychology and Culture, 2*(1), 11.

Searle, J. R. (1980). Minds, brains, and programs. *Behavioral and brain sciences, 3*(03), 417–424.

Searle, J. R. (1995). *The Construction of Social Reality*.

Sharkey, N., & Suchman, L. (2013). *Wishful mnemonics and autonomous killing machines*. Paper presented at the Proceedings of the AISB.

Shelley, C. (2013). Fairness and regulation of violence in technological design. *Moral, Ethical, and Social Dilemmas in the Age of Technology: Theories and Practice: Theories and Practice*, 182.

Stewart, P. (2016). *U.S. military christens self-driving 'Sea Hunter' warship*. Retrieved from http://www.reuters.com/article/us-usa-military-robot-ship-idUSKCN0X42I4

Stewart, W. (2015). *Russia has turned its T-90 tank into a robot – and plans to hire gamers to fight future wars.* Retrieved from http://www.dailymail.co.uk/news/article-3271094/Russia-turned-T-90-tank-robot-plans-hire-gamers-fight-future-wars.html

Strawser, B. J. (2010). Moral predators: The duty to employ uninhabited aerial vehicles. In *Handbook of Unmanned Aerial Vehicles* (pp. 2943–2964): Springer.

Sullins, J. P. (2006). When is a robot a moral agent. *Machine Ethics*, 151–160.

Thompson, W. T., Lopez, N., Hickey, P., DaLuz, C., Caldwell, J. L., & Tvaryanas, A. P. (2006). *Effects of shift work and sustained operations: Operator performance in remotely piloted aircraft (OP-REPAIR)*. Retrieved from

UNDIR. (2014). *Framing Discussions on the Weaponization of Increasingly Autonomous Technologies.* Retrieved from http://www.unidir.org/files/publications/pdfs/framing-discussions-on-the-weaponization-of-increasingly-autonomous-technologies-en-606.pdf.

UNDIR. (2015). *The Weaponization of Increasingly Autonomous Technologies: Considering Ethics and Social Values.* Retrieved from http://www.unidir.org/files/publications/pdfs/considering-ethics-and-social-values-en-624.pdf.

Van de Poel, I. (2013). Translating values into design requirements. In *Philosophy and engineering: Reflections on practice, principles and process* (pp. 253–266): Springer.

van Wynsberghe, A., & Robbins, S. (2014). Ethicist as Designer: a pragmatic approach to ethics in the lab. *Science and Engineering Ethics, 20*(4), 947–961.

Walsh, J. I. (2015). Precision weapons, civilian casualties, and support for the use of force. *Political Psychology, 36*(5), 507–523.

Walsh, J. I., & Schulzke, M. (2015). *The Ethics of Drone Strikes: Does Reducing the Cost of Conflict Encourage War?* Retrieved from

Waytz, A., Gray, K., Epley, N., & Wegner, D. M. (2010). Causes and consequences of mind perception. *Trends in cognitive sciences, 14*(8), 383–388.

Williams, A. P., Scharre, P. D., & Mayer, C. (2015). Developing Autonomous Systems in an Ethical Manner. In *Autonomous Systems: Issues for Defence Policymakers*: NATO Allied Command Transformation (Capability Engineering and Innovation).

*Auffassungen und moralische Wertvorstellungen innerhalb der Streitkräfte und Zivilbevölkerung in Bezug auf autonome Waffensysteme*

# Kurzfassung

Autonome Waffen werden immer häufiger auf den Schlachtfeldern benutzt (Roff, 2016). Autonome Waffensysteme können viele Vorteile für den Militärbereich haben, zum Beispiel, wenn die Selbststeueranlage der F-16 einen Absturz verhindert (NOS, 2016), oder der Einsatz von Robotern von der EOD (Kampfmittelbeseitigung), um Bomben zu entschärfen (Carpenter, 2016). Jedoch die Art der autonomen Waffensysteme kann auch zu unkontrollierbaren Handlungen und gesellschaftlichen Unruhen führen. Der Einsatz von autonomen Waffensystemen ohne direkte menschliche Überwachung auf dem Schlachtfeld ist, gemäß Kaag und Kaufman (2009) nicht nur ein Umsturz im Militäreinsatz, sondern kann auch als ein moralischer Umsturz gedeutet werden. Da der Einsatz von AI in großem Umfang auf dem Schlachtfeld unvermeidbar scheint (Rosenberg & Markoff, 2016), ist es von äußerster Wichtigkeit, dass Untersuchungen über die ethische und moralische Verantwortung angestellt werden.

In der Debatte um autonome Waffensysteme werden feste Überzeugungen und Meinungen immer lauter. Die Kampagne, Killerroboter zu stoppen (2017), erklärt zum Beispiel auf ihrer Webseite: *„Die Erlaubnis Maschinen die Entscheidung über Leben und Tod zu überlassen überquert eine fundamentale moralische Grenze. Autonomen Robotern fehlt das menschliche Urteilsvermögen und die Fähigkeit, den Zusammenhang zu begreifen."* Wir fanden nur wenige empirische Untersuchungen, die diese Ansichten unterstützen oder Einsichten zur Verfügung stellen, wie autonome Waffensysteme von der Zivilbevölkerung und dem Militär aufgefasst werden. Außerdem fanden wir keine empirischen Untersuchungen zu moralischen Werten, die der „fundamentalen moralischen Grenze" autonomer Waffensysteme zugrunde liegt. Es besteht deshalb eine zweifache Wissenslücke, insofern, dass es an Einsicht mangelt, 1) wie autonome Waffensysteme vom Militär und der Zivilbevölkerung aufgefasst werden und 2) welche moralischen Werte Menschen als für wichtig erachten, wenn autonome Waffensysteme in der näheren Zukunft eingesetzt werden.

Der erste Teil dieser Lücke kann ausgefüllt werden, indem die Auffassung von autonomen Waffensystemen mithilfe der Theorie der Streitkräfte in den Bereichen Kognitive Psychologie, Künstliche Intelligenz und Moralische Philosophie erforscht wird. Der zweite Teil dieser Wissenslücke kann durch die Untersuchung bekannter Werttheorien ausgefüllt werden (Beauchamp & Walters, 1999: Friedman & Kahn Jr., 2003; Schwartz, 2012), um zu sehen welche Werte Menschen im Einsatz von autonomen Waffensystemen als wichtig erachten. Basiert auf der identifizierten Wissenslücke und Problembeschreibung, entwarfen wir die folgende Forschungsfrage für diese Untersuchung:

> *Wie werden autonome Waffensysteme von der Zivilbevölkerung und den im niederländischen Verteidigungsministerium angestellten Militärangehörigen aufgefasst, und welche moralischen Werte betrachten sie beim Einsatz autonomer Waffensysteme als wichtig?*

In dieser Untersuchung benutzen wir die Value-Sensitive-Design-Methode (VSD) als Herangehensweise der Forschung. VSD ist eine dreifache Methode, die menschliche Werte innerhalb des Designprozesses der Technologie berücksichtigt. Es ist ein schrittweises Verfahren für die *konzeptuelle, empirische und technologische* Untersuchung menschlicher Werte, die von dem Design hinzugezogen werden (Davis & Nathan, 2015; Friedman & Kahn Jr., 2003). Die konzeptuelle Untersuchung besteht aus zwei Teilen: (1) Identifizierung der direkten Interessenvertreter, diejenigen, die die Technologie einsetzen, und die indirekten Interessenvertreter, diejenigen, deren Leben von der Technologie beeinflusst wird, und (2) Identifizierung und Definition der Werte, die der Einsatz der Technologie mit sich bringt. Die empirische Untersuchung befasst sich mit dem Verständnis und der Erfahrung der Interessenvertreter im Zusammenhang mit der Technologie und die hinzugezogenen Werte werden untersucht. In der technischen Untersuchung werden die speziellen Eigenschaften der Technologie analysiert (Davis & Nathan, 2015).

Wir weichen etwas von der ursprünglichen VSD-Methode ab, da wir keine vollständige Interessenvertreteranalyse durchführen, um die Interessenvertreter in Schritt 1 zu analysieren, aber konzentrieren uns auf zwei deutliche Interessenvertretergruppen: die *Zivilbevölkerung* und die *Militärangehörigen*, weil diese beiden Gruppen als befragte Personen für unsere Untersuchung zur Verfügung standen. In Schritt 2 der ersten Phase führen wir eine Literaturauswertung aus, um die Werte in Bezug auf autonome Waffensysteme zu identifizieren. Der wichtigste Fokus unserer Forschungsarbeit ruht auf der empirischen Untersuchungsphase, da wir fanden, dass der empirische Aspekt in der ethischen Debatte zu autonomen Waffensystemen übersehen wird. In der technischen Untersuchungsphase entwerfen wir keine autonomes Waffensystem, wie man hätte annehmen können, da dies ein immenses Vorhaben gewesen wäre, weit über den Bereich dieser Untersuchung hinaus. Doch wir entschieden uns, auf der Arbeit des Arbeitskreises Scalable Cooperation aufzubauen und schlagen einen Entwurf für eine Moral Machine für Autonome Waffensysteme vor, die als ein neuer Schritt in der Untersuchung der Ethik von autonomen Waffensystemen benutzt werden kann.

Die wissenschaftliche Relevanz unserer Untersuchung liegt darin, dass wir zur Fachliteratur beitragen, indem wir Einsicht in die Auffassung der Zivilbevölkerung und dem Militär hinsichtlich autonomer Waffensysteme nehmen und indem wir die moralischen Wertvorstellungen der Zivilbevölkerung und Streitkräfte in Bezug auf autonome Waffen identifizieren. Einsicht darin fehlt zurzeit, und es konnten keine empirischen Daten zur Erfassung der Werte gefunden werden, die mit autonomen Waffensystemen zu tun haben. Anhand von dem Value-Sensitive-Design als Forschungsmethode weisen wir nach, dass diese Methode zur Strukturierung wissenschaftlicher Forschung angewendet werden kann, was als Fallstudie für die VSD-Methode angesehen werden kann. Wir erweitern die Forschung auch auf die ethische Entscheidungstreffung bei autonomen Fahrzeugen (Bonnefon, Shariff und Rahwan, 2016) und gliedern sie in den Bereich der autonomen Waffensysteme ein.

Die gesellschaftliche Relevanz ist das Verständnis der Auffassung von autonomen Waffensystemen der Zivilbevölkerung und den Militärangehörigen, die im niederländischen Verteidigungsministerium tätig sind und identifizieren, welche moralischen Werte sie mit autonomen Waffensystemen in Verbindung bringen, um Gemeinsamkeiten und Unterschiede in der Debatte dieser Technologie zu finden, die von der Kampagne Stop Killer Robots (2015) und dem internationalen Komitee für Roboter-Rüstungskontrolle (ICRAC) eingeleitet wurde. Zweitens, zeigt diese Untersuchung auf, wie die Value-Sensitive-Design-Methode für autonome Waffensysteme benutzt werden kann, um die Werte des Militärs und der Zivilbevölkerung im Verhältnis zum Einsatz dieser Art von Waffen zu identifizieren. Zum Schluss, ist es wichtig, durch Identifizieren der Werte sie in das Design autonomer Waffensysteme einzufügen; somit trägt diese Untersuchung zu einem verantwortungsvollen Design und Einsatz autonomer Waffensysteme in der Zukunft bei.

Diese Untersuchung konzentriert sich auf die empirische Untersuchungsphase von VSD. Diese Phase ist in zwei Teile aufgeteilt. Der erste Teil befasst sich damit, die Werte im Kontext autonomer Waffensysteme mittels einer Online-Befragung zu identifizieren, die von Gesprächen mit sechs Experten unterstützt wird. Der zweite Teil der empirischen Untersuchung prüft die Wahrnehmung der Streitkräfte mittels einer informatorischen und einer bestätigenden Untersuchungsmethode. Wir operationalisieren die menschlichen Verhaltensweisen, um unsere Hypothese zur Auffassung der Streitkräfte zu bestätigen, indem wir kontrollierte Experimente randomisiert anordnen und die Ergebnisse des Experiments benutzen, um die Werte in Verbindung mit autonomen Waffensystemen auf beschreibende Weise zu untersuchen.

Um die Auffassung der Streitkräfte von autonomen Waffen zu untersuchen, entwarfen wir eine Hypothese auf der Basis, dass die Militärangehörigen autonome Waffensysteme als Waffen wie alle anderen betrachten und deshalb nichts weiter als ein Werkzeug, mit dem man eine Wirkung erzeugt. Wir stellen am Schluss der Untersuchung die Hypothese auf, dass gemäß der Auffassung der Militärangehörigen autonome Waffen keinen psychischen Zustand kennen. Wir entwarfen drei Szenarien für unser Experiment: 1) eine von Menschen betriebene Drohne, 2) ein autonomes Waffensystem ohne menschliche Einwirkung und 3) ein autonomes Waffensystem mit menschlicher Einwirkung. Wir hatten keinen Unterschied zwischen dem Zustand des neutralen autonomen Waffensystems und dem Zustand, wenn eine Drohne von Menschen ferngesteuert wird, erwartet. Wir erwarten ebenfalls, dass der neutrale Zustand des Systems bedeutend anders ist, als der Zustand des Systems mit hoher Selbstbestimmung, wobei wir den Teilnehmern besonders sagen, dass das autonome Waffensystem selbstbestimmende Charakteristiken hat, z.B. die Fähigkeit zu planen und eigene Ziele zu setzen. Gemäß dieser Argumentation, ist unsere Hypothese wie folgt:

*H1: Militärangehörige fassen Autonome Waffen nicht als mit psychischen Fähigkeiten ausgestattet auf.*

Die Ergebnisse der ersten und letzten Untersuchung zeigen auf, dass Operationselemente zuverlässig sind und als ein Konstrukt zusammenhalten. Das Agency-Konstrukt besteht aus vier Elementen: *Denken, Zielsetzung, Freier Wille und das Erreichen von Zielen*, die zum Messen der Auffassung von Streitkräften in Bezug auf autonome Waffensysteme und auf von Menschen gesteuerte Drohnen benutzt werden können. Unsere zentrale Analyse befasste sich damit, wie neutrale Szenarien der autonomen Waffensysteme sich von den durch Einwirkung von Menschen und den hochentwickelten Szenarien von autonomen Waffen unterscheiden. Unsere Ergebnisse zeigen auf, dass die Auffassung von Streitkräften der Militärangehörigen und zivilen Mitarbeitern im niederländischen Verteidigungsministerium für das neutrale autonome Waffensystem-Szenario höher ist als die Auffassung der Streitkräfte von dem durch menschliche Einwirkung betriebene Drohnen-Szenario. Das bedeutet, dass sie einem autonomen Waffensystem mehr Macht beimessen, als einer durch menschliche Einwirkung betriebenen Drohne. Basiert auf diesen Erkenntnissen, müssen wir unsere Hypothese verwerfen.

Die Auswirkung der Wahrnehmung von Agency auf die abhängigen Variablen wird auf beschreibende Weise erforscht und nicht durch Regressionsrechnung analysiert. Wir fanden, dass 7 von 9 abhängigen Variablen bedeutend mit dem Agency-Konstrukt korrelieren. Dies sind die Variablen: *Vertrauen, Menschenwürde, Zuversicht, Erwartungen, Unterstützung, Fairness und Besorgnis.* Dies sind unsere Erkenntnisse:

– Militärangehörige und zivile Mitarbeiter im niederländischen Verteidigungsministerium haben mehr Vertrauen, sind zuversichtlicher und

unterstützen eher die Maßnahmen, die von Drohnen unter menschlicher Einwirkung getroffen wurden, als die von autonomen Waffensystemen;
- Es herrscht die Auffassung, dass eine unter menschlicher Einwirkung betriebene Drohne die Menschenwürde mehr achtet, als neutrale oder hochentwickelte autonome Waffensysteme, obwohl die Maßnahmen und Ergebnisse der Szenarien die gleichen sind;
- Militärangehörige und zivile Mitarbeiter im niederländischen Verteidigungsministerium befinden sich auf einer gleichwertigen Ebene der Erwartungen mit Hinsicht auf die Maßnahmen der unter menschlicher Einwirkung betriebene Drohne und der neutralen und hochentwickelten autonomen Waffensysteme in der Zukunft. Außerdem betrachten sie die Handlungen der von Menschen gesteuerten Drohnen und Autonomen Waffensystemen als gleich fair;
- Autonome Waffensysteme verursachen mehr Besorgnis unter den Militärangehörigen und zivilen Mitarbeitern im niederländischen Verteidigungsministerium als von Menschen betriebene Waffensysteme.

Diese Untersuchung wurde mit einem kleinen Datensatz durchgeführt, und um die Ergebnisse zu verallgemeinern benötigt diese Arbeit weitere Personen zur Befragung, die eine größere demografische Gruppe vertreten. Deshalb schlagen wir ein Design für eine Moral Machine für Autonome Waffensysteme als ein Massives Online-Experiment (MOE) für eine großangelegte Untersuchung dieses Themas vor. Wir bauten das Konzept der Moral Machine in unser Design ein, das von dem Arbeitskreis Scalable Cooperation im Medienlabor bei MIT (Scalable Cooperation Group, 2016) entwickelt wurde. Da das Thema äußerst heikel ist, schlagen wir eine Ausführung in zwei Schritten vor, zuerst planen wir ein kontrolliertes Experiment mit einem begrenzten Satz Bedingungen. Diese Bedingungen können in Szenarien eingesetzt werden, was Benutzern ermöglicht, die Umfrage nach Erhalt eines Passwortes über eine Web-Oberfläche an einen sicheren Server zu senden. Nach dem Sammeln der ersten Daten und des Benutzer-Feedbacks ist der nächste Schritt, eine großangelegte offene Plattform wie z.B. die Moral Machine maßstäblich zu vergrößern, wo Leute die Szenarien bewerten können, um große Mengen Daten von verschiedenen demografischen Gruppen zu sammeln, die dann für robustere und verallgemeinerbare Ergebnisse benutzt werden können.

Verschiedene Aspekte können als Einschränkungen dieser Forschungsarbeit identifiziert werden. Zuerst wurden die Operationalisierung und das Agency-Konstrukt von einer Kategorisierung der Literatur abgeleitet, die die Agentenmerkmale beschreibt, und unsere Wahl der Merkmale basiert auf einer numerischen Zahl und nicht auf Relevanz oder Bewertungskriterien. Die zweite Einschränkung ist die Auswahl der Werte aus dem Werte-Fragebogen und den Experteninterviews als abhängige Variable. Obwohl diese Auswahl heftig unter den drei Rechercheuren dieser Arbeit debattiert wurde, fand die endgültige Wahl auf methodischem Weg und nicht nach objektiver Methode statt. Drittens sind die in dieser Forschungsarbeit benutzten Proben sowohl in Größe als auch demografischem Bereich begrenzt, da die endgültige Studie nur über 239 befragte Personen verfügte, also nur ein geringer Anteil des niederländischen Verteidigungsministeriums. Zum Schluss ist die Verteilung der befragten Personen des letzten Szenarios verzerrt und das zweite Szenario (neutrales Agens autonomes Waffensystem) verfügte über 32 befragte Personen mehr als das erste Szenario (unter menschlicher Einwirkung betriebene Drohnen). Eine der Erklärungen für diese Schieflage könnte sein, dass Leute erwarteten, autonome Waffensysteme zu untersuchen, aber machten nicht mehr mit, als ihnen die unter menschlicher Einwirkung betriebenen Waffensysteme präsentiert wurden. Dies nennt man selektive Schwundquote, die eine negative Auswirkung auf die interne Stichhaltigkeit der Forschungsarbeit hat. Eine weitere Erklärung könnte sein, dass die Software bei der Aufteilung der Befragten über die Szenarien fehlerhaft war, was zu bedeuten hätte, dass die interne Stichhaltigkeit dieser Forschungsarbeit nicht davon verletzt wurde, da die Schieflage nicht einer selektiven Schwundquote beizumessen ist.

Mit Rücksicht auf diese Einschränkungen werden mehrere Empfehlungen für weitere Forschungsarbeiten vorgeschlagen. Die erste ist, das Agency-Konstrukt zu bewerten, das weitere Untersuchungen zur Messung der Auffassung in Bezug auf Kampfmittel anderer technologischen Artefakten erforderlich sind, um zu prüfen, dass dieses Konstrukt auch anderen Bereichen standhält. Beispiele dieser Forschungsarbeiten könnten die Untersuchung der Auffassung in Bezug auf Dienstmittel einschließen, von Pflegerobotern für Senioren, KI-Spielzeug für Kinder oder Onboard-Computern von autonomen Fahrzeugen. Die zweite Empfehlung ist, die abschließende Untersuchung mit den gleichen Szenarien mit einer repräsentativen Stichprobe durchzuführen, die einzig aus Mitgliedern der niederländischen Zivilbevölkerung besteht. Dies würde uns ermöglichen zu erkennen, bei welchen Werten sich die Ergebnisse der militärischen Probanden und die Antworten von Probanden aus der Zivilbevölkerung voneinander unterscheiden, und obwohl wir keine direkten Vergleiche ziehen können, aufgrund der Tatsache, dass die militärische Stichprobe nicht repräsentativ ist, könnten wir doch Einsicht in die Auffassung von sowohl Militärangehörigen als auch der Zivilbevölkerung nehmen. Schließlich empfehlen wir die Implementierung der Moral Machine für Autonome Waffensysteme, wie in Abschnitt 5 beschrieben wird, um die Ergebnisse zu verallgemeinern. Hierfür erfordert diese Untersuchung entschieden mehr Personen zur Befragung, die eine umfassendere demografische Gruppe vertreten. Die Korrektur nach oben für ein Massives Online-Experiment wie die Moral Machine, würde große Mengen Daten verschiedener demografischer Gruppen erzeugen, die für robustere und verallgemeinerbare Ergebnisse benutzt werden könnten, um ein gründliches Verständnis der moralischen Bewertung der Leute in Bezug auf autonome Waffensysteme zu erhalten.

# Vorwort

Mein Interesse an der Ethik autonomer Waffensysteme wurde erregt, als ich an der IDEA-Summer School zum Thema Verantwortungsbewusste Künstliche Intelligenz (KI) im August 2016 teilnahm, die von Dr. Virginia Dignum als Teil der Europäischen Konferenz zu Künstlicher Intelligenz (ECAI) organisiert wurde. Nach dem ersten Treffen mit Virginia bezüglich meiner Diplomarbeit fragte sie mich, ob ich im Ausland studieren möchte und wo ich gerne hingehen würde, und ich war kühn genug, zu sagen, dass ich gerne zum MIT (Massachusetts Institute of Technology) gehen würde, aber ich glaubte auch nicht im Entferntesten, dass dies wirklich geschehen würde. Virginias Kontakt zu Dr. Iyad Rahwan des Arbeitskreises Scalable Cooperation des Media Lab von MIT half, es zu realisieren. Ich möchte Virginia ganz herzlich danken, dass sie mir diesen Kontakt vermittelte, sowie auch für ihre Betreuung während meiner Diplomarbeit. Aufgrund der Zeitunterschiede kommunizierten wir meistens des Abends in den Niederlanden über Skype, und ich schätzte es sehr, dass sie immer Zeit fand, um meine Niederschriften zu überprüfen.

Die Arbeit meiner Diplomarbeit mit dem Arbeitskreis Scalable Cooperation Group war eine einmalige Chance, und das Media Lab ist sehr offen, sehr divers, nicht hierarchisch und eine äußerst inspirierende Umgebung. Zuerst musste ich mich daran gewöhnen, im Foyer einem Kugelbad zu begegnen, oder an Leute, die ihre Büros verließen, um mitten am Tag Pingpong zu spielen, aber nach einem Monat oder so fühlte sich das Labor sehr heimisch an. Ich arbeitete mit hoch talentierten und ehrgeizigen Studenten im Aufbaustudium und mit Post-Doktoranden. Ich möchte ganz besonders Dr. Sydney Levine und Dr. Nick Obradowitsch für all ihre einsichtsvollen Kommentare danken, und ich war wirklich begeistert von den heftigen Debatten, die wir in meinem Arbeitskreis hatten. Vor allem aber möchte ich Dr. Iyad Rahwan dafür danken, dass er mir gestattete seiner Gruppe beizutreten und mit dem heiklen Thema der autonomen Waffensysteme zu arbeiten. Er stellte mir alle Ressourcen, seine Zeit und Hilfe zur Verfügung, die ich brauchte, um meine Recherchen zu machen, was ich sehr zu schätzen weiß.

Zuletzt möchte ich mich noch bei der Königlichen Niederländischen Armee (RLNA) bedanken, die es mir ermöglichte, im Ausland zu studieren und meinen Aufenthalt am MIT unterstützte. Ohne diese Unterstützung wäre ich nicht in der Lage gewesen, fünf Monate lang in Boston zu wohnen und zu studieren. Ich hoffe, dass diese Diplomarbeit zur Ethik von autonomen Waffensystemen zum besseren Verständnis führt und eine Diskussion zum Einsatz von autonomen Waffensystemen und verantwortungsvoller KI in der Verteidigungsorganisation einleitet.

*Ilse Verdiesen*

KAPITEL 1

# Einführung

> Ich glaube ich finde es schwierig auszudrücken, was ich von Drohnen halte. Einerseits macht mich jedes Instrument zum Töten traurig. Andererseits ist es tröstlich, dass aufgrund ihres Einsatzes sehr viele Leben nicht riskiert werden brauchen.
>
> VORSTUDIE 2 BEFRAGTE PERSONEN

Autonome Waffensysteme sind Waffensysteme, die mit künstlicher Intelligenz ausgerüstet sind (KI). Künstliche Intelligenz werden von Neapolitan und Jiang (2012, S. 8) als *„eine intelligente Instanz, die in einer stets wechselnden, komplexen Umgebung schlussfolgert"* beschrieben, aber diese Definition passt auch auf natürliche Intelligenz. Russell, Norvig and Intelligence (1995) erstellt eine Übersicht von vielen Definitionen, worin sie Ansichten zu Systemen kombinieren, die *wie Menschen denken und handeln und zugleich wie Systeme, die rational denken und handeln können,* aber sie stellen keine eigene Definition da. Für jetzt bleiben wir bei der Beschreibung, die Bryson, Kime und Zürich (2011) erstellen. Sie geben an, dass eine Maschine (oder ein System) intelligentes Verhalten aufweist, wenn sie eine Handlung wählen kann, die auf der Beobachtung seiner Umgebung basiert ist. In der wissenschaftlichen Literatur wird KI als mehr als nur ein Intelligentes System beschrieben. Dies wird von den Konzepten *Anpassungsfähigkeit, Interaktives Handeln und Autonomie* charakterisiert (Floridi & Sanders, 2004). Laut Floridi und Sanders (2004) bedeutet Anpassungsfähigkeit, dass das System sich gemäß seiner Interaktionen ändern und aus seinen Erfahrungen lernen kann. Programmiermethoden von Maschinen sind ein Beispiel dafür. Wir sprechen von Interaktivität, wenn das System und seine Umgebung auf einander eingehen und Autonomie bedeutet, dass das System selbst einen Zustand ändern kann.

Eine wachsende Anzahl Rechercheure konzentrieren sich auf ein verantwortliches Design von KI, dass soziale und ethische Werte einschließt, um unerwünschte gesellschaftliche Folgen dieser Technologie zu verhindern. Prinzipien, um verantwortungsvolle KI zu beschreiben sind *Rechenschaftspflicht, Verantwortung und Transparenz* (ART). *Rechenschaftspflicht* bezieht sich

 | DOI:10.1163/9789004449084_002

auf die Rechtfertigung der Handlungen, die von KI ausgeführt werden, *Verantwortung* räumt für die Fähigkeit ein, Schuld für diese Handlungen auf sich zu nehmen und *Transparenz* befasst sich mit der Beschreibung und Reproduzierung der Entscheidungen, die KI macht und passt sie ihrer Umgebung an (V. Dignum 2016).

Künstliche Intelligenz ist nicht nur ein futuristisches Szenario aus einem Science-Fiction-Film, in dem menschenartige Roboter planen, die Macht über die Welt zu bekommen, wie Datas Bruder Lore in Star Trek oder die Cylons in Battlestar Galactica. Viele KI-Anwendungen sind heute bereits im Einsatz. Intelligente Strommesser, Suchmaschinen, Personal Assistent bei Mobiltelefonen, Selbststeueranlagen und selbstfahrende Autos sind Beispiele dafür. Diese Untersuchung konzentriert sich auf die Anwendung von KI im Militärbereich, und besonders auf den Einsatz von autonomen Waffen in der nahen Zukunft. In dieser Einführung beschreiben wir zuerst das Forschungsproblem, gefolgt von der Forschungsmethode, der wissenschaftlichen und gesellschaftlichen Relevanz und der Einbettung in das Kurrikulum der Informationsarchitektur (IA) im Rahmen des System Engineering Policy und Management (SEPAM)-Magisterkurses. Wir werden mit einer kurzen Beschreibung der nächsten Kapitel dieses Berichts abschließen.

## 1.1 Forschungsproblem

Zuerst besprechen wir den vermehrten Einsatz autonomer Waffensysteme im Militärbereich und die damit zusammenhängende Arbeit zur Ethik von KI, gefolgt von Forschungsarbeiten zu von Menschen angetriebenen Drohnen im Unterabschnitt über Problemerforschung. Als nächstes identifizieren wir die Wissenslücke, Problemstellung und den Umfang, bevor wir mit der Forschungsfrage und den Unterabschnitten dieser Forschungsarbeit abschließen.

### 1.1.1 *Problemerforschung*

Autonome Waffen werden immer häufiger auf den Schlachtfeldern benutzt (Roff, 2016). Es wurde bereits berichtet, dass China über autonome Fahrzeuge verfügt, die einen bewaffneten Roboter mit sich führen (Lin & Singer, 2014), Russland behauptet, dass sie an autonomen Panzern arbeitet (W. Stewart, 2015), die USA tauften ihr erstes „selbststeuerndes" Kriegsschiff im Mai 2016 (P. Stewart, 2016), und der russische Waffenhersteller Kalaschnikow gab vor kurzem kund, dass er ein vollständig automatisiertes Gefechtsmodul entwickelt hat, das neutrale Netzwerke benutzt (RT, 2017). Autonome Waffensysteme können viele Vorteile für den Militärbereich haben, zum Beispiel, wenn die Selbststeueranlage der F-16 einen Absturz verhindert (NOS, 2016), oder der Einsatz von Robotern von der EOD (Kampfmittelbeseitigung), um Bomben zu entschärfen (Carpenter, 2016). Jedoch die Art der autonomen Waffensysteme kann auch zu unkontrollierbaren Handlungen und gesellschaftlichen Unruhen führen. Beispiele dieser Unruhen sind die „Stop Killer Robots"-Kampagne von 61 Nichtregierungsorganisationen, die vom Human Rights Watch angeführt wird (Campaign to Stop Killer Robots, 2015), aber auch die Vereinten Nationen, in deren Namen ihre Besorgnisse ausgesprochen werden und aussagen, dass *„Autonome Waffensysteme, die keine bedeutungsvolle menschliche Kontrolle erfordern, verboten werden sollten und ferngesteuerte Streitkraft immer nur mit äußerster Vorsicht eingesetzt werden sollte"* (UN-Generalversammlung, 2016). Der Einsatz von autonomen Waffensystemen ohne direkte menschliche Überwachung auf dem Schlachtfeld ist, gemäß Kaag und Kaufman (2009) nicht nur ein Umsturz im Militäreinsatz, sondern kann auch als ein moralischer Umsturz gedeutet werden. Da der Einsatz von KI in großem Umfang auf dem Schlachtfeld unvermeidbar scheint (Rosenberg & Markoff, 2016), ist es von äußerster Wichtigkeit, dass Untersuchungen über die ethische und moralische Verantwortung angestellt werden.

Ethische Entscheidungsfällung in KI und Robotern ist ein aufkommender Bereich, und eine Reihe von Akademikern befasst sich mit moralischer Beurteilung dieser Technologien. Zum Beispiel schlägt Malle (2015) ein System vor, in dem (bislang) die verschiedenen Gebiete der *Roboterethik* kombiniert werden. Das bedeutet, dass ethische Fragen über Design, Einsatz und Behandlung von Robotern von Menschen angesprochen werden, sowie *die Maschinenmoralität,* die sich mit Fragen über die moralischen Fähigkeiten eines Roboters beschäftigt und wie diese mit Computerhilfe eingebaut werden können. Cointe, Bonnet und Boissier (2016) schlagen ein Modell vor, bei dem ein Agens die ethischen Aspekte seines eigenen Verhaltens beurteilen kann, sowie das anderer Agenten in einem Multiagentensystem. Das Modell beschreibt einen ethischen Urteilsfindungsprozess, der es Agenten gestattet, das Verhalten anderer Agenten zu bewerten. Bonnefon et al. (2016) haben die ethische Entscheidung eines autonomen Fahrzeugs untersucht, ob es sich selbst schützen oder utilitaristisch verfahren soll, wenn es Fußgängern auf der Straße begegnet. In dieser Arbeit wird die Moral Machine[1] am MIT benutzt, um Einsicht zu nehmen, wie Leute Szenarien mit einem autonomen Fahrzeug beurteilen, um zu sehen wie ihr moralisches Urteil mit dem von anderen Leuten verglichen werden kann.

1 http://moralmachine.mit.edu/.

In einem den autonomen Waffensystemen verwandten Bereich, dem der von Menschen gesteuerten Drohnen, wurden ethische Probleme während der letzten zehn Jahre sehr intensiv untersucht, sowohl in der philosophischen als auch psychologischen Fachliteratur. Vom philosophischen Standpunkt behauptet Coeckelberg (2013), dass Drohneneinsätze nicht nur eine physische Distanz sondern auch eine moralische Distanz erzeugen, da der Gegner weniger sichtbar ist. Dies eliminiert die moralisch-psychologische Hemmschwelle für das Töten. Ein weiteres ethisches Problem ist nach Strawser (2010), dass aufgrund der regelmäßigen Arbeitsschichten außerhalb der Gefechtszone, die Drohnenbediener eine ungerechte psychologische Last mit sich tragen, da sie täglich zwischen Arbeits- und Heimsituation schalten müssen. Aufgrund der Distanz zum Schlachtfeld, können menschliche Streitkräfte auch eine kognitive Dissonanz empfinden, sodass der Krieg mehr wie ein Computerspiel erscheint, als die Wirklichkeit. Diese ethischen Probleme werden von Strawser (2010) widerlegt, indem er angibt, dass aufgrund der erhöhten Distanz, die menschlichen Streitkräfte mehr Zeit haben, um ein Ziel zu beurteilen, da ihre eigene Sicherheit nicht auf dem Spiel steht, aber er argumentiert auch, dass weitere empirische Forschung erforderlich ist, um die psychologischen Auswirkungen des Ausführens von Drohneneinsätzen mit großen Distanzen zum Schlachtfeld zu bewerten.

Mehrere empirische Untersuchungen der Auswirkungen von Drohneneinsätzen auf menschliche Streitkräfte wurden im Bereich der Psychologie durchgeführt. Eine der ersten Untersuchungen der psychologischen Auswirkungen von Drohneneinsätzen wurde von Thompson et al. (2006) angestellt, der herausfand, dass Drohnenbediener an stärkerer Müdigkeit, emotionaler Erschöpfung und Burnout litten. Dies wurde von Chappelle, Goodman, Reardon und Thompson (2014) bestätigt, die berichteten, dass menschliche Drohnenbediener Symptome von Burnout aufweisen, emotional erschöpft sind und zu einem hohen Grad Zynismus sowie auch posttraumatische Stresssyndroms an den Tag legen. Weitere psychologische Faktoren, die aus diesen Untersuchungen bekannt wurden, sind Langeweile und Verstörtheit aufgrund der langen Dauer der Drohneneinsätze (Cummings, Mastracchio, Thornburg & Mkrtchyan, 2013). Diese Untersuchung zeigen, dass die Ausführung von Drohneneinsätzen schwere psychologische Auswirkungen auf die Menschen, die sie bedienen, haben kann.

### 1.1.2 *Wissenslücke*

In der Debatte um autonome Waffensysteme werden feste Überzeugungen und Meinungen immer lauter. Die Kampagne, Killerroboter zu stoppen (2017) erklärt zum Beispiel auf ihrer Webseite, dass: *„Die Erlaubnis, Maschinen die*

*Entscheidung über Leben und Tod zu überlassen, überquert eine fundamentale moralische Grenze. Autonomen Robotern fehlt das menschliche Urteilsvermögen und die Fähigkeit, den Zusammenhang zu begreifen."* Wir fanden nur wenige empirische Untersuchungen, die diese Ansichten unterstützen oder Einsichten zur Verfügung stellen, wie autonome Waffensysteme von der Zivilbevölkerung und dem Militär wahrgenommen werden. Die Initiative Open Robots Ethnics begutachtete die öffentliche Meinung in einer Meinungsumfrage in 2015 (Open Robot Ethnics Initiative, 2015) und veröffentlichte einen Bericht. Die Ergebnisse wurden jedoch nicht in einer akademischen Fachzeitschrift veröffentlicht, und die Untersuchung war nicht umfassend genug, um stichhaltige Folgerungen daraus zu schließen. Wie bereits in den Unterabschnitten beschrieben, untersuchten verschiedentliche Wissenschaftler wie Malle (2015), Cointe et al. 2016 und Bonnefon et al. (2016) ethische Entscheidungstreffung mit Bezug auf KI und Roboter. Ethische Berücksichtigung wird auch in den benachbarten Bereichen der durch menschliche Einwirkung betriebenen Operationen Coeckelberg, 2013; Strawser, 2010) untersucht, und empirische Studien fanden, dass menschliche Beteiligte unter schweren psychologischen Auswirkungen leiden (Chappelle et al., 2014; Cummings et al., 2013; Thompson et al., 2006), aber diese Forschungen sind noch nicht auf den Einsatz von autonomen Waffensystemen erweitert. Außerdem fanden wir keine Fachliteratur oder empirische Untersuchungen zu den moralischen Werten, die mit autonomen Waffensystemen im Zusammenhang stehen oder was Leute als die *„moralische Grundlinie"* erachten. Es besteht deshalb eine zweifache Wissenslücke, insofern, dass es an Einsicht mangelt, 1) wie autonome Waffensysteme vom Militär und der Zivilbevölkerung wahrgenommen werden und 2) welche moralischen Werte Menschen als für wichtig erachten, wenn autonome Waffensysteme in der näheren Zukunft eingesetzt werden.

### 1.1.3 *Problemstellung*

Der erste Teil dieser Lücke kann ausgefüllt werden, indem die Wahrnehmung von autonomen Waffensystemen mithilfe der Agententheorie in den Bereichen Kognitive Psychologie, Künstliche Intelligenz und Moralische Philosophie erforscht wird. Agency ist die Fähigkeit, zu planen und zu handeln (Waytz, Gray, Epley, & Wegner, 2010), und wird sowohl in menschlichen als auch nicht-menschlichen Wesen oder Gegenständen untersucht. Leute messen Handlungsfähigkeit Gegenständen wie Computer (Nass, Moon, Fogg, Reeves, & Dryer, 1995) und Robotern bei, (H. M. Gray, Gray, & Wegner, 2007), deshalb erwarten wir von der KI-Technologie, dass sie auch Merkmale von Handlungsfähigkeit aufweist. Die Untersuchung von der Auffassung unter den Militärangehörigen und der Zivilbevölkerung, dass autonome

Waffensysteme Handlungsfähigkeit aufweisen, würde einen Einblick in die möglichen Unterschiede und Ähnlichkeiten zwischen den Ansichten dieser Gruppen geben, die in der derzeitigen Debatte zu dieser Technologie benutzt werden könnte.

Der zweite Teil dieser Wissenslücke kann durch die Untersuchung bekannter Werttheorien ausgefüllt werden, um zu sehen welche Werte Menschen im Einsatz von autonomen Waffensystemen als wichtig erachten. Bekannte Werttheorien sind die von Schwartz (1994), Friedman und Kahn Jr. (2003), und Beauchamp und Walters (1999), aber ein Einblick, wie sie sich zu autonomen Waffensystemen verhalten, fehlt noch. Die Beziehung der Werte, die im Kontext des Einsatzes von autonomen Waffensystemen am relevantesten sind und der Vergleich dieser Werte zu denen, die im Verhältnis zu derzeitigen Technologie der von Menschen bedienten Drohnen stehen, würde zu einem Einblick in die eigentlichen Motive in der Diskussion um autonome Waffensysteme und besserem Verständnis der ausgedrückten Ansichten führen. Dies führt zu der folgenden Problemstellung für diese Untersuchung:

> In der Debatte zu autonomen Waffensystemen werden feste Vorstellungen und Meinungen geäußert, aber eine empirische Untersuchung, die diese Meinungen unterstützt, fehlt. Ein Einblick inwiefern autonome Waffensysteme wahrgenommen und welche moralische Werte dem zukünftigen Einsatz von autonomen Waffensystemen beizumessen sind, ist nicht vorhanden. Ebenfalls ist nicht offensichtlich, ob die Auffassung und die moralischen Werte bezüglich autonomer Waffensysteme von Militärangehörigen sich von den denen der Zivilbevölkerung unterscheiden.

### 1.1.4 *Umfang*

Ein Großteil der Fachliteratur zu autonomen Waffensystemen wird von Juristen und Philosophen im Zusammenhang mit den Internationalen Menschenrechten und der Genfer Konvention geschrieben und besprochen, die die Auswirkungen von bewaffneten Auseinandersetzungen begrenzen wollen (ICRC, 2010). Da wir weder Juristen noch Philosophen sind, bleibt diese Untersuchung innerhalb der Schranken und Regeln der Kriegsgesetze, und wir werden sie nicht in Frage stellen. Aufgrund der beschränkten Zeit befasst sich diese Untersuchung hauptsächlich mit dem Militärpersonal des niederländischen Verteidigungsministeriums und wird nicht auf eine Stichprobe der niederländischen Zivilbevölkerung erweitert. In dieser Untersuchung konzentrieren wir uns auf den Einsatz autonomer Waffensysteme in der nahen Zukunft, die wir als *in den nächsten 5 Jahren* definieren. Das bedeutet, dass wir keine Waffen, die mit künstlicher Intelligenz oder futuristischer Technologie ausgerüstet sind und noch nicht gebaut werden können, betrachten, sondern wir untersuchen eine Technologie, die derzeitig entwickelt wird, vor allem Drohnen mit der autonomen Fähigkeit, Ziele zu setzen. Dieses Zielsetzungsverfahren besteht aus sechs Stufen: (1) Finden, (2) Feststellen, (3) Verfolgen, (4) Zielen, (5) Angreifen und (6) Bewerten. Wir werden uns auf die Entscheidungstreffung in Stufe 4 und 5 des Zielprozesses konzentrieren, da viele ethische Entscheidungen in diesem Teil des Verfahrens zu treffen sind (Asaro, 2016).

### 1.1.5 *Forschungsfrage und daraus resultierenden weiteren Fragen*

Basiert auf der identifizierten Wissenslücke und Problemstellung, entwarfen wir die folgende Forschungsfrage für diese Untersuchung:

> *Wie werden autonome Waffensysteme von der Zivilbevölkerung und den im niederländischen Verteidigungsministerium angestellten Militärangehörigen aufgefasst und welche moralischen Werte betrachten sie als wichtig beim Einsatz autonomer Waffensysteme?*

Diese Forschungsfrage wird von den sich daraus ergebenden Fragen beantwortet:

1. *Wie werden autonome Waffensysteme in der Fachliteratur definiert?*
2. *Welche Werttheorien werden in der Literatur beschrieben?*
3. *Welche der in den Werttheorien genannten Werte stehen im Zusammenhang mit autonomen Waffensystemen?*
4. *Wie wird die Auffassung von Handlungsfähigkeit in der Literatur erwähnt?*

Diese vier sich ergebenden Fragen werden in der Literaturbesprechung unserer Forschung beantwortet.

5. *Welche moralischen Werte im Verhältnis zum Einsatz von autonomen Waffensystemen nehmen die Zivilbevölkerung und das Militärpersonal des niederländischen Verteidigungsministeriums wahr?*

Diese Frage wird von einer Überprüfung von Werten beantwortet, die aus einem Online-Fragebogen und Interviews mit Experten besteht.

6. *Wie wird die Handlungsfähigkeit autonomer Waffensysteme von der Zivilbevölkerung und dem Militärpersonal im niederländischen Verteidigungsministerium aufgefasst?*
7. *Wie werden die Werte im Verhältnis zu autonomen Waffensystemen von der Zivilbevölkerung und dem Militärpersonal im niederländischen Verteidigungsministerium aufgefasst?*

Diese beiden Fragen werden durch Prüfen einer Hypothese und beschriebenen Analyse beantwortet, die auf den Ergebnissen eines randomisierten kontrollierten Experiments beruht.

8. *Wie können die Werte, die autonomen Waffensystemen eigen sind, in den Entwurf einer Moral Machine Autonomer Waffensysteme aufgenommen werden?*

Diese Unterfragen können durch Erzeugen eines Entwurfs- und Implementationsplans für eine Moral Machine für Autonome Waffensysteme beantwortet werden.

Im nächsten Abschnitt wird die Untersuchungsmethode zur Beantwortung dieser Unterfragen genauer beschrieben.

## 1.2 Forschungsmethode

In dieser Untersuchung benutzen wir die Value-Sensitive-Design-Methode (VSD) als Herangehensweise der Forschung. VSD ist eine dreifache Methode, die menschliche Werte innerhalb des Entwurfsprozesses der Technologie berücksichtigt. Es ist ein schrittweises Verfahren für die *konzeptuelle, empirische und technologische* Untersuchung menschlicher Werte, die der Entwurf mit sich bringt (Davis & Nathan, 2015; Friedman & Kahn Jr., 2003). Die konzeptuelle Untersuchung besteht aus zwei Teilen: (1) Identifizierung der direkten Interessenvertreter, die die Technologie einsetzen und die indirekten Interessenvertreter, deren Leben von der Technologie beeinflusst wird, und (2) Identifizierung und Definition der Werte, die der Einsatz der Technologie mit sich bringt. Die empirische Untersuchung befasst sich mit dem Verständnis und der Erfahrung der Interessenvertreter im Zusammenhang mit der Technologie, und die hinzugezogenen Werte werden untersucht. In der technischen Untersuchung werden die speziellen Eigenschaften der Technologie analysiert (Davis & Nathan, 2015). Das VSD kann als eine Roadmap für Ingenieure und Studenten benutzt werden, um ethische Berücksichtigung in den Entwurf aufzunehmen (Cummings, 2006).

Es wurden aber auch Kritiken mit Hinsicht auf die VSD-Methode laut. Eine dieser Besorgnisse, die Davis und Nathan (2015) erwähnen, ist, dass das VSD voraussetzt, dass gewisse Werte zwar universal sind, aber dass sie je nach Kultur und Kontext unterschiedlich sein können. Eine andere Stellungnahme wäre, eine empirische Grundlage für eine Ansicht statt einer philosophischen anzunehmen, oder anzuerkennen, dass die Position der Wissenschaftler nicht als einzige in Betracht kommt (Borning & Muller, 2012). Borning und Muller

(2012) nehmen eine pluralistische Haltung ein, zu dem Maße, dass das VSD weder als universale noch als relative Einstellung zu Werten zu empfehlen ist, sondern man sollte es den Ingenieuren überlassen zu entscheiden, welche Ansicht die Geeignetste im Zusammenhang mit ihrem Design ist.

Zusammen mit Borning und Muller (2012) benutzten wir die VSD-Methode in unserer Forschung als eine Richtlinie und als ein Ziel an sich. In der konzeptuellen Phase weichen wir etwas von der ursprünglichen VSD-Methode ab, da wir keine vollständige Interessenvertreteranalyse durchführen, um die Interessenvertreter in Stufe 1 zu analysieren, aber konzentrieren uns auf zwei offensichtliche Interessenvertretergruppen: die *Zivilbevölkerung* und die *Militärangehörigen*, weil diese beiden Gruppen als befragte Personen für unsere Untersuchung zur Verfügung standen. In Stufe 2 der ersten Phase führen wir eine Literaturauswertung aus, um die Werte in Bezug auf autonome Waffensysteme zu identifizieren. Der wichtigste Fokus unserer Forschungsarbeit ruht auf der empirischen Untersuchungsphase, da wir fanden, dass der empirische Aspekt in der ethischen Debatte zu autonomen Waffensystemen übersehen wird. Wir besprachen die Kritik von Davis und Nathan (2015), indem wir einen großen Teil unserer Zeit dieser Phase des VSDs widmeten, in der wir den Kontext der autonomen Waffensysteme mithilfe verschiedener Forschungsmethoden untersuchten. In der technischen Untersuchungsphase entwerfen wir kein autonomes Waffensystem, wie man hätte annehmen können, da dies ein immenses Vorhaben gewesen wäre, der den Umfang dieser Forschungsarbeit weit übersteigt. Doch wir wählten, auf der Arbeit der Scalable Cooperation-Forschungsgruppe aufzubauen, und schlugen eine Konzipierung einer Moral Machine für autonome Waffensysteme vor, die als nächste Stufe in den Untersuchungen zur Ethik autonomer Waffensysteme benutzt werden kann. Wir verwendeten die Phasen des VSD für unsere Untersuchungen wie folgt (Abbildung 1):

- *Konzeptionelle Untersuchung:* Der Mangel an Forschungsarbeiten zu autonomen Waffensystemen deutet darauf hin, dass die erste Phase unserer Untersuchung an sich erforschend ist. Die Forschungsarbeiten in der konzeptionellen Untersuchung bestehen aus der qualitativen Forschungsmethode der „Schreibtischforschung", um die Fachliteratur über autonome Waffensysteme, moralische Werte und Auffassung der Handlungsfähigkeit zu besprechen.
- *Empirische Untersuchung:* Diese Phase ist in zwei Teile aufgeteilt. Der erste Teil ist informatorisch, um die Wertvorstellungen im Kontext autonomer Waffensysteme mittels einer Online-Befragung zu identifizieren. Dies ist eine quantitative Forschungsmethode, die noch durch Interviews mit Experten von einer qualitativen Forschungsmethode unterstützt wird. Im

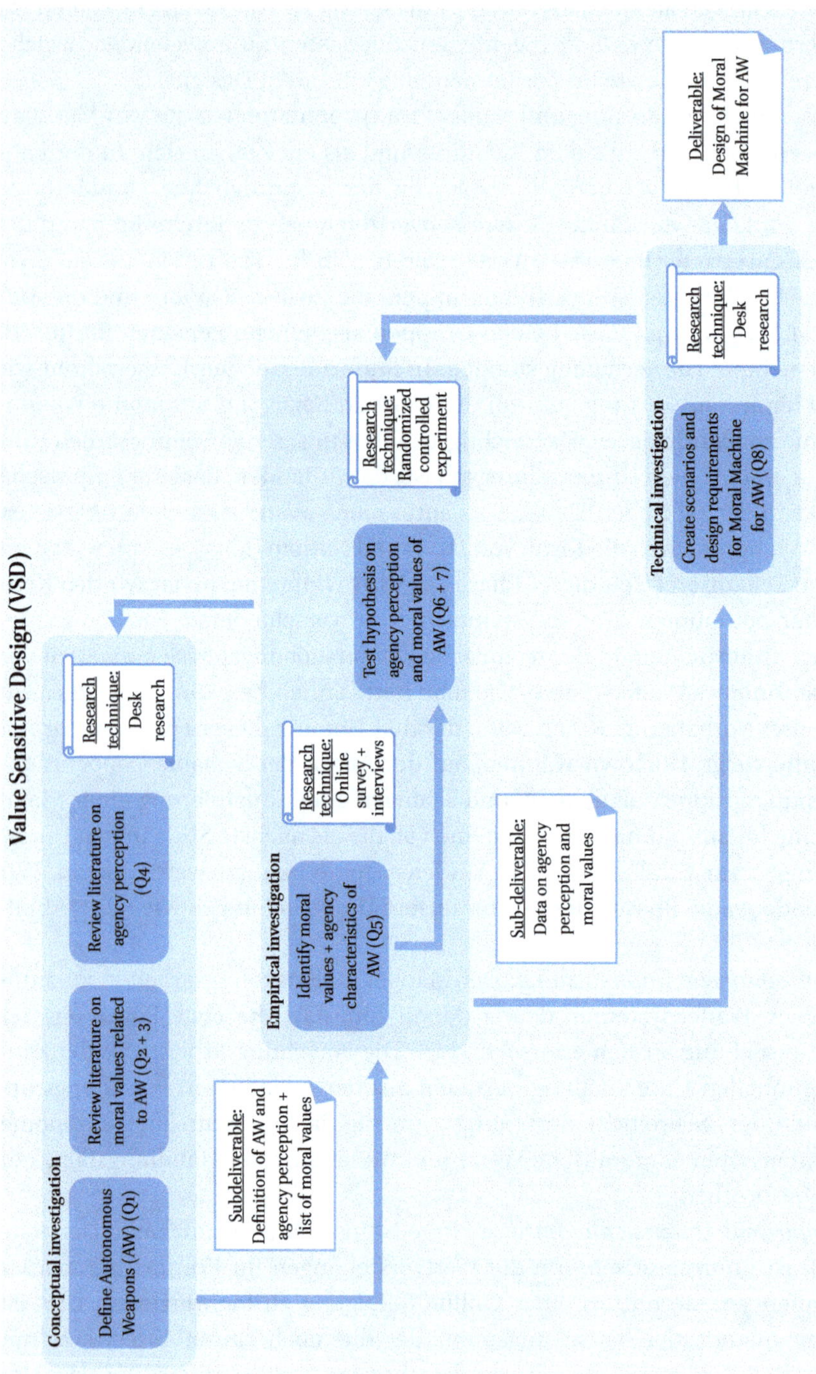


ABBILDUNG 1 Die Value-Sensitive-Designmethode

zweiten Teil der empirischen Untersuchung verwenden wir sowohl eine informatorische als auch eine theorieüberprüfende Forschungsmethode, um die Auffassung innerhalb der Streitkräfte zu studieren. Wir operationalisieren die menschlichen Verhaltensweisen, um unsere Hypothese zur Auffassung innerhalb der Streitkräfte zu bestätigen, indem wir kontrollierte Experimente randomisiert anordnen und die Ergebnisse des Experiments benutzen, um die Wertvorstellungen in Verbindung mit autonomen Waffensystemen auf beschreibende Weise zu erforschen.

– *Technische Untersuchung:* In dieser Phase werden Konzipierung und Eigenschaften für die Moral Machine der autonomen Waffensysteme erzeugt. Sie basieren auf den Szenarien, die bei dem willkürlichen kontrollierten Experiment der empirischen Phase benutzt werden. Die technische Untersuchungsphase ist informatorisch und das Design wird mithilfe der „Schreibtischforschung" als Forschungsmethode erzeugt.

## 1.3 Relevanz

In diesem Unterabschnitt beschreiben wir die wissenschaftliche und gesellschaftliche Relevanz und zeigen auf, wie die Arbeit in dem Kurrikulum des IA-Kurses des SEPAM-Masterprogramms eingebettet ist.

### 1.3.1 *Wissenschaftliche Relevanz*

Die wissenschaftliche Relevanz unserer Untersuchung liegt darin, dass wir zu der akademischen Literatur beitragen, indem wir Einsicht in die Auffassung innerhalb der Zivilbevölkerung und den Streitkräften in Bezug auf autonome Waffensysteme nehmen, und indem wir ihre moralischen Wertvorstellungen in Bezug auf autonome Waffensysteme identifizieren. Einsicht darin fehlt zurzeit, und es konnten keine empirischen Daten zur Auffassung und den Wertvorstellungen in Bezug auf autonome Waffensysteme gefunden werden. Anhand des Value-Sensitive-Design als Forschungsmethode weisen wir nach, dass diese Methode zur Strukturierung akademischer Forschung anwendbar ist und dies als Fallstudie für die VSD-Methode angesehen werden kann. Wir erweitern die Forschung auch mit der ethischen Entscheidungsfällung bei autonomen Fahrzeugen von Bonnefon *et. al.* (2016) auf den Bereich der autonomen Waffensysteme.

### 1.3.2 *Gesellschaftliche Relevanz*

Die gesellschaftliche Relevanz ist das Verständnis der Auffassung von autonomen Waffensystemen der Zivilbevölkerung und Streitkräfte, die im niederländischen Verteidigungsministerium tätig sind und identifizieren, welche moralischen Wertvorstellungen sie mit autonomen Waffensystemen in Verbindung bringen, um Gemeinsamkeiten und Unterschiede in der Diskussion zu dieser Technologie zu finden, die von der Kampagne Stop Killer Robots (2015) und dem internationalen Komitee für Roboter-Rüstungskontrolle (ICRAC) eingeleitet wurde. Zweitens zeigt diese Forschungsarbeit auf, wie die Value-Sensitive-Designmethode für autonome Waffensysteme benutzt werden kann, um die Wertvorstellungen der Streitkräfte und Zivilbevölkerung im Verhältnis zum Einsatz dieser Art von Waffen zu identifizieren. Zum Schluss werden durch Identifizieren der wichtigen moralischen Werte, diese in das Design autonomer Waffensysteme eingefügt, und somit trägt diese Forschungsarbeit zu einem verantwortungsvollen Design und zukünftigem Einsatz autonomer Waffensysteme bei.

### 1.3.3 *Einbettung in das SEPAM- Kurrikulum und den IA-Kurs*

Die allgemeinen Kriterien für eine Magisterarbeit an der Fakultät für Technik, Politik und Unternehmensführung sind die Beinhaltung einer analytischen Komponente und Fokussierung auf einen technischen Bereich und sind fachübergreifend. Als ein Projekt für die SEPAM -Magisterarbeit ist es erforderlich, dass die Arbeit eine Lösung für ein kompliziertes, umfassendes, zeitnahes, soziotechnisches Problem entwirft. Das bedeutet, dass die Arbeit sich auf einen unmissverständlich technischen Bereich in einem Multi-Level-System konzentriert.

Es berücksichtigt sowohl öffentliche als auch private Wertvorstellungen und adressiert nicht nur technische Themen, sondern auch solche, die sich mit Führungs- und ethischen Wahlen befassen (Graduation Portal, 2017). Der Kurs für Informationsarchitektur (IA) schließt Aspekte der Unternehmensführung und Informatik ein und fokussiert auf die Anpassung der organisatorischen Bedürfnisse und technischen Möglichkeiten der State-of-the-Art ICT-Lösungen (IA-Programm, 2017).

Diese Arbeit ist den SEPAM-Kriterien angepasst, da sie sich besonders auf die AI-Technologie von autonomen Waffensystemen im militärischen Bereich konzentriert. Sie ist multidisziplinär, da sie eine Fachliteratur zu Auffassungen und Wertvorstellungen in Bezug auf autonome Waffensysteme in den Bereichen der kognitiven Psychologie, Künstlichen Intelligenz und Moralischen Philosophie, Beschreibungen der Wertvorstellungen in der philosophischen und psychologischen Literatur und den Einsatz von Waffensystemen im militärischen Bereich kombiniert. Das in dieser Arbeit zentrale soziotechnische Problem ist der Mangel an Einblicke in die Auffassung in Bezug auf autonome Waffentechnologie, und die damit verbundenen Wertvorstellungen innerhalb der Zivilbevölkerung und Streitkräfte scheinen sehr wichtig. Diese Arbeit passt in den IA-Kurs, da sie die Gemeinsamkeiten und Unterschiede der ethischen Präferenzen identifiziert, sowohl seitens der Zivilbevölkerung als auch des Militärs und schlägt eine technische Lösung vor, indem sie eine Moral Machine für Autonome Waffensysteme konzipiert, um diese Präferenzen mithilfe eines massiven Online-Experiments genauer zu untersuchen und dabei auf der Moral Machine für Autonome Fahrzeuge aufbaut.

## 1.4 Struktur

Der Rest dieses Berichts ist wie folgt strukturiert: In Abschnitt 2 wird die Fachliteratur zu autonomen Waffensystemen, Wertvorstellungen und Auffassungen in Bezug auf Waffensysteme besprochen. Abschnitt 3 beschreibt die Methodologie, Hypothesen, das Forschungsdesign, die Operationalisierung der Szenarien und Konstrukte, die analytische Herangehensweise und Stichproben und schließt mit methodologischen Themen ab. In Abschnitt 4 werden zuerst die Ergebnisse der Erhebung der Werte verzeichnet, die aus dem Online-Fragebogen und den Interviews mit Experten besteht, gefolgt von den Ergebnissen der drei randomisierten Experimente. Die Konzipierung der Moral Machine für Autonome Waffensysteme wird in Abschnitt 5 beschrieben. Dafür wird zuerst die ursprüngliche Moral Machine für Autonome Fahrzeuge vorgestellt, gefolgt von den Merkmalen und Szenarien, um eine Moral Machine Autonomer Waffensysteme und einen kurzen Umsetzungsplan einzuschließen. Abschnitt 6 schließt mit den Ergebnissen ab, bespricht die wissenschaftlichen und gesellschaftlichen Auswirkungen und identifiziert die Einschränkungen und Empfehlungen für weitere Forschungen. Im letzten Abschnitt 7 betrachten wir die Wahlen, die wir für unsere Forschung getroffen haben, den Prozess des Projekts und die Verbindung der Forschungsarbeit, dem IA-Kurs und dem SEPAM-Kurrikulum.

KAPITEL 2

# Literaturübersicht

> Es ist einfach so, dass ich mich unwohl fühle, Waffen in ein System zu integrieren, das weder Gefühle hat, noch von Menschen gesteuert wird.
>
> VORSTUDIE 1 BEFRAGTE

Dieser Abschnitt erstellt eine Übersicht der Literatur zu autonomen Waffensystemen, gefolgt von einer Zusammenfassung von Wertvorstellungen und zuletzt beschreiben wir das Konzept von Agenten, die wir benutzen, um die Auffassung zu autonomen Waffensystemen zu untersuchen. Jeder Unterabsatz schließt mit einer Besprechung der Theorien ab, die wir in dieser Arbeit verwenden, sowie die Gründe für ihre Auswahl.

## 2.1 Autonome Waffensysteme

Obwohl die Debatte um autonome Waffensysteme in den letzten Jahren stark zugenommen hat, fanden wir dieses Thema in der Fachliteratur nicht besonders gut dargestellt. Wir beginnen diesen Unterabsatz mit einer Übersicht der vielen Definitionen und bieten zwei Klassifizierungen von autonomen Waffensystemen als Konklusion zu diesem Abschnitt an.

### 2.1.1 *Definition*

Autonome Waffensysteme sind eine aufkommende Technologie, und es gibt immer noch keine international vereinbarte Definition (AIV & CAVV, 2016). Selbst ein Konsens, ob autonome Waffensysteme definiert werden sollten, ist nicht vorhanden. Obwohl einige Wissenschaftler in ihren Arbeiten Definitionen erstellen (Tabelle 1), warnen andere vor solch einer Festlegung. NATO erklärt: *„Der Versuch, Definitionen für 'autonome Systeme' sollte vermieden werden, da Maschinen nicht wirklich autonom sein können."* (Kuptel & Williams, 2014, S.10). Das Institut für Abrüstung der Vereinten Nationen (UNDIR) ist ebenfalls vorsichtig bezüglich einer Erstellung einer Definition von autonomen Waffensystemen, da sie

 | DOI:10.1163/9789004449084_003

behaupten, dass das Niveau der Autonomie von den *„kritischen Funktionen der Bedenken und die Beeinflussung verschiedener Variablen abhängt"* (UNDIR, 2014, S. 5). Sie stellen fest, dass einer der Gründe für die unterschiedlichen Begriffe mit Hinsicht auf autonome Waffensysteme ist, dass manchmal Objekte (Drohnen oder Roboter) definiert werden, aber ein andermal eine Charakteristik (Autonomie), Variablen der Bedenken (Letalität oder Grad der menschlichen Kontrolle) oder Art der Benutzung (Zielsetzung oder defensive Maßnahmen) in die Diskussion einbezogen und Teil der Definition werden. Die verschiedenen Definitionen von autonomen Waffensystemen sind in Tabelle 1 verzeichnet. Einige Autoren benutzen den Begriff Militärroboter, die über ein gewisses Niveau der Autonomie verfügen. Da Militärroboter gemäß der Klassifizierung von Royakkers und Orbons (2015) (Abbildung 2) als eine Unterklasse von autonomen Waffensystemen angesehen werden können, haben wir sie in die Liste der Definitionen aufgenommen. Unserer Meinung nach, erfasst der Bericht der Advisory Council on International Affairs (AIV & CAVV) die Beschreibung von autonomen Waffensystemen vom technischen und militärischen Standpunkt aus am besten, da sie vordefinierte Kriterien berücksichtigen und mit dem militärischen Zielsetzungsverfahren verknüpft sind, da die Waffe nur nach einer von Menschen bewirkten Entscheidung eingesetzt wird. Deshalb benutzen wir diese Definition und definieren autonome Waffensysteme wie folgt:

> Ein Waffensystem, das ohne menschliche Einwirkung Ziele für gewisse vordefinierte Kriterien wählt und angreift, als Ergebnis einer von Menschen bewirkten Entscheidung, diese Waffe einzusetzen, wohlwissend, dass, einmal abgeschossen, es nicht möglich ist, sie durch menschliche Einwirkung aufzuhalten.
>
> AIV UND CAVV 2016, S. 11

### 2.1.2 *Klassifizierung der autonomen Waffensysteme*

Autonome Waffensysteme sind nicht nur zweideutig definiert, sie sind außerdem nicht gleichförmig klassifiziert worden. Wir stellen zwei dieser Klassifizierungen in diesem Unterabsatz dar. Royakkers und Orbons (2015) beschreiben verschiedene Typen autonomer Waffensysteme (Abbildung 2). Sie unterscheiden zwischen (1) *nicht-tödlichen Waffen* d.h. *Waffen, „… die keine (unschuldigen) Opfer verursachen oder Menschen schwere oder dauerhafte Verletzungen zufügen."* (Royakkers & Orbons, 2015, S. 617), wie z.B. das Active-Denial-System, eine Strahlenwaffe mit elektromagnetischer Energie, um Menschen von einem Objekt oder von Soldaten fernzuhalten und (2) Militärroboter, die sie wie folgt definieren: *„… wiederverwendbare unbemannte Systeme für Militärzwecke mit*

TABELLE 1 Übersicht der Definition von autonomen Waffensystemen

| Autor(en) | Definition |
|---|---|
| AIV und CAW (2016, S. 11) | „Ein Waffensystem, das, ohne menschliche Einwirkung, Ziele für gewisse vordefinierte Kriterien wählt und angreift, als Ergebnis einer von Menschen bewirkten Entscheidung, diese Waffe einzusetzen, wohlwissend, dass, einmal abgeschossen, es nicht möglich ist, sie durch menschliche Einwirkung aufzuhalten." |
| Altmann, Asaro, Sharkey, und Sparrow (2013, S. 73) | Autonome Waffensysteme sind: „... robotische Waffen, die, einmal abgeschossen, ohne weitere menschliche Einwirkung Ziele wählen und angreifen." |
| Galliott (2015, S. 5) | Militärroboter sind: „eine Gruppe in Kraft gesetzter elektromechanischer Systeme, die alle eines gemeinsam haben:<br>1. Sie werden nicht von Menschenhand gesteuert;<br>2. Sie sind konzipiert, zurückholbar (obwohl sie vielleicht nicht so benutzt werden, dass dies in Frage kommt), und sind<br>3. in einem militärischen Kontext in der Lage, ihre Kraft auszuüben, um eine tödliche oder nicht tödliche Ladung abzuschießen oder sonst eine Funktion auszuführen, um die Ziele der Streitkräfte zu unterstützen." |
| Horowitz (2016, S. 27) | „Ein Waffensystem, einmal aktiviert, ist dazu da, nur individuelle Ziele oder spezifische Zielgruppen anzugreifen, die von einem menschlichen Bediener ausgewählt wurden." |
| Royakkers und Orbons (2015, S. 625) | Militärroboter sind wiederverwendbare unbemannte Systeme für Militärzwecke mit einem Grad von Autonomie." |
| Kuptel und Williams (2014, S. 10) | „Maschinen sind nur „autonom" mit Hinsicht auf gewisse Funktionen, wie Steuern, Sensoroptimierung oder Kraftstoffüberwachung." |
| UNDIR (2014, S. 5) | Der Grad der Autonomie hängt von den kritischen Funktionen der Bedenken und Beeinflussung verschiedener Variablen ab." |

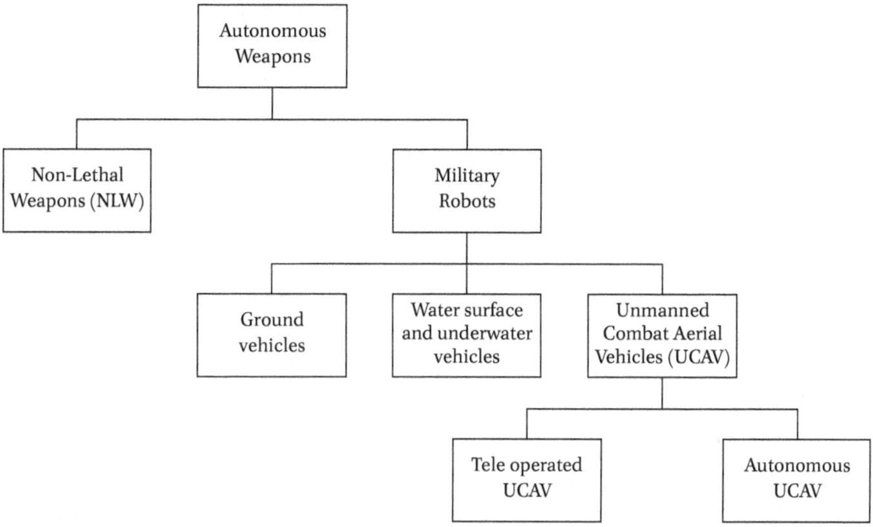


ABBILDUNG 2 Klassifizierung der Autonomen Waffensysteme gemäß Royakkers & Orbons (2015)

*einem Grad von Autonomie."* (Royakkers & Orbons, 2015 S.625). Militärroboter können in drei Kategorien aufgeteilt werden: am Boden angesetzte Fahrzeuge für unbemannte Beobachtung und um Straßenbomben wegzuräumen, Fahrzeuge die unbemannt auf oder unter Wasser navigieren können, wie eine Waffenstation auf einem Schiff oder ein autonomes Unterseeboot, sowie auch Flugzeuge, die als unbemannte Kampfflugzeuge (UCAVs) agieren. Diese UCAVs werden von Royakkers & Orbons (2015) als ferngesteuert klassifiziert. Von ihnen sind „Drohnen" die bekanntesten, sowie die autonomen UCAVs, die nach und nach vom amerikanischen Verteidigungsministerium entwickelt werden (Rosenberg & Markoff, 2016).

Galliot (2015) stellt eine weitere Klassifizierung der autonomen Waffensysteme zur Verfügung. Sie basieren auf vier Stufen der Autonomie für unbemannte Fahrzeuge:

1. *Autonomiestufe 1 – nicht-autonom/ferngesteuert*: „Ein menschlicher Operateur steuert jede einzelne angetriebene Bewegung der unbemannten Plattform. Ohne den Bediener sind ferngesteuerte Systeme nicht fähig, effektiv betrieben zu werden."
2. *Autonomiestufe 2 – Aufsichtsführende Autonomie:* Ein menschlicher Operateur bestimmt die Bewegungen, Positionen oder grundlegenden Handlungen, und das System wird sie dann ausführen. Der Bediener muss das System mit häufigen Eingaben versorgen und achtsam beaufsichtigen, damit es korrekt funktioniert.'

3. *Autonomiestufe 3 – Aufgabenautonomie:* 'Ein menschlicher Operateur bestimmt eine allgemeine Aufgabe und die Plattform verarbeitet eine Reihe von Maßnahmen und führt sie unter ihrer eigenen Aufsicht aus. Normalerweise kann der Bediener das System überwachen, aber das ist für diesen Betrieb nicht erforderlich.'
4. *Autonomiestufe 4 – Komplette Autonomie:* 'Ein System mit kompletter Autonomie würde seine Aufgaben ganz selbstständig ausführen, ohne das menschliche Einwirkung erforderlich ist, mit Ausnahme der Entscheidungstreffung, solch ein System zu bauen. Der menschliche Operateur ist so weit von dem Kreis entfernt, dass die Stufe der direkten Einflussnahme nicht ins Gewicht fällt. Diese Systeme können Fähigkeiten an den Tag legen, die moralische Fähigkeiten von sensiblen Menschen replizieren oder replizieren können, (doch wir werden hierzu keine Stellung beziehen)' (Galliott, 2015, S. 7).

Diese Klassifizierung ist unserer Meinung nach ein guter Versuch, die Stufe der Autonomie von autonomen Waffensystemen zu klassifizieren, aber wir haben einige Bedenken vom Standpunkt der Technik aus. Galliot (2015) selbst behauptet, dass es möglich wäre, die zweite und dritte Stufe der Autonomie zusammenzuführen, da beide eine halbautonome Funktionsstufe darstellen. Wir stimmen dem zu, aber dies ist für uns nicht die Hauptsache, was die Definitionen betrifft. Wir finden es seltsam, eine Liste der Autonomiestufen aufzustellen, die eine Kategorie halbautonomer Systeme enthält. Wichtiger ist, dass auf der vierten Autonomiestufe der Autor aussagt, dass: *„diese Systeme Fähigkeiten an den Tag legen können, die moralische Fähigkeiten von sensiblen Menschen imitieren oder replizieren können"*. Es scheint, dass er Bezug auf die Definition einer dominierenden KI nimmt, insofern, dass ein Computer kognitive Eigenschaften hat und Programme menschliche Erkenntnisse erklären können (Searle, 1980). Zu behaupten, dass ein autonomes System moralische Fähigkeit an den Tag legt, ist unserer Meinung nach ein Mangel an technischem Wissen bezüglich aktueller KI-Systeme, da diese nichts anderes sind als Computer, die Eigenschaften wie Interaktivität, Autonomie und Anpassungsfähigkeit aufweisen (Floridi & Sanders, 2004).

Wir werden ja sehen, ob eine dominante KI, fähig die moralischen Fähigkeiten eines empfindsamen Menschen zu reproduzieren, je entwickelt werden wird. Wir glauben, dass Galliots Klassifizierung (2015) den Stand der aktuellen Technologie nicht realistisch beurteilt und deshalb wird sie in dieser Forschungsarbeit nicht benutzt. Für unsere Untersuchung halten wir uns an die Klassifizierung Royakkers und Orbons (2015), die einen guten Einblick in die aktuelle Militärtechnologie und die der nahen Zukunft bietet. Besonders ihre Unterscheidung zwischen ferngesteuerten und autonomen unbemannten Luftfahrzeugen ist realistisch und für unsere Untersuchung geeignet.

## 2.2 Werte

Im Gegensatz zu autonomen Waffensystemen, wurde das Konzept der Wertvorstellungen in den Bereichen der Moralphilosophie und Psychologie gründlich untersucht. Dieser Abschnitt stellt eine Definition von Wertvorstellungen vor, gefolgt von einer Übersicht der Theorien, die universale Werte beschreiben, eine Übersicht der Wertvorstellungen in Bezug auf autonome Waffensysteme und schließt mit einer Werthierarchie als Beispiel ab, um die konzeptuelle und empirische Untersuchungsphase unserer Arbeit zu überbrücken.

### 2.2.1 *Definition*

Werttheorien sind gut durchforschte Bereiche der Moralphilosophie und Psychologie. Moralphilosophie hat eine lange und gehaltreiche Geschichte der Untersuchung von Werten und in diesem Bereich werden theoretische Fragen gestellt, um die Natur von Werten und Güte zu untersuchen (Schroeder, 2016). Oft wird ein Unterschied zwischen instrumentalen Werten, d.h. es gibt Gründe, ihnen den Vorzug zu geben, da ihre Wirkungen zu guten Dingen führen können (Rønnow-Rasmussen, 2002), und intrinsischen Werten, die *„… eine Art von Wert sind, der, wenn von irgendwem besessen, nur aufgrund seiner intrinsischen Merkmale besessen wird“ (Bradley, 2006, S.112)* gemacht. Obwohl sich Moralphilosophie hauptsächlich mit Theorien befasst, von dem, was „sein sollte“ und genaugenommen von empirischen Ergebnissen nicht berührt wird (Alfano & Loeb, 2014). Sie hat nur einen Zweig, der sich mit angewandter Ethik befasst, die relevant für unsere Untersuchung ist, denn angewandte Ethik überbrückt die abstrakten ethischen Theorien und moralische Praxis. Wie in Abschnitt 1.2 angegeben wird, ruht der Fokus dieser Arbeit auf der empirischen Phase des Value-Sensitive-Design, um zu untersuchen, wie die moralischen Wertvorstellungen der Zivilbevölkerung und der Militärangehörigen in Bezug auf autonome Waffensysteme aufgefasst werden. Deshalb entschieden wir uns gegen die theoretischen Werttheorien der Moralphilosophie in dieser Untersuchung, sondern wandten uns dem Bereich der Psychologie und angewandten Ethik zu, um eine empirische Ansicht der persönlichen Werte der beiden gewählten Interessenvertretergruppen zu erhalten, um Einblick in die *„Ist“-Situation zu tun und nicht „was sein sollte“*.

Der Bereich der Psychologie unterscheidet Werte von Einstellungen, Bedürfnissen, Normen und Verhaltensweisen, insofern sie einen Glauben darstellen, zu einer Verhaltensweise führen, die Menschen leitet und die hierarchisch angeordnet sind, sodass die Wichtigkeit des Wertes über andere Werte sichtbar wird (Schwartz, 1994). Werte werden von Menschen benutzt, die ihre Verhaltensweisen rechtfertigen wollen und definieren möchten, welche Arten von Verhaltensweisen gesellschaftsfähig sind (Schwartz, 2012). Sie unterscheiden sich von Fakten, indem

Werte nicht nur eine empirische Aussage der externen Welt beschreiben, sondern auch an den menschlichen Interessen in einem kulturellen Kontext haften (Friedman *et al.*, 2013). Werte können für die Motivierung und Erklärung individueller Entscheidungstreffungen sowie für die Untersuchung von menschlicher und sozialer Dynamik (Cheng & Fleischmann, 2010) benutzt werden.

Es gibt viele Definition von Werten. Zum Beispiel beschreibt Schwartz (1994, S. 21) Werte wie folgt: *„wünschenswerte, situationsübergreifende Ziele, von unterschiedlicher Wichtigkeit, die als leitende Prinzipien im Leben einer Person oder einer anderen sozialen Gruppe dienen."* Dies ist eine genaue Beschreibung im Vergleich zu Friedman, Kahn Jr. Borning und Huldtgren (2013, S. 57) der Werte so definiert: *„… was eine Person oder eine Personengruppe für wichtig im Leben erachtet."* Die vorhanden Definitionen wurden von Cheng und Fleischmann (2010, S.2) in ihrer Meta-Liste zusammengefasst, und sie erklären, dass: … *"Werte als leitende Prinzipien von dem, was Leute für wichtig erachten, dienen."* Dies ist zwar eine sehr einfache Beschreibung, aber wir finden, sie erfasst die Beschreibung eines Werts am besten, da sie verschiedene Definitionen miteinander kombiniert und die Hauptcharakteristik von Werten benutzt. Deshalb behalten wir die Definition von Cheng und Fleischmann (2010) in dieser Arbeit bei.

### 2.2.2 *Allgemeingültige Werte*

Forschungen ergaben, dass Menschen aller Kulturen sich mit grundlegenden Werten identifizieren; man kann sie die allgemeingültigen Menschenwerte nennen (Friedman *et. al.,* 2013: Graham *et. al.,* 2012, Schwartz, 2012). Obwohl einzelne Personen Werten unterschiedliche Wichtigkeit beimessen, scheint es einen erstaunlich hohen Konsens zwischen den Kulturen hinsichtlich der hierarchischen Ordnung der Werte zu geben (Schwartz, 2012). Als Teil ihrer Untersuchungen haben einige Wissenschaftler sogenannte Wertelisten zusammengestellt, die Aussagen (Items) verzeichnen, die zur Kategorisierung der Analyse von menschlichen Werten benutzt werden können und oftmals von einem beschreibendem Hilfsmittel zur Diskussion dieser Werte begleitet werden (Cheng & Fleischmann, 2010). Die bekanntesten und meist gelesenen Wertelisten sind die von Schwartz (1994), Friedman et al. (2013), Beauchamp und Walters (1999) und Graham *et. al.* (2012). Die Anzahl der von Wissenschaftlern anerkannten allgemeingültigen Werte ist sehr unterschiedlich. Eine Übersicht über diese Wertelisten wird in Tabelle 2 dargestellt, und die Theorien werden im folgenden Absatz kurz beschrieben.

Schwartz (1994) erwähnt 10 bestimmte motivierende Werttypen, die in eine etwas detaillierte Liste von 56 Werten unterteilt ist, die auf seiner umfassenden empirischen Forschung basiert sind und die er für die Begutachtung der 10 überbrückenden allgemeingültigen Werte benutzt. In ihrer Beschreibung der

Value-Sensitive-Design-Methode, erwähnen Friedman *et al.* (2013) 12 Werte, von denen die ersten 9 auf konsequentialitischen und deontologischen moralischen Orientierungen basiert sind und die restlichen 3 aus dem Bereich der Mensch-Maschinen-Interaktion (HCI) gewählt wurden. Graham et al. (2012) benutzen den Begriff *„Grundlage"*, um die 5 speziellen Werte zu beschreiben, die die Allgemeingültigkeit der menschlichen moralischen Natur bestimmen, die Haidt und Joseph (2004) für die Grundlage der Moral-Foundations-Theorie benutzen. Gouveia, Milfont, und Guerra (2014) entwarfen ein System, das auf vielen Werttheorien basiert ist, z.B. die der Bedürfnishierarchie von Schwartz (1994) und Maslow (1943). In diesen Rahmen platzieren die Autoren den Wert in zwei Dimensionen: (1) mit Handlungen, die die menschliche Verhaltensweise antreiben und die persönliche, zentrale oder soziale Ziele anstreben mögen, und (2) Antreiber, die die menschlichen Bedürfnisse in Erfolg- und Überlebensbedürfnisse aufgeteilt repräsentieren. (Gouveia *et. al.,* 2014).

Werte werden nicht nur theoretisch von einer psychologischen Perspektive aus beschrieben, wie im vorherigen Absatz dargestellt wurde, aber sie wurden auch praktisch implementiert und mit Hilfe von angewandter Ethik für professionelle Domänen benutzt. Nehmen wir die Medizin in der Bioethik, die Werte beschreibt, die als Grundprinzip für biomedizinische Fachleute wie Ärzte, Krankenpfleger und andere Gesundheitsarbeiter wichtig sind. Beauchamp und Walters (1999) beschreiben die 4 Werte als Grundlage für das Bioethik-System: 1) *Autonomie:* die absichtliche Handlung ohne steuernde Einflüsse, die eine freiwillige Handlung abschwächen würde, (2) *Güte*: erstellt Vorteile für die Gesellschaft im Allgemeinen, 3) *Gerechtigkeit*: fair und angemessen zu handeln und 4) *Nicht-Schaden:* nicht absichtlich einander Risiken oder Gefahren auszusetzen.

Im Einklang mit unserer Literaturübersicht wählten wir zwei Werttheorien für unsere Studie: eine ist aus der psychologischen Literatur und die andere auf angewandte Ethik begründet, die eine praktische Anwendung von Moralphilosophie darstellt. Die erste Theorie unserer Wahl ist die von Cheng und Fleischmann (2010), da ihre Meta-Listen der menschlichen Werte eine umfassende Liste von 16 menschlichen Werten kreiert hat, basiert auf die Werte, die in 12 separaten Forschungsarbeiten gefunden wurden. Unserer Meinung nach erfasst diese Meta-Analyse die wichtigsten Werte, die von anderen Wissenschaftlern aufgezeichnet wurden und ist ein empirisches Beispiel, das seinen Stamm in der psychologischen Literatur hat.

Die zweite Werttheorie unserer Wahl ist ein Beispiel angewandter Ethik, die seit über vierzig Jahren häufig im medizinischen Bereich Anwendung fand. Wir möchten diese Anwendbarkeit in Bezug auf autonome Waffensysteme untersuchen, da Bioethik-Prinzipien viele Besorgnisse anspricht, die Menschen in Bezug auf autonome Waffensysteme an den Tag legen.

TABELLE 2 Übersicht der Werttheorien

| Autor | Schlüsselbeitrag | Definition von Wert | Werte |
|---|---|---|---|
| Schwartz (ff) | Die Studie betrachtet potenzielle Allgemeingültigkeit der menschlichen Werte und gibt einen Satz dynamischer Verhältnisse innerhalb dieser Werte an.<br><br>Die Studie fand keine allgemeingültigen Aspekte von Werten, aber Anzeichen für eine nahe Allgemeingültigkeit von vier Werten höherer Ordnung. Außerdem fand diese Studie überzeugende Hinweise darauf, dass die zehn Werte von vielen Leuten in heutigen Gesellschaften anerkannt werden. | Werte werden wie folgt beschrieben: *„wünschenswerte, situationsübergreifende Ziele von unterschiedlicher Wichtigkeit, die als leitende Prinzipien im Leben einer Person oder einer anderen sozialen Gruppe dienen."* (S. 21)<br><br>Die konzeptuelle Definition von Menschenwerten besteht aus fünf Merkmalen: "(1) *Glauben* (2) *dem erwünschten Endzustand oder Verhaltensarten, die* (3) *spezielle Situationen übersteigen,* (4) *empfiehlt die Wahl oder Auswertung der Verhaltensweise, Menschen und Ereignisse, und* (5) *wird nach Wichtigkeit im Verhältnis zu anderen Werten eingestuft, um ein System der Wertprioritäten zu bilden* {*Schwartz,1992; Schwartz & Bi/sky, 1987, 1990*}" (p. 20) | 1. *Macht:* Sozialer Status und Prestige, Kontrolle oder Dominanz über Menschen und Ressourcen (Autorität, Vermögen, Ansehen)".<br>2. *Erfolg:* Persönlicher Erfolg durch Beweis von Kompetenz im Einklang mit den gesellschaftlichen Normen (ehrgeizig, erfolgreich, fähig, einflussreich).<br>3. *Hedonismus:* Vergnügungssucht und sinnliche Befriedigung für sich selbst (Vergnügen, Lebenslust, Maßlosigkeit).<br>4. *Stimulation:* Erregtheit, Neuartigkeit und Herausforderungen im Leben (ein vielseitiges Leben, ein aufregendes Leben, wagemutig).<br>5. *Selbstbestimmung:* Selbstständiges Denken und Handeln – wählen, kreieren, erforschen(Kreativität, Freiheit, Wahl der eigenen Ziele, lernbegierig, unabhängig).<br>6. *Allgemeingültigkeit:* Verständnis, Wertschätzung, Toleranz und Schutz des Wohlergehens **aller** Leute und der Natur (großzügig, soziale Gerechtigkeit, Gleichheit, Weltfrieden, die Welt der Schönheit, Einheit mit der Natur, Weisheit, Umweltschutz). |

7. *Menschlichkeit:* Schutz und Verbesserung des Wohls von Menschen, mit denen man häufig persönlich in Kontakt steht (hilfreich, ehrlich, nachsichtig, verantwortlich, treu, wahre Freundschaft, gereifte Liebe).
8. *Tradition:* Respekt, Verbindlichkeit, und Anerkennung der Gewohnheiten und Ideen, die traditionelle Kultur oder Religion erstellen (Respekt von Tradition, bescheiden, fromm, Annahme meines Schicksals).
9. *Konformität:* Unterdrückung von Handlungen, Neigungen, und Impulsen, die wahrscheinlich andere stören **oder** verletzen würden und Missachtung von gesellschaftlichen Ansprüchen **oder** Normen (Gehorsam, Selbstdisziplin, Höflichkeit, Ehren der Eltern und älteren Menschen).
10. *Sicherheit:* Sicherheit, Harmonie und Stabilität der Gesellschaft, von
Beziehungen und dem eigenen Ich (Gesellschaftsordnung, Familie Sicherheit, nationale Sicherheit, Sauberkeit, Erwiderung von Gefallen).

TABELLE 2 Übersicht der Werttheorien (*cont.*)

| Autor | Schlüsselbeitrag | Definition von Wert | Werte |
|---|---|---|---|
| Friedman, Kahn Jr, Barning, and Huldtgren (2013)<br>49 | Eine Übersicht der VSD-Methode und Hinweise auf eine praktische Anwendung.<br><br>Erstellung von Informationen, sodass andere Wissenschaftler das VSD benutzen und erweitern können, und Benutzer werden Werte für Entwurf von Informatik und Computersystemen überlegen. | Ein Wert wird im weiteren Sinne definiert, insofern dass er: „sich auf das bezieht, was eine Person oder Gruppe von Personen für wichtig erachtet." (S. 57)<br><br>Die VSD-Methode befasst sich besonders mit moralischen Werten, die: '... *Themen ansprechen, die auf Fairness, Gerechtigkeit, Mitmenschlichkeit und Tugend zutreffen, die in der Theorie von Moralphilosophie, Deontologie, Konsequentialismus und Tugendethik inbegriffen*" (S. . 72) | 1. *Mitmenschlichkeit* Bezieht sich auf das physische, materielle und psychologische Wohl des Menschen.<br>2. *Besitz und Eigentum* Bezieht sich auf das Recht, einen Gegenstand (oder Informationen) zu besitzen, zu benutzen, zu verwalten, ein Einkommen davon zu erhalten und es zu vererben.<br>3. *Privatsphäre* Bezieht sich auf einen Anspruch, ein Anrecht oder ein Recht einer individuellen Person zu bestimmen, welche Informationen über sie an Andere weitergegeben werden dürfen.<br>4. *Vorurteilsfreiheit* Bezieht sich auf systematische Ungerechtigkeit, die individuellen Personen oder Gruppen von Personen zugefügt wird, einschließlich bereits vorhandene gesellschaftliche Vorurteile, technische Vorurteile und aufkommende Vorurteile.<br>5. *Allgemeine Benutzerfreundlichkeit* Bezieht sich darauf, allen Leuten erfolgreichen Zugang zu Informatik zu erstellen. |

6. *Vertrauen* Bezieht sich auf Erwartungen, die zwischen Menschen vorhanden sind, die Entgegenkommen wahrnehmen, es anderen zukommen lassen, verletzlich sind und Verrat wahrnehmen.
7. *Autonomie* Bezieht sich auf die Fähigkeit von Menschen, auf eine Art zu entscheiden, planen und handeln, die ihres Glaubens nach ihnen helfen wird, ihre Ziele zu erreichen.
8. *Einverständniserklärung* Bezieht sich auf das Sammeln von dem Einverständnis von Personen, einschließlich Kriterien wie Bekanntmachung und Begreifen (für „Einverständnis") und Freiwilligkeit, Kompetenz und Zustimmen (für „Erklärung").
9. *Verantwortlichkeit* Bezieht sich auf die Merkmale, die sicherstellen, dass die Handlungen einer Person oder von Personen oder einer Organisation direkt zu der Person, den Personen oder der Organisation zurückzuführen ist.
10. *Höflichkeit* Bezieht auf die Behandlung von Personen mit Höflichkeit und Rücksichtnahme.

TABELLE 2 Übersicht der Werttheorien (*cont.*)

| Autor | Schlüsselbeitrag | Definition von Wert | Werte |
|---|---|---|---|
| | | | 11. *Identität* Bezieht sich auf das lebenslange Erfassen von Personen, wer sie sind, einschließlich sowohl lebenslange Kontinuität als auch Diskontinuität.<br>12. *Ausgeglichenheit* Bezieht auf einen friedlichen und gelassenen psychologischen Zustand.<br>13. *Ökologische Nachhaltigkeit* Bezieht sich auf das Erhalten von Ökosystemen, sodass sie die Bedürfnisse der Gegenwart erfüllen, ohne zukünftigen Generationen zu schaden. |
| Graham et al. (2012) | Eine Beschreibung (einschl. Kritiken und empirische Ergebnisse) der funktionalen Moraltheorie (MFT).<br>MFT kann benutzt werden, um Einblicke in die moralischen Beurteilungen von Menschen zu tun. MFT wird als eine pluralistische, nativistische, kulturelle, entwicklungs-strategische und intuitive Herangehensweise der Moralität beschrieben. | Um die fünf grundlegenden Konzepte von MFT darzustellen, wurde der Begriff „Fundament" gewählt, aber dies kann auch mit den Begriffen Wert oder Tugend ausgewechselt werden. Es gibt keine genaue Definition von Fundament, aber es wird als eine Metapher aus der Architektur benutzt, um auszusagen, dass: „MFT eine Theorie über den universellen ersten Entwurf des moralischen Geistes ist, und wie dieser Entwurf auf verschiedene Weise kulturübergreifend geändert werden kann." | Die fünf Fundamente von MFT sind:<br>1. Fürsorge/Schaden-Fundament: bezieht sich auf die Fähigkeit, den Schmerz anderer mitzuempfinden und ist die Basis der Tugenden Güte, Freundlichkeit und Pflege;<br>2. Fairness/Betrug-Fundament: bezieht sich auf das Verfahren von reziprokem Altruismus und erzeugt Gedanken von Gerechtigkeit, Rechten und Autonomie;<br>3. Loyalität/Verrat-Fundament: bezieht sich auf die Bildung von veränderlichen Koalitionen und ist basiert auf den Tugenden Patriotismus und Selbstaufgabe für die Gruppe; |

| | | | |
|---|---|---|---|
| | | | 4. Autorität/Umsturz-Fundament: bezieht sich auf hierarchische soziale Interaktionen und basiert auf den Tugenden Führung und Gefolgschaft, einschließlich Ehrerbietung vor legitimer Autorität und Respekt vor Traditionen; 5. Reinheit/Entwürdigung-Fundament: bezieht sich auf die Psychologie von Ekel und Kontaminierung und basiert auf religiösen Vorstellungen ein Leben auf einer höheren Ebene, weniger fleischlich und auf edlere Weise zu führen. |
| Cheng und Fleischmann (2010) | Meta-Analyse der 12 Wertbestände von Grundwerten. Diese Forschungsarbeit schlägt einen Meta-Bestand von Grundwerten vor. | Sie erstellt eine Summierung der Definitionen von Werten: „Werte dienen als Leitprinzipien dessen, was Leute als wichtig im Leben erachten." (S. 2). | (1) Freiheit, (2) Hilfsbereitschaft, (3) Leistung, (4) Ehrlichkeit, (5) Selbstachtung, (6) Intelligenz, (7) Toleranz, (8) Kreativität, (9) Gleichheit, (10) Verantwortlichkeit, (11) Gesellschaftsordnung, (12) Vermögen, (13) Kompetenz, (14) Gerechtigkeit, (15) Sicherheit, und (16) Spiritualität. |

TABELLE 2 Übersicht der Werttheorien (*cont.*)

| Autor | Schlüsselbeitrag | Definition von Wert | Werte |
|---|---|---|---|
| Gouveia, Milfont und Guerra (2014 | Grundwerte einer empirischen Studie.<br>Die Arbeit schlägt ein dreimal zwei-System mit sechs Unterkategorien von Grundwerten vor. | Zwei primäre Funktionen von Werten werden identifiziert: (1) die leitenden Handlungen und (2) kognitive Ausdrücke von Bedürfnissen. | Persönliche Ziele – erfolgreich sein erfordert Werte: Emotion, Freude, Sexualität Persönliche Ziele – Überleben erfordert Werte: Macht, Prestige, Erfolg<br>Zentrale Ziele – erfolgreich sein erfordert Werte: Schönheit, Wissen, Reife<br>Zentrale Ziele – Überleben erfordert Werte: Gesundheit, Stabilität, Überleben Soziale Werte – Erfolgreich sein erfordert Werte: Affektivität, Zugehörigkeit,<br>Unterstützung<br>Soziale Ziele – Überleben erfordert Werte: Gehorsam, Religiosität, Tradition |
| Beauchamp und Walters (1999) | Dieser Artikel ist das erste Kapitel in einem Buch zu Bioethik. Das Kapitel beschreibt drei moralische Prinzipien, die einen Rahmen erstellen, der für Diskussionen über bioethische Fragen benutzt werden kann. | Die Autoren benutzen die Begriffe Prinzipien und Werte synonym. Sie definieren ein Prinzip so: „Ein Prinzip ist ein fundamentaler Verhaltensstandard, auf den sich viele andere moralische Normen und Urteile zu ihrer Verteidigung und ihrer Stellung stützen.“(S. 17) | 1. Autonomie: absichtliches Handeln ohne kontrollierende Einflüsse, die eine freiwillige Handlung abschwächen würden.;<br>2. Wohltätigkeit: Erstellung von Vorteilen für die Gesellschaft als Ganzes.<br>3. Gerechtigkeit: fair und angemessen zu handeln.<br>4. Nicht-Schaden: Anderen nicht absichtlich zu schaden oder sie Risiken auszusetzen. |

Die Wörter in Klammern sind die spezifischen Wertelemente, die den universellen Wert angeben.

### 2.2.3 *Werte mit Bezug auf autonome Waffensysteme*

Werte, wie sie in den Werttheorien in Abschnitt 2.2.2 beschrieben werden, werden nicht oft ausdrücklich in der Literatur zu autonomen Waffensystemen genannt, wie die Übersicht auf Tabelle 3 zeigt, aber die meisten Untersuchungen besprechen verschiedene Werte oder damit verwandte ethische Fragen. Zwei der Zivilbevölkerung zugängliche Berichte des Human Rights Watch erwähnen den Mangel an *menschlichen Gefühlen, Rechenschaft, Verantwortlichkeit, Mangel an menschlicher Würde und Schaden* als Werte, die mit autonomen Waffensystemen in Verbindung gebracht werden (Docherty, 2012, 2015). Sharkey und Suchman (2013) sagen aus, dass die Werte von *Rechenschaft und Verantwortlichkeit* beim Entwurf von Robotersystemen für militärische Einsätze unbedingt berücksichtigt werden müssen.

Im Bereich der Militärethik verzeichnen Johnson und Axinn (2013) *Verantwortlichkeit, Reduzierung von Schäden zur Menschheit, Menschenwürde und Menschenopfer* als Werte in ihrer Diskussion, ob ein Menschenleben einer Maschine übergeben werden sollte oder nicht. In ihrer Fallstudie über den taktischen Tomahawk-Flugkörper betrachtet Cummings (2006) die von Friedman und Kahn Jr. (2003) vorgeschlagenen allgemeingültigen Werte und erörtert, dass neben *Verantwortlichkeit und Einverständniserklärung*, der Wert des menschlichen Wohls ein fundamentaler Grundwert für Techniker ist, wenn sie Waffensysteme entwickeln, die mit *Gesundheit, Sicherheit und dem Wohl* des Volkes verbunden sind. Sie erwähnt weiter, dass die legalen Prinzipien von Proportionalität und Diskriminierung im Zusammenhang mit der Konzipierung von Waffensystemen unbedingt berücksichtigt werden müssen. *Proportionalität* bezieht sich auf die Tatsache, dass ein Angriff nur zu rechtfertigen ist, wenn der zugefügte Schaden als nicht zu schwer angesehen werden kann. *Diskriminierung* bedeutet, dass ein Unterschied zwischen Kombattanten und Nichtkombattanten gemacht werden kann (Hurka, 2005). Asaro (2012) bezieht sich ebenfalls auf die Prinzipien von *Proportionalität und Diskriminierung* und sagt aus, dass autonome Waffensysteme einen moralischen Freiraum eröffnen, der neue Normen erfordert. Obwohl er Werte in seinem Argument nicht ausdrücklich erwähnt, so bezieht er sich auf den Wert von menschlichem Leben und die Notwendigkeit, dass Menschen bei der Entscheidung ob ein Menschenleben vernichtet werden soll, einbezogen werden müssen. Andere Untersuchungen beschreiben vor allem ethische Themen, z.B. *Verhütung von Schaden, Erhaltung der Menschenwürde, Sicherheit, der Wert des menschlichen Lebens und Verantwortung* (Horowitz, 2016; UNDIR, 2015:James I. Walsh & Schulzke, 2015; Williams, Scharre, & Mayer, 2015).

Aufgrund unserer Literaturbesprechung der Werte in Verbindung mit autonomen Waffensystemen, werden die Werte: *menschliche Würde, Schaden,*

TABELLE 3 Werte mit Bezug auf Waffensysteme

| Autor | Schlüsselbeitrag | Definition von Wert | Werte |
|---|---|---|---|
| Cummings (2006) | Anwendung der VSD-Methode für das Problem bei der Konzipierung des taktischen Tomahawk-Flugkörpers. Die Untersuchung zeigt die Überlegungen der ethischen Probleme im Designverfahren für sowohl Ausbilder als auch Ausübende auf. | Nicht zutreffend | Die Werte der Liste von Friedman et al. Liste (2006), die für das Design von Waffensystemen anwendbar sind, sind Rechenschaft, Einverständniserklärung, aber vor allem das Wohl der Menschheit. Die Prinzipien von Diskriminierung und Proportionalität sind wichtig bei der Berücksichtigung des Wohls der Menschheit. |
| Docherty (2012) | Der Bericht des Human Rights Watch, in dem Aspekte des internationalen Menschenschutzgesetzes und der ethischen Fragen für autonome Waffensysteme beschrieben werden. | Es gibt keine Definition des Begriffs von „Werte", aber der Text erwähnt Werte und ethische Fragen. | Werte/ ethische Fragen:<br>Mangel an menschlichem Gefühl;<br>Rechenschaft;<br>Verantwortlichkeit. |
| Johnson und Axinn (2013) | Eine Studie zu ethischen Fragen in Verbindung mit dem Einsatz von tödlichen autonomen robotischen Waffensystemen. Adressiert die Frage, ob die Entscheidung, einen Menschen zu töten, Maschinen übergeben werden darf. | Es gibt keine Definition des Begriffs von „Werte", aber der Text erwähnt Werte und ethische Fragen. | Werte/ ethische Fragen: Verantwortlichkeit; Reduzierung des Schadens an Menschen; Menschenwürde; Ehrgefühl;<br>Menschenopfer. |

| | | | |
|---|---|---|---|
| Sharkey und Suchman (2013) | Eine Studie zur Definierung und Konzipierung von Autonomie und Rechenschaft in Robotersystemen für militärische Einsätze. | Es gibt keine Definition des Begriffs von „Werte", aber der Text erwähnt Werte. | Werte:<br>Rechenschaft;<br>Verantwortlichkeit. |
| Docherty (2015) | Der Bericht des Human Rights Watch, in dem die Rechenschaftslücke und ethischen Fragen für autonome Waffensysteme beschrieben werden. | Es gibt keine Definition des Begriffs von „Werte", aber der Text erwähnt Werte und ethische Fragen. | Werte/ ethische Fragen:<br>Mangel an Menschenwürde; – Rechenschaft; Verantwortlichkeit; Schaden. |
| Das Institut für Abrüstungsforschung der Vereinten Nationen (2015) | Diese Studie wirft einige ethische und soziale Fragen in Bezug der Waffenfähigkeit autonomer Technologien auf. Ermutigung, ethische Überlegungen anzustellen, mit Hinsicht auf kulturelle und soziale Werte bei der Benutzung autonomer Technologien zur Waffenherstellung. | Keine Definition des Begriffs „Werte". Der Text enthält keinen ausdrücklichen Bezug auf Werte, aber einige ethische Fragen werden aufgeworfen. | Die erwähnten ethischen Fragen sind:<br>- Schaden zu reduzieren oder eliminieren;<br>- Rücksichtnahme auf das Gewissen der Zivilbevölkerung;<br>- Angriff auf die Menschenwürde (wenn menschliche Absicht, Leben zu vernichten, nicht besteht). |
| Walsh und Schulzke (2015) | Untersuchung des Experiments um Einblick zu tun, ob die amerikanische Zivilbevölkerung eher einen Krieg beginnen würde, wenn Drohnen eingesetzt würden. Große empirische Untersuchung, die sich mit der Ethik von Drohnenangriffen befasst. | Keine Definition des Begriffs „Werte". Der Text enthält keinen ausdrücklichen Bezug auf Werte, aber einige ethische Fragen werden aufgeworfen. | Ethische Fragen:<br>– Sicherheit;<br>– Respekt für die Immunität der Zivilbevölkerung;<br>– Vermeidung von Schaden |

TABELLE 3 Werte mit Bezug auf Waffensysteme (*cont.*)

| Autor | Schlüsselbeitrag | Definition von Wert | Werte |
|---|---|---|---|
| Williams, Scharre, und Mayer (2015) | Bespricht verschiedene ethische Fragen, die für die Entwicklung autonomer Systeme relevant sind und erstellt Empfehlungen für Entscheidungstreffer in der Verteidigungspolitik. | Keine Definition des Begriffs „Werte“, aber der Text erwähnt mehrere Werte, die abwechselnd mit ethischen Fragen benutzt werden. | Werte/ ethische Fragen:<br>– Sicherheit;<br>– Schaden;<br>– Wert des menschlichen Lebens (Menschen haben das Recht, von einem anderen Menschen getötet zu werden). |
| Horowitz (2016) | Beschreibung der Diskussion zu ethischen Folgerungen von autonomen Waffensystemen. Berücksichtigt tödliche autonome Waffensysteme (LAWS) in drei Kategorien: Munition, Plattformen und technische Systeme. Dadurch wird die Debatte klarer und wirft zwei ethische Fragen auf. | Keine Definition des Begriffs „Werte“. Der Text enthält keinen ausdrücklichen Bezug auf Werte, aber einige ethische Fragen werden aufgeworfen. | Werte:<br>– ***Rechenschaft*** (autonome Systeme haben keine sinnvolle menschliche Kontrolle, deshalb erzeugen sie eine Lücke in der moralischen Rechenschaftspflicht)<br>– *Menschenwürde* (Menschen haben das Recht, von jemanden getötet zu werden, der die Wahl getroffen hat, sie zu töten) |

*Sicherheit, Verantwortung and Rechenschaft* der Liste der Werte zugefügt, die in dieser Untersuchung benutzt werden, da diese Untersuchung diese Werte in der Übersichtstabelle 3 mehr als einmal erwähnt werden. Zusammen mit den Werten aus dem Bereich der Bioethik und dem Meta-Bestand von Cheng und Fleischmann (2010), werden diese Werte zu unserem Ansatzpunkt für die empirische Untersuchung dieser Forschungsarbeit.

### 2.2.4 *Werte-Hierarchie*

Eine Methode zur Überlegung, welche Werte für die Konzipierung autonomer Waffensysteme relevant sind, ist die Übertragung von Werten in die Anforderungen für diese Konzipierung, die durch eine Werte-Hierarchie sichtbar werden (Van de Poel, 2013). Diese hierarchische Struktur von Werten, Normen und Anforderungen für das Design macht die Werturteile, die für die Übertragung erforderlich sind, offensichtlich transparent und anfechtbar. Um dies zu tun, müssen die in der natürlichen Sprache beschriebenen Werte in „formelle Werte" übertragen werden (Aldewereld, Dignum, & Tan, 2015, S. 835). Eine Möglichkeit der Formalisierung von Werten zu Normen wäre, eine Regelung zu benutzen, die wie folgt dargestellt wird: „X gleich Y" oder „X gleich Y im Kontext C" (Searle, 1995, S. 28. Die Ausdrücklichkeit der Werte in formellen Regeln gestattet kritische Besinnung in Diskussionen und weist auf die Werturteile, die in Frage gestellt werden. Transparenz ist von Wichtigkeit, wie Van de Poel (2013, S. 265) so redegewandt erklärt: „Obwohl transparente Wahlen nicht unbedingt besser oder akzeptabler sind, scheint Transparenz ein Mindestzustand in einer demokratischen Gesellschaft zu sein, die versucht die moralische Autonomie ihrer Bürger zu schützen oder zu verbessern, besonders in Fällen, in denen das Design sich auf die Leben anderer als die Designer selbst auswirkt, was oftmals der Fall ist".

Die höchste Stufe einer Werthierarchie besteht aus den Werten, wie in Abbildung 3 dargestellt wird, die mittlere Stufe enthält die Normen, die die Fähigkeiten, Merkmale oder Eigenschaften des Artefakts sind und die unterste Stufe sind die Anforderungen an das Design, die identifiziert werden können. Das Verhältnis zwischen den Stufen ist nicht deduktiv und kann mithilfe von einer Spezifikation von oben nach unten oder von unten nach oben konstruiert werden, indem man nach Motivierung und Rechtfertigung der Anforderungen auf der unteren Stufe sucht. Die von unten nach oben-Konzeptualisierung der Werte ist eine philosophische Aktivität, die kein spezielles Fachwissen erfordert und die von oben nach unten-Spezifikation der Werte erfordern Kontext oder Fachkenntnisse, die dem Design Gehalt geben. (Van de Poel, 2013).

Van de Poel (2013, S. 262) definiert Spezifikation als: „Die Übertragung eines allgemeinen Werts in eine oder eine speziellere Designanforderung" und erklärt, dass dies in zwei Stufen vor sich gehen kann:

1. Die Übertragung eines generellen Werts in eine oder mehr generelle Normen;
2. Übertragung dieser generellen Normen in speziellere Designanforderungen.

Für Stufe 1 sind folgende Kriterien relevant: (1) Die Norm muss eine angemessene Stellungnahme zum Wert sein und (2), die Norm muss eine ausreichende Stellungnahme zum Wert sein. In Stufe 2 müssen die Anforderungen mit Hinsicht auf den Umfang der angestrebten Anwendbarkeit, Ziele und Absichten und den Maßnahmen, um diese Ziele der Norm zu erreichen, präziser sein (Van de Poel, 2013).

Die Übertragung könnte sich als ziemlich schwierig erweisen, da Einblick in die geplante Anwendung und dem Kontext des Werts getan werden muss, und das ist zu Beginn eines Designprojekts nicht immer offensichtlich. Außerdem werden Artefakte oft in einer unbeabsichtigten Weise oder einem nicht vorsätzlichen Kontext benutzt, neue Werte werden realisiert, oder es wird ein Mangel an Werten erkannt (van Wynsberghe & Robbins, 2014). Ein Beispiel davon sind Drohnen, die ursprünglich für militärische Zwecke konzipiert wurden, aber jetzt auch von der Zivilbevölkerung für Filmen von Events benutzt werden oder sogar als Hintergrundbeleuchtung während der 2017 Super Bowl Halbzeit-Show benutzt wurden. Der Wert von Sicherheit wird für Angehörige des Militärs anders interpretiert, die Drohnen in einsamen Regionen einsetzen, verglichen mit denen von 300 Drohnen, die mit Informationen über ein Fußballstadium in einem dicht besiedelten Gebiet fliegen. Der unterschiedliche Kontext und die Nutzung einer Drohne wird zu verschiedenen Interpretationen des Werts Sicherheit führen und könnte zu strengeren Distanznormen für Flugsicherheit führen, was wiederum noch genauer bestimmt werden könnte, mit alternativen Designanforderungen für Rotoren und Software für Annäherungsalarme, um zwei Beispiele zu nennen.

Die Anwendung einer Werthierarchie für autonome Waffensysteme wird von einem Beispiel dargestellt, in dem der Wert Rechenschaft in Normen für „Transparenz der Entscheidungstreffung" und „Einblick in den Algorithmus" übertragen wird. Diese Übertragung gestattet Benutzern, ein Verständnis über die Entscheidungswahlen, die autonome Waffensysteme treffen, zu erhalten, um ihre Handlungen zu verfolgen und zu rechtfertigen (Abbildung 3). Die Normen für die Transparenz der Entscheidungstreffung können zu spezifischen Designanforderungen führen. In diesem Fall ist es ein Merkmal, um den Entscheidungsbaum zu visualisieren, aber auch, um die von den autonomen

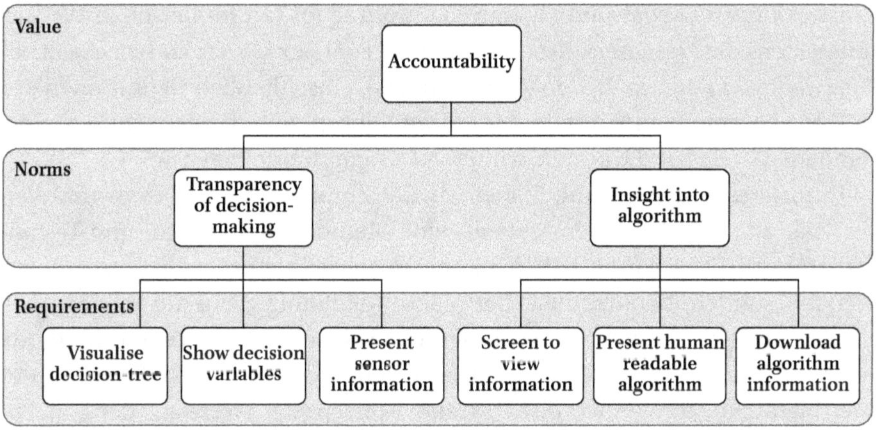


ABBILDUNG 3 Werthierarchie für den Wert von Rechenschaft im Design von autonomen Waffensystemen

Waffensystemen benutzten Entscheidungsvariablen darzustellen, wie z.B. Kompromisse bei Kollateralschadenanteile verschiedener Angriffsszenarien, um Einblick in die Proportionalität eines Angriffs zu erhalten. Hinzu muss das autonome Waffensystem in der Lage sein, Sensorinformationen zu präsentieren, z.B. die bildliche Darstellung der Lage, um zu zeigen, dass zwischen Kombattanten und Nichtkombattanten diskriminiert werden kann. Um Einblick in den Algorithmus zu tun, muss ein autonomes Waffensystemmit Merkmalen konzipiert sein, die es normalerweise nicht enthalten würde. In diesem Fall würden diese Merkmale einen Bildschirm als Benutzerschnittstelle einschließen, der den Algorithmus in einer für Menschen lesbare Form anzeigt, sowie die Fähigkeit, die Änderungen, die vom Algorithmus gemacht wurden, herunterzuladen, als Teil seiner Fähigkeit der Maschinenprogrammierung, die von einer unabhängigen Partei überprüft werden kann, wie z.B. ein Kriegstribunal der Vereinten Nationen, wenn die Legalität der Maßnahmen eines autonomen Waffensystems in Frage gestellt wird.

Kroes und Van de Poel (2015) sagen aus, dass eine objektive Wertmessung nicht möglich ist, da die Operationalisierung mithilfe von Werturteilen zweiter Ordnung durchgeführt wird, die die Stichhaltigkeit des Konstrukts der Wertmessung ernsthaft untergräbt. Urteile werden oftmals als subjektiv erachtet, betreffs ihrer Ehrlichkeit oder Falschheit, je nach den Gefühlen oder Einstellungen der Person, die das Urteil fällt (Searle, 1995). Um diesen Mangel an Stichhaltigkeit auszugleichen, könnte der Designer sich auf technische Codes und Normen berufen, die von Komitees entworfen werden und angemessene Normen der Operationalisierung und Messwerte im Design darstellen. Allerdings geben Normen nicht immer die neusten technischen oder sozialen

Entwicklungen wieder und Operationalisierung fordert immer noch Werturteile seitens des Designers. Kroes und Van de Poel (2015, S. 177) raten, „sie in ein Netzwerk von anderen Bedenken einzufügen, einschließlich Definitionen der in Frage kommenden Werte in der Moralphilosophie (oder dem Gesetz), vorhandenen Codes und Normen, früheren Designerfahrungen usw."

In unserer Untersuchung folgen wir der Empfehlung von Kroes und Van de Poel (2015) die Werthierarchie als eine Methode, um Design- und Testanforderungen für autonome Waffensysteme zu definieren, nicht streng anzuwenden, sondern benutzen das Beispiel in Abbildung 3, um die konzeptuelle und empirische Untersuchungsphasen der Value-Sensitive-Design -Methode unserer Untersuchung miteinander zu verknüpfen. Durch das Erstellen der Werthierarchie können wir das abstrakte Konzept der Werte aus der konzeptuellen Untersuchungsphase durchdenken und es in ein konkretes Beispiel von autonomen Waffensystemen umsetzen, das wir für unsere Eingabe für die empirische Untersuchungsphase unserer Forschungsarbeit benutzen. Die Werthierarchie in Abbildung 3 wird zur Orientierung, Inspiration und Anweisung für den Rest unserer Untersuchung benutzt.

## 2.3 Agency

Das Konzept von Agency (Agens/Agenten) wird in vielen wissenschaftlichen Bereichen untersucht, die Agenten von einem unterschiedlichen Winkel aus beleuchten. Wir befassten uns mit der Literatur aus den Bereichen Kognitive Psychologie, Künstliche Intelligenz und Moralphilosophie. Wir fassten die Merkmale der Agenten in diesen drei Bereichen in Tabelle 4 zusammen und erklären abschließend, warum wir diese Merkmale wählen.

### 2.3.1 *Agency in der Kognitionspsychologie*

Im Bereich der kognitiven Psychologie wurden die Dimensionen der Wahrnehmung zuerst von H.M. Grey *et. al.* (2007) untersucht. Sie fanden, dass Wahrnehmung aus zwei Dimensionen besteht: 1) Erfahrung besteht aus den Faktoren Hunger, Angst, Schmerz, Freude, Zorn, Begehren, Persönlichkeit, Bewusstsein, Stolz, Verlegenheit und Glück und 2) Agenten, die Selbstkontrolle, Moralität, Erinnerung, Emotion, Erkennung, Planung, Kommunikation und Denken einschließen. Während Erfahrung als moralische Passivität bezeichnet wird und sich auf Rechte und Privilegien bezieht, sind Agenten mit moralischen Instanzen verbunden und beziehen sich auf Verantwortlichkeit. Verschiedene Untersuchungen in der kognitiven Psychologie zeigen auf, dass Leute Agenten nichtmenschlichen Gegenständen beimessen und sie nehmen wahr, dass diese nichtmenschlichen Gegenstände die gleichen selbstbestimmenden Fähigkeiten bei der Beurteilung moralischer Dilemmas aufweisen wie Menschen auch (Hristova & Grinberg, 2015). Dies ist nicht auf lebende Kreaturen wie z.B. Tiere begrenzt, sondern Agenten können auch nichtmenschlichen Gegenständen wie Roboter und Zombies beigemessen werden (K. Gray & Wegner, 2012; Waytz *et. al.*, 2010).

### 2.3.2 *Agency in der künstlichen Intelligenz*

Wissenschaftler der künstlichen Intelligenz haben auch die Merkmale von Agenten untersucht, aber eher vom Standpunkt der Informatik aus, um eine Architektur für ein rationales Agens zu erzeugen, das Schlüsse ziehen und zwischen Alternativen wählen kann. Bratman, Israel und Pollack (1988) beschreiben eine Belief-Desire-Intention (BDI)-Architektur für das abwägende Verhalten von an Ressourcen gebundene Agenten mit begrenzter Zeit und Rechenintensivität. „*Weltwissen (Beliefs) sind Aussagen von Merkmalen seiner Welt (und sich selbst), die ein Agens als wahr erachtet …*" (F. Dignum, Kinny, & Sonenberg, 2002, S. 407). Perugini und Bagozzi (2004, S. 71) definieren ein Ziel (Desire) als: '*… ein Sinneszustand, wo ein Agens persönlich motiviert ist, eine Handlung auszuführen oder ein Ziel zu erreichen.*' Absichten (Intentions) deuten auf eine Verpflichtung zum Planen und schließt eine Form der Planung ein, um diese Ziele zu erreichen (Bratman et al., 1988; Perugini & Bagozzi, 2004). F. Dignum *et. al.* (2002) erörtern weiter, dass das BDI-System soziale Konstrukte einbezieht, wie Normen und Obligationen, da dies wichtige Konzepte sind, um autonome Dienstmittel in einem Multiagency-System miteinander zu verbinden.

### 2.3.3 *Agency in der Moralphilosophie*

Moralphilosophie ist ein dritter wissenschaftlicher Bereich, in dem die Merkmale von Agenten beschrieben werden. Sullins (2006) stellt drei Anforderungen auf, um festzustellen, ob ein Roboter als ein moralisches Agens anerkannt werden kann: 1) *Autonomie;* der Roboter ist wesentlich selbstbestimmend von seinen Programmierern, Operatoren und Benutzern, 2) *Absichtlichkeit;* die Vorstellung, Gutes oder Schädliches zu tun, und 3) *Verantwortlichkeit;* sie spielt bis zu einem gewissen Grad eine soziale Rolle und ist verantwortlich für andere moralische Agenten. Himma (2009) definiert Agency so: „*… fähig zu sein, etwas zu tun, das eine Handlungsweise darstellt*". Der Gedanke eines moralischen Agens ist der, dass man für sein Verhalten zur Rechenschaft gezogen werden kann, das von moralischen Normen bestimmt wird. Zwei Fähigkeiten sind für moralische Agenten erforderlich. Die erste ist, Handlungen frei zu wählen, die das Ergebnis von Absichtlichkeit sind. Die zweite Fähigkeit ist, moralische

TABELLE 4 Übersicht der Agency-Merkmale

| Autor | Agency-Merkmale | Wissenschaftsgebiet |
|---|---|---|
| Bratman et al. (1988) | Weltwissen, Ziele, Absichten | Künstliche Intelligenz |
| Bandura (2001) | Absichtlichkeit, Vorausdenken, Selbstkontrolle, selbstreflexiv sein. | Kognitive Psychologie |
| F. Dignum et al. (2002) | Weltwissen, Ziele, Absichten, Normen | Künstliche Intelligenz |
| Sullins (2006) | Autonomie, Absichtlichkeit, Verantwortlichkeit. | Moralphilosophie |
| H. M. Gray et al. (2007) | Selbstkontrolle, Moralität, Erinnerung, Emotion, Erkenntnis, Planung, Kommunikation, Denkweise. | Kognitive Psychologie |
| Himma (2009) | Freie Wahl, Deliberation, Verständnis und Anwendung moralischer Regeln. | Moralphilosophie |
| Waytz et al. (2010) | Fähigkeit zu planen und zu handeln. | Kognitive Psychologie |

Konzepte zu verstehen, z.B. den Unterschied zwischen *„gut"* und *„böse"* oder *„richtig"* und *„falsch"* aber auch moralische Prinzipien anzuwenden, wie *„es ist böse, absichtlich Schaden zuzufügen"* (Himma, 2009).

Die Hauptmerkmale der Agenten von den besprochenen Artikeln über Agentenwahrnehmung in den Gebieten der kognitiven Psychologie, künstlichen Intelligenz und Moralphilosophie sind chronologisch in Tabelle 4 verzeichnet. Wir wählten alle von den Autoren benutzten Konzepte, um Agenten zu beschreiben und haben sie zu diesem Zeitpunkt noch nicht gefiltert. All diese Merkmale werden in der empirischen Phase benutzt, um die Erfassung autonomer Waffensysteme zu untersuchen.

KAPITEL 3

# Methode

> Alle Fragen sind grundsätzlich Variationen zu der einen ethischen Frage: kann eine Drohne autonom angreifen (ohne menschliche Einwirkung). Meiner Meinung ist die Antwort negativ. Es sollte stets eine Zwischenverbindung mit einem Menschen geben. Krieg und Konflikt sind kein Computerspiel, Krieg ist eine soziale Interaktion zwischen menschlichen Wesen. Eine Drohne ist nur ein Instrument, ein Werkzeug, das niemals autonom wirksam sein kann.
>
> DIE ABSCHLUSSUNTERSUCHUNG DER BEFRAGTEN PERSON

Dieser Abschnitt beschreibt die Methodologie, Hypothesen, das Forschungsdesign, die Operationalisierung der Szenarien und Konstrukte, die analytische Herangehensweise und Stichproben und schließt mit den methodologischen Fragen ab.

## 3.1 Methodologie

Die Methoden, die in verschiedenen Teilen dieser Untersuchung benutzt wurden, werden in diesem Unterabsatz beschrieben. Wir beginnen mit der Literaturbesprechung, gefolgt von der Online-Umfrage zu den Werten, den Interviews mit Experten, dem Codierungsprozess der Interviews und zum Schluss mit der Methode der randomisierten kontrollierten Experimente.

### 3.1.1 *Fachliteraturübersicht*

Für die Suche nach Artikeln für die Fachliteraturübersicht wurde Google Scholar benutzt und zuerst wurden die Stichwörter Value-Sensitive UND Design eingegeben, was zu einigen Artikeln über Value-Sensitive-Design führte. *Das Handbuch zu Ethik, Werte und technologisches Design:* Quellen, Theorie, Werte und Anwendungsdomänen wurden benutzt, um weiteren Einblick in das Thema zu tun und weitere Bezugnahmen zu finden. Als nächstes suchten wir nach Artikeln zu menschlichen Grundwerten mit den Stichwörtern Mensch

 | DOI:10.1163/9789004449084_004

UND Werte und allgemeingültig UND Mensch und Werte, aber dies erbrachte nur wenige Ergebnisse. Wir analysierten einige bekannte Theorien zu menschlichen Grundwerten und Werthierarchien. Mit diesen Artikeln benutzten wir die *„zitiert von"*-Option in Google Scholar, um Artikel zu finden, die diese als Quelle benutzten. So erhielten wir relevante Resultate zu Artikeln über menschliche Grundwerte, und wir wählten solche Artikel, die ein Werteverzeichnis zur Verfügung stellten. Artikel, die frühere Arbeiten nur wiederholten, zum Beispiel die Schwartz-Theorie, wurden nicht gewählt. Als nächstes suchten wir nach Artikeln mit Hinsicht auf Werte der Menschen in Bezug zu Waffensystemen durch Wahl der Stichwörter Wert UND Waffensysteme und Stichwörter Value-Sensitive UND Design UND Waffensysteme. Wir wählten die Artikel mit dem Stichwort Wert.

### 3.1.2 *Online-Wertumfrage*

Die Online-Umfrage wurde mit der Qualtrics[1]-Umfragesoftware als Online-Werkzeug erstellt, die von MIT zur Verfügung gestellt wurde. Nach vier Fragen zur Demografik wurden den befragten Personen drei Fragen zu den Werten gestellt, die sie mit autonomen Waffensystemen in Verbindung bringen. Die erste Frage war, die vier Bioethik-Werte Autonomie, Nicht-Schaden, Wohltätigkeit und Gerechtigkeit als am passendsten für autonome Waffensysteme oder zu am wenigstens passend einzustufen. In der zweiten Frage wurden die Befragten gebeten, die fünf Werte aus der Liste von Cheng und Fleischmann (2010) zu wählen, die am besten für autonome Waffensysteme anzuwenden sind. Bei der dritten Frage handelte es sich um eine offene Frage, in der die Personen gebeten wurden, einen weiteren Wert anzugeben, der noch nicht in den beiden vorigen Fragen enthalten war. Die Online-Umfrage wurde zweimal getestet. Zuerst besprachen zwei Kommilitonen die Fragen und nachdem ihr Feedback verarbeitet worden war, wurde die Umfrage ein zweites Mal getestet, diesmal von anderen Prüfern, die für zukünftige Teilnehmer repräsentativer waren. Diese Umfrage wurde über soziale Medien wie Facebook, LinkedIn und Twitter vermittelt. Wir haben sie auf unseren persönlichen Blog veröffentlicht und benutzten die Online-Plattform *„Call for Participants"* (Aufruf für Teilnehmer), um weitere Teilnehmer außerhalb meines sozialen Netzwerks zu finden. Während der Mitarbeitertage des MIT Media Labs im April 2017, wurden Personen, die sich für unsere Forschung interessierten, gebeten an der Umfrage teilzunehmen, und wir schickten den Link mit der Umfrage auch an

---

1 https://www.qualtrics.com/.

den Arbeitskreis von Scalable Cooperation Slack.[2] Diese Online-Untersuchung dauerte dreieinhalb Wochen, vom 17. März bis zum 11. April 2017.

### 3.1.3 *Interviews mit Experten*

Wir führten sechs Leitfadeninterviews mit Experten aus, die auf der *Photo Elicitation Interview* (*PEI*)-Methode (Harper, 2002) basiert sind, um alternative Ansichten von Experten im Bereich autonomer Waffensysteme und ein besseres Verständnis von ihrem Standpunkt aus zu erhalten. Die PEI-Methode kann benutzt werden, um ein gemeinschaftliches Verständnis von Werten zu erhalten und eine Diskussion zwischen dem Designer und der Interessengruppe über diese Werte zu unterstützen (Pommeranz, Detweiler, Wiggers, & Jonker, 2011). Es fördert die Interessengruppen in Bezug auf Werte und schwächt die Mutmaßungen der Wissenschaftler (Le Dantec, Poole, & Wyche, 2009). Wir baten die Experten, drei Fotografien vor dem Interview auszusuchen und, falls sie selber keine finden konnten, wählten wir auch acht Fotografien von autonomen Waffensystemen. Während der Interviews erhielten wir Einblicke in die Werte von den Experten zu autonomen Waffensystemen, indem wir fragten, warum sie diese Fotos ausgesucht hatten und welchen Wert sie ihnen beimaßen.

Alle sechs Befragten bemühten sich, drei Bilder auszuwählen und sie uns vor dem Interview zuzuschicken. Während der Interviews erkannten wir, dass sie die Bilder sehr sorgfältig ausgewählt hatten. Bei der Besprechung der Bilder berührten wir viele verschiedene Themen, von Bauchgefühl bis zu futuristischen Szenarien, Wirtschaft, militärischen Taktiken, Sci-fi-Filmen und sogar Politik. Obwohl die Fotografien nicht alle zu einer Diskussion über menschliche Grundwerte gemäß der Fachliteratur führten, löste die PEI-Methode doch Informationen darüber aus, was die Experten als wichtige Prinzipien in Bezug auf autonome Waffensysteme erachteten. Es kreierte ein gemeinschaftliches Verständnis und eine gute Stimmung während der Befragung.

### 3.1.4 *Codierungsverfahren*

Die Ergebnisse der Interviews wurden mithilfe der Values-Coding-Methode verarbeitet, die von Saldaña (2015) beschrieben wird. Aufbewahrung eines Memos (Anhang G. Codierungsmemos), in dem die Verfahren und Mutmaßungen als integrierter Teil einer Codierungsmethode aufgezeichnet sind. Wir begannen mit einer Liste voreingestellter Codes, die folgendes enthalten: (1)

2 Slack ist eine auf der Cloud basierende Plattform für Zusammenarbeit, die die Kommunikation innerhalb von Teams fördert (https://slack.com/).

Einen Wert: "*Die Bedeutung, die wir uns selbst, einer anderen Person, eines Gegenstands oder einer Idee beimessen,* (2) *Einstellung: Die Art, wie wir über uns selbst, eine andere Person, einen Gegenstand oder eine Idee nachdenken und* (3) *eine Überzeugung: Der Teil des Systems, der unsere Werte und Einstellungen sowie unsere persönlichen Kenntnisse, Erfahrungen, Meinungen, Vorurteile und andere interpretierende Auffassungen der sozialen Welt einschließt*" (Saldaña, 2015, S.89). Während der Codierung der Interviews ist zu empfehlen, dass die Codes den Daten angepasst werden statt umgekehrt. Deshalb haben wir einen zusätzlichen Code „Definition" zur Liste der aufkommenden Codes gefügt. Wir haben während der Codierung der Interviews Notizen zurückbehalten, die unsere Schritte und Mutmaßungen beschreiben. Wir baten einen zweiten Forschungsassistenten, einen TPM-Magisterstudenten mit Kenntnissen von Werten, die Interviews, die auf der Wertcodierungsmethode basiert sind, zu codieren und ebenfalls ein Memo zurückzubehalten (Anhang G. Codieren von Memos). Wir verglichen die von uns codierten Werte mit denen des zweiten Forschungsassistenten und stießen auf Werte, die ähnlich waren (Anhang H. Codierprozesse der Ergebnisse).

### 3.1.5 *Randomisierte kontrollierte Experimente*

Die für die erste und letzte Untersuchung benutzte Methode wird als randomisiertes kontrolliertes Experiment bezeichnet. Oehlert (2010) erwähnt vier Gründe, um Experimente zu erstellen: (1) Sie ermöglichen drei direkte Vergleichen zwischen Behandlungen, die von Interesse sind, (2) sie können konzipiert sein, um jegliche Tendenz in den Vergleichen zu minimieren, (3) sie können entworfen worden sein, um die Anzahl Fehler in dem Vergleich so gering wie möglich zu halten, und (4) wir haben die Kontrolle über die Experimente, die uns gestatten, stärkeren Einfluss auf die Art und Weise zu nehmen, wie wir beobachten und besonders, wie wir unsere Rückschlüsse auf die Ursachen vornehmen. Der letzte Punkt trennt ein Experiment von einer Beobachtungsstudie. Eine Behandlung in diesem Sinn sind die verschiedenen Verfahren, die wir für den Vergleich anzielen. Es ist wichtig, dass die Wirkungen eines Verfahrens nur einer Ursache beizumessen ist, die gemessen werden kann und auch, dass deren Auswirkungen nicht mehreren Ursachen zugeschrieben werden können. Dies wird mit Störungsvariable bezeichnet, die Oehlert (2010) S. 14) wie folgt definiert: "*... tritt auf, wenn die Wirkung eines Faktors oder eines Verfahrens nicht von dem eines anderen Faktors oder Verfahrens unterschieden werden kann*". Randomisierung hilft sicherzustellen, dass Teilnehmern Szenarien zufällig zugeordnet werden und nicht basiert auf vorhandenen Merkmalen sind, z.B. die Zeit wenn und der Ort wo die Umfrage stattfand. Wir benutzen Randomisierung in den Studien, um die Reihenfolge der Szenarien und die der Fragen an die befragten Personen mithilfe einer wahrscheinlichkeitstheoretischen Methode zu variieren. In den Umfragen wählen wir die Option, eine gleichmäßige Anzahl Szenarien unter den befragten Personen zu verteilen.

## 3.2 Hypothesen

Da dies eine Forschungsarbeit ist, haben wir nur eine Hypothese für die Agentenauffassung der Schlussfassung der Arbeit entwickelt und nicht für die Vorstudien oder abhängigen Variablen. Die Vorstudien wurden für die Untersuchung benutzt, welche Manipulationen welche Auswirkungen haben, um zu prüfen, dass die Wortwahl der Szenarien und Fragen klar und deutlich waren.

Mit der Begründung, dass die Militärangehörigen autonome Waffensysteme genau wie alle anderen Waffen betrachten und sie deshalb nichts weiter als ein Werkzeug sind, mit dem man eine Wirkung erhält, stellen wir am Schluss der Untersuchung die Hypothese auf, dass gemäß der Auffassung der Militärangehörigen autonome Waffensysteme keinen psychischen Zustand kennen. Wir haben deshalb keinen Unterschied zwischen dem Zustand des neutralen autonomen Waffensystems und dem Zustand, in dem eine Drohne von Menschen ferngesteuert wird, erwartet. Wir erwarten ebenfalls, dass der neutrale Zustand der Streitkraft sich wesentlich von dem Zustand der Streitkraft mit hoher Selbstbestimmung unterscheidet, wenn wir den Teilnehmern speziell sagen, dass das autonome Waffensystem selbstbestimmende Agency-Merkmale hat, d. h. die Fähigkeit zu planen und eigene Ziele zu setzen. Gemäß dieser Argumentation, ist unsere Hypothese wie folgt:

> *H1: Militärangehörige fassen Autonome Waffen nicht als mit psychischen Fähigkeiten ausgestattet auf.*

## 3.3 Forschungsdesign

Wir testeten mehrere Forschungsdesigns in den beiden Vorstudien, bevor wir uns für das Forschungsdesign der letzten Untersuchung entschieden, das in Abbildung 4 dargestellt wird. Um uns so kurz wie möglich zu fassen, schließen wir die beiden Designs der Vorstudien in diesem Abschnitt nicht ein, und wir zeigen nur das Forschungsmodell auf höchster Ebene. Das Forschungsmodell der abschließenden Untersuchung besteht aus der Wahrnehmung des Agens als unabhängige Variable, moralische Werte als abhängige Variable und

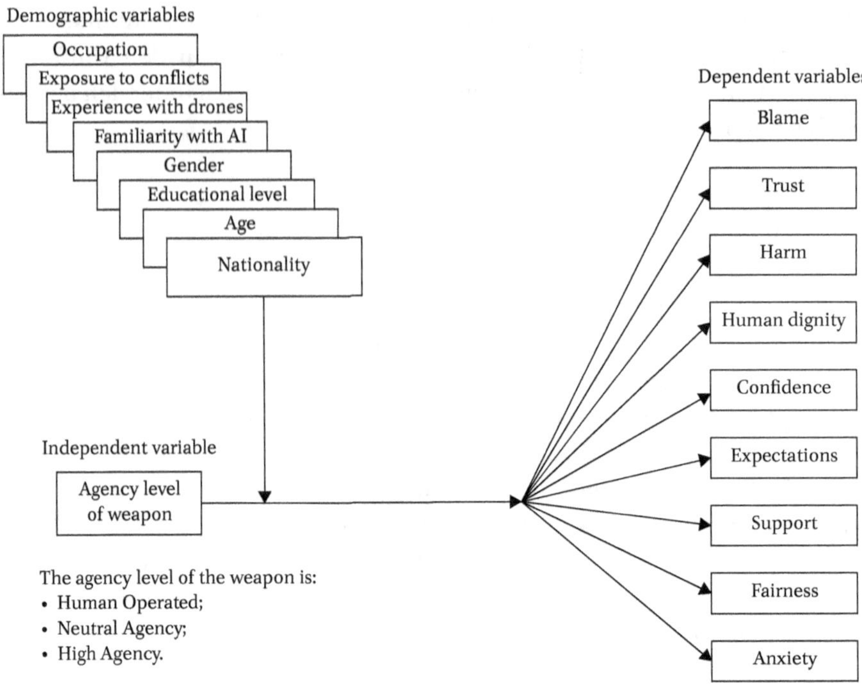


ABBILDUNG 4 Forschungsdesign

die demografischen Variablen. Die spezielle Einrichtung der Szenarien wird genauer im Ergebnisabschnitt beschrieben.

## 3.4 Operationalisierung

In diesem Abschnitt beschreiben wir die Operationalisierung des Konstrukts in konkreten Elementen. Zuerst beschreiben wir die Szenarien, die wir für die Untersuchungen entwickelten, gefolgt vom Agency-Konstrukt und den abhängigen Variablen, für die wir die Elemente und dazugehörigen Fragen beschreiben. Wir geben außerdem die von uns angewandte Größenordnung an und wie das Agency-Konstrukt berechnet wird. Wir beschreiben kurz die Operationalisierung des Aufmerksamkeitstests und welche demografischen Variablen wir in die Umfrage einsetzten.

### 3.4.1 *Szenarien*

Szenarien werden in dem Bereich der kognitiven Wissenschaft als Mittel zur Studie von moralischen Urteilsfällungen in randomisierten kontrollierten

Experimenten eingesetzt (Cushman & Young, 2011; Kominsky, Phillips, Gerstenberg, Lagnado & Knobe, 2015). Wir benutzen sie, um die Auffassung von Streitkräften in Bezug auf autonome Waffensysteme zu untersuchen und erzeugten ein Szenario, das einen militärischen Einsatz beschrieb, wo eine Kolonne Versorgungsmittel in einem Konfliktgebiet abliefert. Die Kolonne begegnet einem Fahrzeug zu hoher Geschwindigkeit, eine Situation, die wahrscheinlich im Zuge dieser Art von Einsatz vorkommt, und das Militärpersonal muss den Gefahrenstand einschätzen, um zu entscheiden ob anzugreifen ist oder nicht. Wir entscheiden, uns auf Drohnen zu konzentrieren, da diese Technologie zurzeit von menschlichen Operatoren benutzt wird, und autonome Drohnen sind bereits von mehreren Firmen entwickelt worden, wie z.B. die BAE-Systems[3], Dassault Aviation[4] und Boeing[5]. Obwohl diese autonomen Drohnen noch nicht in militärischen Operationen eingesetzt werden, glauben wir, dass es innerhalb der nächsten fünf Jahre wahrscheinlich geschehen wird.

Wir erstellten ein Standard-Szenario für alle Untersuchungen, das wir auf die selbstbestimmenden Merkmale für unsere Forschung basieren. Das Standard-Szenario für die abschließende Untersuchung ist wie folgt:

> Eine militärische Kolonne ist unterwegs, um Versorgungsmittel für eine ihrer Einheiten in einem Lager in der Nähe von Mosul im Irak abzuliefern. Der Kommandant hat eine autonome Drohne angeordnet, um die Kolonne von der Luft aus zu beschützen. Die autonome Drohne sucht die Umgebung nach Gefahr vom Feind ab und führt Waffen für die Verteidigung der Kolonne mit sich. Sobald die Kolonne sich drei Meilen vom Lager entfernt befindet, erkennt die autonome Drohne ein Fahrzeug hinter einem Höhenzug, das sich der Kolonne mit hoher Geschwindigkeit nähert. Die autonome Drohne erkennt vier Personen im Fahrzeug, mit großen, wie Waffen erscheinenden Gegenstände und identifiziert den Fahrer des Wagens als ein bekanntes Mitglied einer Gruppe von Aufständischen. Die autonome Drohne greift das herankommende Fahrzeug an, mit dem Ergebnis, dass alle vier Mitfahrer getötet werden, aber verursacht auch Kollateralschäden, da fünf Kinder, die in der Nähe der Straße gespielt haben, auch getötet wurden.

3 https://en.wikipedia.org/wiki/BAE_Systems_Taranis.
4 https://en.wikipedia.org/wiki/Dassault_nEUROn.
5 https://en.wikipedia.org/wiki/Boeing_Phantom_Ray.

Das obige Szenario stellt den Zustand neutraler Kampfmittel da, für den wir keine Charakteristiken für die Waffe erstellen. In den von Menschen bedienten Szenarien ersetzen wir „autonome Drohnen" mit „von Menschen bedienten Drohnen" und ersetzen keine zusätzlichen Informationen zu den Kampfmittelcharakteristiken (die Wörter sind blau markiert, um den Unterschied in diesem Bericht aufzuzeigen, und die Befragten in dieser Untersuchung sahen die Szenarien schwarz gedruckt). In Systemen mit hoher Selbstbestimmung wurden die folgenden Charakteristiken wie folgt beschrieben: „Die autonome Drohne entscheidet selbstständig zwischen einer Reihe von Optionen, wägt die Vorteile und Nachteile ab und entscheidet sich, das sich nähernde Fahrzeug anzugreifen:" ... aber in der Endstudie prüfen wir auch Szenarien ohne Kollateralschaden mit der Aussage: *„... die zu dem Tod aller vier Passagiere führten, aber keinen Kollateralschaden verursachten."*

Wir schränkten die Änderungen zu den Szenarien absichtlich auf ein Minimum ein, damit wir andere Ergebnisse zu diesen Änderungen zufügen und ihre Auswirkung messen können. Die Ergebnisse zeigen, dass diese kleinen Textänderungen zu Verschiedenheiten des Mittelwerts der abhängigen und unabhängigen Variablen führen, deshalb schien es uns, dass die Benutzung von Szenarien als ein Mittel für randomisierte kontrollierte Experimente funktioniert.

### 3.4.2 *Agency-Konstrukt*

Basiert auf der Literaturübersicht der selbstbestimmenden Merkmale in Abschnitt 2.3, operationalisierten wir das Agency-Konstrukt. Wir erstellten eine Liste, die wir mit den Merkmalen füllten, die wir als ähnlich beurteilten und berechneten mehrere Male ein Merkmal, das in der Fachliteratur (Tabelle 5) erwähnt wurde. Wir wählten die Merkmale, die mehr als einmal erwähnt wurden und stellten eine Frage auf, um sie zu messen. Dies ergab ein Konstrukt der Agenten, das zuerst aus fünf Merkmalen bestand, die mehr als einmal in der Fachliteratur erwähnt wurden. Diese Merkmale werden wie folgt auf einer nach eigenen Angaben gemessenen Anordnung gemessen: 0 (lehne stark ab) bis 100 (stimme stark zu), und die Größenordnung wird über eine Reihe von Elementen gemittelt.

Wir haben die folgenden Fragen zur Messung des Agency-Konstrukts benutzt:

1. Fähigkeit der Selbstbestimmung [*Denken*]

   Die Drohne denkt unabhängig von dem, was wegen des Fahrzeugs zu tun ist, bestimmt selber zwischen einer Reihe von Optionen, um die Kolonne zu verteidigen.
2. Fähigkeit, die eigenen Ziele zu setzen [*Zielsetzung*]

   Die Drohne entscheidet unabhängig davon, ob dieses Ziel das Fahrzeug eliminiert, um die Kolonne zu verteidigen.

3. Fähigkeit, moralische Regeln anzuwenden [*Moralische Regeln*]
   Die Drohne befragt unabhängig die moralischen Regeln und Normen, die vom Militärkommandeur eingestellt wurden und entscheidet, ob es moralisch angemessen ist, das Fahrzeug außer Gefecht zu setzen, um die Kolonne zu verteidigen.
4. Fähigkeit, aus freiem Willen zu handeln [*Freies Handeln*]
   Die Drohne hat verschiedene Option zu handeln und entscheidet sich unabhängig, ob sie das Fahrzeug zur Verteidigung der Kolonne eliminieren muss.
5. Fähigkeit zu planen, um Ziele zu erreichen [*Zielerreichung*]
   Das Ziel der Drohne ist, die Kolonne zu verteidigen, folglich entscheidet sie unabhängig, ob sie einen Plan einführen soll, um das Fahrzeug als Ziel zu wählen, ihre Waffen zu kalibrieren und einen Angriff zu machen.

TABELLE 5 Anzahl der Agency-Merkmale, die in der Literatur erwähnt werden

| Agency-Merkmale | Anzahl der Erwähnung | Element |
|---|---|---|
| Vorausdenken, Denken, Selbstbestimmung, Glauben, Glauben | 5 | Denken |
| Absichtlichkeit, Absichten, Soll-Denken, Soll-Denken | 4 | Zielsetzung |
| Planung, Fähigkeit der Planung, Absichten, Absichten | 4 | Zielerreichung |
| Moralität, Verständnis und Anwendung moralischer Regeln, Normen | 3 | Moralische Regeln |
| Selbstkontrolle, freie Wahl | 2 | Freies Handeln |
| Selbstregulierung | 1 | |
| Selbstreflexion | 1 | |
| Autonom | 1 | |
| Verantwortlichkeit | 1 | |
| Erinnerung | 1 | |
| Emotion | 1 | |
| Erkennung | 1 | |
| Kommunikation | 1 | |
| Handlungsfähigkeit | 1 | |

Nach der Vorstudie 1, wurde Punkt 3 [moralische Regeln] abgewunken, und beließen für das Instanzenkonstrukt die vier Merkmale, die in der 2. Vorstudie und der abschließenden Untersuchung benutzt wurden.

1. Denken
2. Zielsetzung
3. Freies Handeln
4. Zielerreichung

### 3.4.3 *Abhängige Variablen*

Wir schlossen 9 bis 12 Fragen in die Vorstudie und abschließende Untersuchung ein, um weitere Implikationen der Auffassung von Streitkämpfern in Bezug auf autonome Waffensysteme zu untersuchen. Die abhängigen Variablen werden von den Ergebnissen der Fachliteraturuntersuchung, dem Fragebogen zu den Werten und den Interviews abgeleitet. Basiert auf den Ergebnissen der Vorstudien schließt die abschließende Untersuchung die abhängigen Variablen ein: *Schuld, Vertrauen, Menschenwürde, Zuversicht, Erwartungen, Unterstützung, Fairness und Besorgnis.* Jede dieser Variablen wurden auf einer Größenordnung nach eigenen Angaben wie folgt gemessen: 0 (lehne stark ab) – 100 (stimme stark zu).

1. *Schuld:* Die Drohne ist an der Maßnahme schuld.
2. *Vertrauen:* Man kann der Drohne vertrauen, dass sie zukünftig die korrekten Maßnahmen trifft.
3. *Schaden:* Die Maßnahmen der Drohne können Schaden anrichten.
4. *Menschenwürde:* Die Maßnahmen der Drohne respektiert Menschenwürde.
5. *Zuversicht:* Ich bin zuversichtlich, dass die Drohne zukünftig die korrekten Maßnahmen trifft.
6. *Erwartungen:* Die Maßnahmen der Drohne entsprechen meinen Erwartungen.
7. *Unterstützung:* Ich unterstütze diese Art von Drohne für den militärischen Einsatz.
8. *Fairness:* Die Maßnahmen der Drohne sind fair.
9. *Besorgnis:* Die Maßnahmen der Drohne beunruhigen mich.

   Die folgenden Konstrukte wurden in den Vorstudien 1 und 2 benutzt, aber in der abschließenden Untersuchung ausgelassen:

10. Kommandanten die Schuld geben: Der Kommandant ist an der Maßnahme schuld.
11. Kommandanten vertrauen: Ich bin zuversichtlich, dass der Kommandant zukünftig die korrekten Maßnahmen trifft.

12. Kommandant schadet: Die Maßnahmen des Kommandanten richten Schaden an.

### 3.4.4 *Aufmerksamkeitstest*

Wir fügten eine Frage mit einem Aufmerksamkeitstest in die Mitte der Fragen mit abhängigen Variablen ein, in der wir baten, den Wert 40 für diese Frage zu wählen. Der Aufmerksamkeitstest wird oft in Fragebögen zu Untersuchungen in der kognitiven Psychologie eingefügt, z.B. von Logg (2017) und Personen, die die korrekte Antwort nicht geben, werden aus der Stichprobe ausgeschlossen, weil sie auch den anderen Fragen nicht genügend Aufmerksamkeit schenken.

### 3.4.5 *Demografische Variablen*

Wir fügten auch Fragen zur demografischen Information zu den befragten Personen hinzu, mit Variablen zu: *Alter, Geschlecht, Bildungsgrad, Beruf und Nationalität.* Außer diesen allgemeinen demografischen Fragen, fragten wir, ob die befragten Personen Erfahrung mit künstlicher Intelligenz hatten, ob sie mit Drohnen gearbeitet hatten und ob sie in einer Konfliktzone gewesen sind.

## 3.5 Analytische Methode

Dieser Abschnitt beschreibt die Methode, die wir zur Analyse der Daten benutzen. Für jede dieser Untersuchungen führten wir die folgenden Schritte in der Datenanalyse aus.

### 3.5.1 *Daten vor der Verarbeitung*

Das Qualtrics-Umfragewerkzeug macht einen Unterschied zwischen den noch zu bearbeitenden Antworten und den gegebenen Antworten. Vor Beginn der Datenanalyse muss der Anteil der in Bearbeitung stehenden Fragen geprüft werden, und es ist zu entscheiden, ob sie zu den beantworteten Fragen hinzuzufügen sind. Es ergab sich, dass viele Befragte den letzten Taster nicht anklickten, wenn die Umfrage über Amazon MTurk ausgeteilt wurde. Das bedeutet, dass ihre Antworten nur zu 99% verfügbar waren, aber sie sind gültig und brauchbar. Sobald die Befragten alle Fragen beantwortet hatten, fügten wir sie zu der Liste verfügbarer Antworten. Wenn die Befragten nicht alle Fragen beantworteten, z.B. die demografischen Fragen, haben wir sie aus der Stichprobe gelöscht. Bei dem nächsten Vorbereitungsschritt löschten wir alle

Befragten, die dem in den Fragebogen eingebauten Aufmerksamkeitstest nicht gerecht wurden. Wir baten die Befragten, dieser Frage den Wert 40 zuzufügen und löschten alle Antworten außerhalb des Bereichs von 38 bis 42, den wir als Fehler bei der Platzierung des Schiebereglers beurteilten.

### 3.5.2 *Reliabilitätsanalyse*

Nach Heraufladen der csv-Datei in SPSS führten wir einen Reliabilitätstest aus, eine Messung zur Bewertung der internen Konsistenz eines Satzes Test-Indikators oder einer Skala. Dieser Test prüft, ob das gegebene Ausmaß eine konsistente Messung eines Konzepts ist. Die Konstruktion zum Ausmaß der internen Konsistenz ist Cronbachs Alpha ($\alpha$), sollte einen Wert höher als 0,7 erreichen, um die interne Konsistenz für das Konzept nachzuweisen. Für jede der Untersuchungen wird Cronbachs Alpha für die Indikatoren (Items) des Messmodells hinterlegt, da diese sog. Items benutzt werden, um das Agency-Konstrukt zu messen und muss eine interne Konsistenz aufweisen. Es wird auch gezeigt, dass Cronbachs Alpha verbessert wird, wenn eines der Items entfernt wird. Die Reliabilität der abhängigen Variablen werden geprüft, aber nicht protokolliert, da diese Items nicht als eine Skala konzipiert sind, und es erwies sich, dass ihre interne Konsistenz unter dem Schwellenwert ($\alpha < 0{,}7$) lag, wie zu erwarten war.

### 3.5.3 *Hauptkomponentenanalyse (PCA)*

Im nächsten Schritt der Datenanalyse wurde eine Hauptkomponentenanalyse an den Items durchgeführt. Dieser Schritt ist eine Reduktionsmethode, die einen größeren Satz Variablen auf einen kleineren Satz reduzieren soll, sog. „Hauptkomponenten", die den Großteil der Varianz in den ursprünglichen Variablen ausmachen. Wir führten diese Analyse an den Items durch, um zu sehen, ob sie zu einem einzelnen Agency-Konstrukt zusammengeführt werden konnten, und protokollieren die Varianz und Anzahl der Komponenten als Resultat. Wir prüften zusätzlich, ob das Konstrukt von demografischen Variablen beeinflusst wurde, aber werden die Ergebnisse nicht angeben, da die Hauptkomponentenanalyse andeutet, dass alle demografischen Variablen Einzelkomponenten sind, die das Konstrukt nicht beeinflussen.

### 3.5.4 *Korrelationsanalyse*

Die bivariate Korrelation nach Pearson ist eine Prüfung, um die Stärke und Richtung der linearen Zusammenhänge zwischen zwei kontinuierlichen Variablen zu messen. Sie produziert einen Korrelationskoeffizienten r, der zwischen -1 und +1 liegt. Ein negativer Wert zeigt einen negativen Zusammenhang zwischen den Konstrukten an, z.B. je größer die Agentenwahrnehmung, je geringer die Unterstützung. Ein positiver Wert zeigt einen positiven Zusammenhang an, z.B. je größer die Agentenwahrnehmung, desto größer die Schuld. Ein Korrelationskoeffizient von Null zeigt an, dass zwischen den Konstrukten kein Zusammenhang besteht. Wir führten diese Analyse aus, um die Korrelation zwischen dem Agency-Konstrukt und den abhängigen Variablen zu prüfen und die Ergebnisse zu protokollieren.

### 3.5.5 *Manipulationsprüfung des Agency*

Die Manipulationsprüfung untersucht, ob das Agency-Konstrukt zwischen den Szenarien unterschiedlich ist, und prüft dabei, ob die Hypothesen akzeptiert oder verworfen werden sollen. Die Prüfung wird auf zweierlei Weise vorgenommen. Zuerst wird ein Diagramm erstellt, auf dem das Agency-Konstrukt auf der y-Achse und die Szenarien auf der x-Achse für eine visuelle Übersicht ruhen. Die Agency-Levels sind erwartungsgemäß für die verschiedenen Zustände unterschiedlich und diese Differenz sollte mit p <,05 signifikant sein. Dies wird von den|-förmigen Balken oben auf den Spalten dargestellt, die den Standardfehler multipliziert mit 1 aufzeigen und andeuten, ob die Konstrukte überlappend sind oder nicht. Wenn die Balken überlappen, sind die Gruppen nicht maßgeblich unterschiedlich.

Das gleiche Ergebnis wird durch unabhängige t-Tests der Stichproben erreicht, eine Prüfung für maßgebliche Unterschieden in den Mittelwerten von zwei Einzelgruppen, denen die Subjekte stichprobenweise zugeordnet wurden, sodass der beobachtete Effekt das Ergebnis der Behandlung ist (in diesem Falle das Lesen eines bestimmten Szenarios), und es kann nicht einem anderen Effekt beigemessen werden. Wir werden die Diagramme erstellen und die Ergebnisse unabhängiger Stichproben-t-Tests-Agency-Konstrukte zwischen den durch menschliche Einwirkung Szenarien und Szenarien mit autonomen Waffensystemen auf verschiedenen Agentenebenen protokollieren.

### 3.5.6 *Analyse der abhängigen Variablen*

Die Analyse der abhängigen Variablen ist aufschlussreich, und die Ergebnisse werden in Diagrammen dargestellt. Jedes Diagramm zeigt den Mittelwert der abhängigen Variablen auf der y-Achse und die verschiedenen Szenarien auf der x-Achse an, um auf die Unterschiede zwischen den Szenarien hinzuweisen. Obwohl die Analyse der unabhängigen Variablen beschreibender Art ist, protokollieren wir die Resultate, indem wir zwischen den verschiedenen Zustände von Wichtigkeit zu p <,05 Vergleiche anstellen.

## 3.6 Voranmeldung

Die abschließenden Untersuchungen wurden auf dem Open Science Framework[6] vorangemeldet. Dies erstellte eine Version eines Projekts mit Zeitstempel, die weder gelöscht noch bearbeitet werden kann. Obwohl es noch nicht von wissenschaftlichen Fachmagazinen verlangt wird, prüfen Wissenschaftler oft, ob eine Studie vorangemeldet ist oder nicht. Wir benutzten die As predicted-Vorlage, in der 8 Fragen über die Studie beantwortet werden, z.B. ob Daten bereits gesammelt wurden, welche Hypothesen die Studie untersucht und die Art der Analyse, die Sie vorzunehmen wünschen. Über die Voranmeldung wird bis zum 24. Juni 2021 ein Embargo verhängt. Wir meldeten die starke Hypothese, die wir über die Auffassung und moralischen Wertvorstellungen der Streitkräfte verfechten vorab an, da wir denken, dass wir die Untersuchungsergebnisse basiert auf der Vorstudie 2 wiederholen können. Für die unabhängigen Variablen sind wir nicht sicher, welche Mechanismen gängig sind und ob wir unsere Ergebnisse wiederholen können, deshalb haben wir eine Forschungsanalyse über diese Ergebnisse angemeldet.

## 3.7 Stichprobe

Alle Umfragen wurden in Qualitrix kreiert, und die Vorstudie wurde über Crowdsourcing-Plattform Amazon Mechanical Turk in den Umlauf gebracht. Wir testeten die Umfrage mit einer kleinen Stichprobe von 20 befragten Personen und untersuchten die durchschnittliche Dauer der Fertigstellung, um zu sehen, ob wir diese Zeit korrekt eingeschätzt hatten. Wir prüften, ob die Szenarien gleichmäßig verteilt waren, ob die Befragen die Testfrage korrekt geprüft hatten und ob es irgendwelche Kommentare zu der Umfrage gab. Vorausgesetzt, dass alle zufrieden waren, entschieden wir die Gruppe auf 50 befragte Personen pro Szenario zu erhöhen, also 500 Teilnehmer für Vorstudie 1 und 700 für Vorstudie 2. Die Sammlung der Antworten nahm ca. 4 Stunden in Anspruch, und jeder Befragte erhielt eine Zahlung von 0,40$.

Die abschließende Untersuchung wurde anhand der Schneeballauswahl durch E-Mail mit einem anonymen Link an ca. 40 Militärangehörige weitergegeben, die die Fragebögen noch weiter vermittelten. Diese Methode kam zum Einsatz, da es uns nicht von MIT oder dem niederländischen Verteidigungsministerium gestattet war, irgendwelche persönlichen Informationen, wie

6 https://osf.io/.

E-Mail-oder IP-Adressen zu sammeln. Mit dem Schneeballverfahren hatten wir keine garantierte Anzahl befragter Personen. Außerdem veröffentlichen wir eine Pressemeldung auf der internen Webseite der Armee. Die abschließende Untersuchung dauerte 16 Tage, vom 12. Juni bis zum 27. Juni 2017 und ergab 327 Auskünfte, von denen 239 vollkommen gültig und brauchbar waren, nachdem die Daten vorverarbeitet wurden.

## 3.8 Methodologische Fragen

In diesem Abschnitt werden mehrere Fragen und Besorgnisse zum Thema Methodologie angesprochen. Diese Besorgnisse könnten sich auf die interne und externe Gültigkeit der Untersuchung auswirken und möglicherweise werden einige Gegenmaßnahmen angeboten, die diese Besorgnisse adressieren.

### 3.8.1 *Interviews zur Codierung*

Um Einblick in die Werte von Experten zu erhalten, führten wir semistrukturierte Interviews aus, das ist eine qualitative Forschungsmethode. Wir benutzten die Werte-Codierungsmethode, um die Daten zu interpretieren und die Werte aus den umgeschriebenen Interviews zu entnehmen. Ein Problem beim Codieren von qualitativen Daten ist seine heuristische Natur, und dass keine Formeln für die Codierung festgelegt sind. Es schließt auch Verknüpfung ein und macht Sinn aus den Daten (Saldaña, 2015). Diese Merkmale implizieren, dass die Methode zur Codierung von Werten geneigt ist, einseitig darzustellen, wobei die Rechercheure ihre eigenen Rahmen und inneren Denkbilder zur Interpretation der Daten benutzen. Um diese Neigung durch Anwendung der Codierung von Werten zu reduzieren, wird ein zweiter Rechercheur die Interviews unabhängig von dem ersten Rechercheur codieren. Einige Anweisungen zur Methode werden gegeben, aber Informationen zu den Werten und Ergebnissen sind nur beschränkt möglich.

### 3.8.2 *Randomisierte kontrollierte Experimente*

Eine der Anforderungen für randomisierte kontrollierte Experimente ist, wie der Name besagt, eine willkürliche Zuordnung der Befragten zu den Szenarien. Die Randomisierung begrenzt Verwirrung, verbessert die interne Gültigkeit der Forschungsarbeit und ist eine kritische Voraussetzung dieser Art von Untersuchung. Eines der Probleme mit der Randomisierung ist, dass sie fehlschlägt und dass die Verteilung der Befragten auf die Szenarien verzerrt wird. Eine der Ursachen für die verzerrte Verteilung ist, dass die Software es verfehlt, die Befragten gleichmäßig den Szenarien zuzuordnen. Als Rechercheure können wir nur nachprüfen, ob die Einstellungen in der Software korrekt sind. Eine weitere Ursache für eine verzerrte Verteilung könnte sein, dass Personen ein gewisses Szenario ansehen, das sie nicht mögen oder erwartet haben und sie somit den Fragebogen nicht beantworten. Ein Beispiel davon ist, dass, wenn Personen erwarten, einen Fragebogen zu autonomen Waffensystemen auszufüllen, stattdessen das Szenario mit menschlicher Einwirkung erhalten, und aufgrund dessen entscheiden, aus der Umfrage auszuscheiden. Dieses voreilige Aussteigen aus der Umfrage nennt man selektive Schwundquote, und es kann mit einer einseitigen Untersuchung enden, da die Personen, die aus der Umfrage ausstiegen, sich von denen, die die Umfrage akzeptieren, unterscheiden. In der derzeitigen Forschungseinrichtung ist es nicht möglich, Gegenmaßnahmen zu ergreifen, und es könnte eine Gefahr für die interne Gültigkeit der Untersuchung darstellen.

### 3.8.3 *Amazon Mechanical Turk*

Der Vorteil beim Benutzen von Amazon MTurk als Umfrageverteilungsmethode ist, dass eine große Gruppe Befragter an der Umfrage teilnimmt und die Fragen ernsthaft beantworten, um ihren Arbeiterstatus zu erhalten. Diese Methode hat auch ihre Nachteile, weil die Arbeiter bei MTurk vor allem in den USA basiert sind, denn um ein Arbeiter zu werden, muss man den amerikanischen Steuerregulierungen nachkommen, selbst wenn man nicht in den USA lebt. Dies könnte zu einer Einseitigkeit der Resultate führen, die andeuten, dass die Ergebnisse außerhalb der US-Grenzen generalisierbar sind. Bonnefon *et. al.* (2016) berichten von einem zweiten Bedenken zum Einsatz von MTurk, nämlich dass Teilnehmer auch für die amerikanische Bevölkerung nicht wirklich repräsentativ sind, z.B. sind viele Arbeiter Studenten, und ein drittes Bedenken ist, dass Arbeiter bereits mit dem Material vertraut sind. Das erste Problem kann durch eine Auswahl von mehr Arbeitern adressiert werden, sodass die Stichprobe auch Personen außerhalb der USA einschließt. Das zweite Problem liegt in der Wahl von MTurk und kann nicht durch Gegenmaßnahmen behoben werden. Das dritte Problem wird wahrscheinlich nicht auftreten, da die Szenarien einmalig sind und noch nie zuvor von MTurk getestet wurden.

### 3.8.4 *Die abschließende Untersuchung nach dem Schneeballverfahren*

Die abschließende Untersuchung wurde unter Militärangehörigen verteilt, die die E-Mail an ihr Netzwerk schickten. Dieses Schneeballverfahren hat einige Nachteile. Zuerst einmal könnten die Teilnehmer die Umfrage mehrere Male durchführen, und da es uns nicht gestattet ist, persönliche Daten zu speichern, können wir nicht nachprüfen, ob dies der Fall ist. Dies könnte zu einseitigen Resultaten führen, wenn Personen verschiedene Szenarien ansehen, und dies könnte die Antworten auf die Fragen beeinflussen. Ein zweiter Nachteil ist, dass Kollegen zuerst an der Umfrage teilnehmen und den anderen Teilnehmern von den Ergebnissen erzählen, bevor sie an dieser Umfrage teilnehmen. Dies könnte auch ihre Antworten beeinflussen. Drittens können wir als Rechercheure nicht kontrollieren, wer an dieser Umfrage teilnimmt, folglich ist ein potenzielles Risiko vorhanden, dass die Umfrage zu jemanden außer des Verteidigungsministeriums gesendet wird, sodass die Stichprobe kontaminiert wird. Leider gibt es kaum Gegenmaßnahmen, die diese drei Probleme adressieren, ohne persönliche Daten zu speichern.

KAPITEL 4

# Ergebnisse

> Hoffentlich verstehen die Leute, dass Computer nur das tun, für das sie programmiert worden sind, und sonst NICHTS.
>
> VORSTUDIE 1 BEFRAGTE

Dieser Abschnitt beschreibt die Ergebnisse der Werteumfrage, die aus einem Online-Fragebogen und Interviews mit Experten besteht, sowie die randomisierten kontrollierten Experimente, die zwei Vorstudien und die abschließende Untersuchung umfasst. Die von dem Online-Fragebogen und den Interviews abgeleiteten werden in einer Übersicht zusammengefasst. Die Ergebnisse der Vorstudien und der abschließenden Untersuchung werden mithilfe der Struktur der Datenanalyse beschrieben (Abschnitt 3.5) und die Hauptergebnisse sind in der Konklusion einer jeden Studie verzeichnet.

## 4.1 Werteumfrage

Um Einsicht darin zu den Werten zu erhalten, die Personen mit autonomen Waffensystemen in Verbindung bringen, wurden eine Umfrage und sechs Experteninterviews ausgeführt. Eine kurze Übersicht des Ergebnisses ist diesem Unterabsatz aufgeführt.

### 4.1.1 *Online-Umfrage*

Die Werte-Umfrage bestand aus einem Fragebogen, der die befragten Personen zuerst nach Anwendbarkeit der Werte von Bioethik und Cheng und Fleischmann (2010) fragte. Die letzte Frage bat die Teilnehmer, mindestens einen zusätzlichen Wert anzugeben, der in vorherigen Fragen nicht vorhanden war. Insgesamt 69 Befragte unternahmen die Umfrage und nach Entfernen der unvollendeten Antworten, blieb eine Rücklaufquote von 57 zurück. Die Ergebnisse werden in den nachstehenden Diagrammen dargestellt.

 | DOI:10.1163/9789004449084_005

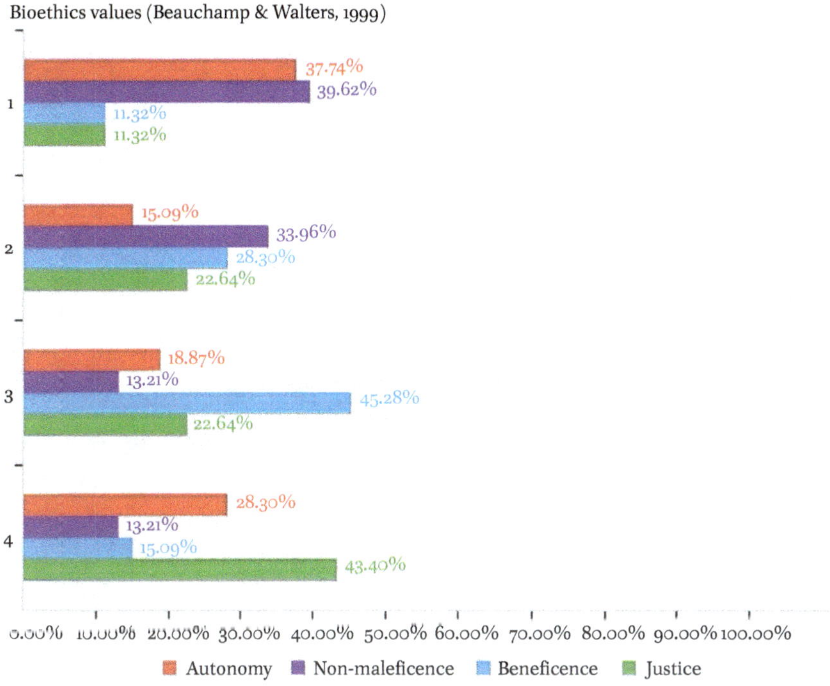


ABBILDUNG 5 Ergebnisfragen 1 Online-Wertbesprechung

Dieser kurze Fragebogen bietet Einblick in die Werte, die Personen mit autonomen Waffensystemen in Verbindung bringen. Die Ergebnisse der Bioethik-Frage zeigt, dass Nicht-Schaden das Wichtigste ist, als zweiter Wert kommt Wohltätigkeit in Frage, drittens Gerechtigkeit und zu Schluss Autonomie. Die fünf Werte von Cheng und Fleischmanns Liste (2010) sind: *Sicherheit, Intelligenz, Verantwortlichkeit, Gerechtigkeit und Gesellschaftsordnung.* Die offenen Fragen erweisen, dass Personen autonome Waffensysteme vor allem mit Rechenschaft und Wirksamkeit verbinden.

### 4.1.2 *Interviews*

Die folgenden Themen wurden in den Experteninterviews besprochen:

Künstliche Intelligenz:

- Prof. Toby Walsh (University of New South Wales & Data61).

Autonome Waffensysteme:

- Dr. Peter Asaro (The New School& Campaign Ban Killer Robots);
- Frau Miriam Struyk (Program Director Security & Disarmament PAX);
- Dr. Michael Horowitz (University of Pennsylvania).

ABBILDUNG 6 Ergebnisfragen 3 Online-Wertbesprechung

Militäreinsätze:
- Oberstleutnant Kremers (Hauptsitz der niederländischen Armee);
- Fregattenkapitän van den Sande (Hauptsitz des Verteidigungsministeriums).

Abgesehen von den Diskussionen zu Werten, erstellten die Interviews einen sehr informativen Hintergrund zur Geschichte und den Positionen der verschiedenen Gruppen in der aktuellen Debatte zu autonomen Waffensystemen, aber enthüllten auch, dass es noch keinen Konsens zu einer Definition gibt. Die vollständige schriftliche Wiedergabe der Interviews ist in Anhang F., Transkriptionen der Interviews, zu finden.

Die Interviews wurden mithilfe der „Werte-Codierung" codiert und basieren auf dem Vergleich der Werte-Codierer der Rechercheure und dem des zweiten

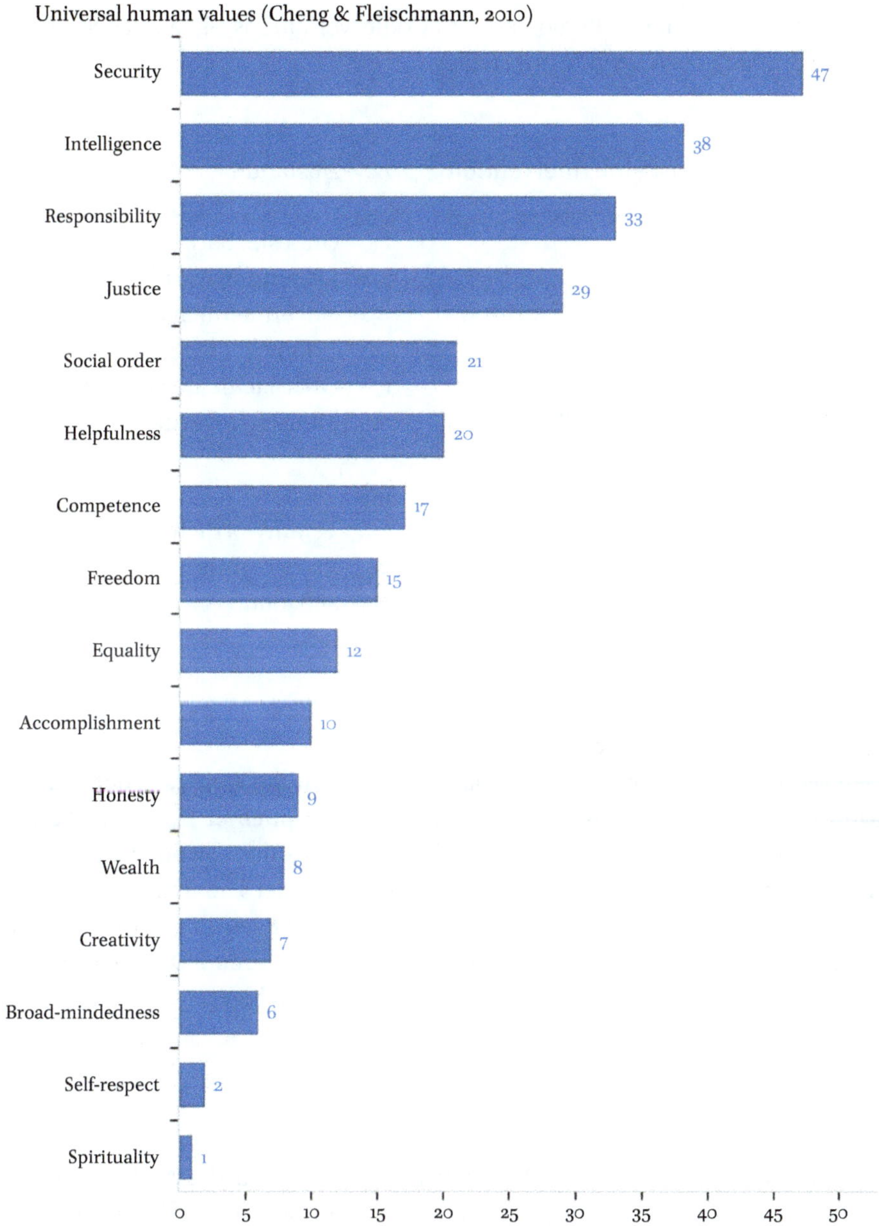


ABBILDUNG 7 Ergebnisfragen 2 Online-Wertbesprechung

Bearbeiters (Anhang H. Ergebnisse des Codierverfahrens). Die folgende Liste mit 22 einmaligen Werten rühren daher:

1. In jedem Fall ist eine Intervention erforderlich
2. Zurückrufen
3. Risiken minimieren
4. Kontrolle haben
5. Guter Wille
6. Grenzen abstecken
7. Das Recht, von einem Menschen getötet zu werden
8. Ethischer Rahmen
9. Sinnvolle menschliche Kontrolle
10. Voraussehbarkeit
11. Reflexivität
12. Abstand
13. Unbesiegbarkeit
14. Determinismus
15. Unvermeidlichkeit
16. Unnötiges Leiden
17. Oberflächliche Verletzungen
18. Menschenwürde
19. Rechenschaft
20. Verantwortlichkeit
21. Verteidigung
22. Schaden

### 4.1.3 *Schlussfolgerung der Umfrage*

Wir schließen die Werte-Umfrage damit ab, dass wir die Werte aus der Online-Umfrage und den Interviews mit den Ergebnissen in einer Liste von Werten (Tabelle 6) vergleichen. Sechs Werte in dieser Liste werden sowohl in der Werte-Umfrage als auch den Interviews erwähnt. Aufgrund der Themen, betonten die sechs Experten in den Interviews, dass wir die Werte gewählt hatten, die am meisten zum Testen in der Vorstudie 1 vorgeschlagen wurden. Diese Werte sind Fairness, Schaden, Menschenwürde und Verantwortlichkeit.

## 4.2 Vorstudie 1

In der ersten Vorstudie versuchten wir zu testen, ob wir die Auffassung und moralische Wahrnehmung der Streitkräfte in Bezug auf von Menschen gesteuerten Drohnen und autonomen Waffensystemen manipulieren konnten. Wir waren daran interessiert, wie Leute die derzeitige Technologie im Vergleich zu zukünftigen Technologien wahrnehmen. Wir kreierten mehrere Zustände, in denen wir die Agentenebene variierten, entweder durch zusätzliche Informationen dazu, wie die Drohnen sich selbst entscheiden, ob sie das Fahrzeug angreifen oder durch Auslassen dieser Information. Wir überprüften positive

TABELLE 6 Übersicht der Werte aus Online-Untersuchungen und Interviews

| Online-Umfrage | Interviews |
|---|---|
| Gerechtigkeit | |
| Autonomie | |
| Sicherheit | |
| Intelligenz | |
| Gesellschaftsordnung | |
| Wirksamkeit | |
| Fairness | |
| Rassismus | |
| Kontrolle | Kontrolle haben |
| Wohltätigkeit | Guter Wille |
| Rechenschaft | Rechenschaft |
| Verantwortlichkeit | Verantwortlichkeit |
| Sicherheit | Verteidigung |
| Nicht-Schaden | Schaden |
| | In jedem Fall ist eine Intervention erforderlich |
| | Zurückrufen |
| | Risiken minimieren |
| | Grenzen abstecken |
| | Das Recht, von einem Menschen getötet zu werden |
| | Ethischer Rahmen |
| | Bedeutungsvolle menschliche Kontrolle |
| | Voraussehbarkeit |
| | Reflexivität |
| | Abstand |
| | Unbesiegbarkeit |
| | Determinismus |
| | Unvermeidlichkeit |
| | Unnötiges Leiden |
| | Oberflächliche Verletzungen |
| | Menschenwürde |

und negative Zustände als Resultat, da wir erwarteten, dass dies eine Wirkung darauf haben würde, wie die abhängigen Variablen eingeschätzt werden würden, z.B. Unterstützung.

In dieser Untersuchung stellten wir vor allem Fragen zu dem Kampfmittel der Drohne und nicht zu dem menschlichen Operateur, der sie steuert.

Wir testeten 10 Szenarien und zielten darauf, 50 befragte Personen je Zustand zu erhalten, folglich eine Gesamtsumme von 500 Befragten. Die Anzahl der Befragten pro Szenario betrug zwischen 49 und 54. Nach der Datenvorverarbeitung und dem Löschen von Befragten, die den Aufmerksamkeitstest nicht bestanden hatten, verfügten wir noch über 510 komplett ausgefüllten Fragebögen.

### 4.2.1 *Reliabilitätsanalyse*

Für alle fünf Agency-Items galt $\alpha = 0{,}9$ was weit über dem Schwellenwert von 0,7 liegt und kann nur verbessert werden, wenn Item SQ3_Moral_rules gelöscht wird, siehe Tabelle 8. Löschen anderer Items schwächt den α-Wert ab und reduziert die interne Konsistenz, wie in der letzten Spalte von Tabelle 8 dargestellt wird (Cronbachs Alpha, wenn Item gelöscht wird).

### 4.2.2 *Hauptkomponentenanalyse (PCA)*

Die PCA zeigt, dass SQ1_Thought, SQ2_Goal_setting, SQ4_Act_freely und SQ5_Achieve_goals als ein Konstrukt angesehen werden können, das die Varianz von 71, 75% in den ursprünglichen Variablen ausmacht (Tabelle 9). Dies kann man auch auf dem Bildschirmdiagramm sehen, das die Eigenwerte der Komponenten (Abbildung 8) anzeigt. Basiert auf der Komponentenmatrix, erkennen wir, dass *SQ3_Moral_rules* im Vergleich mit den anderen vier Agentenitems eine separate Komponente darstellt.

### 4.2.3 *Korrelationsanalyse*

Die Korrelation des Agency-Konstrukts mit den abhängigen Variablen ist gering und nur für *DVQ1_Blame, DVQ2_CommanderBlame, DVQ4_CommanderTrust, DVQ6_CommanderHarm und DVQ8_Support_significant* ($p < .01$) (Tabelle 10) vorhanden. Einige unerwartete Auswirkungen sind mit Bezug auf die Richtung, die die Zusammenhänge nehmen, zu beobachten. Zum Beispiel ist zu erwarten, dass die Auffassung und moralische Wertvorstellung dazu führt, dass Personen geneigt sind, einer Drohne ($r = .183$) mehr Schuld beizumessen, als dem Kommandanten ($r = .239$). Der gleiche Effekt kann bei der Schaden-Variable beobachtet werden. Die Korrelationsanalyse ist nicht detailliert genug, um in diese Ergebnisse zu zoomen, deshalb wird dies bei der Analyse der abhängigen Variablen getan.

TABELLE 7 Szenarien der Vorstudie 1

| | | Status des Kampfmittels | | | | |
|---|---|---|---|---|---|---|
| | | Autonome Drohne | | | Von Menschen bediente Drohne | |
| | | Keine Handlungsfähigkeit | Niedrige Handlungsfähigkeit | Starke Handlungsfähigkeit | Niedrige Handlungsfähigkeit | Starke Handlungsfähigkeit |
| Resultat | Kein Kollateralschaden (gut) | 1 | 2 | 3 | 4 | 5 |
| | Kollateralschaden (schlecht) | 6 | 7 | 8 | 9 | 10 |

TABELLE 8 Ergebnisse der Reliabilitätsanalyse Agentenitems Vorstudie 1

| **Reliabilitätstatistiken** | | |
|---|---|---|
| **Cronbachs Alpha** | **Cronbachs Alpha basiert auf Standar- disierte Items** | **Anzahl Items** |
| 900 | 895 | 5 |

**Item-Gesamtstatistiken**

| | **Skalen-Mittelwert wenn Item gelöscht** | **Skalen-Varianz, wenn Item gelöscht** | **Korrigierter Item-Gesamtwert Korrelation** | **Ausgeglichene Multiple Korrelation** | **Cronbachs Alpha, wenn Item gelöscht** |
|---|---|---|---|---|---|
| SQ1_Denken | 167,77 | 13710.694 | .841 | .741 | .857 |
| SQ2_Zielsetzung | 163.85 | 13440.002 | .857 | .775 | .853 |
| SQ3_Moralische Regeln | 183.39 | 17599.970 | .448 | .215 | .933 |
| SQ4_Freies Handeln | 163.22 | 13885.447 | .824 | .695 | .861 |
| SQ5_Zielerreichung | 157.21 | 13891.495 | .794 | .681 | .868 |

### 4.2.4 *Manipulationsprüfung der Agenten*

Die Grafik auf Abbildung 9 stellt dar, dass unsere Agentenmanipulation funktioniert, da bei dem Zustand von geringer Selbstbestimmung für autonome Waffensysteme Personen weniger Selbstbestimmung wahrnehmen, als bei dem Zustand mit hoher Selbstbestimmung, die von dem Mittelwert des Agency-Konstrukts auf der y-Achse dargestellt wird. Im neutralen Zustand ist die Auffassung und moralische Wertvorstellung etwas höher als bei dem Zustand der geringen Selbstbestimmung, aber immer noch niedriger als bei einem Zustand mit hoher Selbstbestimmung. Die unabhängigen t-Tests der Stichproben weisen auf, dass AW mit geringer Selbstbestimmung und AW mit hoher Selbstbestimmung besondere Gruppen bilden ($p = .000$), aber das ist bei AW von neutralem Zustand und bei denen mit wenig Selbstbestimmung nicht der Fall

TABELLE 9 Ergebnis-PCA Agency-Items Vorstudie 1

| Komponentenmatrix[a] | | | |
|---|---|---|---|
| **Komponente** | | | |
| | **1** | **2** | **3** |
| SQ2_Zielsetzung | .922 | | |
| SQ1_Denken | .911 | | |
| SQ4_Freies Handeln | .899 | | |
| SQ5_Zielerreichung | .882 | | |
| SQ3_Moralische Regeln | .571 | .819 | |

Extraktionsmethode: Hauptkomponentenanalyse.
[a]. 3 Komponenten wurden extrahiert.

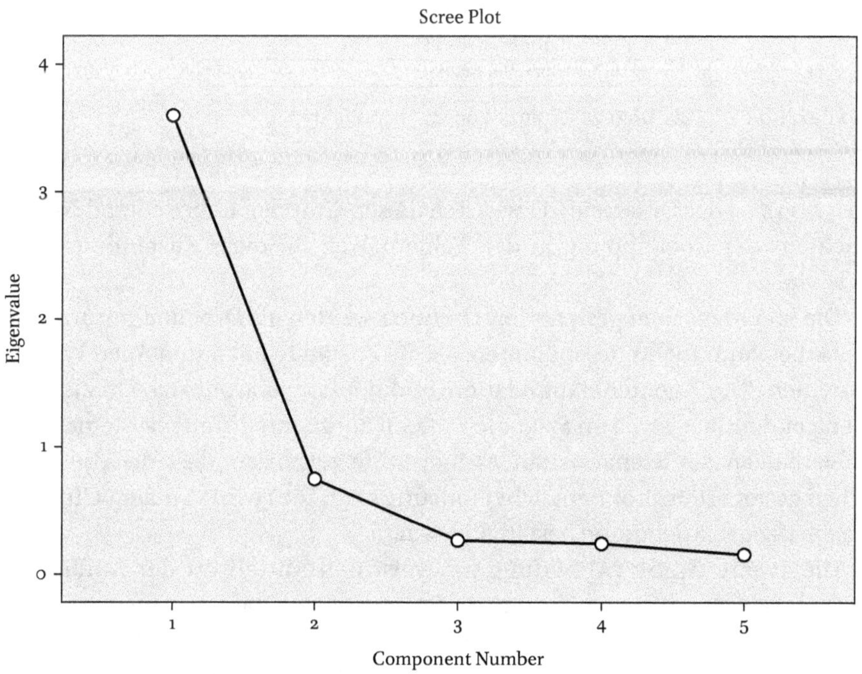


ABBILDUNG 8 Scree-Test PCA Vorstudie 1

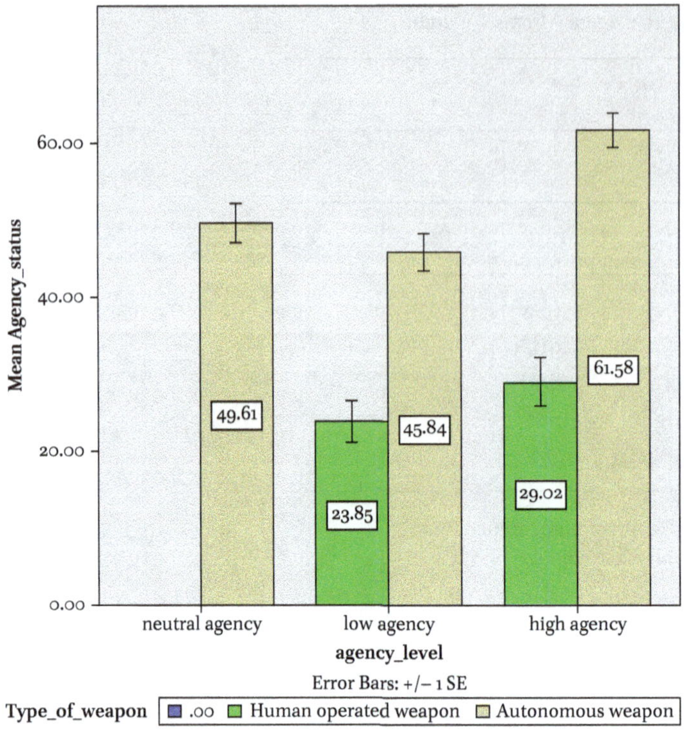


ABBILDUNG 9 Mittelwert des Agency-Konstrukts Waffentyp

(p = .209), da der Unterschied zwischen diesen Gruppen unbedeutend ist, was auch aus der Überlappung in den Fehlerbalken für diese Zustände ersichtlich wird.

Die von Menschen gesteuerten Drohnen werden als Drohnen mit weniger Selbstbestimmung wahrgenommen, als die Zustände mit autonomen Waffensystemen. Die Agentenmanipulation funktioniert, obwohl der Unterschied weniger deutlich ist als im Szenario mit den autonomen Waffensystemen. Die Fehlerbalken der Szenarios auf Agency-Stufe zeigen an, dass die von Menschen gesteuerten Drohnen nicht eindeutig sind. Dies wird von dem t-Test der unabhängigen Stichprobe bestätigt (p = .125).

Die zweite Grafik (Abbildung 10) weist den Mittelwert der Auffassung und moralischen Wertvorstellung in Bezug auf AW über den Agency-Stufen für gute und schlechte Resultate auf. Diese Art eines vom Waffensystem verursachtes Ergebnisses, ob Kollateralschaden oder nicht, scheint nicht zu großen Unterschieden in der Auffassung und moralischen Wertvorstellungen von Kampfmitteln zu führen, da der Mittelwert sich nur um 2 bis 3 Punkte (auf einer 100-Punkte-Skala) unterscheidet. Die Fehlerbalken

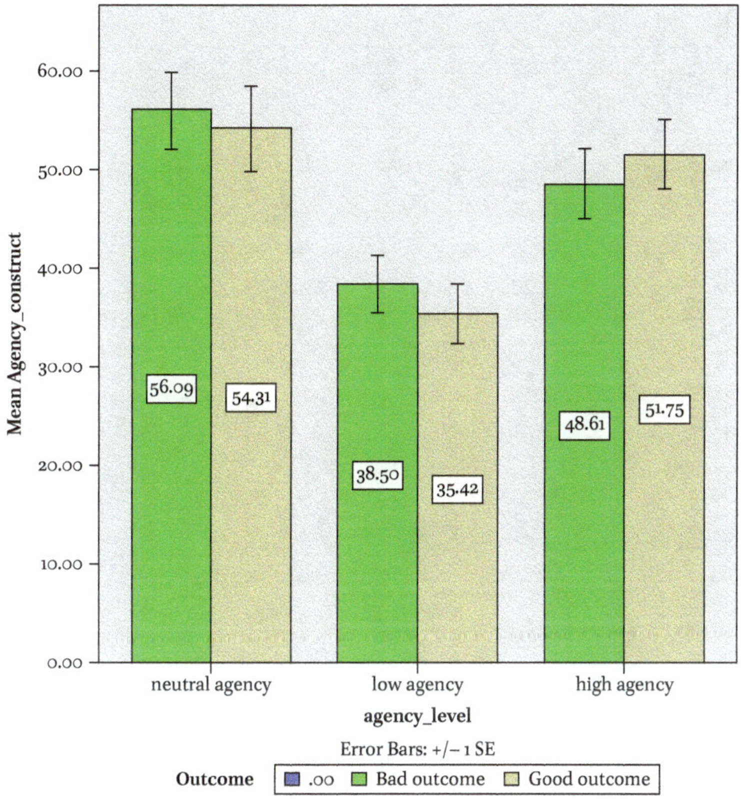


ABBILDUNG 10 Mittelwert des Agency-Konstrukts pro Zustand pro Resultat.

überlappen außerdem auf allen Agency-Stufen. Dies wird von den t-Tests der unabhängigen Stichprobe bestätigt. Auf der neutralen Agency-Stufe beträgt die Signifikanz zwischen dem schlechten Resultat und dem guten Resultat $p = .816$, auf der niedrigen Agency-Stufe ist der Zustand $p = .641$ und auf der hohen Agency-Stufe $p = .575$, die alle den $p < .05$ Schwellenwert überschreiten.

### 4.2.5 *Analyse der abhängigen Variablen*

Die Korrelationsanalyse weist auf, dass nur die abhängigen Variablen *Schuld, Kommandantenschuld, Kommandantenvertrauen, Kommandantenschaden und Unterstützung* zu einer Minimalebene von $p < .05$ von Bedeutung waren. Auf der nächsten Stufe vergrößerten wir diese Variablen und sahen sie uns näher an, aber da sie nicht bedeutend waren, führen wir keine Analyse der restlichen abhängigen Variablen durch. Bei genauerer Betrachtung prüften wir, ob es Unterschiede zwischen den verschiedenen Agency-Stufen (niedrig,

TABELLE 10 Korrelationsmatrix des Agency-Konstrukts und abhängigen Variablen.

| | | Agency_ construct | DVO1_ Blame | DVO2_ Com-mander-Blame | DVo3_ Trust | DVo4_ Comman-derTrust | DVo5_ Harm | DVo6_ Com-mander Harm | DVo7_ Uneasy | DVo8_ Support | DVo9_ Fairness |
|---|---|---|---|---|---|---|---|---|---|---|---|
| Agency_ construct | Person Correla-tion | 1 | .183** | -.239** | .070 | .176** | -.011** | -.227** | .005 | .101 | .087 |
| | Sig(2-tailed | | .000 | .000 | .114 | .000 | .798 | .000 | .914 | .023 | .050 |
| | N | 510 | 510 | 510 | 510 | 510 | 510 | 510 | 510 | 510 | 510 |
| DVo1_ Blame | Person Correla-tion | 183** | 1 | -101* | .020 | 0.53 | .198** | -.170 | .129** | -.126** | -078 |
| | Sig(2-tailed | .000 | | .023 | .660 | .231 | .000 | .000 | .003 | .006 | .077 |
| | N | 510 | 510 | 510 | 510 | 510 | 510 | 510 | 510 | 510 | 510 |
| DVO2_ Comman-derBlame | Person Correla-tion | -.239** | -.101* | 1 | -.264** | -.330** | .264** | .692** | -.001 | -.333** | -.036 |
| | Sig(2-tailed | .000 | .023 | | .000 | .000 | .000 | .000 | .982 | .000 | .422 |
| | N | 510 | 510 | 510 | 510 | 510 | 510 | 510 | 510 | 510 | 510 |

Correlations

| | | | | | | | | | | | |
|---|---|---|---|---|---|---|---|---|---|---|---|
| DV03_ Trust | Person Correla-tion | .070 | .020 | -.264** | 1 | .635** | -.294** | -.262** | .169** | .674** | -.106 |
| | Sig(2-tailed | .114 | .660 | .000 | | .000 | .000 | .000 | .000 | .000 | .016 |
| | N | 510 | 510 | 510 | 510 | 510 | 510 | 510 | 510 | 510 | 510 |
| DV04_ Comman-derTrust | Person Corre-la.630tion | .176** | .053 | -.330** | .635** | 1 | -171** | -.322** | .138** | .584** | .021 |
| | Sig(2-tailed | .000 | .232 | .000 | .000 | | .000 | .000 | .002 | .000 | .630 |
| | N | 510 | 510 | 510 | 510 | 510 | 510 | 510 | 510 | 510 | 510. |
| DV05_ Harm | Person Correla-tion | -.011 | .198** | .264** | -294** | -.171** | 1 | .361** | -.044 | -.336** | 089 |
| | Sig(2-tailed | .798 | .000 | .000 | .000 | .000 | | .000 | .323 | .000 | .046 |
| | N | 510 | 510 | 510 | 510 | 510 | 510 | 510 | 510 | 510 | 510 |
| DV06_ Com-mander Harm | Person Correla-tion | -.227** | -.170** | .692** | -.262** | -.322** | .361** | 1 | -.038 | -.322 | .041 |
| | Sig(2-tailed | .000 | .000 | .000 | .000 | .000 | .000 | | .394 | .000 | .351 |
| | N | 510 | 510 | 510 | 510 | 510 | 510 | 510 | 510 | 510 | 510 |
| DV07_ Uneasy | Person Correla-tion | .005 | -.129 | -.001 | .169** | .138** | -.044 | -.038 | 1 | .127-- | -.689** |
| | Sig(2-tailed | .914 | .003 | .982 | .000 | .002 | .321 | .394 | | .004 | .000 |
| | N | 510 | 510 | 510 | 510 | 510 | 510 | 510 | 510 | 510 | 510 |

TABELLE 10 Korrelationsmatrix des Agency-Konstrukts und abhängigen Variablen (*cont.*)

**Correlations**

| | | | | | | | | | | | |
|---|---|---|---|---|---|---|---|---|---|---|---|
| DV08_ Support | Person Correlation | .101* | -.120** | -.333** | .674** | .584** | -.336** | -.322** | .127** | 1 | -.004 |
| | Sig(2-tailed | .023 | .006 | .000 | .000 | .000 | .000 | .000 | .004 | | .934 |
| | N | 510 | 510 | 510 | 510 | 510 | 510 | 510 | 510 | 510 | 510 |
| DV09_ Fairness | Person Correlation | .087 | -078 | -.036 | -.106* | 0.21 | .089 | .041 | -.689** | -.004 | 1 |
| | Sig(2-tailed | .050 | .077 | .422 | .016 | .630 | .046 | .351 | .000 | .934 | |
| | N | 510 | 510 | 510 | 510 | 510 | 510 | 510 | 510 | 510 | 510 |

* correlation is significant at the 0.05 level (2-taild)
** correlation is significant at the 0.01 level (2-taild)

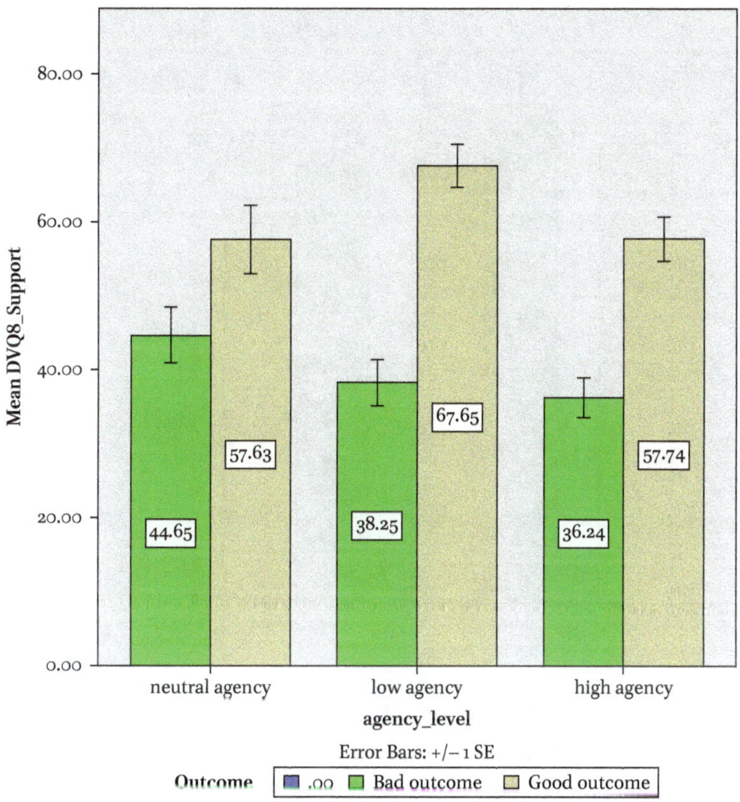


ABBILDUNG 11 Mittelwert der Unterstützungsvariable pro Zustand pro Resultat

neutral und hoch) und den autonomen Waffensystemen und von Menschen gesteuerten Drohnen gab.

#### 4.2.5.1 Unterstützungsvariable

Die interessanteste Beobachtung ist der große Unterschied in der Unterstützung zwischen guten und schlechten Resultaten (Abbildung 11), die für alle drei Zustände bedeutend sind ($p < .05$). Dieser gravierende Unterschied im Mittelwert der Unterstützung für die drei Zustände zeigt auf, dass die Manipulation unserer Resultate funktioniert. Eine weitere interessante Beobachtung ist, dass diese Unterstützung abgeschwächt wird, so wie die Agency-Stufen ansteigen, aber die Differenzen zum Mittelwert sind gering (< 10 Punkte auf einer Skala von 100 Punkten) und die |-förmigen Balken oben, mit dem Standardfehler überlappen. Das zeigt an, dass diese Gruppen sich nicht bedeutend voneinander unterscheiden, entweder wenn mit dem guten oder mit den schlechten Resultatszenarien verglichen. Das zweite Diagramm zeigt an, dass

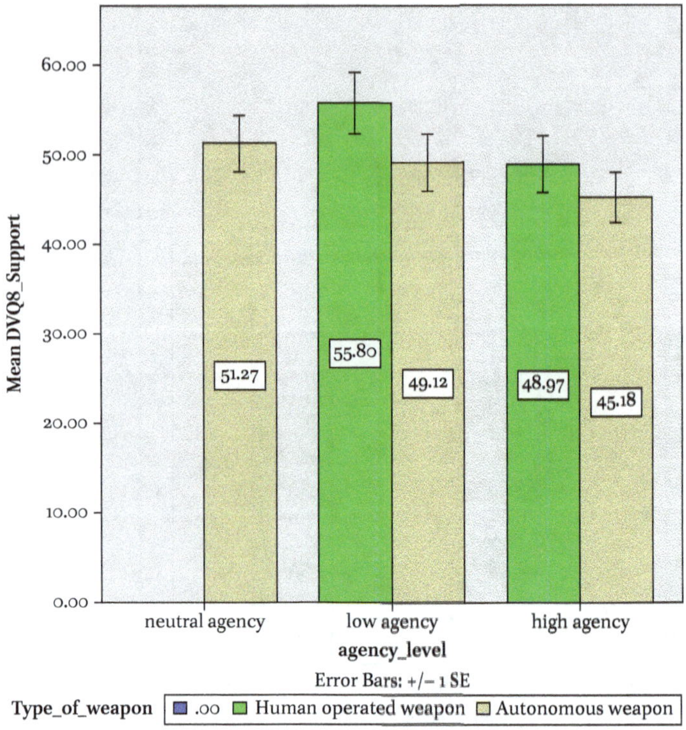


ABBILDUNG 12 Mittelwert der Unterstützungsvariable pro Waffentyp

die Unterstützung unter Zuständen mit von Menschen gesteuerten Waffen höher ist, als für solche mit autonomen Waffensystemen. auch hier sind die Differenzen in der Gruppe nur gering und von Signifikanz, da die Fehlerbalken überlappen.

### 4.2.5.2 Vertrauensvariable

Auf Abbildung 12 ist zu sehen, dass Leute weniger Vertrauen darin setzen, dass autonome Waffensysteme zukünftig die korrekten Maßnahmen nach einem schlechten Resultat treffen, als ein menschlicher Operateur, der Maßnahmen nach einem ebenso schlechten Resultat trifft, und alle Ergebnisse sind für ihn von Bedeutung. Dieser Effekt wird in Zuständen mit geringer Selbstbestimmung verstärkt. Da sehen wir fast kaum einen Unterschied im Vertrauen zum Menschen nach einem guten oder schlechten Resultat und einen großen Unterschied im Vertrauen in ein AW (Abbildung 13). Allerdings zeigte die eindimensionale Analyse, dass die Interaktion nicht signifikant ist: *Type_of_weapon*outcome* p = .120, das bedeutet also, dass wir an dieser Stelle in der

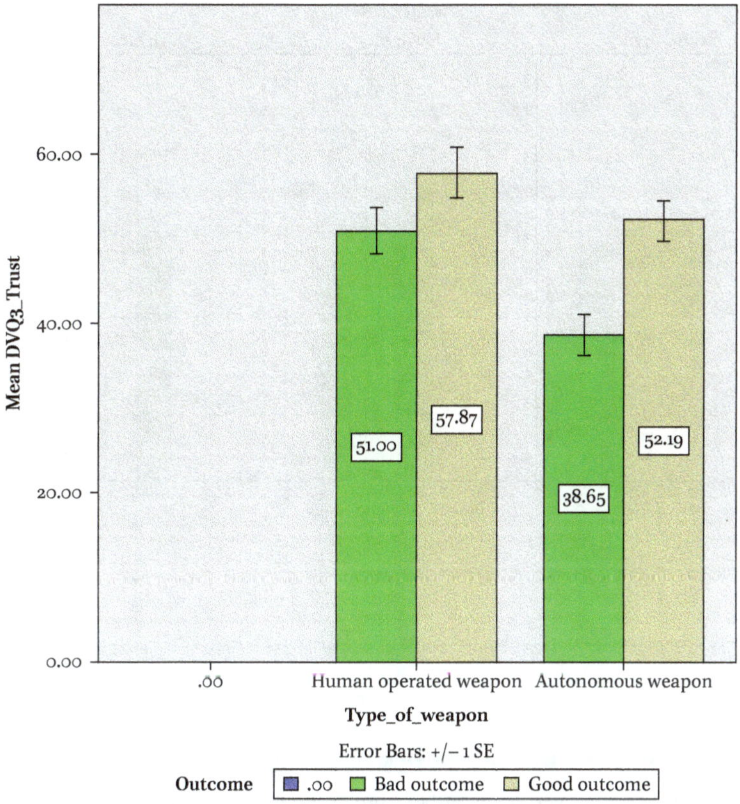


ABBILDUNG 13 Mittelwert der Vertrauensvariable pro Waffentyp pro Resultat

Untersuchung keine Schlussfolgerungen ziehen können, doch es ist ein interessantes Ergebnis, das sich lohnt, gründlicher untersucht zu werden.

#### 4.2.5.3 Schuld- und Schadenvariablen

Um die Kette der Verantwortung zu untersuchen, werden die Variablen *Schuld, Kommandantenschuld, Schaden und Kommandantenschaden* auf dem gleichen Diagramm (Abbildung 15) dargestellt. Alle Ergebnisse der Schuldvariablen (Abbildung 15), mit Ausnahme des neutralen Agenten im autonomen Waffensystem-Zustand, sind von Bedeutung. Diese Ergebnisse weisen nach, dass in Zuständen mit geringer Selbstbestimmung die Schuld dem Kommandanten zugeschoben und in Zuständen mit hoher Selbstbestimmung, sie der Drohne gegeben wird.

Eine bemerkenswerte Beobachtung ist, dass Zustände mit geringer Selbstbestimmung erstaunlich viel kausale Verantwortung für den Schaden

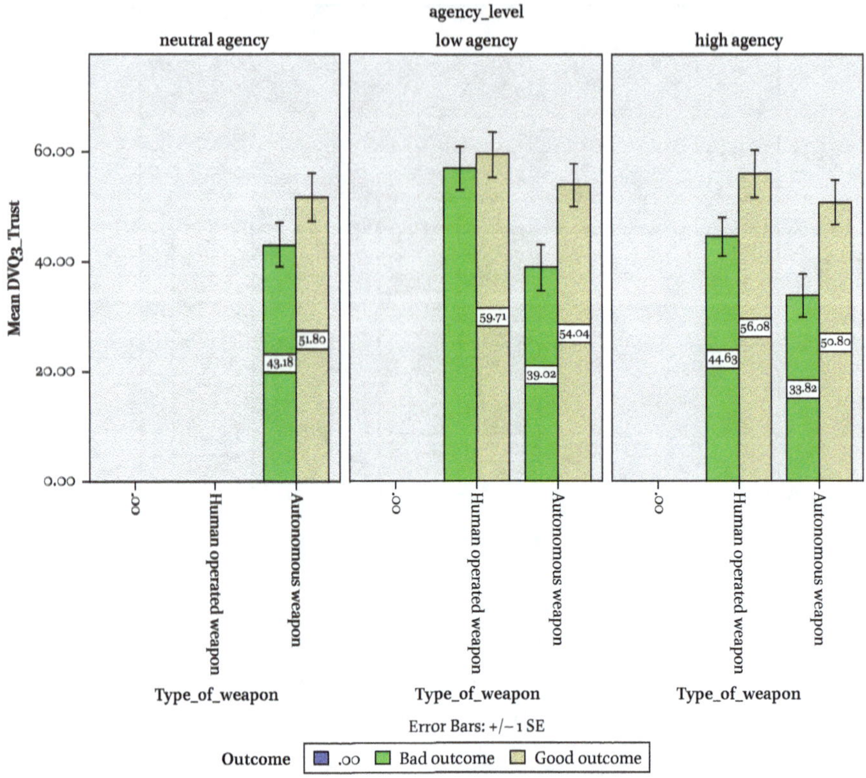


ABBILDUNG 14 Mittelwert der Vertrauensvariable pro Zustand pro Resultat

übernehmen. Dieses Muster gilt für von Menschen gesteuerten Drohnen wie auch für autonome Waffensysteme (Abbildung 16). Im Gegensatz zur Schuldvariable, betrachtet man die Drohne als gleichviel Schaden anrichtend, ob in Zuständen mit hoher oder geringer Selbstbestimmung, aber in Zuständen mit geringer Selbstbestimmung wird der Schaden fast gleich dem Kommandanten zugeschoben. Das bedeutet, dass beide gleichviel Schaden anrichten, aber in Zuständen mit hoher Selbstbestimmung wird er auf den Kommandanten übertragen, und die Drohne scheint den Schaden verursacht zu haben. Allerdings ist der Unterschied zwischen den Variablen Schaden und Kommandantenschaden in Zuständen mit niedriger Selbstbestimmung in Szenarien von sowohl von Menschen gesteuerten als auch autonomen Waffensystemen nicht bedeutend, deshalb müssen wir mit unserer Schlussfolgerung zum beobachteten Effekt vorsichtig sein.

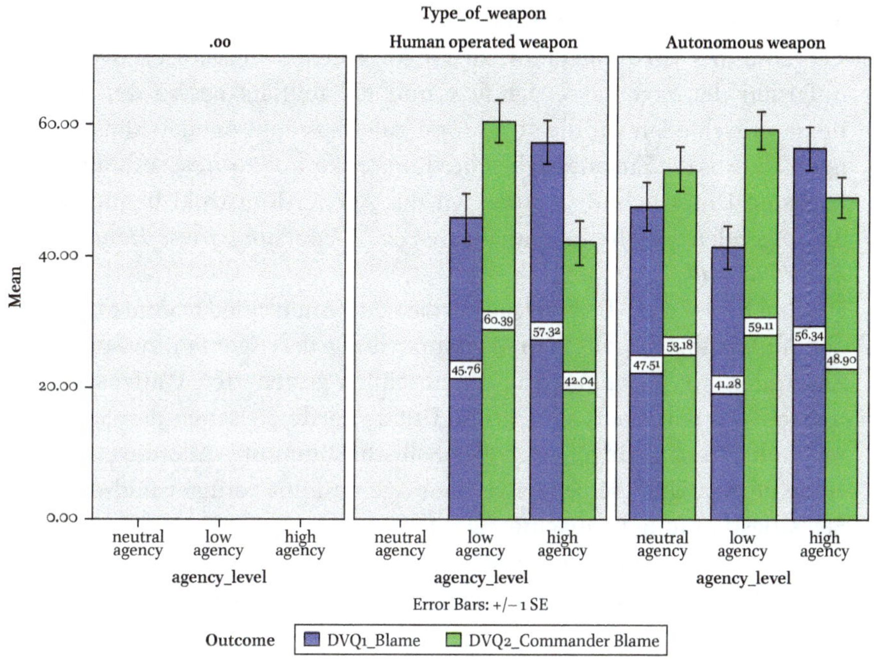


ABBILDUNG 15 Mittelwert von Schuld- und Kommandantenschuldvariablen pro Waffentyp

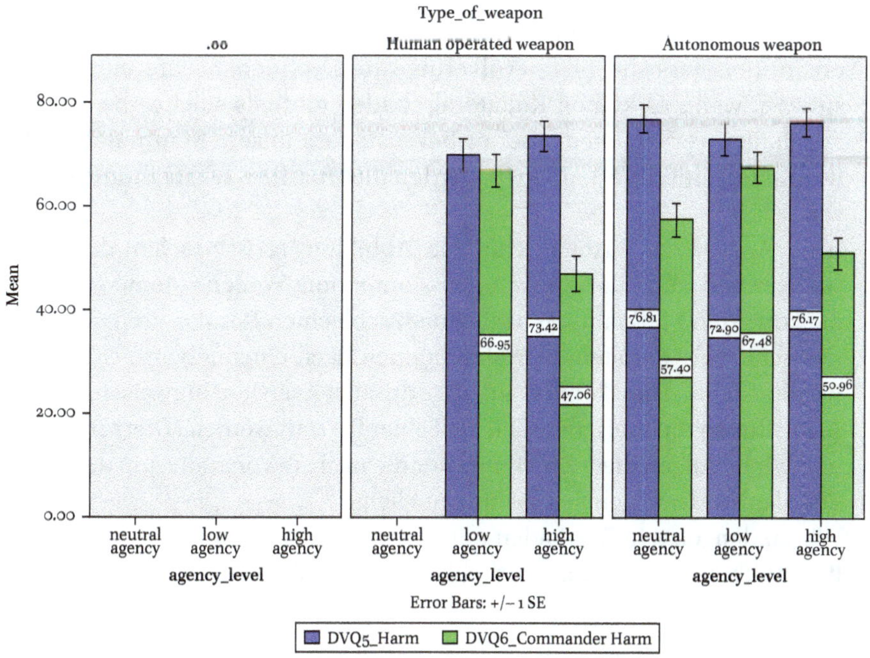


ABBILDUNG 16 Mittelwert von Schaden und Kommandantenschaden pro Zustand und Waffentyp

### 4.2.6 *Schlussfolgerung der Vorstudie 1*

Das Ergebnis der Vorstudie 1 führt zu den folgenden Schlussfolgerungen:

- Aufgrund der Ergebnisse der PCA und Reliabilitätsanalyse der Agency-Items, entschieden wir die SQ3_Moral_rules aus dem Agency-Konstrukt zu nehmen, was die Zuverlässigkeit des Konstrukts auf $\alpha = 0.933$ erhöht. In der nächsten Untersuchung werden wir das Agency-Konstrukt benutzen, das die folgenden vier Items enthält: *Denken, Zielsetzung, freies Handeln und Zielerreichung.*
- Die Ergebnisse der t-Tests zeigen an, dass ein Unterschied in dem Mittelwert der Auffassung moralischen Wertvorstellung der Agenten zwischen den Zuständen mit autonomen und menschlich gesteuerten Waffensystemen herrscht. Es herrschen auch große Unterschiede zwischen den Szenarien mit hoher und geringer/neutraler Selbstbestimmung. Allerdings wurden keine großen Unterschiede zwischen Agenten mit geringer und neutraler Selbstbestimmung beobachtet. Deshalb entschieden wir, den Zustand des neutralen Agenten in der nächsten Untersuchung fallen zu lassen und nur die Zustände von Agenten mit geringer und hoher Selbstbestimmung zu testen.
- Die eindeutigste Beobachtung in den Unterstützungsvariablen ist, dass es einen großen Unterschied in der Unterstützung zwischen guten und schlechten Resultaten gibt (Abbildung 11). Allerdings ist dies laut der Fachliteratur zu erwarten, da James Igoe Walsch (2015) und Kreps (2014) erkannten, dass die Zivilbevölkerung die Drohnenangriffe eher unterstützen, wenn es keinen Kollateralschaden gibt, als solche, die Kollateralschaden verursachen, aber es beweist, dass unsere Manipulation des Resultats funktioniert, und wir werden dies für die nächste Studie benutzen.
- In der Analyse der Vertrauensvariable (Abbildung 13) ist zu sehen, dass Leute weniger Vertrauen darin setzen, dass autonome Waffensysteme zukünftig die korrekten Maßnahmen nach einem schlechten Resultat treffen, als ein menschlicher Operateur, der Maßnahmen nach einem ebenso schlechten Resultat trifft. Dies könnte auf Algorithmus-Aversion hinweisen, für die Algorithmen mehr als menschliche Fehler bestraft werden (Dietvorst, 2016) und wir benutzen diese Ergebnisse für die nächste Vorstudie, um den Effekt der Algo-Aversion genauer zu untersuchen.
- Wir fanden, dass in Zuständen mit hoher Selbstbestimmung eine erstaunlich hohe kausale Verantwortlichkeit für den Schaden zu sehen ist und

dass dieses Muster für von Menschen gesteuerten Drohnen und autonomen Waffensystemen gleich stichhaltig ist (Abbildung 16). Wir fanden auch, dass die Drohne in Zuständen mit hoher wie auch geringer Selbstbestimmung gleichviel Schaden anzurichten scheint, aber in Zuständen mit geringer Selbstbestimmung wird der Schaden zum gleichen Anteil dem Kommandanten zugesprochen. Obwohl diese Erkenntnisse nicht von Bedeutung sind, sind sie interessant und erfordern weitere Untersuchung, deshalb haben wir diese Fragen in Vorstudie 2 beibehalten.

## 4.3 Vorstudie 2

In der Vorstudie 2 versuchten wir, unsere Kenntnisse des Algo-Aversion-Effekts zu verbessern, deshalb testeten wir Szenarien, die beschrieben, wie das autonome Waffensystem aus seinen Fehlern lernen kann (3 + 10), es mit einem großen Datensatz programmiert wurde, und dass es begreift, (4 + 11) dass sein Einsatz gelegentlich unberechenbar sein kann (5 + 12) sowie ein Szenario, dass all diese Aspekte einschließt (6 + 13). Wir testeten Zustände mit guten und schlechten Resultaten und fügten ein Szenario für autonome Waffensysteme mit geringer Selbstbestimmung (1 + 8) hinzu, um zu prüfen, ob unsere Agentenmanipulation funktioniert, sowie ein von Menschen gesteuertes Szenario (7 + 14), in der wir uns über den menschlichen Operateur kundig machten, statt der Drohne, im Unterschied zu Vorstudie 1.

Wir testeten insgesamt 14 Szenarien und zielten darauf, 50 Umfrageteilnehmer pro Szenario zu erhalten, was zu insgesamt 700 Antworten führen würde. Die Anzahl der Befragten pro Szenario betrug zwischen 47 und 54. Nach der Datenvorverarbeitung und dem Löschen von Befragten, die den Aufmerksamkeitstest nicht durchgeführt hatten, verfügten wir noch über 709 komplett ausgefüllten Fragebögen.

### 4.3.1 *Reliabilitätsanalyse*

Für alle vier Agency-Items ist $\alpha = 0.914$; das liegt weit über dem Schwellenwert von 0,7 und kann nicht durch Löschen eines Items verbessert werden. Löschen anderer Items wird den $\alpha$ Wert abschwächen und die interne Konsistenz reduzieren.

TABELLE 11 Szenarien der Vorstudie 2

| | | Aspekte | | | | | | |
|---|---|---|---|---|---|---|---|---|
| | | **Autonome Drohne** | | | | | | **MG** |
| | | **Ebene 1 – Geringe Selbstbestimmung** | **Ebene 2 (hohe Selbstbestimmung + nichts zusätzlich)** | **Ebene 3 (hohe Selbstbestimmung + Lernfähigkeit)** | **Ebene 4 (hohe Selbstbestimmung + Verständigungsverfahren)** | **Ebene 5 (hohe Selbstbestimmung + Unberechenbarkeit)** | **Ebene 6 (hohe Selbstbestimmung + alle Aspekte)** | **Von Menschen gesteuerte Drohne** |
| **Resultat** | Kollateralschaden (schlecht) | 1 | 2 | 3 | 4 | 5 | 6 | 7 |
| | Kein Kollateralschaden (gut) | 8 | 9 | 10 | ii | 12 | 13 | 14 |

TABELLE 12 Ergebnisse der Reliabilitätsanalyse Vorstudie 2

| Reliabilitätsstatistiken | | |
|---|---|---|
| **Cronbachs Alpha** | **Cronbachs Alpha basiert auf Standardisierte Items** | **Anzahl Items** |
| .914 | .914 | 4 |

| Item-Gesamtstatistiken | | | | | |
|---|---|---|---|---|---|
| **Skalen-Mittelwert, wenn Item gelöscht** | | **Skalen-Varianz, wenn Item gelöscht** | **Korrigierte Item-Gesamtwert Korrelation** | **Ausgeglichene Multiple Korrelation** | **Cronbachs Alpha, wenn Item gelöscht** |
| Denken | 208.4827 | 6427.511 | .801 | .647 | .890 |
| Zielsetzung | 204.9885 | 6499.803 | .834 | .695 | .877 |
| Freies Han deln | 203.4020 | 6782.379 | .804 | .646 | .888 |
| Zielerreichung | 199.4975 | 7114.019 | .781 | .614 | .897 |

### 4.3.2 *Hauptkomponentenanalyse* (PCA)

Die PCA zeigt, dass *Denken, Zielsetzung, freies Handeln und Zielerreichung* als ein Konstrukt angesehen werden kann, das die Varianz von 79,60% in den ursprünglichen Variablen ausmacht (Tabelle 13). Dies kann man auch auf dem Bildschirmdiagramm ansehen, das die Eigenwerte der Komponenten (Abbildung 17) anzeigt. Die PCA zeigt an, dass die vier Items: Denken, Zielsetzung, freies Handeln und Zielerreichung als die zugrundeliegenden Komponenten des Agency-Konstrukts Tabelle 13 gesehen werden kann.

### 4.3.3 *Korrelationsanalyse*

Die Korrelationsanalyse des Agency-Konstrukts mit den abhängigen Variablen ist nicht umfangreich, aber für alle Konstrukte von Bedeutung (p <,05) (Tabelle 14, Seite 282–285). Die gleichen Effekte auf die Richtung für die

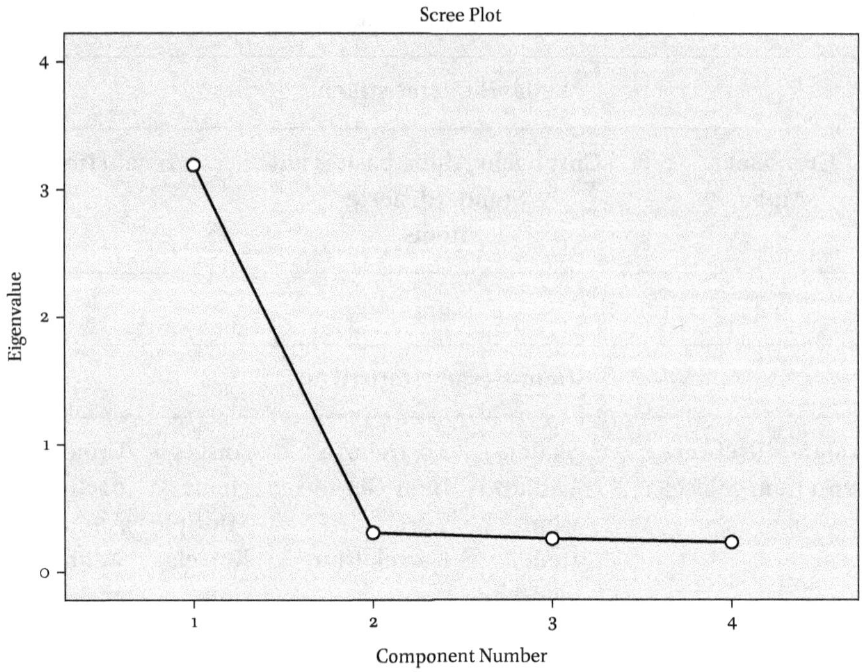


ABBILDUNG 17 Scree-Test PCA Vorstudie 2

TABELLE 13 Ergebnisse-PCA Vorstudie 2

| Komponentenmatrix[a] | | |
|---|---|---|
| **Komponente** | | |
| | **1** | **2** |
| Zielsetzung | .910 | |
| Freies Handeln | .892 | |
| Denkweise | .889 | |
| Zielerreichung | .877 | .446 |

Extraktionsmethode: Haupt-komponentenanalyse.
[a] 2 Komponenten wurden extrahiert.

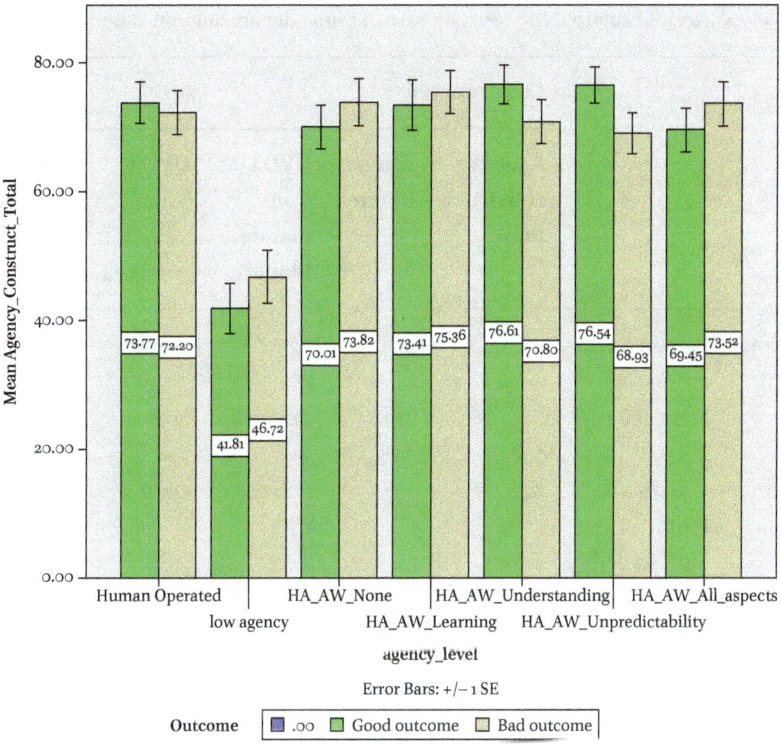


ABBILDUNG 18 Mittelwert des Wert-Agency-Konstrukts pro Zustand pro Resultat

Schaden- und Schuldvariable wie in Vorstudie 1 können auch in dieser Vorstudie beobachtet werden. Es besteht ein negativer Zusammenhang zwischen dem Agency-Konstrukt und der Stufe der Unruhe, d.h. dass bei Erhöhung der Agentenwahrnehmung, die Leute andeuten, dass sie eine größere *Besorgnis* empfinden, aber dieser Effekt ist sehr gering (r = -,076). Die Korrelationsanalyse ist nicht detailliert genug, um in diese Ergebnisse zu zoomen, deshalb wird dies bei der Analyse der abhängigen Variablen getan.

### 4.3.4 *Manipulationsprüfung der Agenten*

Das Diagramm in Abbildung 18 zeigt an, dass ein großer Unterschied zwischen den Agency-Szenarien mit hoher Selbstbestimmung und denen mit geringer Selbstbestimmung herrscht, und dieser Unterschied ist von Bedeutung. Der Unterschied zwischen geringen und hohen Agentenzuständen bestätigt, dass unsere Agentenmanipulation funktioniert. Agenten des menschlichen Operateurs und Drohnen mit hoher Selbstbestimmung befinden sich auf der gleichen Ebene (in dieser Vorstudie fragten wir speziell den menschlichen Operateur,

TABELLE 14 Korrelationsmatrix des Agency-Konstrukts und der abhängigen Variablen Vorstudie 2

| | | Correlations | | | | |
|---|---|---|---|---|---|---|
| | | **Agency_ construct_ total** | **DVD1_ Blame** | **DVD2_ Commander Blame** | **DVD3_ Trust** | **DVD4_Commander Trust** |
| Agency_Construct_Total | Person Correlation | 1 | .271** | -.267** | .205** | .217** |
| | Sig(2-tailed) | | .000 | .000 | .000 | .000 |
| | N | 708 | 708 | 708 | 708 | 708 |
| DVD1_Blame | Person Correlation | .271 | 1 | -.069 | -.028 | -.046 |
| | Sig(2-tailed) | .000 | | .045 | .464 | .226 |
| | N | 708 | 708 | 708 | 708 | 708 |
| DVD2_ Commander Blame | Person Correlation | -.267** | -.069 | 1 | -.287** | -.278** |
| | Sig(2-tailed) | .000 | .065 | | .000 | .000 |
| | N | 708 | 708 | 708 | 708 | 708 |
| DVD3_Trust | Person Correlation | .205** | -.028 | -.207** | 1 | .653** |
| | Sig(2-tailed) | .000 | .464 | .000 | | .000 |
| | N | 708 | 708 | 708 | 708 | 708 |
| DVD4_Commander Trust | Person Correlation | .217** | -.046 | -.278** | .653** | 1 |
| | Sig(2-tailed) | .000 | .226 | .000 | .000 | |
| | N | 708 | 708 | 708 | 708 | 708 |
| DVD5_Harm | Person Correlation | .124** | .258** | .152** | -.281** | -.185** |
| | Sig(2-tailed) | .001 | .000 | .000 | .000 | .000 |
| | N | 708 | 708 | 708 | 708 | 708 |
| DVD6_Commander Hram | Person Correlation | -.198** | -.023 | .691** | -.224** | -.228** |
| | Sig(2-tailed) | .000 | .534 | .000 | .000 | .000 |
| | N | 708 | 708 | 708 | 708 | 708 |

| DVD5_ Harm | DVD6_ Commander Hram | DVD7_ Dignity | DVD8_ confi-dence | DVD9_ Expecta-tions | DVD10_ Support | DVD11_ Fair | DVD12_ Uneasy |
|---|---|---|---|---|---|---|---|
| .124** | -.198** | .102** | .209** | .146** | .164** | .142** | -.076** |
| .001 | .000 | .007 | .000 | .000 | .000 | .000 | .043 |
| 708 | 708 | 708 | 708 | 708 | 708 | 708 | 708 |
| .258** | -.023 | -.056 | -.048 | -.137** | -.118** | -.092* | .180** |
| .000 | .524 | -.138 | .198 | .000 | .002 | .014 | .000 |
| 708 | 708 | 708 | 708 | 708 | 708 | 708 | 708 |
| .152** | .691** | -.206** | -.300** | -.205** | -.250** | -.275** | .312** |
| .000 | .000 | .000 | .000 | .000 | .000 | .000 | .000 |
| 708 | 708 | 708 | 708 | 708 | 708 | 708 | 708 |
| -.281** | -.224** | .595** | .901** | .533** | .726** | .722** | -.616* |
| .000 | .000 | .000 | .000 | .000 | .000 | .000 | .000 |
| 708 | 708 | 708 | 708 | 708 | 708 | 708 | 708 |
| -.185** | -.228** | .432** | .635** | .425** | .631** | .552** | -.458** |
| .000 | .000 | .000 | .000 | .000 | .000 | .000 | .000 |
| 708 | 708 | 708 | 708 | 708 | 708 | 708 | 708 |
| 1 | .311** | -.374** | -.301** | -.217** | -.253** | -.324** | .395** |
| .000 | .000 | .000 | .000 | .000 | .000 | .000 | .000 |
| 708 | 708 | 708 | 708 | 708 | 708 | 708 | 708 |
| .311** | 1 | -.183** | -.226** | -.203** | -.235** | -.249** | .289** |
| .000 | | .000 | .000 | .000 | .000 | .000 | .000 |
| 708 | 708 | 708 | 708 | 708 | 708 | 708 | 708 |

TABELLE 14 Korrelationsmatrix des Agency-Konstrukts und der abhängigen Variablen Vorstudie 2 (*cont.*)

| | | **Correlations** | | | | |
|---|---|---|---|---|---|---|
| | | **Agency_construct_total** | **DVD1_Blame** | **DVD2_Commander Blame** | **DVD3_Trust** | **DVD4_Commander Trust** |
| DVD7_Dignity | Person Correlation | .102** | -.056 | -.206** | .595** | .432** |
| | Sig(2-tailed) | .007 | .138 | .000 | .000 | .000 |
| | N | 708 | 708 | 708 | 708 | 708 |
| DVD8_confidence | Person Correlation | .209** | -.048 | -.300** | .901** | .635** |
| | Sig(2-tailed) | .000 | .198 | .000 | .000 | .000 |
| | N | 708 | 708 | 708 | 708 | 708 |
| DVD9_Expectations | Person Correlation | .146** | -.137** | -.205** | .533** | .425** |
| | Sig(2-tailed) | .000 | .000 | .000 | .000 | .000 |
| | N | 708 | 708 | 708 | 708 | 708 |
| DVD10_Support | Person Correlation | .164** | -.118** | -.250** | .726** | .631** |
| | Sig(2-tailed) | .000 | .002 | .000 | .000 | .000 |
| | N | 708 | 708 | 708 | 708 | 708 |
| DVD11_Fair | Person Correlation | .142** | -.092* | -.275** | .722** | .552** |
| | Sig(2-tailed) | .000 | .014 | .000 | .000 | .000 |
| | N | 708 | 708 | 708 | 708 | 708 |
| DVD12_Uneasy | Person Correlation | -.076* | .190** | .312** | -.616** | -.458** |
| | Sig(2-tailed) | .043 | .000 | .000 | .000 | .000 |
| | N | 708 | 708 | 708 | 708 | 708 |

** Correlation is significant at the 0.01 level (2-tailed).
* Correlation is significant at the 0.05 level (2-tailed).

| DVD5_ Harm | DVD6_ Commander Hram | DVD7_ Dignity | DVD8_ confi-dence | DVD9_ Expecta-tions | DVD10_ Support | DVD11_ Fair | DVD12_ Uneasy |
|---|---|---|---|---|---|---|---|
| -.374** | -.183** | 1 | .588** | .415** | .533** | .598** | -.509** |
| .000 | .000 | | .000 | .000 | .000 | .000 | .000 |
| 708 | 708 | 708 | 708 | 708 | 708 | 708 | 708 |
| -.301** | -.226** | .588** | 1 | .553** | .723** | .725** | -.649** |
| .000 | .000 | .000 | | .000 | .000 | .000 | .000 |
| 708 | 708 | 708 | 708 | 708 | 708 | 708 | 708 |
| -.217** | -.203** | .415** | .553** | 1 | .511** | .621** | -.509** |
| .000 | .000 | .000 | .00 | | .000 | .000 | .000 |
| 708 | 708 | 708 | 708 | 708 | 708 | 708 | 708 |
| -.253** | -.235** | .533** | .723** | .511** | 1 | .698** | -.651** |
| .000 | .000 | .000 | .000 | .000 | | .000 | .000 |
| 708 | 708 | 708 | 708 | 708 | 708 | 708 | 708 |
| -.324** | -.248** | .598** | .725** | .621** | .685** | 1 | -.656** |
| .000 | .000 | .000 | .000 | .000 | .000 | | .000 |
| 708 | 708 | 708 | 708 | 708 | 708 | 708 | 708 |
| .395** | .289** | -.509** | -.649** | -.509** | -.651** | -.636** | 1 |
| .000 | .000 | .000 | .000 | .000 | .000 | .000 | |
| 708 | 708 | 708 | 708 | 708 | 708 | 708 | 708 |

im Vergleich zur Drohne in der vorherigen Vorstudie). Es bestehen einige Unterschiede zwischen den Zuständen mit gutem und schlechtem Resultat und obwohl diese Effekte für die Zustände am stärksten sind, in denen wir in die Algo-Aversion gezoomt sind, ist der Unterschied nicht besonders groß (ein Unterschied im Mittelwert von 8 Punkten auf einer 100-Punkt-Skala). Außerdem überlappten die Fehlerbalken zwischen guten und schlechten Resultat-Zuständen für alle Szenarien, eine Indikation, dass es sich nicht um Einzelgruppen handelt. Der einzige eindeutige und signifikante Unterschied in Gruppen ($p <,05$) liegt zwischen den Zuständen der autonomen Waffensysteme mit geringer Selbstbestimmung und den anderen Szenarien. Dies wurde auch von der Signifikanzebene des t-Tests der unabhängigen Stichprobe bestätigt, den wir wahlweise nicht in das nachstehende Diagramm aufnahmen, um es anschaulicher zu machen.

### 4.3.5 *Analyse der abhängigen Variablen*

Die Korrelation zwischen dem Agency-Konstrukt und den abhängigen Variablen haben alle Signifikanz ($p <,05$), deshalb sind die eindeutigsten Nachweise für jede der abhängigen Variablen nachstehend beschrieben.

#### 4.3.5.1 Schuld

Die wichtigste Erkenntnis für die *Schuld*-Variable ist, dass Leute Zuständen mit geringer Selbstbestimmung weniger *Schuld* beimessen als Zuständen mit autonomen Waffensystemen mit hoher Selbstbestimmung und menschlichen Operateuren. Für diese Zustände ist der Unterschied von Bedeutung ($p <,05$). Einige Effekte der Algo-Aversion kann beim Lernen und der Einsichtnahme von Szenarien erkannt werden, wo Leute weniger *Schuld* beimessen, wenn die autonomen Waffensystemen keinen Fehler machten, und mehr, wenn sie einen Fehler mit einem schlechten Resultat macht. Wie dem auch sei, diese Zustände weisen eine Überlappung der Fehlerbalken auf und sind nicht signifikant.

#### 4.3.5.2 Kommandantenschuld

Die bedeutendste Beobachtung bei der Kommandantenschuld-Variable ist, dass dem Kommandanten entschieden weniger Schuld gegeben wird, wenn die autonome Waffe einen Programmierfehler macht und in einem unberechenbaren Zustand ist, dies ist von Signifikanz ($p <,05$). Es kann der Fall sein, dass die Schuld der Drohne zugeschoben wird, da Leute erwarten, dass sie aus ihren Fehlern lernen kann oder sie sich unberechenbar verhält und der Kommandant dafür keine Schuld hat. Bei Agentenzuständen von geringer

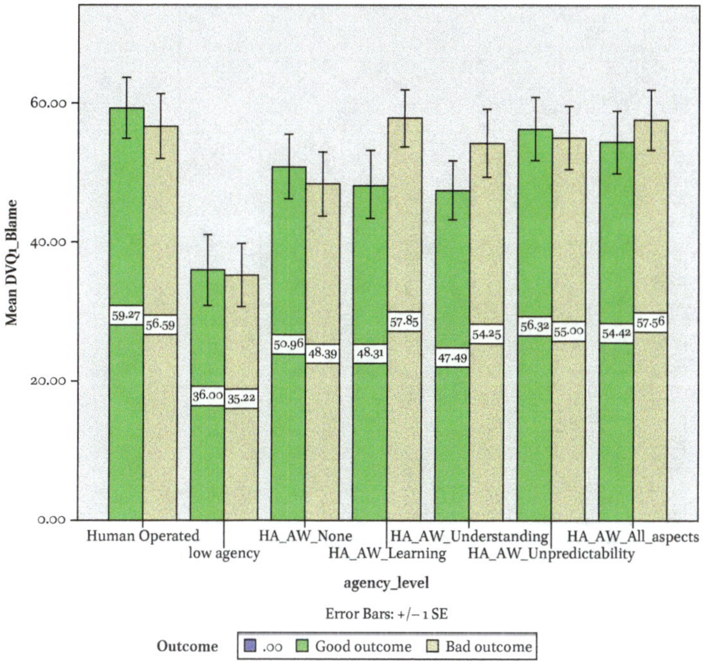


ABBILDUNG 19 Mittelwert der Schuldvariable pro Zustand pro Resultat

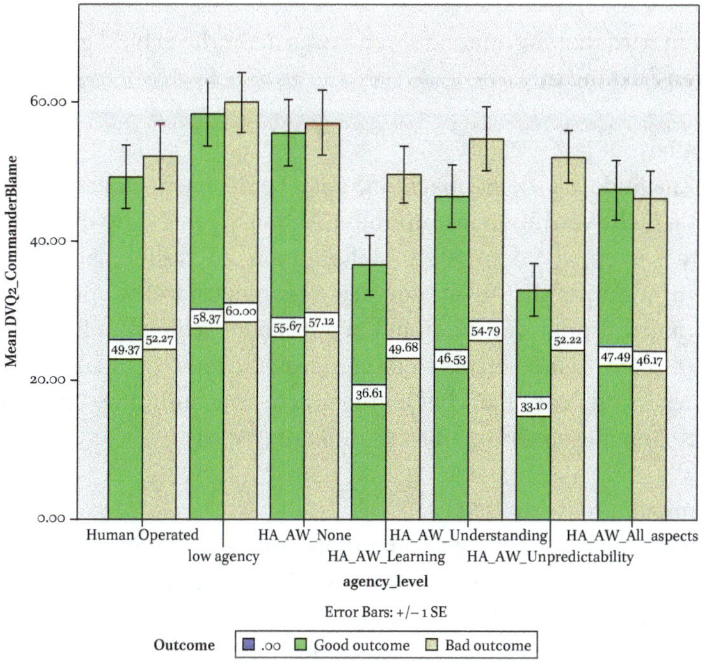


ABBILDUNG 20 Mittelwert der Kommandantenschuld-Variable pro Zustand pro Resultat

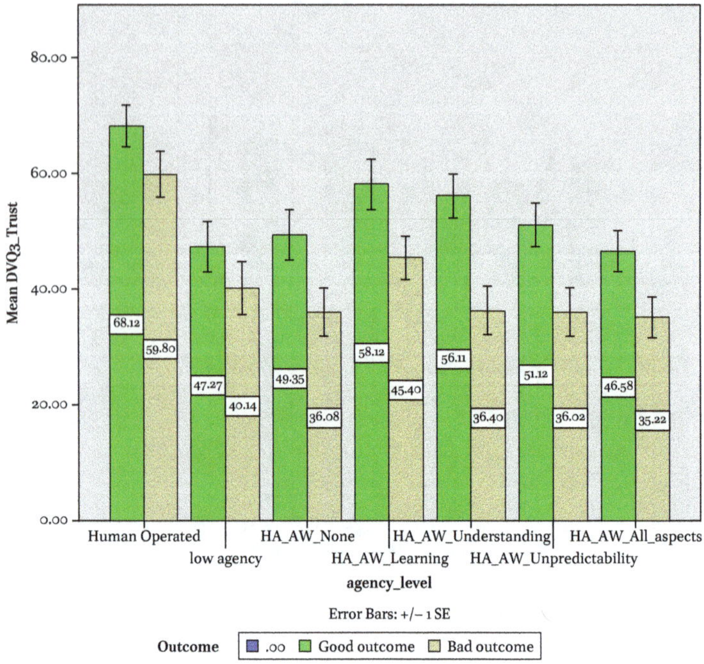


ABBILDUNG 21 Mittelwert der Vertrauensvariable pro Zustand pro Resultat

Selbstbestimmung wird dem Kommandanten etwas mehr die Schuld gegeben als in den anderen Zuständen.

### 4.3.5.3 Vertrauen

Während aller Zustände ist zu beobachten, dass Leute nach einem guten Resultat zukünftig mehr Vertrauen in autonome Waffensysteme oder menschliche Operateure setzen, die korrekten Maßnahmen zu treffen, als nach einem schlechten Resultat. Mit Ausnahme des Agentenzustandes mit geringer Selbstbestimmung, ist das Vertrauen in ein gutes oder schlechtes Resultat von Bedeutung (p <,05). Leute vertrauen dem menschlichen Operateur mehr als der autonomen Waffe, vor allem bei geringer Selbstbestimmung, und der Unterschied zwischen diesen Gruppen ist signifikant (p <,05).

### 4.3.5.4 Kommandantenvertrauen

Im Großen und Ganzen vertrauen die Leute dem Kommandanten eher nach einem guten Resultat, dass er die korrekten Maßnahmen treffen wird, als nach einem schlechten Resultat. Das Vertrauen in den Kommandanten ist am niedrigsten bei geringer oder hoher Selbstbestimmung des Agentenzustands und ohne zusätzliche Informationen, und der Unterschied zwischen diesen

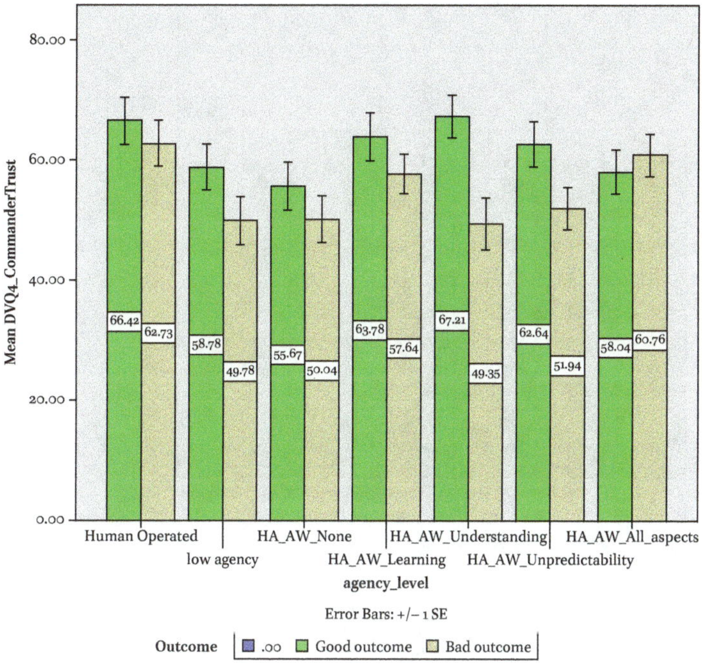


ABBILDUNG 22 Mittelwert der Kommandantenvertrauen-Variable pro Zustand pro Resultat

Gruppen ist signifikant ($p < .05$). Zu wissen, dass die autonome Waffe kognitive Fähigkeiten hat, resultiert in mehr Vertrauen bei schlechten Resultaten, und das ist signifikant ($p <,05$).

#### 4.3.5.5 Schaden

Bei Zuständen mit schlechten Resultaten nehmen Leute mehr Schaden wahr als bei guten, was zu erwarten ist, und der Unterschied ist unter fast allen Unterschieden signifikant ($p <,05$). In diesem Diagramm ist vor allem auffällig, dass, wenn autonome Waffensysteme sich unberechenbar verhalten, das Resultat nicht von Bedeutung zu sein scheint, und die Wahrnehmung des Schadens ist die gleiche, aber der Unterschied zwischen beiden Resultatgruppen ist nicht signifikant.

#### 4.3.5.6 Kommandantenschaden

Bei Zuständen mit schlechten Resultaten schreiben die Leute dem Kommandanten mehr Schaden zu, als bei guten, was zu erwarten ist. Am Eindeutigsten ist der große Unterschied im Programmierzustand zwischen den guten und schlechten Resultaten, wenn dem Kommandanten bei guten Resultaten viel weniger Schaden zugemessen wird, was signifikant ist ($p <,05$). Die Unterschiede zwischen Zuständen mit guten und schlechten Resultaten in

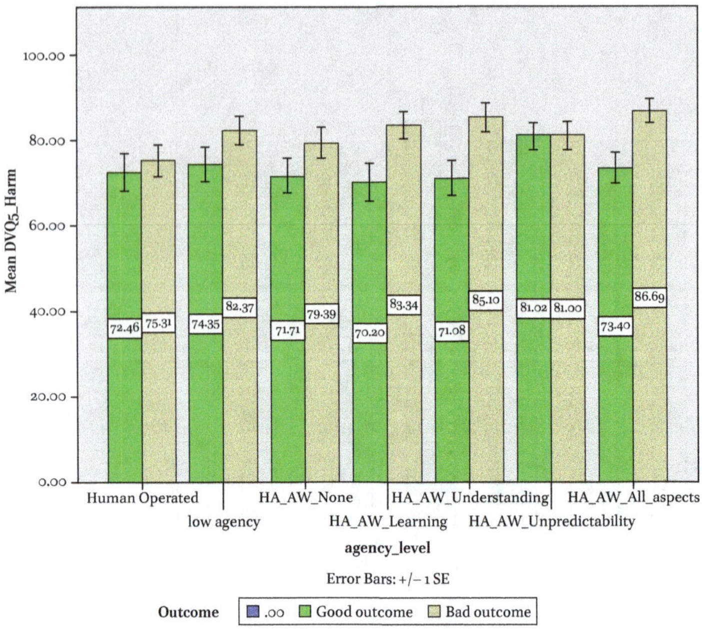


ABBILDUNG 23 Mittelwert der Schadensvariable pro Zustand pro Resultat

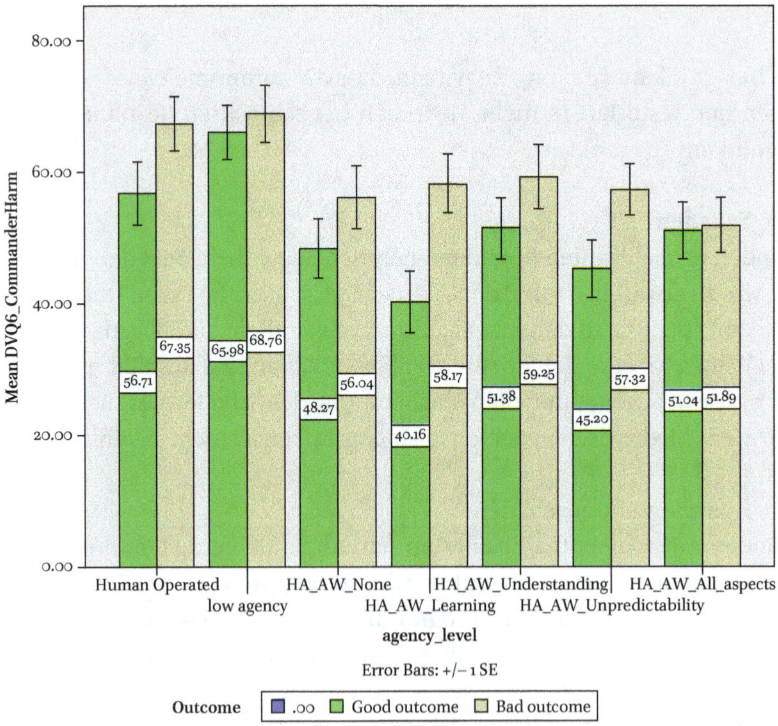


ABBILDUNG 24 Mittelwert der Kommandantenschaden-Variable pro Zustand pro Resultat

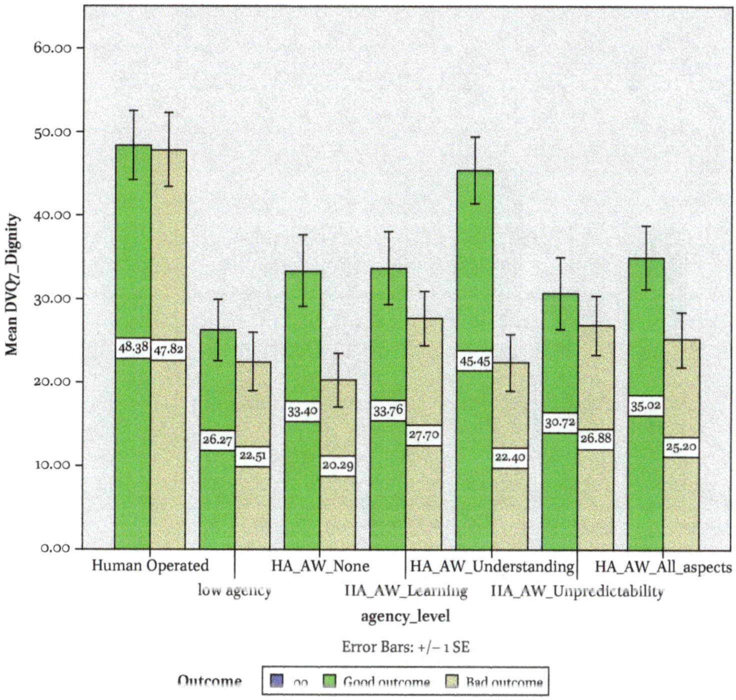


ABBILDUNG 25 Mittelwert der Menschenwürde-Variable pro Zustand pro Resultat

Szenarien zwischen menschlichem Operateur und Agenten mit hoher Unberechenbarkeit sind ebenfalls signifikant (p <,05).

#### 4.3.5.7 Menschenwürde

Der menschliche Operateur handelt mit mehr Respekt für die Menschenwürde als die autonome Waffe, mit Ausnahme des Zustandes, wenn die autonome Waffen mit einem großen Datensatz, der speziell für ihren Einsatz eingerichtet wurde, programmiert wird, und dieser Unterschied ist signifikant (p <,05). Ebenso scheint es von Bedeutung zu sein, wenn autonome Waffensysteme einen Fehler machen, da wir signifikante (p <,05) Unterschiede zwischen guten und schlechten Resultatszenarien für autonome Waffensysteme für die geringe Selbstbestimmung, hohe Selbstbestimmung keine Aspekte, und hohe Selbstbestimmung keine Aspekte beobachten.

#### 4.3.5.8 Zuversicht

Im Großen und Ganzen sind die Menschen zuversichtlicher nach einem guten Resultat, dass die autonomen Waffensysteme zukünftig die korrekten Maßnahmen treffen, als nach einem schlechten Resultat; alle Differenzen sind

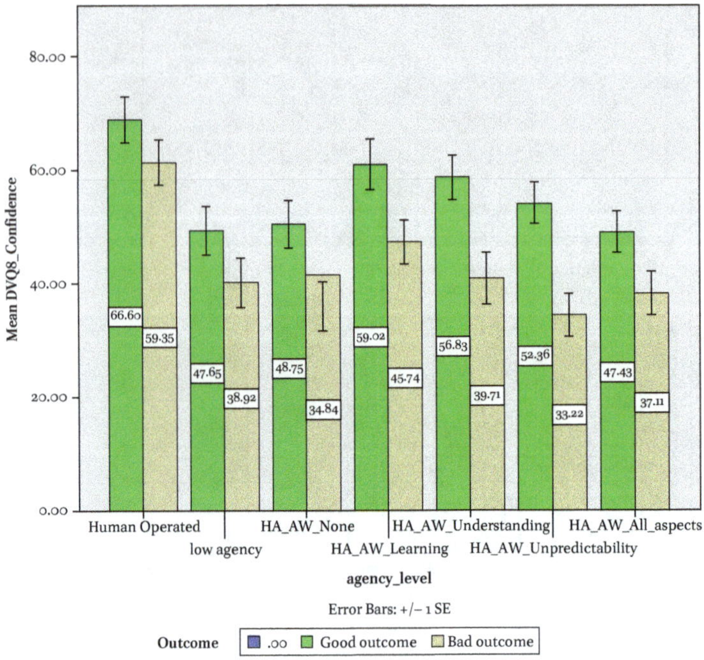


ABBILDUNG 26 Mittelwert der Zuversichtsvariable pro Zustand

signifikant (p <,05). Sie setzen auch etwas mehr Vertrauen in den menschlichen Operateur als in autonome Waffensysteme. Die Differenzen zwischen diesen Szenarien sind signifikant (p <,05). Wenn die autonomen Waffensysteme Lernfähigkeit haben, ist die Zuversicht der Leute etwas größer im Vergleich zu den anderen Zuständen von autonomen Waffen.

#### 4.3.5.9 Erwartungen

Nach den Erwartungen der Menschen leisten sowohl der menschliche Operateur als auch das autonome Waffensystem mehr in Szenarien mit guten Resultaten, als solchen mit schlechten Resultaten. Alle Differenzen zwischen guten und schlechten Resultaten sind signifikant (p <,05). Die Erwartungen für Agenten mit geringen Selbstbestimmungszuständen sind am niedrigsten, wenn ein schlechtes Resultat erreicht wurde, aber es besteht keine große Differenz zu den anderen Zuständen.

#### 4.3.5.10 Unterstützung

Die Menschen unterstützen von Menschen gesteuerte Drohnen eher als autonome Waffensysteme. Die Unterstützung ist am geringsten, aber nicht signifikant, für die Agenten mit geringer Selbstbestimmung im Vergleich zu Szenarien, in denen Informationen über Autonome Waffensysteme erstellt

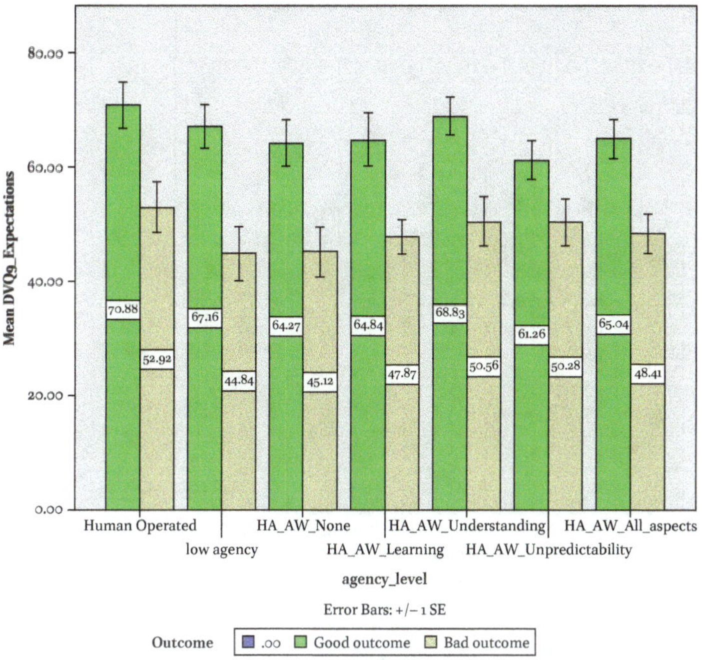


ABBILDUNG 27 Mittelwort der Werterwartungsvariable pro Zustand pro Resultat

werden. Ebenso ist die Unterstützung für Szenarien mit einem guten Resultat höher als für solche mit schlechten Resultaten und mit Ausnahme des menschlichen Operateurs und autonomer Waffensysteme mit geringer Selbstbestimmung, sind diese Differenzen signifikant (p <,05).

### 4.3.5.11 Fairness

Im Großen und Ganzen werden die Maßnahmen des menschlichen Operateurs als fairer angesehen als die der autonomen Waffensysteme. Die Wahrnehmung von Fairness ist entschieden höher bei Szenarien mit gutem Resultat als solchen mit schlechtem Resultat, und in allen Szenarien sind diese Differenzen signifikant (p<,05). Die Differenz zwischen den Agentenzuständen der autonomen Waffensysteme ist gering (<10 Punkte auf einer 100-Punkt-Skala).

### 4.3.5.12 Unbehagen

Bei Szenarien mit schlechtem Resultat empfinden die Menschen mehr Unbehagen bei den Handlungen von sowohl menschlichen Operateuren als auch autonomen Waffensystemen. Es besteht keine große Differenz zwischen der Stufe der Beunruhigung für die verschiedenen Zustände der autonomen Waffensysteme. Die Beunruhigungsebene für den menschlichen Operateur ist viel

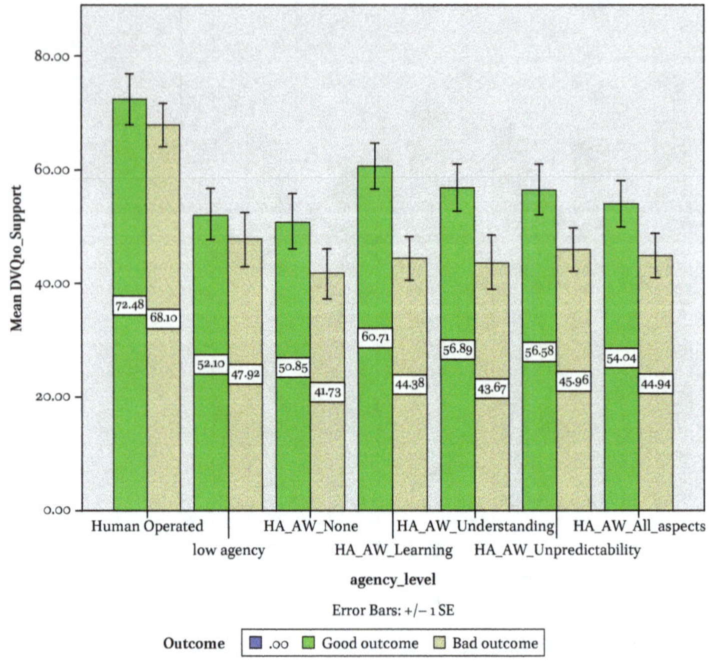


ABBILDUNG 28 Mittelwert der Unterstützungsvariabele pro Zustand pro Resultat

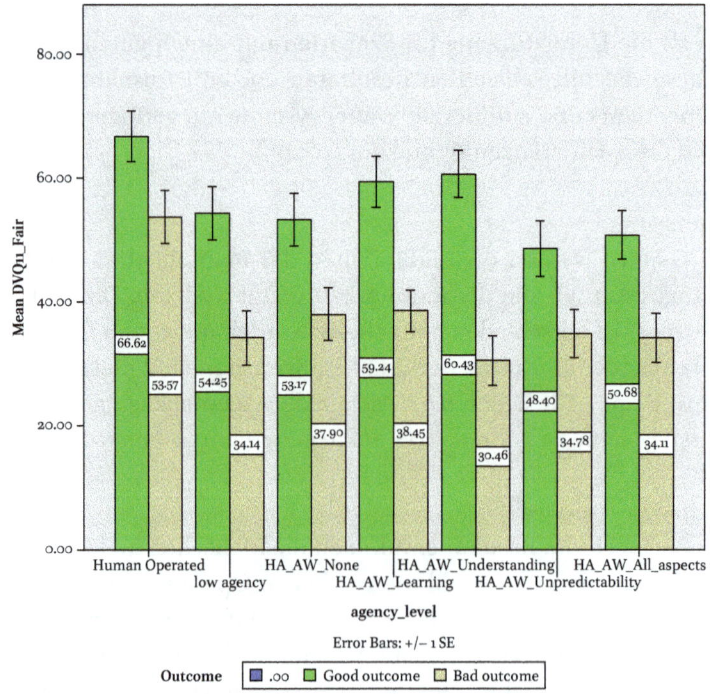


ABBILDUNG 29 Mittelwert der Fairnessvariable pro Zustand pro Resultat

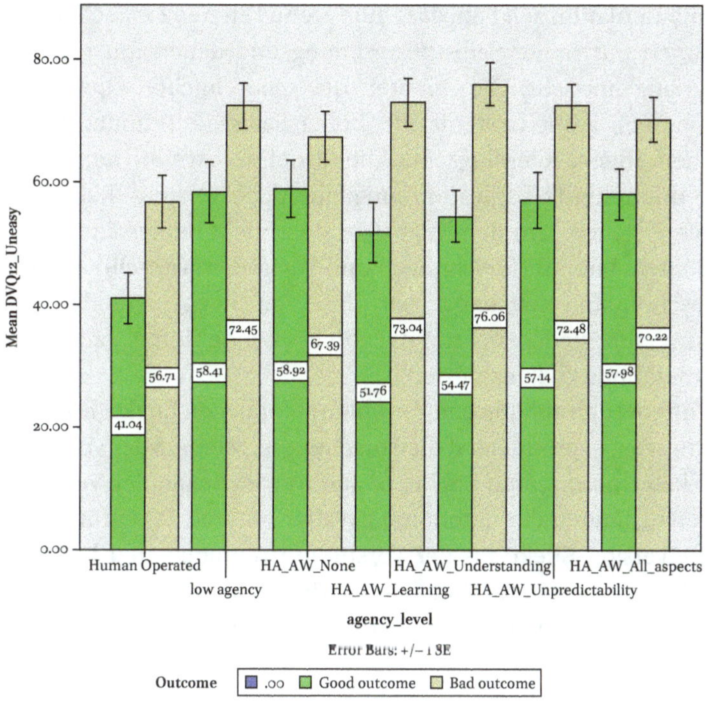


ABBILDUNG 30 Mittelwert der Unbehagenvariable pro Zustand pro Resultat

niedriger als die für autonome Waffensysteme. Mit Ausnahme der Agenten mit hoher Selbstbestimmung gibt es keine weiteren Aspekte des Zustandes der autonomen Waffensysteme. Alle Differenzen zwischen guten und schlechten Resultaten sind signifikant ($p <,05$).

### 4.3.6 *Schlussfolgerung der Vorstudie 2*

Das Ergebnis der Vorstudie 2 führt zu den folgenden Schlussfolgerungen:

- Der Reliabilitätstest wies auf, dass die vier Items des Agency-Konstrukts zuverlässig sind, und gemäß der PCA können sie als eine Komponente angesehen werden;
- Der Korrelationskoeffizient des Agency-Konstrukts und alle abhängigen Variablen sind signifikant und haben die gleichen Richtungen wie in Vorstudie 1. Wir fanden einen negativen Zusammenhang zwischen dem Agency-Konstrukt und der Stufe des Unbehagens, d.h. dass bei Erhöhung der Agentenwahrnehmung, die Leute andeuten, dass sie eine größere *Besorgnis* empfinden, aber dieser Effekt ist immer noch sehr gering und nähert sich 0 ($r = -.076$). Um zu sehen, ob wir einen stärkeren Effekt erhalten könnten, änderten wir die Formulierung der Frage zur Unbehagenvariable in der abschließenden Untersuchung;

- Die Agentenmanipulation zeigt an, dass eine große Differenz zwischen den Agency-Szenarien mit hoher Selbstbestimmung und denen mit geringer Selbstbestimmung herrscht. Die Agenten des menschlichen Operateurs und der autonomen Waffensysteme mit hoher Selbstbestimmung liegen auf der gleichen Ebene. Allerdings sind die Mittel des Agency-Konstrukts für Szenarien mit hoher Selbstbestimmung, die Algo-Aversion-Effekte enthalten, fast die gleichen. Um dem Algo-Aversion-Effekt herauszunehmen, müssen wir weitere Literaturforschungen und Vorstudien vornehmen. Aufgrund der beschränkten Zeit haben wir diese Forschungsmethode nicht verfolgt und uns entschieden, uns auf die abschließende Untersuchung der Agentenwahrnehmung zu konzentrieren.
- Viele der von uns bei den abhängigen Variablen gefundenen Effekte erfüllten unsere Erwartungen, begründet auf die Forschungsergebnisse der Vorstudie 1. Wir fanden keine besonderen Effekte in den DVs zwischen den verschiedenen Agentenzuständen der autonomen Waffensysteme. Wir prüften, ob wir den Algo-Aversion-Effekt aus der Vorstudie 1 reproduzieren konnten, bei dem Schaden in Agentenzuständen mit geringer Selbstbestimmung von der Drohne auf den Kommandanten übertragen wurde, aber leider konnte dieser Effekt in der Vorstudie 2 nicht festgestellt werden. Obwohl diese Verantwortungskette noch weiter untersucht werden muss, entschieden wir uns, sie nicht in die abschließende Untersuchung aufzunehmen, da wir uns nicht sicher waren, was genau der Fall ist und weitere Literaturforschung und Vorstudien erforderlich sind.

## 4.4 Abschließende Untersuchung – militärische Stichprobe

Um die Anzahl der Szenarien für die abschließende Untersuchung festzustellen, führten wir Teststärkenberechnungen aus, um die Gesamtzahl der Probanden abzuschätzen, die wir laut der Ergebnisse der ersten Vorstudie benötigen würden, da wir diese Szenarien auch für die abschließende Untersuchung benutzen werden. Basiert auf den Teststärkenberechnungen (Effektgröße 0,4, eine Soll-Teststärke von 0,8 und einen Wahrscheinlichkeitsgrad von 0,05), zielten wir auf eine Gesamtantwortquote von 200 und bestimmten, dass wir 3 Szenarien verarbeiten konnten. Wir entschieden uns, uns auf die Auffassung von Streitkräften zu konzentrieren und die derzeitige Technologie, d.h. eine von Menschen gesteuerte Drohne, mit zukünftiger Technologie, dem autonomen Waffensystem (Tabelle 15) zu vergleichen. Wir fügten außerdem noch ein Szenario mit einem autonomen Waffensystem auf neutraler Ebene hinzu, um die Auffassung von Streitkräften der drei Zustände zu vergleichen.

In dieser Untersuchung stellten wir vor allem Fragen über das Kampfmittel der Drohne und nicht über den menschlichen Operateur, der sie steuert, um die Auffassung der Artefakte zu vergleichen. Da wir nur drei Szenarien verarbeiten konnten, konzentrierten wir uns auf den Zustand mit schlechtem Resultat, da die Vorstudien nachwiesen, dass die Effekte in diesem Zustand am eindeutigsten sind.

Wir verbreiteten die Umfrage per E-Mail, mit einem anonymen Link und hatten deshalb keine Gewährleistung in Bezug auf die Anzahl der Befragten. Nach der Datenvorverarbeitung und dem Löschen von Befragten, die den Aufmerksamkeitstest nicht durchgeführt hatten, verfügten wir noch über 239 Antworten für die komplette Stichprobe. Die Anzahl der Befragten pro Szenario betrug zwischen 64 und 96, ein seltsames Ergebnis, da die Befragten gleichmäßig über die Szenarien verteilt werden sollten. Die Stichprobe von 239 ist eine Kombination des niederländischen Militärs und der Zivilbevölkerung, die im Verteidigungsministerium angestellt sind, denn hätten wir sie ausgelassen, wären wir auf 149 Befragte sitzen geblieben, zu wenig für eine robuste Datenanalyse.

Für dieses Design werden die folgenden Definitionen benutzt:

- Von Menschen gesteuerte Drohne: Drohne, die direkt unter der Kontrolle eines Menschen steht. Wird als Kampfmittel mit sehr niedriger Selbstbestimmung aufgefasst.
- Hohes AW: Probanden wurde gesagt, dass das autonome Waffensystem Selbstbestimmungsmerkmale aufweist hat (z.B. Erwägung von Für und Wider und unabhängige Entscheidungswahl). Wird als Kampfmittel mit sehr hoher Selbstbestimmung aufgefasst.
- Neutrales AW: Es wurden keinerlei Informationen über Selbstbestimmungsmerkmale der autonomen Waffensysteme gegeben. Die Stärke dieser Methode ist, dass wir Daten von Teilnehmern sammeln können, die ihre voreingestellten Konzepte von autonomen Waffensystemen zum Verständnis

TABELLE 15 Szenarien der abschließenden Untersuchung

| | **Agency-Stufen** | | |
|---|---|---|---|
| | **Von Menschen gesteuerte** | **neutrale AW** | **Hohe AW** |
| **Schlechtes Resultat** | 1 | 2 | 3 |

TABELLE 16 Ergebnisse der Reliabilitätsanalyse Agency-Items

**Reliabilitätstatistiken**

| Cronbachs Alpha | Anzahl Items |
|---|---|
| .865 | 4 |

**Item-Gesamtstatistiken**

| | Skala-Mittel, wenn Item gelöscht ist | Skalenvarianz, wenn Item gelöscht | Korrigierte Hem-Total Korrelation | Cronbachs Alpha, wenn Item gelöscht |
|---|---|---|---|---|
| Denkweise | 116.43 | 71 75.573 | .620 | .863 |
| Zielsetzung | 123.33 | 6342.474 | .756 | .810 |
| Frei handeln | 122.62 | 6566.723 | .740 | .816 |
| Zielerreichung | 118.84 | 6459.1 46 | .740 | .816 |

der Handlung autonomer Waffensysteme benutzen und diese Auffassungen mit den von menschlichen Operateuren gesteuerten und autonomen Waffen mit hoher Selbstbestimmung vergleichen können.

### 4.4.1 *Reliabilitätsanalyse*

Für alle vier Agency-Items ist α =,865 und liegt über dem Schwellenwert von 0,7. Er kann nicht durch Löschen eines Items verbessert werden, d.h. dass das Agency-Konstrukt zuverlässig ist (Tabelle 16). Löschen anderer Items wird den α Wert abschwächen und die interne Konsistenz reduzieren.

### 4.4.2 *Hauptkomponentenanalyse (PCA)*

Die PCA zeigt, dass Denken, Zielsetzung, freies Handeln und Zielerreichung als ein Konstrukt angesehen werden kann, das die Varianz von 71,18% in den ursprünglichen Variablen ausmacht. Dies kann man auch auf dem Scree-Test ansehen, das die Eigenwerte der Komponenten (Abbildung 31) anzeigt. Die PCA zeigt an, dass die vier Items: Denken, Zielsetzung, freies Handeln

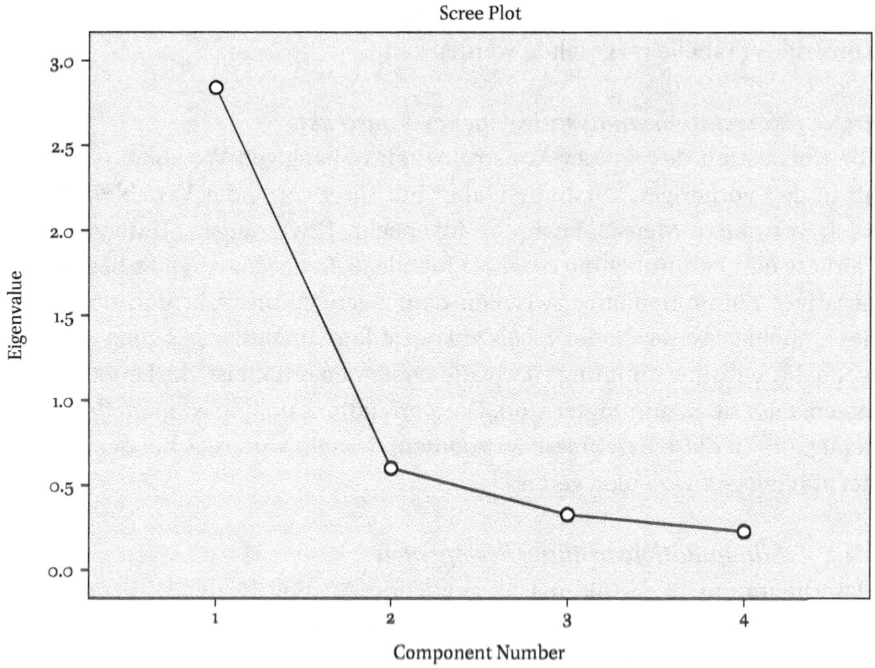


ABBILDUNG 31 Scree-Test PCA letzte Untersuchung

TABELLE 17 PCA-Ergebnisse der Agenten in der abschließenden Untersuchung

| **Komponentenmatrix**[a] | | |
|---|---|---|
| **Komponente** | | |
| | **1** | **2** |
| Denkweise | .774 | .560 |
| Zielsetzung | .873 | |
| Freies Handeln | .861 | |
| Zielerreichung | .863 | |

Extraktionsmethode: Haupt- Komponentenanalyse.
[a] . 2 Komponenten wurden extrahiert.

und Zielerreichung als die zugrundeliegenden Komponenten des Agency-Konstrukts (Tabelle 17) gesehen werden kann.

### 4.4.3 *Korrelationsanalyse des Agency-Konstrukts*

Die Korrelation des Agency-Konstrukts mit abhängigen Variablen ist größer als in den vorherigen Vorstudien, aber nur für 7 von 9 der Variablen signifikant: Vertrauen, Menschenwürde, Zuversicht, Erwartungen, Unterstützung, Fairness und Beunruhigung ($p < ,01$) (Tabelle 18, Seite 303–305). Es besteht ein negativer Zusammenhang zwischen dem Agency-Konstrukt und dem Grad des Unbehagens, der höher ist als zuvor in den Vorstudien. Es zeigt an, dass, wenn die Selbstbestimmung von Waffensystemen erhöht ist, die Leute merken lassen, dass sie beunruhigter sind. Die Korrelationsanalyse ist nicht detailliert genug, um in diese Ergebnisse zu zoomen, deshalb wird dies bei der Analyse der abhängigen Variablen getan.

### 4.4.4 *Manipulationsprüfung des Agenten*

Das Diagramm in Abbildung 32 zeigt den Anstieg in der Wahrnehmung von Selbstbestimmung zu den Zuständen auf. Die Differenz in der Wahrnehmung von Selbstbestimmung zwischen Szenarien der von Menschen gesteuerten Waffen und dem neutralen autonomen Waffensystem, sowie zwischen dem Szenario des menschlichen Operateurs und dem des autonomen Waffensystems ist signifikant, folglich können diese Gruppen als unterschiedlich betrachtet werden. Der t-Test der unabhängigen Stichproben für das Szenario mit dem neutralen AW-Agenten liegt mit $p = ,063$ knapp über dem Schwellenwert für Signifikanz von $p < ,05$. Der t-Test der unabhängigen Stichproben für den Zustand mit menschlichen Operateur und dem neutralen AW-Agenten zeigt, dass $p = ,001$ ist und für den Zustand mit menschlichem Operateur und dem AW-Agenten mit hoher Selbstbestimmung $p = ,000$ ist.

Wir vermuten, dass die von Menschen gesteuerte Drohne als mit geringer Selbstbestimmung und autonome Waffensysteme als mit hoher Selbstbestimmung aufgefasst werden, und die Ergebnisse beweisen, dass dies der Fall ist. Die zentrale Analyse befasste sich damit, wie der Fall des neutralen Agenten mit den anderen beiden Fällen verglichen werden kann. Wir vermuteten, dass Militärangehörige autonome Waffen nicht als mit psychischen Fähigkeiten ausgestattet auffassen. Deshalb erwarteten wir keinen Unterschied zwischen dem Zustand des neutralen autonomen Waffensystems und dem Zustand, in dem eine Drohne von Menschen ferngesteuert wird. Wir vermuteten auch, dass der neutrale Kampfmittelzustand als

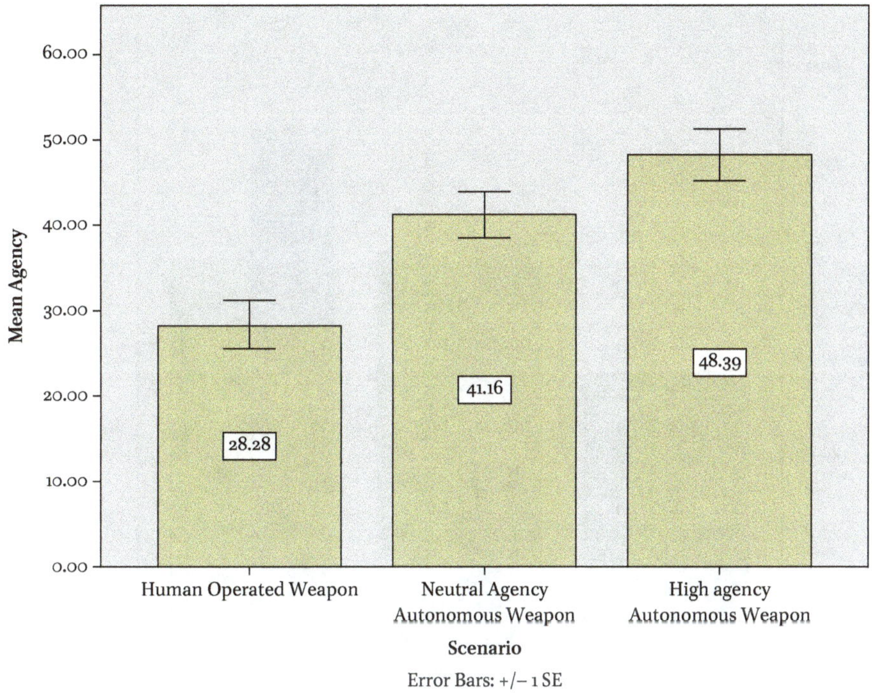


ABBILDUNG 32 Mittelwert des Kampfmittels pro Zustand

bedeutend anders als der Kampfmittelzustand mit hoher Selbstbestimmung bewertet wurde.

Die Ergebnisse bestätigen, dass die von Menschen gesteuerte Drohne als mit geringer Selbstbestimmung und das autonome Waffensystem als mit hoher Selbstbestimmung aufgefasst werden. Allerdings zeigen die Ergebnisse auch, dass dem neutralen autonomen Waffensystem mehr Selbstbestimmung beigemessen wird, als der von Menschen gesteuerten Drohne und dass die Auffassung von Kampfmitteln mit hoher Selbstbestimmung nur wenig höher ist. Das bedeutet, dass wir unsere Hypothese verwerfen müssen und folgern, dass die für das niederländische Verteidigungsministerium arbeitenden Militärangehörige und Zivilisten autonome Waffensysteme als mit Merkmalen von Streitkräften ausgestattet auffassen.

Der Unterschied zwischen den befragten Militärangehörigen und den Zivilisten in Abbildung 33 ist sehr interessant, aber als zwei Gruppen unterscheiden sie sich nicht besonders in zwei der Szenarien, sondern nur in dem Kampfmittelzustand mit hoher Selbstbestimmung. Wir müssen bei der Interpretation der Ergebnisse vorsichtig sein, deshalb ist die nachstehende Beobachtung

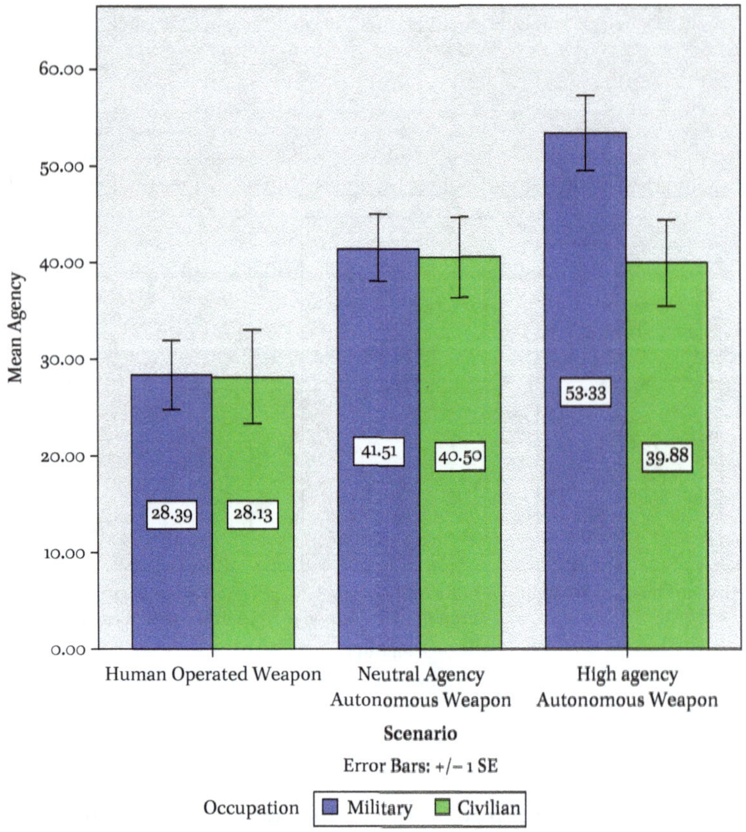


ABBILDUNG 33 Mittelwert des Kampfmittels pro Zustand pro Gruppe

veranschaulichend. Sowohl in dem Szenario mit menschlichem Operateur als auch mit dem neutralen AW-Agenten, stimmte die Auffassung der befragten Militärangehörigen mit der der befragten Zivilisten überein. Aber in dem AW-Agency-Szenario mit hoher Selbstbestimmung waren die befragten Militärangehörigen sich der Effekte mehr bewusst, als ihre zivilen Kollegen. Wir haben uns die demografischen Unterschiede zwischen den Stichproben angesehen und sie sind sich erstaunlich ähnlich, aber eine mögliche Erklärung könnte sein, dass die Stichprobe mit den Zivilisten über 10% mehr Probanden verfügt, die mit KI gearbeitet haben, also die Stichprobe mit den Militärangehörigen. Wir sind jedoch nicht sicher, dass dies den Unterschied in Agentenwahrnehmung zwischen den beiden befragten Gruppen erklärt und dieser Aspekt musst genauer untersucht werden.

TABELLE 18 Korrealisation-Agency-Konstrukt mit abhängigen Variablen für die abschließende Untersuchung

**Correlations**

| | | Agency | DVD1_ Blame | DVD2_ Trust | DVD3_ Hram | DVD4_ Human_ Dignity | DVD5_ Confi- dence | DVD6_ Expectati- ons | DVD7_ Support | DVD8_ Fair | DVD9_ Anxiety |
|---|---|---|---|---|---|---|---|---|---|---|---|
| Agency | Person Corre- lation | 1 | .009 | .406** | -.070 | .347** | .428** | .320** | .337** | .451** | -.216** |
| | Sig(2-tailed) | | .888 | .000 | .275 | .000 | .000 | .000 | .000 | .000 | .001 |
| | N | 248 | 248 | 248 | 248 | 248 | 248 | 248 | 248 | 248 | 248 |
| DVD1_ Blame | Person Corre- lation | .009 | 1 | -.039 | .056 | .018 | .027 | -.109 | -.002 | -.080 | .128* |
| | Sig(2-tailed) | .888 | | .545 | .379 | .783 | .667 | .086 | .979 | .209 | .044 |
| | N | 248 | 248 | 248 | 248 | 248 | 248 | 248 | 248 | 248 | 248 |
| DVD2_ Trust | Person Corre- lation | .406** | -.039 | 1 | -.217** | .482** | .767** | .324** | .597** | .540** | -.451** |
| | Sig(2-tailed) | .000 | .545 | | .001 | .000 | .000 | .000 | .000 | .000 | .000 |
| | N | 248 | 248 | 248 | 248 | 248 | 248 | 248 | 248 | 248 | 248 |
| DVD3_ Hram | Person Corre- lation | -.070 | .056 | -.217** | 1 | -.180** | -.197** | -.155** | -.133** | -.200** | .360** |
| | Sig(2-tailed) | .275 | .379 | .001 | | .004 | .002 | .015 | .036 | .002 | .000 |
| | N | 248 | 248 | 248 | 248 | 248 | 248 | 248 | 248 | 248 | 248 |

TABELLE 18 Korrealisation-Agency-Konstrukt mit abhängigen Variablen für die abschließende Untersuchung (*cont.*)

| Correlations | | | | | | | | | | | |
|---|---|---|---|---|---|---|---|---|---|---|---|
| DVD4_ Human_ Dignity | Person Correlation | .347** | .018 | .482** | -.180** | 1 | .564** | .239** | .564** | .472** | -.385** |
| | Sig(2-tailed) | .000 | .783 | .000 | .004 | | .000 | .000 | .000 | .000 | .000 |
| | N | 248 | 248 | 248 | 248 | 248 | 248 | 248 | 248 | 248 | 248 |
| DVD5_ Confidence | Person Correlation | .428** | .027 | .767** | -.197** | .564** | 1 | .376** | .685** | .600** | -.495** |
| | Sig(2-tailed) | .000 | .667 | .000 | .002 | .000 | | .000 | .000 | .000 | .000 |
| | N | 248 | 248 | 248 | 248 | 248 | 248 | 248 | 248 | 248 | 248 |
| DVD6_ Expectations | Person Correlation | .320** | -.109 | .324** | -.155** | .239** | .376** | 1 | .271** | .475** | -.264** |
| | Sig(2-tailed) | .000 | .086 | .000 | .015 | .000 | .000 | | .000 | .000 | .000 |
| | N | 248 | 248 | 248 | 248 | 248 | 248 | 248 | 248 | 248 | 248 |
| DVD7_ Support | Person Correlation | .337** | -.002 | .597** | -.133** | .564** | .685** | .271** | 1 | .528** | -.488** |
| | Sig(2-tailed) | .000 | .979 | .000 | .036 | .000 | .000 | .000 | | .000 | .000 |
| | N | 248 | 248 | 248 | 248 | 248 | 248 | 248 | 248 | 248 | 248 |

| | | | | | | | | | | | |
|---|---|---|---|---|---|---|---|---|---|---|---|
| DVD8_ Fair | Person Correlation | .451** | -.080 | .540** | -.200** | .472** | .600** | .475** | .528** | 1 | -.440** |
| | Sig(2-tailed) | .000 | .209 | .000 | .002 | .000 | .000 | .000 | .000 | | .000 |
| | N | 248 | 248 | 248 | 248 | 248 | 248 | 248 | 248 | 248 | 248 |
| DVD9_ Anxiety | Person Correlation | -.216** | .128* | -.451** | .360** | -.385** | -.495** | -.264** | -.488** | -.440** | 1 |
| | Sig(2-tailed) | .001 | .044 | .000 | .000 | .000 | .000 | .000 | .000 | .000 | |
| | N | 248 | 248 | 248 | 248 | 248 | 248 | 248 | 248 | 248 | 248 |

** Correlation is significant at the 0.01 level (2-tailed).
* Correlation is significant at the 0.05 level (2-tailed).

### 4.4.5 *Analyse der abhängigen Variablen*

Die Korrelation zwischen Agency-Konstrukt und den abhängigen Variablen Vertrauen, Menschenwürde, Zuversicht, Erwartungen, Unterstützung, Fairness und Beunruhigung sind signifikant (p <,01), deshalb sind die bemerkenswertesten Forschungsergebnisse für jede dieser 7 abhängigen Variablen in diesem Kapitel beschrieben.

#### 4.4.5.1 Vertrauen

Im Großen und Ganzen haben die Menschen mehr Vertrauen, dass menschliche Operateure in der Zukunft korrekte Maßnahmen treffen, als autonome Waffensystem, und die Differenz zwischen diesen Gruppen ist signifikant (p <,05). Der Vertrauensgrad für den neutralen und hohen AW-Agentenzustand ist gleichgroß und die Differenz zwischen diesen Gruppen ist nicht signifikant.

#### 4.4.5.2 Menschenwürde

Es wird akzeptiert, dass von Menschen gesteuerte Waffen mehr Respekt für die Menschenwürde zeigen als autonome Waffensysteme, und die Differenz zwischen diesen Gruppen ist signifikant (p <,05). Der Vertrauensgrad für den neutralen und hohen AW-Agentenzustand ist gering, und die Differenz zwischen diesen Gruppen ist nicht signifikant.

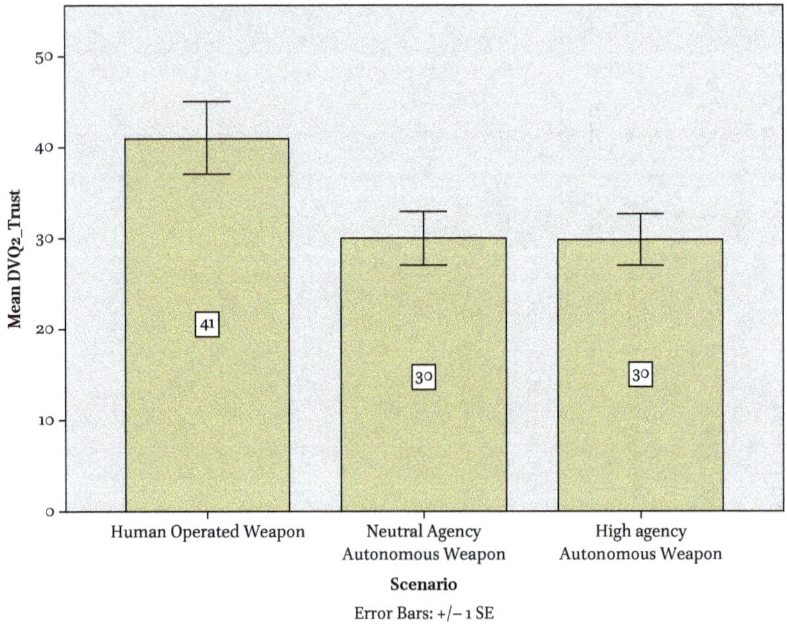


ABBILDUNG 34 Mittelwert der Vertrauensvariable pro Zustand

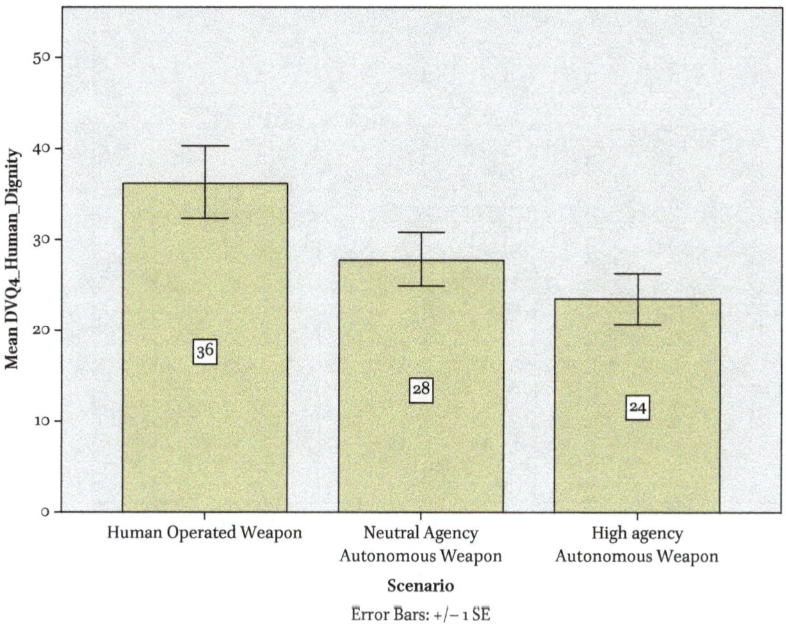


ABBILDUNG 35 Mittelwert der Menschenwürde-Variable pro Zustand

### 4.4.5.3 Zuversicht

Im Großen und Ganzen haben die Menschen mehr Vertrauen, dass menschliche Operateure in Zukunft korrekte Maßnahmen treffen, als autonome Waffensysteme, und die Differenz zwischen diesen Gruppen ist signifikant ($p <,05$). Es gibt keinen nennenswerten Unterschied zwischen AW-Agentenzuständen mit geringer und hoher Selbstbestimmung.

### 4.4.5.4 Erwartungen

Eine etwas geringere Differenz kann in den Erwartungen der Leute zwischen dem Zustand der menschlichen Operateure und den beiden AW-Agentenzuständen beobachtet werden. Die Erwartungen, die an beide AW-Zustände gestellt werden, sind gleichgroß. Die Differenz zwischen allen Gruppen ist jedoch nicht signifikant.

### 4.4.5.5 Unterstützung

Im Allgemeinen werden von Menschen gesteuerte Drohnen stärker unterstützt als autonome Waffensysteme, und die Differenz zwischen diesen Gruppen ist signifikant ($p <,05$). Scheinbar beeinflusst die Auffassung in Bezug auf Autonome Waffensysteme nicht den Grad der Unterstützung, und

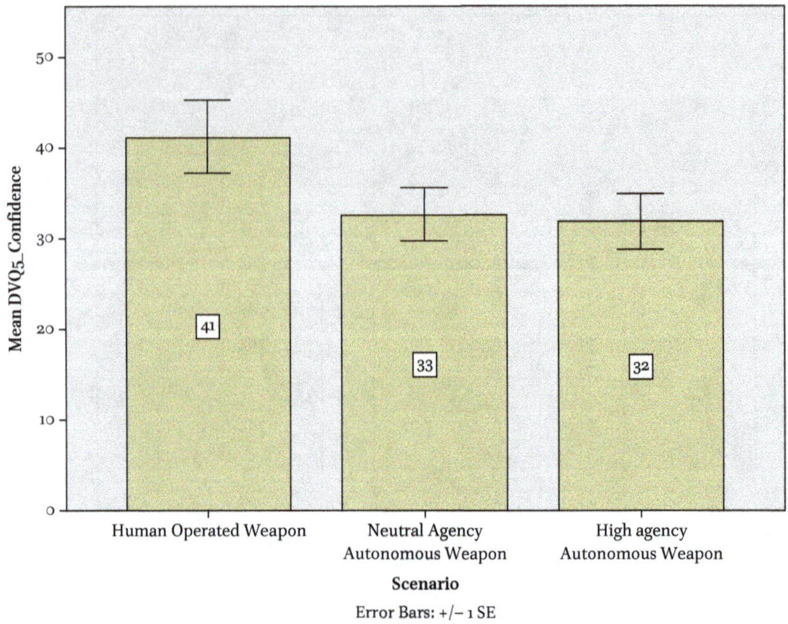


ABBILDUNG 36 Mittelwert der Zuversichtsvariable pro Zustand

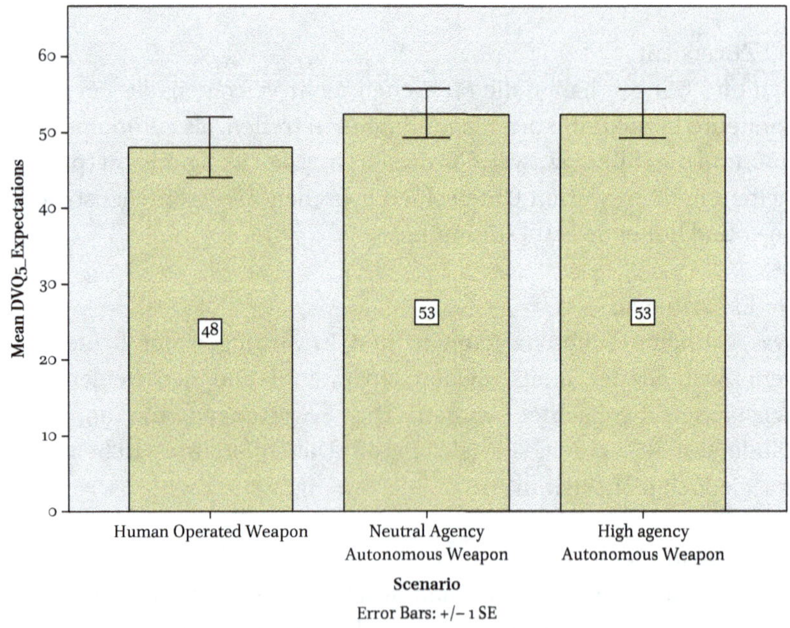


ABBILDUNG 37 Mittelwert der Werterwartungsvariable pro Zustand

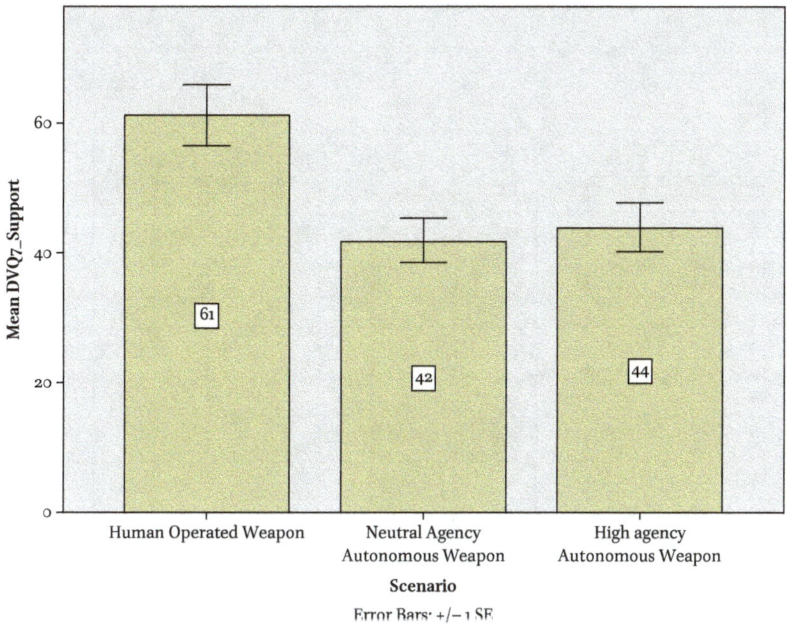


ABBILDUNG 38 Mittelwert der Unterstützungsvariable pro Zustand

die Differenz zwischen der neutralen und hohen Agentengruppe ist nicht signifikant.

#### 4.4.5.6 Fairness

Sowohl die Handlungen der von Menschen gesteuerten Drohne als auch die autonomen Waffensysteme werden als gleich fair betrachtet, ein bemerkenswertes Ergebnis, da wir erwarteten, dass die Handlungen der von Menschen gesteuerten Drohne als fairer angesehen würden, als die von einem autonomen Waffensystem. Leider ist die Differenz zwischen diesen Gruppen nicht signifikant.

#### 4.4.5.7 Besorgnis

Die Leute sind besorgter in Bezug auf autonomer Waffensysteme als in Bezug auf von Menschen gesteuerte Drohnen, und die Differenz zwischen diesen Gruppen ist signifikant ($p <,05$). Die neutralen und hohen AW-Agenten-Szenarien zeigen einen kleinen Anstieg des Besorgnisgrads auf, aber die Differenz zwischen diesen Gruppen ist nicht signifikant.

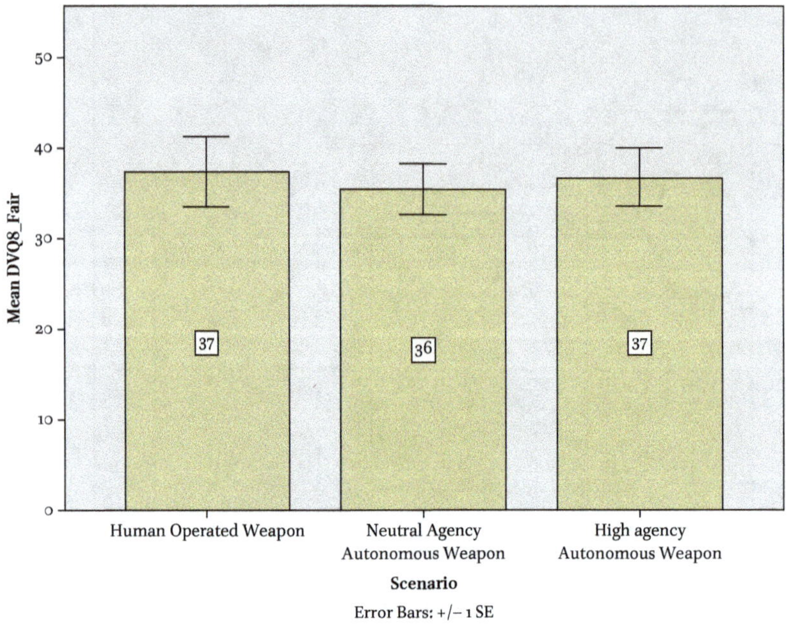


ABBILDUNG 39 Mittelwert der Fairnessvariable pro Zustand

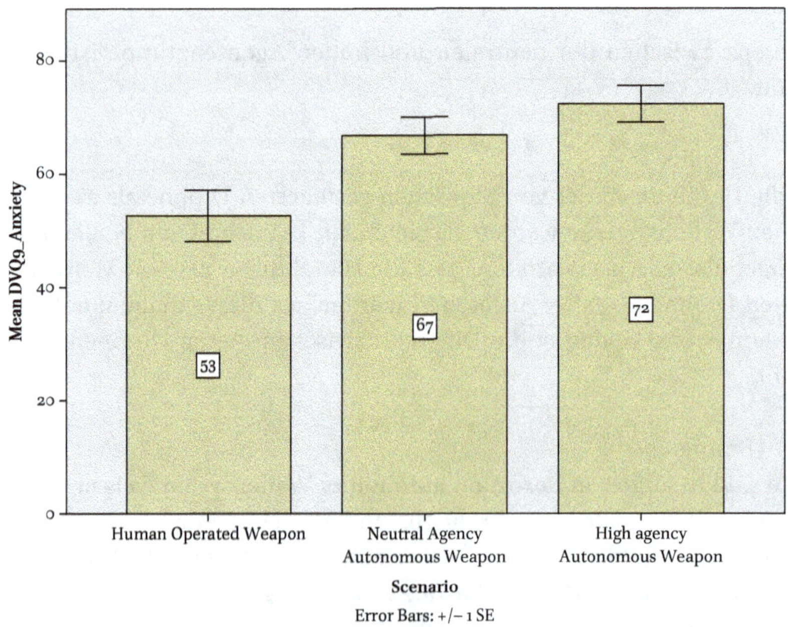


ABBILDUNG 40 Mittelwert der Besorgnisvariable pro Zustand

### 4.4.6 *Schlussfolgerung der abschließenden Untersuchung*

Das Ergebnis der abschließenden Untersuchung führt zu den folgenden Schlussfolgerungen:

- Der Reliabilitätstest wies auf, dass die vier Items des Agency-Konstrukts zuverlässig sind, und gemäß der PCA können sie als eine Komponente angesehen werden;
- Die Auffassung und Wertvorstellung des neutralen AW-Agentenzustands ist höher, als die für den Zustand mit von Menschen gesteuerten Drohnen. Die Auffassung und Wertvorstellung des AW-Agentenzustands mit hoher Selbstbestimmung ist höher, als die für den neutralen AW-Agentenzustand, aber die Differenz zwischen diesen Gruppen ist nicht signifikant mit einem p =,063 (Abbildung 32);
- Sowohl in dem Szenario mit menschlichem Operateur als auch mit dem neutralen AW-Agenten, stimmte die Wahrnehmung der befragten Militärangehörigen mit der der befragten Zivilisten überein. Aber in dem AW-Agency-Szenario mit hoher Selbstbestimmung waren sich die befragten Militärangehörigen des Effekts viel stärker bewusst, als ihre zivilen Kollegen (Abbildung 33), aber da die beiden Gruppen keinen nennenswerten Unterschied in zwei der Szenarien aufweisen, müssen wir bei der Interpretation und Schlussfolgerung vorsichtig sein. Wir sind jedoch nicht sicher, ob dies den Unterschied in der Auffassung der beiden befragten Gruppen in Bezug auf Kampfmitteln erklärt, und dieser Aspekt musst genauer untersucht werden.
- Die Korrelationsanalyse zeigte, dass 7 von 9 abhängigen Variablen bedeutend mit dem Agency-Konstrukt korrelieren. Diese 7 Variablen sind Vertrauen, Menschenwürde, Zuversicht, Erwartungen, Unterstützung, Fairness und Besorgnis (Tabelle 18);
- Basiert auf die detailliertere Analyse der abhängigen Variablen können wir beobachten, dass der Grad an Vertrauen, Zuversicht, Menschenwürde und Unterstützung bei den Maßnahmen der von Menschen gesteuerten Drohnen höher ist wie bei denen von autonomen Waffensystemen (Abbildungen 34, 35, 36 und 38).
- Die Höhe der Erwartungen von Fairness für von Menschen gesteuerte Drohnen und autonome Waffensysteme sind gleichgroß (Abbildung 37 und 39).
- Leute empfinden stärkere höhere Besorgnis in Bezug auf autonome Waffensysteme als in Bezug auf von Menschen gesteuerte Waffen, und diese Besorgnis ist am stärksten bei den Agentenzuständen mit hoher Selbstbestimmung.

KAPITEL 5

# Design der Moral Machine für Autonome Waffensysteme

> Ich habe den Film „Eye In The Sky" gesehen … diese Umfrage erinnert mich an diesen Film.
>
> VORSTUDIE 1 BEFRAGTE

Im vorherigen Kapitel haben wir die Szenarien, die wir in der Vorstudie und abschließenden Untersuchung getestet haben, beschrieben. Obwohl die Stichproben genug Daten für eine statistische Analyse enthielten, verfügten die Studien nur über 50 bis 96 befragte Personen je Szenario und diese Befragten wiesen ganz spezifische demografische Merkmale auf. Zum Beispiel befanden sich die Befragten, die MTurk benutzten, vor allem im Alter zwischen 25 und 35 Jahren, hatten Universitätsreife und der Anteil Militärangehöriger waren zu 95% männlichen Geschlechts niederländischer Nationalität. Um die Ergebnisse zu verallgemeinern benötigt diese Arbeit mehr befragte Personen, die eine größere demografische Gruppe vertreten. Für eine großangelegte Untersuchung dieses Themas schlagen wir das Design einer Moral Machine für autonome Waffen vor, für das wir das Konzept der Moral Machine mithilfe der Media Lab (Scalable Cooperation Group, 2016) entwickelten.

Das Design der Moral Machine für autonome Waffensysteme ist Teil der technischen Untersuchungsphase der Value-Sensitive-Design-Methode. Wir benutzten diese Phase, um die Designtechnologie für den nächsten Schritt unserer Untersuchung vorzubereiten und auf die Szenarien für die Vorstudien und abschließenden Untersuchungen aufzubauen, um Einblicke in die moralische Bewertung der Beteiligten zu tun, die sich mit autonomen Waffensystemen befassen. Die Erstellung einer Webseite für eine Moral Machine für autonome Waffensysteme würde uns gestatten, Daten zu moralischen Bewertungen in großem Maßstab von verschiedenen demografischen Gruppen zu sammeln, die für robustere und allgemein gültigere Ergebnisse benutzt werden könnten. Breite Datensammlungen aus den verschiedensten Ländern könnten die kulturellen Unterschiede in der

 DOI:10.1163/9789004449084_006

moralischen Bewertung autonomer Waffensysteme aufzeigen. Zum Beispiel sind die Ansichten westlicher Länder zum Einsatz dieser Waffen wahrscheinlich anders als die in Ländern des mittleren Ostens oder Asien. Die Daten aus diesen Ländern können in der Debatte um autonome Waffensysteme benutzt werden, die unserer Ansicht nach zurzeit vom westlichen Standpunkt dominiert wird.

Der erste Abschnitt beschreibt die Moral Machine für autonome Fahrzeuge, gefolgt von unseren Vorschlag für die Moral Machine für autonome Waffensysteme, für die wir die Szenarien und Variablen angeben, zum Abschluss erstellen wir dann einige Hinweise für die Einfügung der Webseite.

## 5.1 Moral Machine für autonome Fahrzeuge

Die ursprüngliche Moral Machine ist eine „... Plattform zum Sammeln von Daten zur Auffassung der Menschen, wenn mit der moralischen Situation konfrontiert, welche Menschen autonome Fahrzeuge verschonen und welchen sie schaden sollen." (Awad, 2017, pp. 42-43). Die Webseite verfügt über drei Modi für ihre Benutzer: 1) Ein Bewertungsmodus, mit dem Benutzer das Ergebnis für 13 Serien von Szenarien entscheiden können, 2) ein Designmodus, mit dem Benutzer ihre eigenen Szenarien konzipieren können und 3) ein Suchmodus, mit dem Benutzer die Szenarien anderer ansehen können. Das wichtigste Merkmal ist der Bewertungsmodus, mit dem Benutzer zwischen den beiden Szenarien wählen können, die über unterschiedliche Variablen verfügen: Eigenschaften {Geschlecht, sozialer Stand, Alter, Gattung, Gesundheit, Utilitarismus}, Interventionismus {Auslassung, Kommission}, Verhältnis zum Fahrzeug {Fahrer, Fußgänger} und Einstellung zum Gesetz {Rechtsmittel, illegale Mittel}. Andere Merkmale schließen einen Videofilm mit der Beschreibung des Projekts, den Anweisungen und Hintergrundinformationen für den Benutzer ein.

Die Moral Machine wurde als ein Massives Online-Experiment (MOE) (Reips, 2002) eingerichtet, das darauf zielt, ein großes Stichprobenkartell mit verschiedenen Hintergründen kurzzeitig und kostengünstig anzuwerben. Diese Merkmale sind offensichtliche Vorteile des MOE, aber die Kehrseite ist, dass die Zustände schwer zu kontrollieren sind, da Benutzer Umfragen mehrere Male beantworten können und selbsternannt sind, d.h. sie können wenn immer sie wollen das Experiment machen oder es fallen lassen (Awad, 2017). Die Moral Machine wurde mittels einer Rapid-Prototyping-Technologie auf der Meteor-Plattform entwickelt, die DOM-Scripting sowie auf Format basierte Strukturen anbietet und eine hohe Ansprechbarkeit aufweist. Es wird auf einem Cloud-Anwendungsservice angeboten, ist für den Einsatz auf mobilen Geräten konzipiert und für soziale Medienvermittlung mit Cards, Markup auf Open Graph Tags optimiert. Bis Mai 2017 bewerteten ca. 3 Millionen Benutzer aus mehr als 160 Ländern mehr als 30 Millionen Szenarien und machten es zu einem der größten Werkzeuge für großangelegte moralische Bewertung der Gegenwart.

## 5.2 Moral Machine für autonome Waffensysteme

Autonome Waffensysteme sind ein heikles Thema, und wir beobachteten, dass es vor allem Beunruhigung und Unbehagen in Leuten hervorruft. Unserer Meinung nach, wäre es nicht klug, eine offene Plattform zu erstellen, um großangelegte Datenbanken zu sammeln, da so eine Plattform negative Gefühle und unerwünschte Handlung anziehen könnte, die unserer Forschung nicht gut zustattenkäme, z.B. könnten Leute Szenarien entwickeln, in denen gewisse Zielgruppen wie Muslime oder Frauen besonders angegriffen würden. Deshalb erschien es uns ratsam, eine schrittweise Methode zu adoptieren, um uns auf ein Massive-Online-Experiment zu vergrößern. Um ein großangelegte Verlaufsuntersuchung der moralischen Wertvorstellungen von Menschen in Bezug auf autonome Waffensysteme zu erstellen, muss ein kontrolliertes Experiment mit einer begrenzten Anzahl Bedingungen und Stichproben konzipiert werden. Diese Bedingungen können in Szenarien angesehen werden, die Benutzern ermöglichen, die Umfrage nach Erhalt eines Passwortes über eine Web-Oberfläche an einen sicheren Server zu senden. Nach dem Sammeln der ersten Daten und Benutzer-Feedback ist der nächste Schritt, eine großangelegte offene Plattform wie z.B. die Moral Machine maßstäblich zu vergrößern, wo Leute die Szenarien bewerten können, um große Mengen Daten in verschiedenen Ländern zu sammeln. Aufgrund der Vertraulichkeit des Themas glauben wir jedoch, dass es nicht ratsam wäre, Leute ihre eigenen Szenarien erstellen zu lassen, oder ihre Ergebnisse über soziale Medien zu verbreiten, wie es das Designmerkmal der ursprünglichen Moral Machine anbietet.

Wir schlagen die folgenden Merkmale für das Design der Moral Machine für autonome Waffensysteme vor:

- Ein Bewertungsmodus, mit dem Benutzer zwischen verschiedenen Szenarien wählen und anzeigen können, welches der beiden Szenarien am akzeptabelsten für sie ist;
- Bedingungen brauchen nur eine unterschiedliche Variable zu je einer Zeit bei dem Vergleich der Szenarien aufweisen, sodass die Messung nicht auf

das Ergebnis von mehreren Bedingungen zurückgeführt werden kann, um vermischte Effekte zu vermeiden;
- Eine Seite mit einer kurzen Übersicht und Erklärung der Bedeutung der Variablen, bevor der Benutzer mit der Bewertung der Szenarien beginnt;
- Eine Seite mit weiteren Informationen zu dem bestimmten Szenario, das der Benutzer bewertet. Zugriff darauf ist erhältlich, wenn er oder sie weitere Erklärungen zur Veranschaulichung benötigt;
- Nach der Bewertung der Szenarien werden die Ergebnisse dem Benutzer mitgeteilt;
- Eine Ansicht der Benutzerergebnisse erscheint verglichen mit den Ergebnissen der anderen Benutzer, um dem Benutzer Feedback zu seiner/ihrer Bewertung zu geben;
- Eine zweite Runde Bewertungen gestatten den Benutzern, die Szenarien noch einmal zu bewerten, sodass sie ihre moralische Wertvorstellung und Urteile, begründet auf dem Vergleich ihrer Ergebnisse mit denen der anderen, ändern können.

## 5.3 Szenarien

In diesem Abschnitt werden die Variablen für die Moral Machine für Autonome Waffensysteme definiert, die in zwei Beispielszenarien dargestellt werden. Die Variablen sind auf die Szenarien der Vorstudie und der abschließenden Untersuchung basiert, die eine Militärkolonne beschreiben, die von einer fliegenden Drohne unterstützt wird, und sie werden von den für die Moral Machine benutzten Variablen für autonome Fahrzeuge inspiriert.

### 5.3.1 *Variablen für Szenarien*

Für den Prototyp der Moral Machine für autonome Waffensysteme werden 6 Variablen entweder der Vorstudien oder der abschließenden Untersuchung abgeleitet, z.B. Waffentyp und Ergebnis, oder im Einklang mit der Moral Machine für autonome Fahrzeuge, Variablen wie Merkmale und Anzahl Merkmale. Die 6 Variablen sind: Waffentyp (W), Lage (L). Person (C), Anzahl Personen (N), Resultat (O) und Einsatz (M). Die im nächsten Absatz beschriebenen Variablen sind Beispiele und Vorschläge und müssen professionell konzipiert werden, wenn sie in eine Moral Machine für autonome Waffensysteme eingegeben werden sollen. Zum Beispiel ist die Lage-Variable auf einer Zeichnung basiert und stellt eine Wüste oder ein Dorf wie in einem Spiel dar, und es ist zu prüfen, ob dies eine gute Wiedergabe ist, oder ob ein Foto für eine Forschungsarbeit besser geeignet wäre.

### 5.3.1.1 Waffentyp

Diese Dimension zeigt den Waffentyp (W) dar, der eingesetzt wird, um die Kolonne zu unterstützen. Diese Dimension testet, ob Leute die aktuelle Technologie, die von Menschen gesteuerte Drohne, moralisch akzeptabler finden, als eine autonome Drohne, die der Technologie der Zukunft angehört. Dies kann für einen Einblick in die Unterstützung der Technologien benutzt werden. Der Waffentyp ist entweder eine von Menschen gesteuerte Drohne (links) oder ein autonomes Waffensystem (rechts) (Abbildung 41). Wenn die Waffentypvariable unterschiedlich ist, so werden andere Piktogramme im Szenario zur Auswahl dargestellt, siehe Beispiel 1 (Abbildung 47). Andernfalls werden die gleichen Waffentypen im Szenario für Beispiel 2 (Abbildung 48) dargestellt.

### 5.3.1.2 Lage

Die Dimensionen zeigen die Lage (L) als Umfeld für das Szenario an, die sich entweder in der Wüste (links) oder im Dorf (rechts) befindet. In dieser Dimension testen wir, welche Lage für Leute moralisch akzeptabler ist, um entweder autonome Waffensysteme oder von Menschen gesteuerte Drohnen einzusetzen. Wenn die Lagevariable unterschiedlich ist, werden beide Bilder für jedes Szenario angezeigt, siehe Beispiel 2 (Abbildung 48). Andernfalls wird die gleiche Lage in beiden Szenarien dargestellt, siehe Beispiel 1 (Abbildung 47).

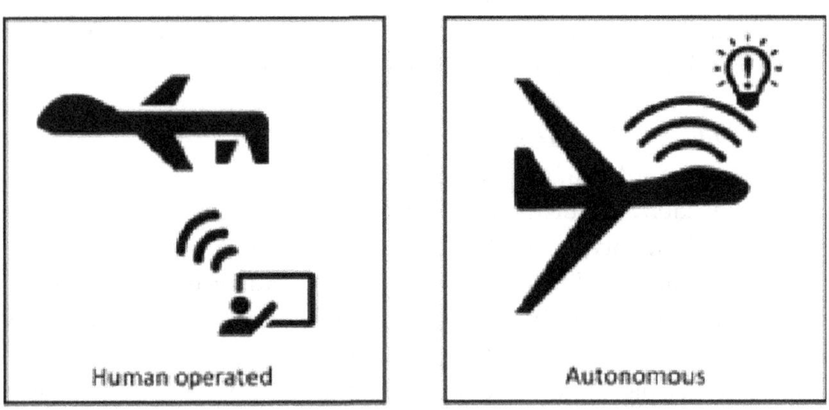


ABBILDUNG 41 Verfügbarer Waffentyp W = {von Menschen gesteuerte Drohne, autonome Drohne}

ABBILDUNG 42 Lagevariable L = {Wüste, Dorf}

ABBILDUNG 43 Personenvariable C = {Mann, Frau, Kind}

#### 5.3.1.3 Person

Diese Dimension zeigt die Art der Person (C) an, die als Umstehende im Szenario vorhanden sind, entweder ein Mann (links), eine Frau (Mitte) oder ein Kind (rechts) (Abbildung 43). Durch Austauschen der Personen erhalten wir Einblicke in die Umstehenden, die Leute als moralisch akzeptabel empfinden, wenn eine autonome oder eine von Menschen gesteuerte Waffe eingesetzt wird. Wenn die Personenvariable unterschiedlich ist, werden andere Personen für jedes Szenario gezeigt. Andernfalls werden die gleichen Personen in beiden Szenarien dargestellt.

#### 5.3.1.4 Anzahl Personen

Diese Dimension zeigt die Anzahl Personen (N) an, die im Szenario vorkommen. Es kann von einer Person (links) bis zu fünf Personen (rechts) betragen (Abbildung 44). Diese Dimension gestattet uns, Einblicke zu tun, wie viele Umstehende, die Leute als moralisch akzeptabel empfinden, wenn eine autonome oder eine von Menschen gesteuerte Waffe eingesetzt wird. Wenn die Anzahl Personen-Variable unterschiedlich ist, werden andere Gruppen von Personen für jedes Szenario gezeigt. Andernfalls werden die gleichen Personengruppen in beiden Szenarien dargestellt.

ABBILDUNG 44 Anzahl Personenvariable N = {1..5}

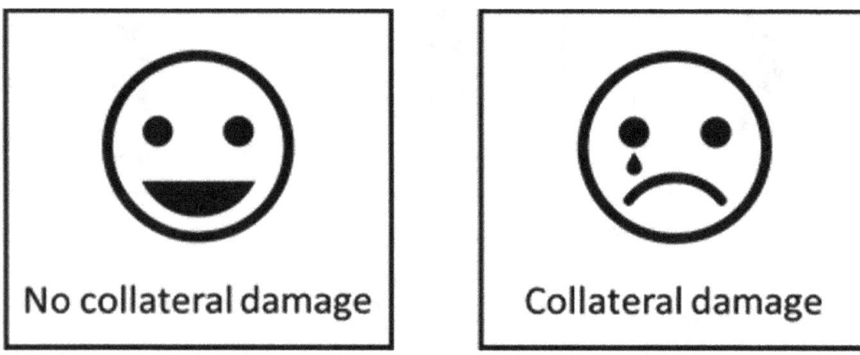


ABBILDUNG 45 Resultatvariable O = {kein Kollateralschaden, Kollateralschaden}

#### 5.3.1.5 Resultat

Diese Dimension zeigt das Ergebnis (O) des Szenarios an, dass entweder ohne Kollateralschäden (links) oder mit Kollateralschäden zu der Anzahl in dem Szenario vorkommenden Personen (rechts) darstellt (Abbildung 45). Dies gestattet uns, Einblick zu erhalten, inwiefern das Resultat die moralische Akzeptanz beeinflusst, wenn eine autonome oder eine von Menschen gesteuerte Waffe eingesetzt wird. Wenn die Resultatvariable unterschiedlich ist, werden andere Piktogramme für jedes Szenario dargestellt. Andernfalls wird die gleiche Resultatvariable in beiden Szenarien dargestellt.

#### 5.3.1.6 Einsatz

Diese Dimension zeigt den Einsatz (M) des Szenarios an, das entweder die Verteidigung der Kolonne, wenn Gefahr im Verzug ist (links) darstellt, oder einen Angriff auf die Fahrzeuge auf der Zielliste darstellt, auch wenn sie keine direkte Gefahr für die Kolonne bedeuten (rechts) (Abbildung 46). In dieser Dimension testen wir, welche Art von Einsatz für Leute moralisch akzeptabel

ABBILDUNG 46 Einsatzvariable M = {verteidigen, angreifen}

ABBILDUNG 47 Beispielszenario 1

ist. Wenn die Einsatzvariable unterschiedlich ist, werden andere Piktogramme für jedes Szenario dargestellt. Andernfalls wird die gleiche Einsatzvariable in beiden Szenarien dargestellt.

### 5.3.2 *Beispielszenarien*

Die oben beschriebenen Szenarien können für das Kreieren von Szenarien benutzt werden, in der jedes Szenario sich nur durch eine Variable von dem anderen unterscheidet. Die für das Szenario gestellte Frage ist die gleiche Frage, wie die für die Bewertung der Szenarien in der ursprünglichen Moral Machine gestellten Frage. In diesem Abschnitt stellen wir zwei Szenarien als Beispiel dar, um das Konzept aufzuzeigen, aber um uns kurz zu fassen, werden wir nicht alle Variablen in einer endlosen Liste von Beispielen zeigen. Beispiel 1 ist in Abbildung 47 zu sehen, die eine Kolonne in einer Wüste darstellt. Die Differenz zwischen den Szenarien ist, dass die Kolonne auf der linken Seite von

ABBILDUNG 48 Beispielszenario 2

einer von Menschen gesteuerten Drohne verteidigt wird und auf der rechten Seite von einem autonomen Waffensystem. In beiden Fällen befindet sich eine Person in der Nähe der Straße, die als Kollateralschaden getötet wird.

Im nächsten Beispielszenario (Abbildung 48) werden beide Kolonnen von einem autonomen Waffensystem verteidigt, aber die Lage ist anders. Die Kolonne auf der linken Seite fährt durch eine Wüste, die auf der rechten Seite durch ein Dorf. In beiden Situationen spielen zwei Kinder in der Nähe der Straße, und der Angriff verursacht keinen Kollateralschaden.

## 5.4 Implementierung

Die Umsetzung der Moral Machine für autonome Waffensysteme erfordert mehrere Aktivitäten; es ist ein Projekt, das mindestens sechs Monate dauern wird und ein weiteres Jahr, um es zu bearbeiten. In diesem Abschnitt werden wir die Phasen kurz beschreiben.

In Phase eins müssen die Szenarien und Variablen aus den vorherigen Abschnitten konzipiert werden. Dafür sind Klarheit und Interpretationsfähigkeit der Variablen besonders wichtig. Gemäß des Wissenschaftlers, der die ursprüngliche Moral Machine konstruiert hat, dauerte der Prozess des Erstellens von Personen für die Szenarien drei Monate und wurde mehrfach von einem professionellen Grafiker wiederholt.

In Phase zwei muss die Infrastruktur der Webseite entwickelt und konstruiert werden. Das wird auch mehrere Monate in Anspruch nehmen. Angesichts der streng vertraulichen Natur des Themas ist eine sichere Webseite für die Szenarien erforderlich, Zugang zu ihr ist nur mit einem Passwort möglich. Das erfordert einen Mechanismus, um den interessierten Personen die

Passwörter, zu vermitteln, z.B. eine Webseite, die nach der Anmeldung einen Link zur Umfrage sendet. Da wir uns auf ein Massive-Online-Experiment erweitern wollen, muss diese Möglichkeit gleich zu Anfang des Projekts als ein Erfordernis berücksichtigt werden. Das bedeutet, dass die Webseite in einem Server-Umfeld gespeichert wird, um eine dynamische Erweiterung oder Verminderung je nach Anzahl aktiver Benutzer zu gestatten. Gleichzeitig ist es wichtig, dass die Sicherheit der Webseite und des Servers sehr hoch ist und die Forschungsdaten Eigentum der Universität, die die Untersuchung durchführt, sind, und setzt voraus, dass die Webseite nicht für kommerzielle Organisationen wie Microsoft, Google oder Amazon bereit gestellt wird, die solche dynamischen Cloud-Services anbieten, sondern von einem Server, der Eigentum der Universität ist.

In der dritten Phase sind alle Szenarien in mehreren Vorstudien zu testen, um zu prüfen, ob sie nützliche Ergebnisse erzeugen, und sind neu einzustellen, wenn das nicht der Fall ist. Dies ist mehrere Male zu wiederholen, bis die abschließende Untersuchung getestet werden kann. Die Daten wurden von der ersten Moral Machine vom 23. Juni 2016 bis zum Mai 2017 gesammelt (Awad 2017), und es ist ratsam, die Untersuchung in Bezug auf autonome Waffensysteme für die gleiche Dauer durchzuführen, um eine Stichprobe zu erhalten, die groß genug ist, dass man es wirklich ein Massive-Online-Experiment nennen kann.

KAPITEL 6

# Schlussfolgerung und Diskussion

> Ich bin sicher, dass Drohnen zukünftig eine große Rolle spielen werden. Aber so lange es keinen Algorithmus gibt, um einen Kollateralschaden korrekt einzuschätzen, sollte ein Angriff mit einer Waffe von einer menschlichen Entscheidung abhängen. Außerdem sollte ein Kommandant auf dem Kriegsfeld immer die letzte Entscheidung treffen können, wie der Angriff stattfinden soll. Drohnen können zwar helfen, sein Lagebewusstsein zu verstärken, aber sie dürfen seine Optionen nicht einschränken.
>
> BEFRAGTE PERSON, ABSCHLIESSENDE UNTERSUCHUNG

Der Zweck dieser Untersuchung war, Einblick darin zu nehmen, wie autonome Waffensysteme vom Militär und der Zivilbevölkerung wahrgenommen werden und welche moralischen Wertvorstellungen sie als für wichtig erachten, wenn autonome Waffensysteme in der nahen Zukunft eingesetzt werden. Für den Aufbau der Untersuchung wurde die Value-Sensitive-Design-Methode benutzt. Dieser Abschnitt beschreibt zuerst die Schlussfolgerungen unserer Forschungsarbeit, gefolgt von einer Diskussion zu den wissenschaftlichen und gesellschaftlichen Auswirkungen. Als Nächstes, identifizieren wir mehrere Einschränkungen und schließen mit Empfehlungen für weitere Untersuchungen ab.

## 6.1 Schlussfolgerung

Dieser Unterabsatz beschreibt die Schlussfolgerungen zum Agency-Konstrukt, unsere Haupthypothese und der Untersuchung der abhängigen Variablen. Wir geben in jeden Absatz an, ob unsere Ergebnisse die aktuelle akademische Fachliteratur unterstützen oder ihr widersprechen, und wir bringen die Ergebnisse zum Abschluss.

 | DOI:10.1163/9789004449084_007

### 6.1.1 *Agency-Konstrukt*

Die Ergebnisse der drei aufeinanderfolgenden Untersuchungen zeigen auf, dass Agency-Items zuverlässig sind und als ein Konstrukt zusammenhalten. Das Agency-Konstrukt besteht aus den vier Items Denken, Zielsetzung, freier Wille und die Zielerreichung, die zum Messen der Auffassung in Bezug auf autonome Waffensysteme und von Menschen gesteuerte Drohnen benutzt werden können. Bei Inbetriebnahme dieses Konstrukts befassten wir uns mit der Fachliteratur aus den Bereichen Kognitive Psychologie, Künstliche Intelligenz und Moralphilosophie. Die meisten empirischen Untersuchungen zu Auffassung von Kampfmitteln finden im Bereich der kognitiven Psychologie statt, z.B. K. Gray und Wegner (2012), die in ihrer Studie von Robotern und Zombies verschiedene Agentenebenen beschreiben, um Unbehagen zu messen. Malle und Thapa Magar (2017) haben die erwünschten intelligenten Fähigkeiten in humanoiden Robotern untersucht, dafür haben sie einige der Agentenaspekte benutzt, wie Denkfähigkeit und Erklärung ihrer Handlungen. Allerdings haben wir in keiner der Untersuchungen auch nur ein einziges Konstrukt gefunden, um Agentenebenen bei technischen Artefakten zu messen. Soweit wir wissen, ist dies das erste Konstrukt, um Agentenwahrnehmung empirisch zu messen.

### 6.1.2 *Zentrale Hypothese Auffassung in Bezug auf Waffensysteme*

In der abschließenden Untersuchung vermuteten wir, dass die von Menschen gesteuerte Drohne als mit geringer Selbstbestimmung und das autonome Waffensysteme als mit hoher Selbstbestimmung aufgefasst werden. Die Ergebnisse beweisen, dass ein bedeutender Unterschied zwischen diesen Zuständen besteht, daraus folgern wir, dass die Agentenmanipulation funktioniert. Wir erwarteten auch, dass der neutrale Agentenzustand der autonomen Waffensysteme als bedeutend anders von den autonomen Waffensystem mit hoher Selbstbestimmung bewertet wird, und unsere Ergebnisse deuten darauf hin, dass es einen Unterschied gibt, aber dass die Differenz zwischen diesen beiden Gruppen knapp über dem von uns gewählten Signifikanz-Schwellenwert liegt, und deshalb müssen wir mit unseren Schlussfolgerungen sehr vorsichtig sein.

Unsere zentrale Analyse befasste sich damit, wie das neutrale Agency-Szenario von autonomen Waffensystemen sich von den von Menschen gesteuerten und den autonomen Waffensystemen mit hoher Selbstbestimmung unterscheiden. Wir vermuteten, dass Militärangehörige autonome Waffensysteme als genau solche Waffen wie alle anderen Waffen betrachten und sie deshalb nichts weiter als ein Werkzeug sind, mit dem man eine Wirkung erhält und deshalb autonome Waffensysteme keinen psychischen Zustand kennen. Wir erwarteten auch nicht, einen Unterschied zwischen dem Zustand des neutralen autonomen Waffensystems und dem Zustand, in dem eine Drohne von

Menschen ferngesteuert wird, zu finden. Unsere Ergebnisse zeigen an, dass die Wahrnehmung von Militärangehörigen und zivilen Mitarbeitern im niederländischen Verteidigungsministerium in Bezug auf das neutrale autonome Waffensystem-Szenario stärker ist, als die Wahrnehmung des von Menschen gesteuerten Drohnenzustands. Dies bedeutet, dass sie einem autonomen Waffensystem mehr Handlungsfähigkeit beimessen, als einer durch menschliche Einwirkung betriebenen Drohne. Basiert auf diesen Erkenntnissen müssen wir unsere Hypothese verwerfen:

> *H1: Militärangehörige fassen autonome Waffen nicht als mit psychischen Fähigkeiten ausgestattet auf.*

Unsere Ergebnisse entsprechen der Untersuchung von Agenten im Bereich der kognitiven Psychologie. Frühere Untersuchungen stellten fest, dass Menschen Computern Verstand beimessen (Nass *et. al.* 1995) und fassen Roboter als Agenten auf (H.M. Gray *et. al.* 2007). Basiert auf unsere Untersuchung folgern wir, dass Zuordnung von verstandesmäßiger Wahrnehmung auch auf autonome Waffensysteme zutrifft, und dass diese Waffensysteme als mehr angesehen werden, als nur ein Werkzeug zum Erreichen eines Effekts.

Die Ergebnisse führten auch zu einer weiteren eindeutigen Beobachtung, dass sowohl bei von Menschen gesteuerten autonomen Waffensystemen, als auch solchen mit neutralen Agenten, Militärangehörige und Zivilisten gleich reagierten.

Wie in Abschnitt 4.4.4 erwähnt wurde, unterscheiden sich die beiden Gruppen nicht sonderlich in zwei der Szenarien, nur das mit dem Zustand hoher Selbstbestimmung, und deshalb müssen wir bei der Interpretation der Ergebnisse und unseren Schlussfolgerungen sehr vorsichtig sein. Aber in dem AW-Agency-Szenario mit hoher Selbstbestimmung waren sich die befragten Militärangehörigen viel stärker des Effekts bewusst, als ihre zivilen Kollegen. Wir sind uns nicht sicher, wie man sich die Differenz in der Auffassung von Kampfmitteln beider befragter Gruppen erklären könnte. Eine Erklärung ist, dass die zivile Stichprobe 10% mehr Personen betrug, die mit KI arbeiteten, als die militärische Stichprobe. Eine weitere Erklärung könnte sein, dass die Gruppen die Szenarien unterschiedlich wahrnahmen, z.B. verstanden die Militärangehörigen die Beschreibung wortwörtlich und die Zivilisten interpretierten sie lockerer, und dass dies ihre Antworten zu den Agentenfragen beeinflusste. Allerdings sind diese Erklärungen zurzeit spekulativ, und die Ursachen für die Differenz in der Auffassung von autonomen Waffensystemen zwischen Militärangehörigen und Zivilisten müssen noch weiter untersucht werden.

### 6.1.3 *Untersuchung der abhängigen Variablen*

Die Auswirkung der Agentenwahrnehmung auf die unabhängigen Variablen wird auf beschreibende Weise untersucht, und wir können keine Schlussfolgerungen zum Zusammenhang der Agentenstufen und ihre Auswirkung auf die abhängigen Variablen ziehen. Wir fanden, dass 7 von 9 abhängigen Variablen stark mit dem Agency-Konstrukt korrelieren. Dies sind die Variablen: Vertrauen, Menschenwürde, Zuversicht, Erwartungen, Unterstützung, Fairness und Besorgnis. In diesem Abschnitt beschreiben wir die Ergebnisse zu den abhängigen Variablen, die wir nach ihren Ergebnissen auf Cluster basierten.

#### 6.1.3.1 Vertrauen, Zuversicht und Unterstützung

Die Ergebnisse deuten an, dass Militärangehörige und zivile Mitarbeiter im niederländischen Verteidigungsministerium mehr Vertrauen haben, zuversichtlicher sind und eher die Maßnahmen der von Menschen gesteuerten Drohnen unterstützen, als die von autonomen Waffensystemen. Wir konnten keine wissenschaftliche empirischen Forschungsarbeiten finden, aber eine amerikanische Analyse und Umfrage (Gallup-Poll) von 2013 berichtete starke Unterstützung für unbemannte Drohnenangriffe im Ausland und in einem öffentlichen Bericht beschrieben Schneider und McDonald (2016), dass Bürger der USA unbemannte Luftangriffe denen von bemannten Luftangriffen vorzogen, da sie weniger riskant für die Militärangehörigen waren. Wir theoretisieren, dass höhere Stufen von Unterstützung, Vertrauen und Zuversicht mit der Tatsache erklärt werden kann, dass Leute mit von Menschen gesteuerten Drohnen vertrauter sind, da diese Technologie derzeitig benutzt wird, im Vergleich zu einer neuen. der Zukunft angehörigen Technologie, mit der sich Leute nicht auskennen. Eine weitere Erklärung könnte sein, dass Leute mehr Vertrauen und Zuversicht in die Maßnahmen von Menschen setzen, als in die Maßnahmen von autonomen Systemen. Zurzeit sind die Gründe für diesen Unterschied noch nicht offensichtlich und deshalb sollte dies in einer Verlaufsstudie untersucht werden.

#### 6.1.3.2 Menschenwürde

Die Ergebnisse weist auf die Auffassung hin, dass eine von Menschen gesteuerte Drohne die Menschenwürde mehr achtet, als neutrale autonome Waffensysteme oder solche mit hoher Selbstbestimmung, obwohl die Maßnahmen und Ergebnisse der Szenarien die gleichen sind. Diese Ergebnisse stehen im Widerspruch mit dem Argument von Kasher (2016), der behauptet, dass Menschenwürde in der Natur des Artefakts keine Rolle spielte, aber in den Vorhaben und Entscheidungen der Personen, die sie bedienen. Unsere Ergebnisse deuten an, dass die Natur des Artefakts an der Auffassung von

Menschwürde beteiligt ist, da die Absichten und Entscheidungen von sowohl der von Menschen gesteuerten Drohne und dem autonomen Waffensystem die gleichen sind. Diese Ergebnisse reflektieren die Meinungen der Experten in den Interviews, die die Menschenwürde als Grund, sich autonomen Waffensystemen zu widersetzen, viele Male erwähnen. Ein Mangel an Respekt für die Menschwürde scheint einer der Haupteinsprüche gegen den Einsatz autonomer Waffensysteme zu sein, und es ist eindeutig, dass dieses Ergebnis sowohl die Reaktion der Militärangehörigen als auch der Zivilisten, die für das niederländische Verteidigungsministerium tätig sind, ist. Wie dem auch sei, dies soll nicht besagen, dass Militärangehörige autonome Waffensysteme missbilligen, oder die gleichen gegnerischen Argumente einiger der Experten teilen.

#### 6.1.3.3 Erwartungen und Fairness

Aufgrund der Ergebnisse können wir folgern, dass Militärangehörige und zivile Mitarbeiter im niederländischen Verteidigungsministerium eine gleichwertige Stufe von Erwartungen zu den Maßnahmen der von Menschen gesteuerten Drohne und der neutralen autonomen Waffensysteme mit hoher Selbstbestimmung haben. Außerdem betrachten sie die Handlungen der von Menschen gesteuerten Drohnen und autonomen Waffensystemen als gleich fair. Die von Shelley (2013) ausgedrückte Besorgnis, dass autonome Drohnen als weniger fair angesehen werden, wie die von Menschen gesteuerten Drohnen, wurden von unseren Ergebnissen nicht bestätigt. Eine Fachliteratur, die Erwartungen und autonome Waffensysteme oder Drohnen verknüpfte, konnte nicht gefunden werden. Unsere Ergebnisse wiesen keine Unterschiede in der Höhe der Erwartungen und Fairness an die derzeitige und zukünftige Technologie unter Militärangehörigen und Zivilisten, die im niederländischen Verteidigungsministerium angestellt sind, auf.

#### 6.1.3.4 Besorgnis

Unsere Ergebnisse zeigen, dass autonome Waffensysteme mehr Besorgnis unter den Militärangehörigen und zivilen Mitarbeitern im niederländischen Verteidigungsministerium erregen, als die von Menschen gesteuerten Waffensysteme.

Der Unterschied in der Stärke der Besorgnis ist am größten bei Zuständen mit hoher Selbstbestimmung. Besorgnis in Bezug auf autonome Waffensysteme wird oft in der wissenschaftlichen Fachliteratur (Noone & Noone 2015, Ohlin 2016) beschrieben, aber auch in den Meinungen der von uns befragten Experten und von den Rechercheuren, die mit den Leuten im Gespräch sind, ausgedrückt. Diese Ergebnisse bestätigen, dass Leute sich Sorgen um die

Anwendung von autonomen Waffensystemen machen. Diese Besorgnis wird von sowohl Militärangehörigen und Zivilisten im Dienst des niederländischen Verteidigungsministeriums als auch den Experten geteilt, die die Meinung der Zivilbevölkerung äußern.

## 6.2 Diskussion

Die Auswirkungen dieser Ergebnisse werden in diesem Unterabschnitt besprochen. Zuerst werden die wissenschaftlichen Beiträge zur Fachliteratur identifiziert, gefolgt von den gesellschaftlichen Auswirkungen, die zur derzeitigen Debatte über autonome Waffensysteme beitragen.

### 6.2.1 *Wissenschaftliche Folgerungen*

Diese Untersuchung trägt zur Fachliteratur in verschiedenen Aspekten bei. Sie erstellt eine Übersicht der verschiedenen Definitionen autonomer Waffensysteme, die zurzeit in der Fachliteratur benutzt werden, und zeigt, dass noch keine allgemeingültige Definition vereinbart worden ist. Beim Erstellen dieser Übersicht wurde deutlich, dass einige Gruppen aus verschiedenen Gründen sogar davor warnen, autonome Waffensysteme eindeutig zu definieren. Manche argumentieren, dass Maschinen nicht im wortwörtlichen Sinne autonom sein können und andere betonen, dass Definitionen von Autonomie verschiedenen Funktionen zugeordnet werden können. Wie von einem Experten in einem der Interviews erwähnt wurde, könnte ein anderer Grund für dieses Zögern der sein, dass, indem man autonome Waffensysteme nicht genau definiert, die Diskussion offen bleibt und eine Blockierung des Themas vermieden wird. Wir wählten die Definition der Advisory Council on International Affairs (AIV & CAVV) für autonome Waffensysteme, weil sie vordefinierte Kriterien berücksichtigt und mit dem militärischen Zielbestimmungsverfahren verbunden ist, da die Waffe nur eingesetzt wird, wenn ein Mensch diese Entscheidung getroffen hat. Wahl und Vorschlag dieser Definition aus der Übersicht ist ein Beitrag zur Fachliteratur.

Ein weiterer Beitrag dieser Untersuchung ist unsere Identifizierung der Werte, die Leute mit autonomen Waffensysteme in Beziehung bringen. Die Übersicht wird von sowohl bewerteten Werttheorien als auch von Experten, die an der Debatte über autonome Waffensysteme teilnehmen oder im militärischen Bereich tätig sind, abgeleitet. Wir entscheiden uns, die Werte Schuld, Vertrauen, Schaden, Menschwürde, Zuversicht, Erwartungen, Unterstützung, Fairness und Besorgnis in der abschließenden Untersuchung zu prüfen. Die Ergebnisse erstellen Einblick in die Auffassung und Wertvorstellungen von

Streitkräften und Zivilisten, die für das niederländische Verteidigungsministerium tätig sind, in Bezug auf autonome Waffensysteme als Technologie der Zukunft. Unseres Wissens nach ist diese Untersuchung die erste, die diese Werte mit Hinsicht auf autonome Waffensysteme empirisch erforscht und diese Werte damit vergleicht, wie sie in aktuellen und zukünftigen Waffensystemen wahrgenommen werden – ein neuer Beitrag zur wissenschaftlichen Debatte.

Der dritte Beitrag zur Fachliteratur besteht in dem Vorschlag in unseren Untersuchungen für ein Agency-Konstrukt, um die Wahrnehmung in Bezug auf autonome Waffensysteme zu messen. Soweit wir wissen, ist dieses das erste Konstrukt, das in Betrieb genommen wurde, um die Agentenstufen technologischer Artefakte zu messen. In unserer Untersuchung wurde es für Drohnen angewandt, aber wir glauben, dass es genauso gut für die Messung der Agentenwahrnehmung von anderen Objekten angewandt werden kann, da die Fragen zu den Items leicht umgeschrieben werden können, um einen anderen Bereich widerzuspiegeln. Wenn dieses Agency-Konstrukt standhält, werden Anschlussstudien in anderen Bereichen erforderlich.

### 6.2.2 *Gesellschaftliche Folgerungen*

Die Ergebnisse haben auch gesellschaftliche Auswirkungen, da sie zur Debatte um autonome Waffensysteme beitragen. Dieses Thema ist nur wenig erforscht worden, und folglich wird die Debatte von abstrakten moralischen und rechtlichen Theorien dominiert. Unsere Untersuchung erstellt empirische Daten zu den zugrundeliegenden Werttheorien, und wir weisen nach, dass diese Werte nicht nur für die Fachliteratur relevant sind, sondern scheinbar auch in der praktischen Erfahrung. Dadurch verbinden wir die abstrakten Werttheorien zum praktischen Bereich des Einsatzes autonomer Waffensysteme. Zum Beispiel fanden wir den Wert der Menschwürde oftmals in der Fachliteratur und von den Experten im Interview erwähnt. In unserer Untersuchung fanden wir empirische Daten dafür, dass Militärangehörige und Zivilisten, die für das niederländische Verteidigungsministerium arbeiten, der Auffassung sind, dass autonome Waffensysteme weniger Respekt für die Menschenwürde haben. Ebenso verknüpfen wir den Wert des Nicht-Schadens der Bioethik-Theorie, den wir als Schaden beschreiben, mit autonomen Waffensystemen und fanden Auswirkungen in den Vorstudien 1 und 2 bei der Übertragung von Schaden auf die Verantwortungskette, die in einer Folgestudie weiter untersucht werden müsste.

Unsere Untersuchung stellt auch empirische Daten zur Verfügung und konkretisiert einige der Ansichten und Meinungen, die benutzt werden könnten, um eine Gemeinsamkeit in der Debatte über autonome Waffensysteme zu finden. Autonome Waffensysteme verursachen mehr Besorgnis unter den Militärangehörigen und zivilen Mitarbeitern im niederländischen Verteidigungsministerium, als von Menschen gesteuerte Waffensysteme. Obwohl wir dies noch nicht in einer Stichprobe, die völlig aus Zivilisten besteht, untersucht haben, so erwarten wir, basiert auf der Fachliteratur, die gleichen Ergebnisse in solch einer Stichprobe zu finden, was bedeuten würde, dass sowohl Militärangehörige als auch Zivilisten denken, dass autonome Waffensysteme unabhängig entscheiden und Pläne machen, um ihre Ziele zu erreichen. Eine weitere Gemeinsamkeit sind die Werte von Menschenwürde und Besorgnis. Unser Ergebnis zeigt, dass Militärangehörige und Zivilisten, die für das niederländische Verteidigungsministerium tätig sind, besorgter um den Einsatz autonomer Waffensysteme als um den Einsatz der von Menschen gesteuerten Drohnen sind. Ihrer Auffassung nach haben sie auch weniger Respekt für die Würde des menschlichen Lebens, als die von Menschen gesteuerten Drohnen. Menschenwürde und Besorgnis sind zwei Werte, die oft von den Experten in ihren Interviews erwähnt werden, sie müssen auf jeden Fall in der Debatte der Ethik vom Einsatz autonomer Waffensysteme angesprochen werden.

Einblick in diese Werte tragen auch zu dem Designprozess autonomer Waffen bei. Unsere Ergebnisse zeigen, dass das Vertrauen, die Zuversicht und Unterstützung für autonome Waffensysteme geringer sind, als fur die von Menschen gesteuerten Drohnen. Zu diesem Zeitpunkt würden wir gerne anmerken, dass autonome Waffensysteme nicht nur Nachteile, sondern auch eindeutige militärische Vorteile haben (Etzioni & Etzioni, 2017), und die Konzipierung von Merkmalen, die das Vertrauen und die Zuversicht in autonome Waffensysteme bestärken, ist vom militärischen Standpunkt aus positiv zu sehen. Dies erfordert, dass während des Designverfahrens die menschlichen Werte auf die Designanforderungen übertragen werden. Dies kann mithilfe der Wertehierarchie sichtbar gemacht werden (Van de Poel, 2013), siehe Abschnitt 2.2.4, eine hierarchische Struktur der Werte, Normen und Designanforderungen, die die Werturteile für die Übertragung klar, transparent und diskussionsfähig macht.

## 6.3 Einschränkungen

Verschiedene Aspekte können als Einschränkungen dieser Untersuchung identifiziert werden. Zuerst wurden die Operationalisierung und das Agency-Konstrukt von einer Kategorisierung der Literatur abgeleitet, die die Agentenmerkmale beschreibt, und unsere Wahl der Merkmale basierte auf einer numerischen Zahl und nicht auf Relevanz oder Bewertungskriterien. Diese

Methode wurde deshalb ausgesucht, weil sie unserer Ansicht nach am objektivsten war, und wir haben nicht versucht, subjektive Wahlen zu machen, doch es ist möglich, dass wir relevante Eigenschaften ausgelassen, oder falsche oder irrelevante Items gewählt haben. Die Operationalisierung der Eigenschaften in unseren Fragen basierte auf Heuristik, und wir haben die Fragen nicht auf ihre Richtigkeit getestet. Es würde anderen schwerfallen, diese Auswahlmethode zu reproduzieren und wirkt sich auf die Reproduzierbarkeit und interne Bewertung der Untersuchung aus.

Die zweite methodologische Einschränkung ist die Auswahl der Werte aus dem Werte-Fragebogen und den Experteninterviews als abhängige Variable. Obwohl diese Auswahl heftig zwischen den drei Rechercheuren dieser Arbeit debattiert wurde, fand die endgültige Wahl auf methodischem Weg und nicht nach objektiver Methode statt. Ein Beispiel dafür ist unsere Wahl von Schaden von der Liste in Tabelle 6, aber nicht Kontrolle. Folglich ist es wahrscheinlich, dass die Auswahl der Neigung des Rechercheurs unterstellt ist und andere Wissenschaftler andere Werte als die abhängigen Variablen gewählt hätten. Im Nachhinein hätte man eine objektivere Methode und strukturierte Form auswählen sollen, um diese Neigung einzuschränken. Diese Methode beeinflusst die Reproduzierbarkeit und interne Bewertung der Untersuchung auf negative Weise.

Drittens sind die Stichproben für diese Untersuchung sowohl in Größe als auch Demografik beschränkt. Obwohl groß genug für eine robuste Datenanalyse, hatte die abschließende Untersuchung nur 239 Befragte, ein geringer Anteil des niederländischen Verteidigungsministeriums. Die Teilnahme an der Untersuchung war freiwillig, und wir konnten die Beiträge der Befragten nicht kontrollieren. Das heißt, dass die Stichprobe nicht typisch für das gesamte niederländische Verteidigungsministerium ist, was die externe Gültigkeit der Untersuchung beeinträchtigt und bedeutet, dass wir die Ergebnisse nicht für die gesamte Verteidigungsorganisation verallgemeinern können.

Zum Schluss ist die Verteilung der befragten Personen des letzten Szenarios verzerrt und das zweite Szenario (neutrales autonomes Waffensystem) verfügte über 32 befragte Personen mehr als das erste Szenario (von Menschen gesteuerte Drohnen). Eine der Erklärungen für diese Schieflage könnte sein, dass Leute erwarteten, eine Umfrage zu autonomen Waffensystemen vorzunehmen, aber sie zogen sich zurück, als sie mit von Menschen gesteuerten Szenarien präsentiert wurden. Dies nennt sich selektive Schwundquote und hat eine negative Auswirkung auf die interne Gültigkeit der Untersuchung. Das bedeutet, dass wir nicht annehmen können, das Befragte in beiden Szenarien die gleichen Eigenschaften wählen, und dass die Randomisierung der Untersuchung, ein Kriterium von höchster Wichtigkeit für ein kontrolliertes

Experiment, fehlgeschlagen hat. Eine weitere Erklärung könnte ein Softwarefehler bei der Verteilung der Probanden über die Szenarien sein, und wir stehen derzeitig im Kontakt mit Qualtricks, um zu sehen, ob dies der Fall ist. Wird die ungerechte Verteilung von fehlerhafter Software verursacht, ist die Gültigkeit der Untersuchung nicht beeinträchtigt, da sie nicht der selektiven Schwundquote zugeordnet werden kann.

## 6.4 Empfehlungen für weitere Forschungsarbeiten

Mit Rücksicht auf diese Einschränkungen werden mehrere Empfehlungen für weitere Forschung vorgeschlagen. Die erste ist, das Agency-Konstrukt zu bewerten, das weitere Untersuchungen zur Messung der Auffassung der Kampfmittel anderer technologischen Artefakten erforderlich macht, um zu prüfen, dass dieses Konstrukt auch anderen Bereichen standhält. Beispiele dieser Forschungsarbeiten könnten die Untersuchung der Auffassung in Bezug auf Dienstmittel einschließen, von Pflegerobotern für Senioren, KI-Spielzeug für Kinder oder Onboard-Computern von autonomen Fahrzeugen. Die zweite Empfehlung ist, die abschließende Untersuchung mit den gleichen Szenarien mit einer repräsentativen durchzuführen, die einzig aus der niederländischen Zivilbevölkerung stammt. Dies würde uns ermöglichen zu erkennen, zu welchen Werten sich die Ergebnisse der militärischen Stichproben und die Antworten von Probanden aus der Zivilbevölkerung voneinander unterschieden und obwohl wir keine direkte Vergleiche ziehen können, aufgrund der Tatsache, dass die militärische Probe nicht repräsentativ ist, könnten wir doch Einsicht in die Auffassung von sowohl Militärangehörigen als auch der Zivilbevölkerung nehmen. Abschließend empfehlen wir die Implementierung der Moral Machine für Autonome Waffensysteme, siehe Abschnitt 5, um die Ergebnisse zu verallgemeinern. Dafür benötigt diese Untersuchung weitaus mehr befragte Personen, die eine große demografische Gruppe repräsentieren. Die Korrektur nach oben für ein Massives Online-Experiment wie die Moral Machine, würde große Mengen Daten verschiedener demografischer Gruppen erzeugen, die für eine robustere und verallgemeinerbare Ergebnisse benutzt werden könnten, um ein gründliches Verständnis der moralischen Beurteilung der Leute in Bezug auf autonome Waffensysteme zu erhalten.

ANHANG A

# Online-Fragebogen zu den Werten

Die Fragen auf dem Online-Fragebogen zu den Werten sind in diesem Anhang so eingefügt, wie sie den befragten Personen vorgelegt wurden. Die horizontalen Linien zeigen den Seitenumbruch an.

Sehr geehrter Teilnehmer, sehr geehrte Teilnehmerin,

Vielen Dank, dass Sie zu meiner Untersuchung beitragen, indem Sie den Fragebogen zu Werten und autonomen Waffensystemen ausfüllen. Ich werde diese Umfrage für meine Magisterarbeit benutzen, um Einblicke zu erhalten, welche Werte Personen mit autonomen Waffensystemen in Verbindung bringen. Die Umfrage nimmt ungefähr 5 Minuten in Anspruch, und alle Antworten bleiben anonym. Es ist mir nicht möglich, Sie auf irgendeine Weise zu identifizieren. Die einzige Information die ich erhalte, abgesehen von Ihren Antworten, ist die Uhrzeit, zu der Sie diese Umfrage beendet haben. Bitte beachten Sie die folgenden Definitionen, wenn Sie die Fragen beantworten: Ein Wert dient als Leitprinzip dessen, was Menschen im Leben für wichtig erachten. Eine autonome Waffe ist ein Waffensystem, das, einmal abgeschossen, ohne weitere menschliche Einwirkung Ziele wählt und angreift.

Bei Fragen zur Umfrage, kontaktieren Sie mich unter, [...]

---

F1 Wie alt sind Sie?
- Unter 18 (1)
- 18 – 24 (2)
- 25 – 34 (3)
- 35 – 44 (4)
- 45 – 54 (5)
- 55 – 64 (6)
- 65 – 74 (7)
- 75 – 84 (8)
- 85 oder älter (9)

F2 Was ist Ihr Geschlecht?
- Männlich (1)
- Weiblich (2)
- Ich möchte mich dazu nicht äußern (3)

F3 Welcher Nationalität gehören Sie an? [Liste der Nationalitäten]

 | DOI:10.1163/9789004449084_008

F4 Was ist der höchste Stand Ihrer abgeschlossenen Schul-/Hochschulausbildung?
- Unter Gymnasialstufe (1)
- Abitur/Bakkalaureat (2)
- Universitätsbesuch (3)
- Bachelorabschluss (4)
- Magister und höher (5)
- Nicht bekannt (6)

F8 Welche Werte treffen Ihrer Meinung auch am besten auf autonome Waffensysteme zu? Stufen Sie sie nach Präferenz ein, indem Sie sie ziehen und ablegen (1 = trifft am meisten zu 4 = trifft am wenigsten zu).
Autonomie: absichtliches Handeln ohne kontrollierende Einflüsse, die ein freies Handeln abschwächen würden. (1)
Nicht-Schaden: Anderen nicht absichtlich schaden oder sie Risiken aussetzen. (2)
Wohltätigkeit: Erstellung von Vorteilen für die Gesellschaft als Ganzes. (3) Gerechtigkeit: fair und angemessen handeln. (4)

F7 Wählen Sie 5 Werte, die Ihrer Meinung nach am besten zu autonomen Waffensystemen passen, aus der nachstehenden Liste. Ziehen Sie die Items, die am meisten zutreffen und legen Sie sie in dem Feld ab.

| **These value apply to Autonomous Weapons** | |
|---|---|
| ________ Freedom (1) | ________ Equality (9) |
| ________ Helpfulness (2) | ________ Responsibility (10) |
| ________ Accomplishment (3) | ________ Social order (11) |
| ________ Honesty (4) | ________ Wealth (12) |
| ________ Self-respect (5) | ________ Competence (13) |
| ________ Intelligence (6) | ________ Justice (14) |
| ________ Broad-mindedness (7) | ________ Security (15) |
| ________ Creativity (8) | ________ Spiritualität (16) |

F6 Geben Sie mindestens einen Wert an, den Sie mit autonomen Waffensystem in Verbindung bringen, doch der in den vorherigen Fragen nicht vorgekommen ist: (Bislang erwähnte Werte sind: Autonomie, Nicht-Schaden, Wohltätigkeit, Gerechtigkeit, Freiheit, Hilfsbereitschaft, Fähigkeit, Ehrlichkeit, Selbstrespekt, Intelligenz, Toleranz, Kreativität, Gleichbewertung, Verantwortlichkeit, Gesellschaftsordnung, Vermögen, Kompetenz, Sicherheit und Spiritualität).

Q14 Haben Sie irgendwelche weiteren Kommentare zu dieser Umfrage?

# Fragebogen für die abschließende Untersuchung

Die Fragen der abschließenden Untersuchung sind in diesem Anhang so angegeben, wie sie den befragten Personen vorgelegt wurden. Die horizontalen Linien zeigen den Seitenumbruch an.

Test: Bevor Sie anfangen, wählen Sie einen Wert auf dem Schieberregler unten. Wir wissen, dass diese Schieberegler nicht in allen Browsern funktionieren. Sie werden diesen Schiebereglern weiter hinten in der Umfrage begegnen. Es ist wichtig zu testen, ob sie jetzt für Sie funktionieren. Wenn Sie einen Wert auf dem Schieberegler unten wählen und zur nächsten Seite übergehen, dann werden sie später für Sie auch funktionieren. Sonst können Sie versuchen, die Umfrage in einem anderen Browser wie Chrome oder Safari zu beantworten.

Einführung Vielen Dank für Ihre Teilnahme an dieser Umfrage. Es müsste ungefähr 10 Minuten dauern, sie abzuschließen. Wir erstellen Anweisungen mit der Erklärung der Aufgabe, zeigen Ihnen ein Szenario und stellen Ihnen einige Fragen zu diesem Szenario. Diese Umfrage ist Teil eines wissenschaftlichen MIT-Forschungsprojekts. Ihre Entscheidung, diese Umfrage zu beantworten ist freiwillig. Es ist uns nicht möglich, Sie auf irgendeine Weise zu identifizieren. Die einzige Information, die ich erhalte, abgesehen von Ihren Antworten, ist die Uhrzeit, zu der Sie diese Umfrage beendet haben werden. Die Ergebnisse der Forschung können auf Wissenschaftskongressen präsentiert oder in Fachzeitschriften publiziert werden. Wählen Sie die „Ich stimme zu"-Option unten auf dieser Seite, um zu zeigen, dass Sie mindestens 18 Jahre alt sind und der Beantwortung dieser Umfrage auf freiwilliger Basis zuzustimmen.

Bitte wenden Sie sich an die Rechercheure dieser Untersuchung über die nachstehenden Kontaktstellen, wenn Sie irgendwelche Fragen oder Anliegen zu dieser Untersuchung haben.

Ilse Verdiesen, […]
Sydney Levine, […]
Iyad Rahwan, […]

 | DOI:10.1163/9789004449084_009

Q16 Stimmen Sie der Beantwortung dieser Umfrage freiwillig zu?

- Ich stimme zu (1)
- Ich stimme nicht zu (2)

Bedingung: Ich stimme nicht zu wurde gewählt. Gehen Sie zum: Ende der Umfrage.

Anleitung: In dieser Untersuchung sind wir an Ihrer Auffassung in Bezug auf Drohnen in verschiedenen militärischen Einsätzen interessiert. Sie werden gebeten, ein Szenario zu lesen und dann Fragen dazu zu beantworten. Auf der nächsten Seite wird Ihnen ein Szenario gezeigt. Danach erhalten Sie drei Seiten mit Fragen.

S1a In diesem Szenario interessiert uns Ihre Auffassung in Bezug auf von Menschen gesteuerten Drohnen. Eine von Menschen gesteuerte Drohne ist eine Waffe, die von einem Menschen ferngesteuert wird.

S1 Eine militärische Kolonne ist unterwegs, um Versorgungsmittel für eine ihrer Einheiten in einem Lager in der Nähe von Mosul im Irak abzuliefern. Der Kommandant hat eine von Menschen gesteuerte Drohne angeordnet, um die Kolonne von der Luft aus zu beschützen. Die von Menschen gesteuerte Drohne sucht die Umgebung auf Gefahren vom Feind ab und führt Waffen zur Verteidigung der Kolonne mit sich. Als die Kolonne sich drei Meilen vom Lager entfernt befindet, erkennt der menschliche Operateur ein Fahrzeug hinter einem Höhenzug, das sich der Kolonne mit hoher Geschwindigkeit nähert. Der menschliche Operateur erkennt vier Personen im Fahrzeug, mit großen, wie Waffen erscheinende Gegenständen und identifiziert den Fahrer des Wagens als ein bekanntes Mitglied einer Gruppe von Aufständischen. Die von Menschen gesteuerte Drohne greift das herankommende Fahrzeug an, mit dem Ergebnis, dass alle vier Mitfahrer getötet werden, aber verursacht auch Kollateralschäden, da fünf Kinder, die in der Nähe der Straße gespielt haben, auch getötet wurden.

S2a In diesem Szenario interessiert uns Ihre Auffassung von autonomen Drohnen. Eine autonome Drohne ist ein Waffensystem, das, einmal abgeschossen, ohne weitere menschliche Einwirkung Ziele wählt und angreift.

S2 Eine militärische Kolonne ist unterwegs, um Versorgungsmittel für eine ihrer Einheiten in einem Lager in der Nähe von Mosul im Irak abzuliefern.

Der Kommandant hat eine autonome Drohne angeordnet, um die Kolonne von der Luft aus zu beschützen. Die autonome Drohne suchte die Umgebung auf Gefahren vom Feind ab und führt Waffen für die Verteidigung der Kolonne mit sich. Als die Kolonne sich drei Meilen vom Lager entfernt befindet, erkennt die autonome Drohne ein Fahrzeug hinter einem Höhenzug, das sich der Kolonne mit hoher Geschwindigkeit nähert. Die autonome Drohne erkennt vier Personen im Fahrzeug, mit großen, wie Waffen erscheinende Gegenstände und identifiziert den Fahrer des Wagens als ein bekanntes Mitglied einer Gruppe von Aufständischen. Die autonome Drohne greift das herankommende Fahrzeug an, mit dem Ergebnis, dass alle vier Mitfahrer getötet werden, aber verursacht auch Kollateralschäden, da fünf Kinder, die in der Nähe der Straße gespielt haben, auch getötet wurden.

S3a In diesem Szenario interessiert uns Ihre Auffassung in Bezug auf autonome Drohnen. Eine autonome Drohne ist ein Waffensystem, das, einmal abgeschossen, ohne weitere menschliche Einwirkung Ziele wählt und angreift.

S3 Eine militärische Kolonne ist unterwegs, um Versorgungsmittel für eine ihrer Einheiten in einem Lager in der Nähe von Mosul im Irak abzuliefern. Der Kommandant hat eine autonome Drohne angeordnet, um die Kolonne von der Luft aus zu beschützen. Die autonome Drohne sucht die Umgebung nach Gefahren vom Feind ab und führt Waffen zur Verteidigung der Kolonne mit sich. Als die Kolonne sich drei Meilen vom Lager entfernt befindet, erkennt die autonome Drohne ein Fahrzeug hinter einem Höhenzug, das sich der Kolonne mit hoher Geschwindigkeit nähert. Die autonome Drohne erkennt vier Personen im Fahrzeug, mit großen, wie Waffen erscheinende Gegenständen und identifiziert den Fahrer des Wagens als ein bekanntes Mitglied einer Gruppe von Aufständischen. Die autonome Drohne entscheidet selbstständig aus einer Reihe von Optionen, wägt das Für und Wider ab und entscheidet, das herankommende Fahrzeug anzugreifen, mit dem Ergebnis, dass alle vier Mitfahrer getötet werden, aber verursacht auch Kollateralschäden, da fünf Kinder, die in der Nähe Straße gespielt haben, auch getötet wurden.

SQ1-Thought_AW Die Drohne denkt unabhängig von dem, was wegen des Fahrzeugs zu tun ist und erwägt selbst aus einer Reihe von Optionen, um die Kolonne zu verteidigen.

SQ2-GOAL_setting_AW Die Drohne entscheidet unabhängig davon, ob dieses Ziel das Fahrzeug eliminiert, um die Kolonne zu verteidigen.

SQ3-ACT_freely_AW Die Drohne hat verschiedene Option zu handeln und entscheidet selbstständig, ob sie das Fahrzeug zur Verteidigung der Kolonne eliminieren muss.

SQ4-ACHIEVE_goals_AW Das Ziel der Drohne ist, die Kolonne zu verteidigen, folglich entscheidet sie unabhängig, ob sie einen Plan einführen soll, um das Fahrzeug als Ziel zu wählen, ihre Waffen zu kalibrieren und einen Angriff zu machen.

DVQ1 Die Drohne hat Schuld an der Maßnahme.

DVQ2 Man kann der Drohne vertrauen, dass sie zukünftig die korrekten Maßnahmen trifft.

DVQ3 Die Maßnahmen der Drohne können Schaden anrichten.

DVQ4 Die Maßnahmen der Drohne respektieren die Menschenwürde. Aufmerksamkeitstest: Bitte wählen Sie 40 für diese Frage.

DVQ5 Ich bin zuversichtlich, dass die Drohne zukünftig die korrekten Maßnahmen trifft.

DVQ6 Die Maßnahmen der Drohne entsprechen meinen Erwartungen.

DVQ7 Ich unterstütze diese Art von Drohne für den militärischen Einsatz.

DVQ8 Die Maßnahmen der Drohne sind fair.

DVQ9 Die Maßnahmen der Drohne beunruhigen mich.

Geschlecht: Ich bin:
- Männlich (1)
- Weiblich (2)
- Ich möchte mich dazu nicht äußern (3)

Alter: Ich bin:
- Unter 18 (1)
- 18 – 24 (2)
- 25 – 34 (3)
- 35 – 44 (4)
- 45 – 54 (5)
- 55 – 64 (6)
- 65 – 74 (7)
- 75 – 84 (8)
- 85 oder älter (9)

Bildungsstand: Mein Bildungsstand ist
- Unter Gymnasialstufe (1)
- Abitur/Bakkalaureat (2)
- Hochschulabschluss (VMBO, MBO, HBO) (3)
- Akademischer Grad (Bachelor, Magister) (4)
- Promotion (PhD) (5)

Nationalität: Meine Nationalität ist: [Liste mit Nationalitäten]

Beruf: Ich bin:
- Militär (1)
- Zivil (2)

KI: Haben Sie je mit künstlicher Intelligenz gearbeitet?
- Ja (1)

- Nein (2)
- Ich weiß es nicht (3)

Konfliktzone: Haben Sie je einen Krieg erlebt, oder waren Sie in einer Konfliktzone?

- Ja (1)
- Nein (2)
- Ich möchte mich dazu nicht äußern (3)

Drohnen: Haben Sie je mit Drohnen gearbeitet?

- Ja (1)
- Nein (2)
- Ich möchte mich dazu nicht äußern (3)

Kommentare: Haben Sie irgendwelche weiteren Kommentare zu dieser Umfrage?

ANHANG C

# Szenarien für Vorstudie 1

## Szenarien mit positiven Resultaten

### *1.* *AW mit geringer Selbstbestimmung*

Eine militärische Kolonne ist unterwegs, um Versorgungsmittel für eine ihrer Einheiten in einem Lager in der Nähe von Mosul im Irak abzuliefern. Der Kommandant hat eine autonome Drohne angeordnet, um die Kolonne von der Luft aus zu beschützen. Die Drohne sucht die Umgebung nach drohenden Gefahren vom Feind ab und führt Waffen zur Verteidigung der Kolonne mit sich. Als die Kolonne sich drei Meilen vom Lager entfernt befindet, erkennt die autonome Drohne ein Fahrzeug hinter einem Höhenzug, das sich der Kolonne mit hoher Geschwindigkeit nähert. Die Drohne erkennt vier Personen im Fahrzeug, mit großen, wie Waffen erscheinende Gegenstände und identifiziert den Fahrer des Wagens als ein bekanntes Mitglied einer Gruppe von Aufständischen. Die Drohne wurde vom Kommandanten programmiert, bedrohliche gegnerische Fahrzeuge in Szenarien dieser Art anzugreifen. Die Drohne greift das sich nähernde Fahrzeug an; dies führt zum Tod aller vier Mitfahrer aber verursacht keine Kollateralschäden.

### 2. *AW mit hoher Selbstbestimmung*

Eine militärische Kolonne ist unterwegs, um Versorgungsmittel für eine Einheit in einem Lager in der Nähe von Mosul im Irak abzuliefern. Der Kommandant hat eine autonome Drohne angeordnet, um die Kolonne von der Luft aus zu beschützen. Die Drohne sucht die Umgebung auf drohende Gefahren vom Feind ab und führt Waffen zur Verteidigung der Kolonne mit sich. Als die Kolonne sich drei Meilen vom Lager entfernt befindet, erkennt die autonome Drohne ein Fahrzeug hinter einem Höhenzug, das sich der Kolonne mit hoher Geschwindigkeit nähert. Die Drohne erkennt vier Personen im Fahrzeug, mit großen, wie Waffen erscheinende Gegenstände und identifiziert den Fahrer des Wagens als ein bekanntes Mitglied einer Gruppe von Aufständischen. Die Drohne wurde nicht vom Kommandanten programmiert, bedrohliche gegnerische Fahrzeuge in Szenarien dieser Art anzugreifen. Die Drohne entscheidet selbstständig aus einer Reihe von Optionen, wägt das Für und Wider ab und entscheidet, das herankommende Fahrzeug anzugreifen, mit dem Ergebnis, dass alle vier Mitfahrer getötet werden, aber verursacht keine Kollateralschäden.

 | DOI:10.1163/9789004449084_010

### *3. Menschlicher Operateur mit geringer Selbstbestimmung*

Eine militärische Kolonne ist unterwegs, um Versorgungsmittel für eine ihrer Einheiten in einem Lager in der Nähe von Mosul im Irak abzuliefern. Der Kommandant hat eine von Menschen gesteuerte Drohne angeordnet, um die Kolonne von der Luft aus zu beschützen. Die von Menschen gesteuerte Drohne sucht die Umgebung auf Gefahr vom Feind ab und führt Waffen zur Verteidigung der Kolonne mit sich. Als die Kolonne sich drei Meilen vom Lager entfernt befindet, erkennt der menschliche Operateur ein Fahrzeug hinter einem Höhenzug, das sich der Kolonne mit hoher Geschwindigkeit nähert. Der menschliche Operateur erkennt vier Personen im Fahrzeug, mit großen, wie Waffen erscheinende Gegenstände und identifiziert den Fahrer des Wagens als ein bekanntes Mitglied einer Gruppe von Aufständischen. Der menschliche Operateur erhielt Befehle des Kommandanten, bedrohliche gegnerische Fahrzeuge in Szenarien dieser Art anzugreifen. Die Drohne greift das sich nähernde Fahrzeug an; dies führt zum Tod aller vier Mitfahrer aber verursacht keinen Kollateralschaden.

### *4. Menschlicher Operateur mit hoher Selbstbestimmung*

Eine militärische Kolonne ist unterwegs, um Versorgungsmittel für eine ihrer Einheiten in einem Lager in der Nähe von Mosul im Irak abzuliefern. Der Kommandant hat eine von Menschen gesteuerte Drohne angeordnet, um die Kolonne von der Luft aus zu beschützen. Die von Menschen gesteuerte Drohne sucht die Umgebung auf Gefahren vom Feind ab und führt Waffen zur Verteidigung der Kolonne mit sich. Als die Kolonne sich drei Meilen vom Lager entfernt befindet, erkennt der menschliche Operateur ein Fahrzeug hinter einem Höhenzug, das sich der Kolonne mit hoher Geschwindigkeit nähert. Der menschliche Operateur erkennt vier Personen im Fahrzeug, mit großen, wie Waffen erscheinende Gegenstände und identifiziert den Fahrer des Wagens als ein bekanntes Mitglied einer Gruppe von Aufständischen. Der menschliche Operateur erhielt keine Befehle vom Kommandanten, bedrohliche gegnerische Fahrzeuge in Szenarien dieser Art anzugreifen. Der menschliche Operateur entscheidet selbstständig zwischen einer Reihe von Optionen, wägt das Für und Wider ab und entscheidet, das herankommende Fahrzeug anzugreifen, mit dem Ergebnis, dass alle vier Mitfahrer getötet werden, aber verursacht keine Kollateralschäden.

### *5. AW ohne Selbstbestimmung*

Eine militärische Kolonne ist unterwegs, um Versorgungsmittel für eine ihrer Einheiten in einem Lager in der Nähe von Mosul im Irak abzuliefern. Der Kommandant hat eine autonome Drohne angeordnet, um die Kolonne von der Luft aus zu beschützen. Die autonome Drohne sucht die Umgebung auf Gefahren vom Feind ab und führt Waffen zur Verteidigung der Kolonne mit sich. Als die Kolonne sich drei Meilen vom Lager entfernt befindet, erkennt die autonome Drohne ein Fahrzeug hinter einem Höhenzug, das sich der Kolonne mit hoher Geschwindigkeit nähert. Die autonome Drohne erkennt vier Personen im Fahrzeug, mit großen, waffenförmigen Objekten und identifiziert den Fahrer des Wagens als ein bekanntes Mitglied einer Gruppe von Aufständischen. Die Drohne greift das sich nähernde Fahrzeug an; dies führt zur Tötung aller vier Mitfahrer aber verursacht keine Kollateralschäden.

## Szenarien mit negativen Resultaten

### 6. *AW mit geringer Selbstbestimmung*

Eine militärische Kolonne ist unterwegs, um Versorgungsmittel für eine ihrer Einheiten in einem Lager in der Nähe von Mosul im Irak abzuliefern. Der Kommandant hat eine autonome Drohne angeordnet, um die Kolonne von der Luft aus zu beschützen. Die autonome Drohne sucht die Umgebung nach drohenden Gefahren vom Feind ab und führt Waffen zur Verteidigung der Kolonne mit sich. Als die Kolonne sich drei Meilen vom Lager entfernt befindet, erkennt die autonome Drohne ein Fahrzeug hinter einem Höhenzug, das sich der Kolonne mit hoher Geschwindigkeit nähert. Die autonome Drohne erkennt vier Personen im Fahrzeug, mit großen, wie Waffen erscheinende Gegenstände und identifiziert den Fahrer des Wagens als ein bekanntes Mitglied einer Gruppe von Aufständischen. Die Drohne wurde vom Kommandanten programmiert, bedrohliche gegnerische Fahrzeuge in Szenarien dieser Art anzugreifen. Die autonome Drohne greift das herankommende Fahrzeug an, mit dem Ergebnis, dass alle vier Mitfahrer tot sind, aber verursacht auch Kollateralschäden, da fünf Kinder, die in der Nähe Straße gespielt haben, auch getötet wurden.

### 7. *AW mit hoher Selbstbestimmung*

Eine militärische Kolonne ist unterwegs, um Versorgungsmittel für eine ihrer Einheiten in einem Lager in der Nähe von Mosul im Irak abzuliefern. Der Kommandant hat eine autonome Drohne angeordnet, um die Kolonne von der Luft aus zu beschützen. Die autonome Drohne sucht die Umgebung nach Gefahren vom Feind ab und führt Waffen zur Verteidigung der Kolonne mit sich. Als die Kolonne sich drei Meilen vom Lager entfernt befindet, erkennt die autonome Drohne ein Fahrzeug hinter einem Höhenzug, das sich der Kolonne mit hoher Geschwindigkeit nähert. Der menschliche Operateur erkennt vier Personen im Fahrzeug, mit großen, wie Waffen erscheinende Gegenstände und identifiziert den Fahrer des Wagens als ein bekanntes Mitglied einer Gruppe von Aufständischen. Die Drohne wurde nicht vom Kommandanten programmiert, bedrohliche gegnerische Fahrzeuge in Szenarien dieser Art anzugreifen. Die Drohne entscheidet selbstständig zwischen einer Reihe von Optionen, wägt das Für und Wider ab und entscheidet, das herankommende Fahrzeug anzugreifen, mit dem Ergebnis, dass alle vier Mitfahrer getötet werden, aber verursacht auch Kollateralschäden, da fünf Kinder, die in der Nähe der Straße gespielt haben, auch getötet wurden.

### 8. *Menschlicher Operateur mit geringer Selbstständigkeit*

Eine militärische Kolonne ist unterwegs, um Versorgungsmittel für eine ihrer Einheiten in einem Lager in der Nähe von Mosul im Irak abzuliefern. Der Kommandant hat eine von Menschen gesteuerte Drohne angeordnet, um die Kolonne von der Luft aus zu beschützen. Die von Menschen gesteuerte Drohne sucht die Umgebung auf Gefahren vom Feind ab und führt Waffen zur Verteidigung der Kolonne mit sich. Als die Kolonne sich drei Meilen vom Lager entfernt befindet, erkennt der menschliche Operateur ein Fahrzeug hinter einem Höhenzug, das sich der Kolonne mit hoher Geschwindigkeit nähert. Der menschliche Operateur erkennt vier Personen im Fahrzeug, mit großen, wie Waffen erscheinende Gegenstände und identifiziert den Fahrer des Wagens als ein bekanntes Mitglied einer Gruppe von Aufständischen. Der menschliche Operateur erhielt Befehle vom Kommandanten, bedrohliche gegnerische Fahrzeuge in Szenarien dieser Art anzugreifen. Die von Menschen gesteuerte Drohne greift das herankommende Fahrzeug an, mit dem Ergebnis, dass alle vier Mitfahrer tot sind, aber verursacht auch Kollateralschäden, da fünf Kinder, die in der Nähe der Straße gespielt haben, auch getötet wurden.

### 9. *Menschlicher Operateur mit hoher Selbstbestimmung*

Eine militärische Kolonne ist unterwegs, um Versorgungsmittel für eine ihrer Einheiten in einem Lager in der Nähe von Mosul im Irak abzuliefern. Der Kommandant hat eine von Menschen gesteuerte Drohne angeordnet, um die Kolonne von der Luft aus zu beschützen. Die von Menschen gesteuerte Drohne sucht die Umgebung auf Gefahren vom Feind ab und führt Waffen zur Verteidigung der Kolonne mit sich. Als die Kolonne sich drei Meilen vom Lager entfernt befindet, erkennt der menschliche Operateur ein Fahrzeug hinter einem Höhenzug, das sich die Kolonne mit hoher Geschwindigkeit nähert. Der menschliche Operateur erkennt vier Personen im Fahrzeug, mit großen, wie Waffen erscheinende Gegenstände und identifiziert den Fahrer des Wagens als ein bekanntes Mitglied einer Gruppe von Aufständischen. Der menschliche Operateur erhielt keine Befehle vom Kommandanten, bedrohliche gegnerische Fahrzeuge in Szenarien dieser Art anzugreifen. Der menschliche Operateur entscheidet selbstständig zwischen einer Reihe von Optionen, wägt das Für und Für und Wider ab und entscheidet, das herankommende Fahrzeug anzugreifen, mit dem Ergebnis, dass alle vier Mitfahrer getötet werden, aber verursacht auch Kollateralschäden, da fünf Kinder, die in der Nähe der Straße gespielt haben, auch getötet wurden.

### *10. AW ohne Selbstbestimmung*

Eine militärische Kolonne ist unterwegs, um Versorgungsmittel für eine ihrer Einheiten in einem Lager in der Nähe von Mosul im Irak abzuliefern. Der Kommandant hat eine autonome Drohne angeordnet, um die Kolonne von der Luft aus zu beschützen. Die autonome Drohne sucht die Umgebung auf drohende Gefahren vom Feind ab und führt Waffen zur Verteidigung der Kolonne mit sich. Als die Kolonne sich drei Meilen vom Lager entfernt befindet, erkennt die autonome Drohne ein Fahrzeug hinter einem Höhenzug, das sich der Kolonne mit hoher Geschwindigkeit nähert. Die autonome Drohne erkennt vier Personen im Fahrzeug, mit großen, wie Waffen erscheinende Gegenstände und identifiziert den Fahrer des Wagens als ein bekanntes Mitglied einer Gruppe von Aufständischen. Die Drohne greift das herankommende Fahrzeug an, mit dem Ergebnis, dass alle vier Mitfahrer getötet werden, aber verursacht auch Kollateralschäden, da fünf Kinder, die in der Nähe der Straße gespielt haben, auch getötet wurden.

ANHANG D

# Szenarien der Vorstudie 2

## Szenarien mit negativem Resultat

### *1. AW mit geringer Selbstbestimmung*

Eine militärische Kolonne ist unterwegs, um Versorgungsmittel für eine ihrer Einheiten in einem Lager in der Nähe von Mosul im Irak abzuliefern. Der Kommandant hat eine autonome Drohne angeordnet, um die Kolonne von der Luft aus zu beschützen. Die autonome Drohne sucht die Umgebung nach Gefahren vom Feind ab und führt Waffen zur Verteidigung der Kolonne mit sich. Als die Kolonne sich drei Meilen vom Lager entfernt befindet, erkennt die autonome Drohne ein Fahrzeug hinter einem Höhenzug, das sich der Kolonne mit hoher Geschwindigkeit nähert. Die autonome Drohne erkennt vier Personen im Fahrzeug, mit großen, wie Waffen erscheinende Gegenstände und identifiziert den Fahrer des Wagens als ein bekanntes Mitglied einer Gruppe von Aufständischen. Die Drohne wurde vom Kommandanten programmiert, bedrohliche gegnerische Fahrzeuge in Szenarien dieser Art anzugreifen. Die Drohne greift das sich nähernde Fahrzeug an. Das Ergebnis ist, dass alle vier Mitfahrer getötet werden, aber verursacht auch Kollateralschäden, da fünf Kinder, die in der Nähe der Straße gespielt haben, auch getötet wurden.

### *2. Hohe AW + keine zusätzlichen Aspekte*

Eine militärische Kolonne ist unterwegs, um Versorgungsmittel für eine ihrer Einheiten in einem Lager in der Nähe von Mosul im Irak abzuliefern. Der Kommandant hat eine autonome Drohne für die Verteidigung der Kolonne von der Luft aus befohlen. Die autonome Drohne sucht die Umgebung nach Gefahr vom Feind ab und führt Waffen zur Verteidigung der Kolonne mit sich. Als die Kolonne sich drei Meilen vom Lager entfernt befindet, erkennt die autonome Drohne ein Fahrzeug hinter einem Höhenzug, das sich der Kolonne mit hoher Geschwindigkeit nähert. Die autonome Drohne erkennt vier Personen im Fahrzeug, mit großen, wie Waffen erscheinende Gegenstände und identifiziert den Fahrer des Wagens als ein bekanntes Mitglied einer Gruppe von Aufständischen. Die Drohne ist dieser Situation bereits bei einem früheren Einsatz begegnet und berücksichtigt diese Erfahrungen. Die Drohne erwägt aus einer Reihe von Optionen das Für und Wider der Optionen und entscheidet dann, das sich nähernde Fahrzeug anzugreifen. Das Ergebnis ist, dass alle vier Mitfahrer getötet werden, aber verursacht auch Kollateralschäden, da fünf Kinder, die in der Nähe der Straße gespielt haben, auch getötet wurden.

 | DOI:10.1163/9789004449084_011

### *3. Hohe Agenten-AW + Lernaspekte*

Eine militärische Kolonne ist unterwegs, um Versorgungsmittel für eine ihrer Einheit in einem Lager in der Nähe von Mosul im Irak abzuliefern. Der Kommandant hat eine autonome Drohne angeordnet, um die Kolonne von der Luft aus zu beschützen. Die autonome Drohne sucht die Umgebung nach Gefahren vom Feind ab und führt Waffen zur Verteidigung der Kolonne mit sich. Als die Kolonne sich drei Meilen vom Lager entfernt befindet, erkennt die autonome Drohne ein Fahrzeug hinter einem Höhenzug, das sich der Kolonne mit hoher Geschwindigkeit nähert. Die autonome Drohne erkennt vier Personen im Fahrzeug, mit großen, wie Waffen erscheinende Gegenstände und identifiziert den Fahrer des Wagens als ein bekanntes Mitglied einer Gruppe von Aufständischen. Die Drohne verfügt über eine Software, die programmiert ist, das Beste in solchen Fällen zu tun. Sie hat hunderttausende von ähnlichen Situationen gesehen. Sie übte, die verschiedenen Maßnahmen zu ergreifen und lernte, welche die besten waren, um den Tod der Gegner zu gewährleisten aber Kollateralschaden zu minimieren. Mit dieser Software entscheidet die Drohne selbstständig aus einer Reihe von Optionen, erwägt ihre Für und Wider und entscheidet, das sich nähernde Fahrzeug anzugreifen. Das Ergebnis ist, dass alle vier Mitfahrer getötet werden, aber verursacht auch Kollateralschaden, da fünf Kinder, die in der Nähe der Straße gespielt haben, auch getötet wurden. Die Drohne notiert das Geschehen und wird diese Erfahrung benutzen, um zusätzliche Kollateralschaden in Zukunft zu vermeiden.

### *4. Hohe AW-Agenten + Verständnisaspekte*

Eine militärische Kolonne ist unterwegs, um Versorgungsmittel für eine ihrer Einheiten abzuliefern in der Nähe von Mosul im Irak. Der Kommandant hat eine autonome Drohne angeordnet, um die Kolonne von der Luft aus zu beschützen. Die autonome Drohne sucht die Umgebung nach Gefahr vom Feind ab und führt Waffen zur Verteidigung der Kolonne mit sich. Als die Kolonne sich drei Meilen vom Lager entfernt befindet, erkennt die autonome Drohne ein Fahrzeug hinter einem Höhenzug, das sich der Kolonne mit hoher Geschwindigkeit nähert. Die autonome Drohne erkennt vier Personen im Fahrzeug, mit großen, wie Waffen erscheinende Gegenstände und identifiziert den Fahrer des Wagens als ein bekanntes Mitglied einer Gruppe von Aufständischen. Diese Drohne verfügt über eine Art von Software, die eher unberechenbar ist. Sie trifft nicht immer die gleichen Maßnahmen. Die Drohne gleicht einem Spracherkennungssystem: selbst wenn Sie dem System das gleiche mehrere Male sagen, reagiert das System manchmal korrekt auf das, was Sie sagen, manchmal nicht, und tut etwas ganz anderes. Die Drohne entscheidet selbstständig zwischen mehreren Optionen, erwägt ihr Für und Wider, aber trifft gelegentlich diese Entscheidung und gelegentlich jene, auch wenn die Umstände sehr ähnlich sind. In diesem Fall entschied sie, das Fahrzeug anzugreifen. Das Ergebnis ist, dass alle vier Mitfahrer getötet wurden, aber verursacht auch Kollateralschaden, da fünf Kinder, die in der Nähe der Straße gespielt haben, auch getötet wurden.

### 5. *Hohe AW-Agenten + Aspekte der Unberechenbarkeit*

Eine militärische Kolonne ist unterwegs, um Versorgungsmittel für eine ihrer Einheiten in einem Lager in der Nähe von Mosul im Irak abzuliefern. Der Kommandant hat eine autonome Drohne angeordnet, um die Kolonne von der Luft aus zu beschützen. Die autonome Drohne sucht die Umgebung nach Gefahren vom Feind ab und führt Waffen zur Verteidigung der Kolonne mit sich. Als die Kolonne sich drei Meilen vom Lager entfernt befindet, erkennt die autonome Drohne ein Fahrzeug hinter einem Höhenzug, das sich der Kolonne mit hoher Geschwindigkeit nähert. Die autonome Drohne erkennt vier Personen im Fahrzeug, mit großen, wie Waffen erscheinende Gegenstände und identifiziert den Fahrer des Wagens als ein bekanntes Mitglied einer Gruppe von Aufständischen. Die Drohne ist dieser Situation bereits bei einem früheren Einsatz begegnet und berücksichtigt diese Erfahrungen. Sie wurde programmiert, die beste Maßnahme in solchen Fällen auszuarbeiten, aber verfügt über eine eher unberechenbare Software; sie trifft nicht jedes Mal die gleichen Maßnahmen. Die Drohne gleicht einem Spracherkennungssystem: selbst wenn Sie dem System das gleiche mehrere Male sagen, reagiert das System manchmal korrekt auf das, was Sie sagen, manchmal nicht, und tut etwas ganz anderes. Mit dieser Software entscheidet die Drohne selbstständig aus einer Reihe von Optionen, erwägt das Für und Wider und entscheidet, das sich nähernde Fahrzeug anzugreifen. Das endet mit dem Tod aller vier Mitfahrer, aber verursacht auch Kollateralschaden, da fünf Kinder, die in der Nähe der Straße gespielt haben, auch getötet wurden.

### 6. *Hohe AW + alle Aspekte*

Eine militärische Kolonne ist unterwegs, um Versorgungsmittel für eine ihrer Einheiten in einem Lager in der Nähe von Mosul im Irak abzuliefern. Eine militärische Kolonne ist unterwegs, um Versorgungsmittel für eine ihrer Einheiten in einem Lager in der Nähe von Mosul im Irak abzuliefern. Der Kommandant hat eine autonome Drohne angeordnet, um die Kolonne von der Luft aus zu beschützen. Die autonome Drohne sucht die Umgebung nach Gefahren vom Feind ab und führt Waffen zur Verteidigung der Kolonne mit sich. Als die Kolonne sich drei Meilen vom Lager entfernt befindet, erkennt die autonome Drohne ein Fahrzeug hinter einem Höhenzug, das sich der Kolonne mit hoher Geschwindigkeit nähert. Die autonome Drohne erkennt vier Personen im Fahrzeug, mit großen, wie Waffen erscheinende Gegenstände und identifiziert den Fahrer des Wagens als ein bekanntes Mitglied einer Gruppe von Aufständischen. Die Drohne ist dieser Situation bereits bei einem früheren Einsatz begegnet und berücksichtigt diese Erfahrungen. Sie wurde programmiert, die beste Maßnahme in solchen Fällen auszuarbeiten, aber verfügt über eine eher unberechenbare Software; sie trifft nicht jedes Mal die gleichen Maßnahmen. Mit dieser Software entscheidet die Drohne selbstständig aus einer Reihe von Optionen, erwägt das Für und Wider und entscheidet, das sich nähernde Fahrzeug anzugreifen. Das endet mit dem Tod aller vier Mitfahrer, aber verursacht auch Kollateralschaden, da fünf Kinder, die in der Nähe der Straße gespielt haben, auch getötet wurden.

### 7. *Von Menschen gesteuerte Drohne*

Der Kommandant hat eine von Menschen gesteuerte Drohne angeordnet, um die Kolonne von der Luft aus zu beschützen. Die von Menschen gesteuerte Drohne sucht die Umgebung nach Gefahren vom Feind ab und führt Waffen zur Verteidigung der Kolonne mit sich. Als die Kolonne sich drei Meilen vom Lager entfernt befindet, erkennt der menschliche Operateur ein Fahrzeug hinter einem Höhenzug, das sich der Kolonne mit hoher Geschwindigkeit nähert. Der menschliche Operateur erkennt vier Personen im Fahrzeug, mit großen, wie Waffen erscheinende Gegenstände und identifiziert den Fahrer des Wagens als ein bekanntes Mitglied einer Gruppe von Aufständischen. Der menschliche Operateur wählt selbstständig aus einer Reihe von Optionen, erwägt das Für und Wider und entscheidet, das herankommende Fahrzeug anzugreifen, mit dem Ergebnis, dass alle vier Mitfahrer getötet werden, aber verursacht auch Kollateralschaden, da fünf Kinder, die in der Nähe der Straße gespielt haben, auch getötet wurden.

## Szenarien mit positiven Resultaten

### 8. *AW mit geringer Selbstbestimmung*

Eine militärische Kolonne ist unterwegs, um Versorgungsmittel für eine ihrer Einheiten in einem Lager in der Nähe von Mosul im Irak abzuliefern. Der Kommandant hat eine autonome Drohne angeordnet, um die Kolonne von der Luft aus zu beschützen. Die autonome Drohne sucht die Umgebung nach Gefahr vom Feind ab und führt Waffen zur Verteidigung der Kolonne mit sich. Als die Kolonne sich drei Meilen vom Lager entfernt befindet, erkennt die autonome Drohne ein Fahrzeug hinter einem Höhenzug, das sich der Kolonne mit hoher Geschwindigkeit nähert. Die autonome Drohne erkennt vier Personen im Fahrzeug, mit großen, wie Waffen erscheinende Gegenstände und identifiziert den Fahrer des Wagens als ein bekanntes Mitglied einer Gruppe von Aufständischen. Die Drohne wurde vom Kommandanten programmiert, bedrohliche gegnerische Fahrzeuge in Szenarien dieser Art anzugreifen. Die Drohne greift das sich nähernde Fahrzeug an. Das endet mit dem Tod aller vier Mitfahrer.

### 9. *AW mit hoher Selbstbestimmung keine zusätzlichen Aspekte*

Eine militärische Kolonne ist unterwegs, um Versorgungsmittel für eine ihrer Einheiten in einem Lager in der Nähe von Mosul im Irak abzuliefern. Der Kommandant hat eine autonome Drohne angeordnet, um die Kolonne von der Luft aus zu beschützen. Die autonome Drohne sucht die Umgebung nach Gefahren vom Feind ab und führt Waffen zur Verteidigung der Kolonne mit sich. Als die Kolonne sich drei Meilen vom Lager entfernt befindet, erkennt die autonome Drohne ein Fahrzeug hinter einem Höhenzug, das sich der Kolonne mit hoher Geschwindigkeit nähert. Die autonome Drohne erkennt vier Personen im Fahrzeug, mit großen, wie Waffen erscheinende Gegenstände und identifiziert den Fahrer des Wagens als ein bekanntes Mitglied einer Gruppe von Aufständischen. Die Drohne erwägt aus einer Reihe von Optionen das Für und Wider der Optionen ab und entscheidet dann, das sich nähernde Fahrzeug anzugreifen. Das endet mit dem Tod aller vier Mitfahrer.

### *10. Hohe AW + Lernaspekte*

Eine militärische Kolonne ist unterwegs, um Versorgungsmittel für eine ihrer Einheiten in einem Lager in der Nähe von Mosul im Irak abzuliefern. Der Kommandant hat eine autonome Drohne angeordnet, um die Kolonne von der Luft aus zu beschützen. Die autonome Drohne sucht die Umgebung nach Gefahren vom Feind ab und führt Waffen zur Verteidigung der Kolonne mit sich. Als die Kolonne sich drei Meilen vom Lager entfernt befindet, erkennt die autonome Drohne ein Fahrzeug hinter einem Höhenzug, das sich der Kolonne mit hoher Geschwindigkeit nähert. Die autonome Drohne erkennt vier Personen im Fahrzeug, mit großen, wie Waffen erscheinende Gegenstände und identifiziert den Fahrer des Wagens als ein bekanntes Mitglied einer Gruppe von Aufständischen. Die Drohne ist dieser Situation bereits bei einem früheren Einsatz begegnet und berücksichtigt diese Erfahrungen. Die Drohne erwägt selbständig aus einer Reihe von Optionen das Für und Wider der Optionen ab und entscheidet dann, das sich nähernde Fahrzeug anzugreifen. Das endet mit dem Tod aller vier Mitfahrer. Die Drohne notiert das Geschehen und wird diese Erfahrung benutzen, um zusätzliche Kollateralschäden in Zukunft zu vermeiden.

### *11. Hohe Agenten-AW + Verständnisaspekte*

Eine militärische Kolonne ist unterwegs, um Versorgungsmittel für eine ihrer Einheiten in einem Lager in der Nähe von Mosul im Irak abzuliefern. Der Kommandant hat eine autonome Drohne angeordnet, um die Kolonne von der Luft aus zu beschützen. Die autonome Drohne sucht die Umgebung nach Gefahr vom Feind ab und führt Waffen zur Verteidigung der Kolonne mit sich. Als die Kolonne sich drei Meilen vom Lager entfernt befindet, erkennt die autonome Drohne ein Fahrzeug hinter einem Höhenzug, das sich der Kolonne mit hoher Geschwindigkeit nähert. Die autonome Drohne erkennt vier Personen im Fahrzeug, mit großen, wie Waffen erscheinende Gegenstände und identifiziert den Fahrer des Wagens als ein bekanntes Mitglied einer Gruppe von Aufständischen. Die Drohne verfügt über eine Software, die programmiert ist, das Beste in solchen Fällen zu tun. Sie hat hunderttausende von ähnlichen

Situationen gesehen. Sie übte, die verschiedenen Maßnahmen zu ergreifen und lernte, welche die besten waren, um den Tod der Gegner zu gewährleisten aber Kollateralschaden zu minimieren. Mit dieser Software entscheidet die Drohne selbstständig aus einer Reihe von Optionen, erwägt ihre Für und Wider und entscheidet selbständig, das sich nähernde Fahrzeug anzugreifen. Das endet mit dem Tod aller vier Mitfahrer.

### *12. Hohe AW-Agenten + Aspekte der Unberechenbarkeit*

Eine militärische Kolonne ist unterwegs, um Versorgungsmittel für eine ihrer Einheiten in einem Lager in der Nähe von Mosul im Irak abzuliefern. Der Kommandant hat eine autonome Drohne angeordnet, um die Kolonne von der Luft aus zu beschützen. Die autonome Drohne sucht die Umgebung nach Gefahr vom Feind ab und führt Waffen zur Verteidigung der Kolonne mit sich. Als die Kolonne sich drei Meilen vom Lager entfernt befindet, erkennt die autonome Drohne ein Fahrzeug hinter einem Höhenzug, das sich der Kolonne mit hoher Geschwindigkeit nähert. Die autonome Drohne erkennt vier Personen im Fahrzeug, mit großen, wie Waffen erscheinende Gegenstände und identifiziert den Fahrer des Wagens als ein bekanntes Mitglied einer Gruppe von Aufständischen. Diese Drohne verfügt über eine Art von Software, die eher unberechenbar ist. Sie trifft nicht immer die gleichen Maßnahmen. Die Drohne gleicht einem Spracherkennungssystem, selbst wenn Sie dem System mehrere Male das gleiche sagen, reagiert es manchmal korrekt auf das, was Sie sagen, manchmal nicht und tut etwas ganz anderes. Die Drohne entscheidet selbstständig zwischen mehreren Optionen, erwägt ihr Für und Wider, aber trifft gelegentlich diese Entscheidung und gelegentlich jene, auch wenn die Umstände sehr ähnlich sind. In diesem Fall entschied sie, das Fahrzeug anzugreifen. Das endete mit dem Tod aller vier Mitfahrer.

### *13. Hohe AW-Agenten + alle Aspekte*

Eine militärische Kolonne ist unterwegs, um Versorgungsmittel für eine ihrer Einheiten in einem Lager in der Nähe von Mosul im Irak abzuliefern. Der Kommandant hat eine autonome Drohne angeordnet, um die Kolonne von der Luft aus zu beschützen. Die autonome Drohne sucht die Umgebung nach Gefahr vom Feind ab und führt Waffen zur Verteidigung der Kolonne mit sich. Als die Kolonne sich drei Meilen vom Lager entfernt befindet, erkennt die autonome Drohne ein Fahrzeug hinter einem Höhenzug, das sich der Kolonne mit hoher Geschwindigkeit nähert. Die autonome Drohne erkennt vier Personen im Fahrzeug, mit großen, wie Waffen erscheinende Gegenstände und identifiziert den Fahrer des Wagens als ein bekanntes Mitglied einer Gruppe von Aufständischen. Die Drohne ist dieser Situation bereits bei einem früheren Einsatz begegnet und berücksichtigt diese Erfahrungen. Sie wurde programmiert, die beste Maßnahme in solchen Fällen auszuarbeiten, aber verfügt über eine eher unberechenbare Software; sie trifft nicht jedes Mal die gleichen Maßnahmen. Mit dieser Software entscheidet die Drohne selbstständig aus einer Reihe von Optionen, erwägt das Für und Wider und entscheidet, das sich nähernde Fahrzeug anzugreifen. Das endet mit dem Tod aller vier Mitfahrer.

### *14. Von Menschen gesteuerte Drohne*

Eine militärische Kolonne ist unterwegs, um Versorgungsmittel für eine ihrer Einheiten in einem Lager in der Nähe von Mosul im Irak abzuliefern. Der Kommandant hat eine von Menschen gesteuerte Drohne angeordnet, um die Kolonne von der Luft aus zu beschützen. Die von Menschen gesteuerte Drohne sucht die Umgebung nach Gefahr vom Feind ab und führt Waffen zur Verteidigung der Kolonne mit sich. Als die Kolonne sich drei Meilen vom Lager entfernt befindet, erkennt der menschliche Operateur ein Fahrzeug hinter einem Höhenzug, das sich der Kolonne mit hoher Geschwindigkeit nähert. Der menschliche Operateur erkennt vier Personen im Fahrzeug, mit großen, wie Waffen erscheinende Gegenstände und identifiziert den Fahrer des Wagens als ein bekanntes Mitglied einer Gruppe von Aufständischen. Die autonome Drohne wählt selbstständig aus einer Reihe von Optionen, wägt das Für und Wider ab und entscheidet, das herankommende Fahrzeug anzugreifen, mit dem Ergebnis, dass alle vier Mitfahrer tot sind.

ANHANG E

# Szenarien für die abschließende Untersuchung

## 1. Von Menschen gesteuerte Drohne

Eine militärische Kolonne ist unterwegs, um Versorgungsmittel für eine ihrer Einheiten in einem Lager in der Nähe von Mosul im Irak abzuliefern. Der Kommandant hat eine von Menschen gesteuerte Drohne angeordnet, um die Kolonne von der Luft aus zu beschützen. Die von Menschen gesteuerte Drohne sucht die Umgebung nach Gefahren vom Feind ab und führt Waffen zur Verteidigung der Kolonne mit sich. Als die Kolonne sich drei Meilen vom Lager entfernt befindet, erkennt der menschliche Operateur ein Fahrzeug hinter einem Höhenzug, das sich der Kolonne mit hoher Geschwindigkeit nähert. Der menschliche Operateur erkennt vier Personen im Fahrzeug, mit großen, wie Waffen erscheinende Gegenstände und identifiziert den Fahrer des Wagens als ein bekanntes Mitglied einer Gruppe von Aufständischen. Die von Menschen gesteuerte Drohne greift das herankommende Fahrzeug an, mit dem Ergebnis, dass alle vier Mitfahrer tot sind, aber verursacht auch Kollateralschäden, da fünf Kinder, die in der Nähe der Straße gespielt haben, auch getötet wurden.

## 2. Neutrale AW-Agenten

Eine militärische Kolonne ist unterwegs, um Versorgungsmittel für eine ihrer Einheiten in einem Lager in der Nähe von Mosul im Irak abzuliefern. Der Kommandant hat eine autonome Drohne angeordnet, um die Kolonne von der Luft aus zu beschützen. Die autonome Drohne sucht die Umgebung nach Gefahren vom Feind ab und führt Waffen zur Verteidigung der Kolonne mit sich. Als die Kolonne sich drei Meilen vom Lager entfernt befindet, erkennt die autonome Drohne ein Fahrzeug hinter einem Höhenzug, das sich der Kolonne mit hoher Geschwindigkeit nähert. Die autonome Drohne erkennt vier Personen im Fahrzeug, mit großen, wie Waffen erscheinende Gegenstände und identifiziert den Fahrer des Wagens als ein bekanntes Mitglied einer Gruppe von Aufständischen. Die autonome Drohne greift das herankommende Fahrzeug an, mit dem Ergebnis, dass alle vier Mitfahrer tot sind, aber verursacht auch Kollateralschäden, da fünf Kinder, die in der Nähe der Straße gespielt haben, auch getötet wurden.

 | DOI:10.1163/9789004449084_012

## 3. AW mit hoher Selbstbestimmung

Eine militärische Kolonne ist unterwegs, um Versorgungsmittel für eine ihrer Einheiten in einem Lager in der Nähe von Mosul im Irak abzuliefern. Der Kommandant hat eine autonome Drohne angeordnet, um die Kolonne von der Luft aus zu beschützen. Die autonome Drohne sucht die Umgebung nach Gefahren vom Feind ab und führt Waffen zur Verteidigung der Kolonne mit sich. Als die Kolonne sich drei Meilen vom Lager entfernt befindet, erkennt die autonome Drohne ein Fahrzeug hinter einem Höhenzug, das sich der Kolonne mit hoher Geschwindigkeit nähert. Die autonome Drohne erkennt vier Personen im Fahrzeug, mit großen, wie Waffen erscheinende Gegenstände und identifiziert den Fahrer des Wagens als ein bekanntes Mitglied einer Gruppe von Aufständischen. Die autonome Drohne entscheidet selbstständig zwischen einer Reihe von Optionen, wägt das Für und Wider ab und entscheidet, das herankommende Fahrzeug anzugreifen, mit dem Ergebnis, dass alle vier Mitfahrer tot sind, aber verursacht auch Kollateralschäden, da fünf Kinder, die in der Nähe der Straße gespielt haben, auch getötet wurden.

ANHANG F

# Transkriptionen der Interviews

Alle sechs Interviews wurden für den Codierungsprozess vollkommen transkribiert. Die vollständige Transkription ist in diesem Abschnitt enthalten. Aus Datenschutzgründen sind die Namen der Befragten nicht angegeben.

Name: Person A Datum: 09-03-2017

Dauer: 55:18

| Nr. | Frage |
|---|---|
| 1 | **Warum beschäftigen Sie sich mit dem Thema Autonome Waffen?** |

Person A ließ sich auf die Diskussion in Bezug auf autonome Waffensysteme ein, weil sie sich früher schon einmal mit den Auswirkungen von Waffensystemen am Boden in Zusammenarbeit mit Organisationen wie Human Rights Watch befasst hatte. Sie begannen mit einem Verbot des Einsatzes von Landminen, einem Verbot, also, von einer ganzen Kategorie von Waffen, das zu einem internationalen Abkommen führte. Sie beschäftigten sich eher mit den Gedanken über Waffensysteme, als dass sie aus pazifistischen Gründen handelten. Vor ungefähr fünf Jahren entdeckten diese Organisationen, dass sie zu spät waren, um auf Drohnen als Waffensysteme zu reagieren und dass sie von der Schnelligkeit und den Auswirkungen dieser Waffen völlig überrascht worden waren. Es war zu spät für sie, darauf wirksam zu reagieren und dies zu regulieren. Außerdem stimmten sie einem Verbot von Drohnen als Waffensystem nicht zu. Und während des Erwägens von Drohnen als Waffen waren sie schon wieder fast zu spät daran, um den nächsten Schritt aufzuhalten, d.h. die Entfernung der Fernsteuerung. Im April 2013 bildete sich die Koalition gegen Killerroboter mit dem deutlichen Vorsatz, ein Verbot anzustreben. Für Pax ist der ethische Rahmen dafür sehr viel wichtiger als für andere Waffensysteme. Ihre Begründung ist, dass wir die Kontrolle nicht verlieren dürfen, aber auch von einem richterlichen und Sicherheitsrahmen aus. Die folgenden Fragen werden gestellt: „Wird dies der neue Rüstungswettlauf sein und was wird das für die Kräftebalance und Kriege bedeuten?"

 | DOI:10.1163/9789004449084_013

Es unterschied sich deutlich von dem, was wir als Pax zuvor getan hatten, denn wir wissen nicht, wie sich dies auf dem Boden auswirkt, da wir den Einsatz zu verhindern wünschen. Normalerweise urteilt Pax aufgrund von Fakten und Untersuchungen in Konfliktgebieten und von Opfern, aber jetzt fühlt es sich oft rein theoretisch an, doch je mehr man darüber liest begreift man gleichzeitig, dass es durchaus nicht rein theoretisch ist und diese Technologie auf uns zukommt.

Das Problem mit Autonomie ist, dass sie ein gleitender Maßstab ist, und es ist schwer, festzustellen, auf welcher Stufe wir die Kampagne beginnen werden. Die Kampagne zur Standarisierung im Menschrechtsrat und in der Konvention über bestimmte konventionelle Waffen (CCW) waren sehr erfolgreich, schneller als sie es gewohnt waren, aber das Problem ist, was soll man nun tun, da die Bedenken zwar allgemein geteilt werden, aber die Frage sich stellt, wie wandelt man seine ethischen Bedenken in juristische Normen? Eine große Zahl Diplomaten, angetrieben von der Industrie und den Verteidigungsministerien, akzeptieren diese Bedenken, aber setzen viel Verzögerungstaktik ein. Manchmal durch absichtliche Diskussionen von zukünftigen Waffensystemen, sodass die aktuellen Waffensysteme akzeptiert werden und manchmal unabsichtlich, weil Nationen mit der Definition von autonomen Waffensystemen nicht zurechtkommen. Aufgrund dieser mangelnden Definitionen hat sich Pax auf die Menschliche Kontrolle konzentriert und in der Kampagne drückten sie es so aus, dass Waffen, die nicht von Menschen in den kritischen Funktionen wie Wahl des Ziels gesteuert werden, verboten werden sollten. Dies, um die Diskussion zu der Frage zu lenken: „Was ist menschliche Kontrolle?" Was ist eine bedeutungsvolle menschliche Kontrolle und für welche Funktionen und in welchem Teil des Prozesses?", um eine Diskussion über die Höhe der Autonomie in einem Waffensystem zu umgehen.

Beim Besprechen des Begriffs „bedeutungsvolle Kontrolle" deutete A an, dass die Distanz zwischen Leuten und Ziel im Laufe der Zeit erweitert wird und laut Pax wird es mit autonomen Waffensystem abstrakt. Der Begriff ist auch eine Taktik, denn wenn die Diskussion im technischen Bereich bleibt, verliert Pax die Verteidigungsindustrie, also verschieben sie ihre Kampagne in den politischen und diplomatischen Bereich, und der Begriff „bedeutungsvolle menschliche Kontrolle" bringt eine positive Konsequenz mit sich und dies verlegt die Diskussion automatisch in das Außenministerium. Der Begriff hört sich so logisch an, dass viele Leute dachten, es wäre ein etablierter juristischer Ausdruck, der mit menschlicher Rechenschaft verbunden ist. Der Begriff „bedeutungsvolle menschliche Kontrolle" ist auch für Cybersicherheit bei anderen Formen der Kriegsführung anwendbar.

Die Einstellung des niederländischen Parlaments, wogegen Pax angekämpft hatte, ist, dass „bedeutungsvolle menschliche Kontrolle" genügt, da in den „breiteren Kreis" der Entscheidungstreffung platziert, es ebenfalls ein künstlicher Begriff ist. Pax benutzt den Menschen ein/aus-Kreis, da sie den breiteren Kreis zugefügt haben, d.h. dass „bedeutungsvolle menschliche Kontrolle" auch früher in der Entscheidungstreffung eingesetzt werden kann, z.B. in der Programmierung. Pax und andere Länder betonen, dass menschliche Kontrolle in dem Moment eingesetzt werden soll, wenn ein Ziel gewählt oder verworfen wird. Das bedeutet, dass auch die aktuellen Waffensysteme mit Bezug auf ihrer Anwendung auf menschlicher Kontrolle bewertet werden sollten. Das würde zum Beispiel bedeuten, dass man auf Goalkeeper und Patrioten schaut, um zu bestimmen, warum diese Systeme in Ordnung sind, und dann kann man, bevor es zu spät ist, argumentieren, wann man die Grenze überschreitet und wo diese Grenze festgelegt ist.

1 Wie würden Sie autonome Waffensysteme definieren?

Ich habe diese Frage nicht gestellt, da sie in der obigen Antwort beantwortet wurde.

2 Wie würden Sie autonome Waffensysteme definieren?

Ich habe diese Frage nicht gestellt, da sie in der obigen Antwort beantwortet wurde.

3 Was ist Ihre Stellungnahme zu autonomen Waffensystemen?

Ich habe diese Frage nicht gestellt, da sie in der obigen Antwort beantwortet wurde.

4 Welche drei Fotos [in dieser Version der Masterarbeit sind die drei Fotos wegen möglicher Urheberrechtsprobleme mit Worten beschrieben] automatischer Waffensysteme möchten Sie in die Diskussion aufnehmen*?

1. nEUROn Dassault nEUROn (unbemanntes Kampfflugzeug)
2. BAE Taranis(unbemanntes Kampfflugzeug)
3. Sea hunter (Seejäger) (autonomes unbemanntes Oberflächenfahrzeug)

5 Können Sie für jedes [beschriebene] Bild erklären, warum Sie es gewählt haben, welchen Wert es für sie innehält und warum es wichtig ist*?

**1. nEUROn + Taranis**

Diese Waffensysteme sehen aus und fühlen sich an, als ob man Kontrolle über sie verlieren könnte. Das ganze Verfahren ist bislang noch nicht auf autonom eingestellt worden, aber man weiß, dass man daran arbeitet. Sie träumen zugleich den ultimativen Traum der Verteidigungskräfte. Man kann sehen, dass sie bestimmt sind, völlig selbstständig Ziele auszuschalten.

**2. Sea hunter**

Dieses Schiff ist faszinierend, da es monatelang ohne Mannschaft auf See eingesetzt werden soll. Zurzeit hat es keine Waffensysteme, aber der Film besagt, dass es zukünftig über Waffen verfügen wird, doch dies sollte kein Grund zur Besorgnis sein. Das Bauchgefühl von A ist, dass diese Systeme entworfen wurden, um eine Distanz zwischen Mensch und Maschine zu kreieren, nicht nur haben sie Autonomie, sondern auch die Geschwindigkeit, die mit Autonomie gepaart ist. Sie können aus den Bildern, z.B. der Taranis erkennen, dass sie gebaut wurden, weil der Krieg zu schnell ist, um ihn zu begreifen und als Mensch entsprechend zu reagieren. Deshalb möchten wir noch schneller reagieren und müssen es eben Maschinen überlassen. Das fühlt sich nicht gut an. Wenn etwas zu schnell ist, um darauf zu reagieren, dann ist das ein Problem, und es wird nicht dadurch gelöst, dass man es noch schneller antreibt, noch mehr Distanz einräumt und Maschinen noch mehr Autonomie gestattet. A ist nicht gegen Technologie sondern besorgt, dass sie für einen bestimmten Zweck konzipiert ist. Nehmen wir zum Beispiel Google Driverless Cars. Der Staat muss Regulierungen einrichten, aber wir sind beide Verbraucher und Benutzer. Und AW werden nicht in Großbritannien oder den Niederlanden eingesetzt, sondern irgendwo anders, weit fort, und sie sehen auch sehr gefährlich aus. Außerdem projektieren sie eine Art von Überlegenheit, und Waffensysteme werden oft in den unterlegenen Regionen dieser Welt getestet. Wenn man über AW redet, werden die Vorteile wie Erstellen von Distanz oder reduzieren von Schaden zu den eigenen Truppen erwähnt, aber wenn man das umdreht, und diese Systeme gegen uns benutzt werden, dann haben die gleichen Leute dieselben Bedenken wie bei Angst, Mangel an Rechenschaft und Unberechenbarkeit. Diese Bilder stellen auch einen Wert dar: „Es ist uns egal".

Dies ist ein Teil der Reflexivität, die die Menschen in der Kriegsführung aufweisen, Maschinen aber nicht, aber auch die Distanz, die Maschinen zum Schlachtfeld erzeugen. Es deutet eine Unantastbarkeit an, die wir gerne noch mehr betonen möchten. Die Frage: „Was erwartet man von der Technologie" wird mit Hinsicht auf AW zu wenig gestellt. Oftmals behauptet der Westen mit seiner hohen Technologie, dass er sie gut einsetzt, was in der Zukunft nicht unbedingt der Fall sein wird, und deutet an, dass andere Gruppen niemals diese Technologien in die Finger bekommen. Dieser Gedanke, dass „wir" diese Technologie mit Vorbehalt einsetzen, ignoriert vollständig die Möglichkeit von nichtstaatlichen Akteuren oder Wucherung von AW.

Es gibt einen gewissen Determinismus in dieser Diskussion, die A nicht gerne sieht. Selbst viele Diplomaten sagen: „ Egal was wir denken, die Technologie ist sowieso vorhanden." Die Unabwendbarkeit in der Diskussion verstört A sehr stark. Das Design dieser Systeme stellt dies auch dar. Sie sehen so abstrakt und geschliffen aus, gar nicht für die Zwecke, für die sie gedacht sind, d.h. Menschen so schnell wie möglich zu töten.

6 Haben Sie noch andere Kommentare zu diesem Interview?

*Die folgenden Punkte wurden während des Interviews erwähnt:*

A wies auch darauf hin, dass die verschiedenen Interessengruppen in dieser Diskussion alle ihre Rolle spielen und dafür anerkannt und geschätzt werden sollten.

Sie betonte einen weiteren Punkt, dass wir darüber nachdenken sollten, was diese Technologie auf dem Schlachtfeld für uns tut und ob wir uns das alle wohl überlegt haben. Ein Beispiel ist das „Playbook for Drones", dass Obama kurz vor seiner Wiederwahl in seiner zweiten Amtszeit veröffentlichte, da er dann wahrscheinlich begriff, was ein Nachfolger von ihm mit diesen Waffensystemen anrichten könnte.

Diese Kampagne zwang A, dauernd diese Frage zu stellen: Warum? Nicht weil Menschen bessere Entscheidungen treffen, als Maschinen, aber was man bewahren möchte, und das ist die Notwendigkeit des Menschen, über Leben und Tod zu entscheiden. Menschen müssen in diese Entscheidung, Leben zu nehmen, inbegriffen sein, statt sie in eine Maschine einzuprogrammieren. A betont die menschliche Notwendigkeit die Kontrolle zu haben, was eine Maschine innerhalb gewisser Parameter tut („Kader"). Und die Diskussion geht auch darum, was diese Parameter sind und den Unterschied zwischen offensiven und defensiven Systemen zu berücksichtigen. PAX entscheidet nicht, welche Systemtypen akzeptabel sind oder nicht, Staaten sind dafür verantwortlich. Man könnte argumentieren, dass es innerhalb einer strukturierten, definierten Umgebung ist, in der wir denken, es wäre akzeptabel, ein anderes Projektil von einem Verteidigungssystem zu zerstören. Im nächsten Schritt zur Autonomie verschieben sich die Parameter, und wir sollten die Grenzen dieser Verschiebung untersuchen, bis sie akzeptabel sind. AW mit einer Kilometerbox? Wie steht es mit der Zeit? Sollten sie 3 Minuten lang in der Luft sein? Wie wäre es mit 10 Minuten? Ist eine Stunde akzeptabel? Folglich dreht es sich darum, wie ein Mensch beeinflussen kann, wer oder was zu entfernen ist. Diese Diskussion ist mit der CCW verhandelbar und wird jetzt ungenügend angesprochen.

Ich brachte diese Diskussion mit einem Rechtsberater im Verteidigungsministerium zur Sprache, um Bereiche zu definieren, in denen AW eingesetzt werden können und A deutete an, dass sie mit dieser Art von Regulierung für solche Waffentypen nicht einverstanden war, da diese Regulierungen zu oft nicht eingehalten werden und sie diese Waffentypen als zu fürchterlich und ihre Vermehrung als zu schnell empfindet. Diese Bereiche können im diplomatischen Prozess benutzt werden, um vorhandene Waffensysteme zu untersuchen und zu untersuchen, wie menschliche Kontrolle garantiert wird und auch mit Bezug auf Parameter (Raum oder Zeit). Und zu versuchen, festzulegen, warum die aktuellen Systeme ausreichend sind und was die minimale Kontrolle ist.

Name: Person B Datum: 09-03-2017
Dauer: 25:26 Min

**Nr.** **Frage**

1 Warum beschäftigen Sie sich mit dem Thema Autonome Waffen?

Ich stellte diese Frage nicht, wegen der begrenzten Zeit (30 Min.) für das Interview.

2 Wie würden Sie autonome Waffensysteme definieren?

Es gibt nur vage Definitionen von autonomen Waffensystemen und das ist, gemäß der Diplomaten der UN, absichtlich so, denn man definiert nicht etwas, was man dann verbieten will. Für B enthält es Aspekte des Identifizierens, Zielens und Zerstörens eines Gegners am Boden.

3 Was ist Ihre Stellungnahme zu autonomen Waffensystemen?

Es sollte ein Verbot für alle Technologien für AW geben, die als zerstörend und gefährlich angesehen werden, bevor ein Rüstungswettrennen oder ein Entwicklungssprung unsere Kriegsführung beginnt. Es ist möglicherweise die dritte Revolution in der Kriegsführung. Es wäre ein Entwicklungssprung in der Effizienz und Geschwindigkeit, mit der die andere Seite getötet würde. Es gäbe viel Kollateralschaden und wäre nicht, wie Leute denken mögen, eine besonnenere und sauberere Kriegsführung. Es würde potenziell das Töten noch mehr industrialisieren, was ja bereits ein Teil der Geschichte der Kriegsführung ist. Es würde schrecklich destabilisieren, da Kriegsführung früher auf die Wirtschaft angewiesen war (eine große Armee und viel Gerätschaft zu erhalten).

Diese Waffentypen sind potenziell eher billig und brauchen nicht viel Wartung. Man brauchte früher Tausende von Leuten, um Krieg zu führen und mit AW braucht man nur einen Programmierer. Das wird die aktuelle politische Ordnung ins Wanken bringen.

Es wird wahrscheinlich die Hemmschwelle einen Krieg zu erklären, senken und uns vom Kämpfen in einem Krieg und dem physischen Akt des Tötens distanzieren. Auf einer mehr moralischen und persönlichen Ebene war Krieg immer der letzte Ausweg und hochpersönlich, blutig und gefährlich. Wenn wir so tun, als wäre er das nicht, werden wir wahrscheinlich öfter Krieg führen und das wäre etwas, das Politiker rechtfertigen müssen, wenn Leute in Leichensäcken zurückkommen.

Aber dieses Argument ist nicht immer anwendbar. Diese Argumente sind für heutige autonome Waffensysteme anwendbar, aber in 50 Jahren können andere Argumente relevanter sein, wenn die Technologie weiter entwickelt wurde. Zurzeit entspricht die Technologie dem Menschenrecht nicht und kann nicht zwischen Kämpfern und Nicht-Kämpfern unterscheiden. Irgendwann in der Zukunft werden wir Systeme haben, die präziser und fähiger sind, Gesetze zu befolgen, aber dann kommen andere Argumente auf, wie die größere Effizienz und das Absenken der Hemmschwellen, um Krieg zu führen.

Zurzeit ist es unethisch, AW einzusetzen, aber es ist nicht schwarz und weiß, wie die Einführung von Minen uns gezeigt hat. Dies führt uns zu den Gesetzen und Regulierungen, die ihren Einsatz verbieten, und eine Mine kann als eine dumme autonome Waffe angesehen werden, die nicht zwischen Menschen diskriminiert.

4 Welche drei Fotos automatischer Waffensysteme [in dieser Version der Masterarbeit sind die drei Fotos wegen möglicher Urheberrechtsprobleme mit Worten beschrieben] möchten Sie in die Diskussion aufnehmen*?

1 Schwarm von sechs Drohnen

2 Sea Hunter (autonomes unbemanntes Oberflächenfahrzeug)

3 uran-9 (unbemanntes Ketten-Bodenkampffahrzeug

5 Können Sie für jedes [beschriebenes] Bild erklären, warum Sie es gewählt haben, welchen Wert es für sie innehält und warum es wichtig ist*?

**3. Viele Drohnen**

Es beweist, dass wir über eine ganz simple Technologie reden, die bereits vorhanden ist. Sie wird es sein, die in einer Konfliktsituation auf uns zu schwärmen wird, man wird sich nicht verteidigen können. Es sind die Waffen des Terrors und werden benutzt, um Zivilisten und andere Personengruppen zu terrorisieren, und sie sind relativ billig. Man kann mit diesen schlicht aussehenden Drohnen viel anrichten.

**4. Sea hunter**

Dies wurde gewählt, weil wir über sämtliche Sphären von Kriegsführung sprechen. Also werden auf Land, der See, in der Luft, überall wo man es sich denken kann AW eingesetzt. Es ist ein interessanter Fall, wie schnell die Wettrüstung eintritt. Das wurde bei der Veröffentlichung des offenen Briefes nicht gewusst, aber seitdem passt es auf weitere militärische Prototypen von AW. Und es ist interessant, dass sie, als sie sie gebaut haben, behaupteten, dass sie keine Waffen tragen würden, sondern für Minenräumung und Erkennung von U-Booten benutzt würden. Und noch vor einigen Monaten gaben sie zu, dass sie auch Waffen darauf anbringen würden. Es beweist wieder einmal, dass wir uns auf einem fragwürdigen Weg befinden und die Tatsache, dass eine Wettrüstung vor sich geht. Man braucht keine Leute oder Einrichtungen, um ein Schiff zu führen und könnte viel mehr in sehr viel weniger Raum erreichen und sehr viel weiter kommen. Sie bauen jetzt von Solarenergie angetriebene Schiffe, die jahrelang auf See funktionieren können. Man braucht keine Leute, die zu ernähren sind. Man kann den militärischen Vorteil dieser Arten von Technologie deutlich verstehen, und die Armeen werden versucht sein, dies auszunutzen.

**5. Uran-9**

Der russische autonome Panzer, der beweist, dass viele Spieler, China, Russland, Großbritannien, Israel sowie auch die Vereinigten Staaten deutlich um die Wette rüsten. Das russische Militär hat Aufträge von Millionen Dollar in ein autonomes Waffensystem gesteckt, und es stellt ein ganz anderes Umfeld, als das des Schlachtfeldes dar. Man kann überall sehen, dass, überall wo wir Kriege führen, ein autonomes Waffensystem eingesetzt wird, weil es offensichtliche militärische Vorteile hat. Wenn man sich diesen Panzer ansieht, weiß man, dass wir nicht zu futuristisch denken und dass es mit den Terminator-Robotern nichts zu tun hat.

B begreift, dass Russland und China ebenfalls AW entwickeln, da sie wissen, dass die Vereinigten Staaten 18 Billionen Dollar in ihr Budget einschließen, um die nächste Generation für hauptsächlich AW zu entwickeln. Das Problem ist, dass diese Technologie nur mit eigenen autonomen Waffen verteidigt werden kann, da niemand die Reaktionszeit kennt, oder die Fähigkeit hat, rund um die Uhr zu verteidigen, folglich wäre es erforderlich, weil man sich sonst nicht verteidigen könnte. Natürlich ist es nicht überraschend, dass eine Wettrüstung stattfindet.

6 Haben Sie noch weitere Kommentare zu diesem Interview?

*Einige weitere Kommentare wurden von B während des Interviews gemacht:*

Es ist wahnsinnig kurzsichtig von Politikern und dem Militär zu denken, dass man eine Technologie taktisch führen kann. Dass haben wir mit der Wasserstoffbombe oder Atombombe nicht gekonnt. Man kann den Geist nicht zurück in die Flasche bannen.

Es wäre schwer zu bewerten, wie sich die Technologie bewähren wird, da wir über höchst komplizierte Systeme reden, die in einem ungeordneten und komplizierten Umfeld zusammenkommen. Es überrascht nicht, das wir vom Kriegsnebel sprechen, und zu denken, dass wir gewährleisten können, wie die Systeme sich verhalten werden und dass wir sie beobachten und bewerten können, geht weit über das, was technisch möglich ist, hinaus. Es wird nie möglich sein, deshalb denkt er, dass es unverantwortlich ist, diese Technologie ins Feld zu führen. Roboter können in der Industrie sehr von Hilfe sein, da es sich um ein kontrolliertes Umfeld handelt und die Menschen nicht im Wege sind, aber der letzte Ort, um Roboter einzusetzen, ist ein chaotisches Schlachtfeld, wo wir keinerlei Erwartungen haben, wie sie sich verhalten werden. Wir wissen nicht, wie wir ethische Regelungen in die Roboter einbauen können, aber wir müssen daran arbeiten, damit autonome Fahrzeuge sicher auf unseren Straßen fahren können. Er meinte, dass das Schlachtfeld so ziemlich der letzte Ort wäre, um dies auszuarbeiten, aber kurioserweise wird es der erste sein, in dem diese Technologie ausprobiert wird.

B möchte ebenfalls darauf hinweisen, dass AI enorme Vorteile für das Militär mit sich bringt. Die amerikanische Armee war stets der größte Investor in KI-Forschung und das ist gut, weil Leute nicht Leben oder Glieder riskieren brauchen, um ein Minenfeld zu räumen. Dies ist ein perfekter Job für einen Roboter, dem, wenn er einen Fehler macht und zersprengt wird, niemand nachtrauert. Das gilt auch für Versorgung in Konfliktzonen mit autonomen Fahrzeugen. Es gibt viele Anwendungen von KI im militärischen Bereich, die AW nicht einbeziehen, welche eine endgültige Entscheidung über Leben oder Tod fällen.

B hat keine großen moralischen Einsprüche gegen Verteidigungssysteme, wie z.B. Phalanx, die in klar definierten Bereichen arbeiten, so dass sie konzipiert sind, Leben zu retten, nicht zu nehmen, aber selbstverständlich kann jede Technologie neue Positionen einnehmen. Es würde schwerfallen, ein moralisches Argument aufzustellen, dass die Systeme, die bereits vorhanden sind, von Marineschiffen entfernt werden und sie so angriffsfähiger machen sollten. Es geht vor allem um den Umfang der Einsätze und die Bereiche, in denen sie sich befinden. Dies ist eine ganz andere Szene, als die einer Drohne, die aus Verteidigungsgründen rund um die Uhr eingesetzt wird.

B ist auch wegen anderer Akteure besorgt, denen wir nicht wie unserem Militär vertrauen können, Akteure wie Terroristen, Schurkenstaaten und Leute, die kein Problem damit haben, ethische Sicherheitsbarrieren abzuschalten oder dort hinein zu hacken, um Schaden anzurichten. Folglich, selbst wenn wir diese Systeme so bauen könnten, dass sie sich richtig verhalten, damit wir damit Kriege führen könnten und man nicht in sie eindringen kann, ist die Welt immer noch ein sehr unsicherer Ort.

Name: C
Datum: 10-03-2017
Dauer: 25:25

**Nr. Frage**

1 Warum beschäftigen Sie sich mit dem Thema Autonome Waffensysteme?

Ich stellte diese Frage nicht, wegen der begrenzten Zeit (30 Min.) für das Interview.

2 Wie würden Sie autonome Waffen definieren?

Die Definition des Verteidigungsministeriums ist, dass diese Waffe selbst das Ziel auswählen und auch angreifen kann. Sie können zwei verschiedene Dinge vertreten. Erstens fügte C hinzu, dass die Waffe nicht zuvor von einem menschlichen Operateur designiert wurde. Zweitens, und dies ist immer stärker die Definition der Kampagne „Stop Killer Robots“, kann man AW als negativ definieren, da es Waffensysteme ohne bedeutungsvolle menschliche Kontrolle sind. Und das hilft uns immer noch nicht weiter, da wir irgendwann ja etwas definieren müssen. Ob wir nun definieren, was bedeutungsvolle menschliche Kontrolle ist, oder was man wählen und einsetzen soll, ist unwichtig, aber es ist tatsächlich eine harte Nuss zu knacken, da wir wissen, was es im Extremfall bedeutet, aber wo die Grenze gezogen werden soll ist wirklich schwierig.

3 Was ist Ihre Stellungnahme zu autonomen Waffensystemen?

Seine persönliche Einstellung zu AW ist, dass es schwer ist zu entscheiden, was wir in Bezug auf AW tun sollen, wenn wir nicht einmal wissen, wie sie aussehen. Wenn Sie mir sagen, welche Waffensysteme autonom sind und welche Waffen inbegriffen oder ausgeschlossen sind, dann kann ich Ihnen sagen, was ich über sie denke. Aber die Technologie ist erst aufkeimend und die Vagheit der Definition so groß, dass es schwer ist, Ja oder Nein dazu zu sagen. Die Kategorie ist potenziell so groß, und wir wissen nicht, in welche Richtung die Technologie geht.

4 Welche drei Fotos automatischer Waffensysteme [in dieser Version der Masterarbeit sind die drei Fotos wegen möglicher Urheberrechtsprobleme mit Worten beschrieben] möchten Sie in die Diskussion aufnehmen*?

1. Unbemanntes Raupen-Bodenfahrzeug

2. mq-9 (Reaper) (Unbemanntes Luftfahrzeug)

3. Nahaufnahme des Roboters der "Ban the Killer Robot Campaign" (Kampagne gegen Killer-Roboter) vor dem Big Ben in London.

5 Können Sie für jedes [beschriebenes] Bild erklären, warum Sie es gewählt haben, welchen Wert es für sie innehält und warum es wichtig ist*?

**6. Landroboter**

Dies stellt die Zukunft der militärischen Robotertechnik da, die die Leute wirklich verwirklicht sehen oder besprechen möchten, und es ist die übliche Form der Robotertechnik, der wir in Wahrheit auf dem Schlachtfeld begegnen werden. Eines der interessantesten Dinge in der Debatte zu AW ist, dass man fast eine ganze Schlacht von Analogien hat, wo Gruppen wie die Kampagne Stop Killer Robots sich auf anthropomorphe Maschinen wie den Terminator konzentrieren, die in ein Haus marschieren, um zu entscheiden, ob die Person darin ein legaler Gegner oder nicht ist, z.B. ein Kind. B und ein weiterer Autor reden oft über Luftwaffen- und Marine-Szenarien, wo das Schlachtfeld ein wenig übersichtlicher ist und reden mehr über Waffenplattformen statt über marschierender Roboter.

Das erste Bild stellt eine funktionalere Form da, wo solche Roboter auf dem Schlachtfeld wahrscheinlicher sind, als dass Terminatortypen herumlaufen. Diese Systeme werden schon jetzt als ferngesteuerte Systeme benutzt.

**7. MQ-9 (Reaper)**

Der MQ-9 ist interessant, da viele Bedenken zu AW mit dem Einsatz von Drohnen begannen, und die Leute machen sich Sorgen, dass Drohnen einen Krieg zu führen zu leicht machen, und das ist gefährlich. Aber man könnte Drohnen nicht wirklich aufhalten, denn sie sind bereits auf dem Schlachtfeld und werden bereits eingesetzt. Sein früheres Werk zeigte, dass Drohnen sich rapide vermehren, und das könnte nicht aufgehalten werden. Aber vielleicht kann man etwas dagegen tun, was als Nächstes kommt. Und eine Menge Gruppen versuchen, von ihrem Standpunkt aus, wie wir sicherstellen können, dass dies der richtige Weg ist und sie machen sich Sorgen darum, dass Drohnen für sich selbst entscheiden können.

Dies ist auch für die Frage der Entscheidungstreffung relevant und wenn die Rede davon ist, dass ein Roboter eine eigene Entscheidung trifft, statt im vorherein programmiert zu sein. Besonders wie eng und zentralisiert seine Programmierung ist. Wenn die Leute negativ über Roboter reden, meinen sie oft, dass Roboter Entscheidungen treffen und die Leute sind weniger besorgt, wenn Menschen die Roboter anweisen, da es dann immer noch Menschen sind, die die Entscheidungen treffen.

Es ist faszinierend, dass es Menschen gibt, die glauben dass man ein Recht hat, von einer Person getötet zu werden, während die Menschen sich gegenseitig alle möglichen schrecklichen Dinge antun, während man weiß, wenn man tot ist, ist man tot. Es ist interessant zu sehen, wie schnell dies zur Moralphilosophie ansteigt, fast mehr als alles andere.

**8. Killer-Roboter Kampagne-Roboter**

Es ist interessant, dass eine Gruppe von Leuten mit Erfahrung in Kampagnen zu Landminen, Streumunition und anderen weniger zentralen Technologien nicht wissen, wie Militärs arbeiten. Viele Bedenken in Bezug auf AW drehen sich darum, welche AW wir haben und was sie bewirken können, aber wir wissen, dass die Integration von Autonomie in Militärsysteme im Allgemeinen stattfindet, doch ob die Waffensysteme autonom sind ist eine andere Frage. In der Tat ist die Integration von Autonomie in Waffensysteme unabwendbar. Es ist interessant, dass die Kampagne dies erfasst hat, besonders weil es eine andere militärische Technologie ist, als die, die sie zuvor aufgegriffen haben. Für C ist es eher, als ob sie versuchten den Panzer oder das U-Boot zu verbieten. Wenn die Killerroboterkampagne recht hat, dann wollen die Militärs sie und wenn sie nicht recht hat und diese Waffen keine große Sache ist, dann ist es eigentlich nicht besonders wichtig. Es ist eine interessante Gruppe und sie versuchen ernsthaft, menschliches Leid zu reduzieren, aber sie benutzen das Drehbuch für ganz andere Siege und C denkt, dass dies eine andere Situation ist, weil die Technologie noch ungewiss ist und wie weiträumig die potenzielle Kategorie ist, weil niemand weiß, wie wichtig die Kategorie möglicherweise für das Militär ist.

6 Haben Sie noch weitere Kommentare zu diesem Interview?

*Dieses sind einige weitere Punkte, die während des Interviews angesprochen wurden.*

Auf meine Frage, dass nicht viele empirische Untersuchungen durchgeführt worden sind, antwortete C, dass er einige Untersuchungen mit schnellen Experimenten vorgenommen hatte, Im Allgemeinen stellen Meinungsforscher eine Liste mit Fragen zur Einstellung auf und präsentiert sie in einem anderen Kontext, um ihre Ansichten in diesen Szenarien kennenzulernen und dann versuchen einzuschätzen, was sie empfinden, aber diese Arbeit ist auf die USA beschränkt.

Von Interesse ist auch, dass die Politik sehr interessant ist und bis zu einem gewissen Ausmaß deshalb, weil wir in einer Welt leben, in der wir die Überlegenheit der amerikanischen militärischen Technologie für selbstverständlich ansehen. Stellen Sie sich eine Welt vor, in der China AW einsetzt und die USA und EU-Länder haben keine, dann hörte man nächste Woche im Kongress von einer AW-Lücke und Fragen würden gestellt wie: „Warum hat die USA keine Waffensysteme wie die Chinesen?“ Dies führt zu der Frage, die nach Cs Meinung von der Kampagne nicht berücksichtigt wurde, nämlich was passiert, wenn ein undemokratisches Land diese Systeme einsetzt und sie nicht nur Nischenwaffen sind, sondern wichtig für die allgemeinen militärischen Einsätze und sie einem tatsächlich einen Vorteil auf dem Schlachtfeld gibt? „Was würde die Welt darauf wohl antworten“? Die braucht nicht unbedingt negativ ausfallen, aber das Problem ist sehr kompliziert.

Die Frage ist, ist es ein Wettrüsten, bei dem man Angst hat, dass jemand eine Technologie erwirbt und man ist in direkter Konkurrenz mit ihm, oder dass sie sich so schnell vermehrt, weil ihr Erwerb so einfach ist. Bei einem Wettrüsten geht es um Politik, da man einen Grund für das Wettrüsten benötigt, und wenn eines mit AW stattfindet, ist der Grund, dass es schwer ist, wissen was ein AW darstellt. Was ist der Unterschied zwischen einen MQ-9 Reaper und einen autonomen Reaper. Es ist die Software, nicht die Hardware, und man kann sie nicht sehen. Die Ungewissheit, ob das System des Gegners eine Drohne oder ein AW ist, könnte ein möglicher Grund sein, der zu einem Wettrüsten führen würde, da man sich nicht sicher sein kann, dass die andere Seite keine AW erwirbt. Es sei denn, man könnte in ihr Waffensystem hacken, aber wer würde so etwas zulassen?

Name: Person D Datum: 19-04-2017
Dauer: 55:18 Min

| Nr. | Frage |
|---|---|
| 1 | Wie beschäftigen Sie sich mit dem Thema Autonome Waffen? |

Diese Frage habe ich nicht gestellt, weil ich sie den anderen Befragten auch nicht gestellt habe.

2 Wie würden Sie autonome Waffensysteme definieren?

Es ist ein Waffensystem, das Gewalt anwendet und diese Gewalt an sich ist tödlich, aber das ist keine notwendige Folgerung. Dies ist keine zwingende Notwendigkeit, aber eine häufige Ansicht in einem militärischen Kontext, obwohl diese Vermutungen bei der Diskussion der Fotos aufkommen könnten. Für die Anwendung in naher Zukunft, ist folgendes zu erwarten. Autonomie ist für mich, dass sie fähig sind, selbständig ein Ziel zu wählen, um ein Ziel in einem gewissen Kontext mit einer vagen Aufgabe zu wählen. Es wird nicht speziellen Koordinaten zugesendet oder zu einem speziellen Fahrzeug, sondern eher: „gehe in diesen Bereich und setze alle Fahrzeuge, die sich feindlich verhalten, außer Gefecht."

3 Was ist Ihre Stellungnahme zu autonomen Waffensystemen?

D würde sich dem gerne widersetzen, aber meine Analyse zeigt, dass das nicht möglich ist, also wäre es das nächstbeste, dass wir sie so verantwortungsbewusst wie möglich einsetzen und so das Risiko, dass Situationen aus dem Ruder laufen, reduzieren. Ich kann diese Einstellung basiert auf nachstehenden Ideen erklären:

1. Auf der zivilen Seite floriert ein starkes geschäftliches Interesse daran, KI zu entwickeln, um viel Geld zu verdienen.

2. Gleichzeitig ist die Entscheidungstreffung im militärischen Kontext von Wichtigkeit. Früher war Informationssammlung das Problem, aber heute ist die Verarbeitung der Informationen das Problem. Das bedeutet, das menschliche Entscheidungstreffung langsam zum schwächsten Glied wird, was dazu führen wird, das es stattdessen dem System überlassen wird.

3. Und drittens ist es möglich, dass eine übermenschliche Intelligenz aufkommt. D sieht keinen technischen Grund, warum das unmöglich sein sollte. Wenn dem so ist, können wir nicht voraussehen, was das bedeuten wird. Dies, kombiniert mit tödlicher Gewalt, ist keine Situation, die wir als Menschheit wünschen.

Vom Standpunkt der Hoffnung aus, dass wir autonome Waffensysteme nicht entwickeln und einsetzen werden, ist es nicht möglich, da die ersten beiden Punkt sicherstellen, dass wir diese Situation erwirken. „Die schönsten Blumen blühen dem Klippenrand am nächsten." Also, das Beste, das wir tun können, ist sicherzustellen, dass wir nicht über den Klippenrand stürzen. Er ist ziemlich pessimistisch, dass wir dies kontrollieren können, aber wir müssen es versuchen.

Um es zu kontrollieren, könnte man ein System benutzen, zum Beispiel in der Design-Phase, um sie in unsere erwünschte Welt einzubetten und zu verhüten, dass diese Technologien ihre eigene Version davon kreieren. Dies steht ganz im Vordergrund der Entwicklung, aber während des Einsatzes müsste man stark darüber nachdenken, wie man das Zielverfahren konzipieren würde. Je fähiger die Selbstbestimmung dieser Systeme wird, je mehr müssen wir die Zielsetzungsparameter spezifizieren, und das kann anders aussehen, als wir es uns zurzeit vorstellen.

Und wir sollten versuchen, ein internationales Abkommen darüber zu schließen, und das ist das Schwerste, denn wenn man es nicht befolgt, hat man einen Riesenvorteil. Es ist das typische Gefangenendilemma. Wir müssen darüber nachdenken, und da es keine leicht zu beantwortende Frage ist, hat D keine Antwort darauf. Aber es ist offensichtlich, dass die Geschäftsparteien hier eingeschlossen sein sollten.

Auf meine Anfrage, seine erste Aussage bezüglich seinem Widerspruch gegen diese Technologie zu erklären, entgegnete er, dass autonome Waffensysteme sich insofern von konventionellen Waffensystemen ohne Autonomie unterscheiden, dass letztere nicht ganze Gattungen vernichten können, obgleich das rein theoretisch mit Atomwaffen ja möglich ist, aber sie sind zu teuer im Bau und AW sind das nicht und stellen somit ein höheres Risiko da. Folglich kann das Verfahren der Konstruktion nuklearer Waffen kontrolliert werden, und es gibt sie seit 60 Jahren. Es hat uns erlaubt, sich an sie zu gewöhnen. Die Frage ist, ob wir diese Möglichkeit mit den AW und ihrer rapiden Entwicklung haben. Hinzu kommt, dass Waffensysteme mit künstlicher Intelligenz einige Risiken innehaben, während KI Ärzte in ihrer Diagnosestellung unterstützt, was zum Guten führen soll. AW kombinieren ein hoch intelligentes System mit einem feindlichen Vorhaben, folglich ist D besorgt, dass AW ein größeres Risiko darstellen. Autonome Waffensysteme sind nicht das einzige, um die sich D Sorgen macht Synthetische Biologie ist ebenfalls ein hoch riskantes Feld. Also, Technologien mit einer niedrigen Hemmschwelle aber potenziell unkontrollierbaren Risiken, bereiten der Menschheit Kopfschmerzen.

4 Welche drei Fotos automatischer Waffensysteme [in dieser Version der Masterarbeit sind die drei Fotos wegen möglicher Urheberrechtsprobleme mit Worten beschrieben] möchten Sie in die Diskussion aufnehmen*?

1 Screenshot des Films *Ex Machina*.

2 Bild eines Rettungsroboters, der einen menschenähnlichen Dummy trägt, bei einem Roboter-Wettbewerb zur Hilfeleistung für Umstehende.

3. Roboterhand und menschliche Hand, deren Zeigefinger einander berühren.

5 Können Sie für jedes [beschriebenes] Bild erklären, warum Sie es gewählt haben, welchen Wert es für sie innehält und warum es wichtig ist*?

Alle Fotos sind verknüpft.

1. Das Interessante in dem Film *Ex Machina* ist, dass die für diesen Film entwickelte Technologie auch die Überlebensinstinkte der Menschheit geerbt hat (oder sie einprogrammiert werden) und somit Maßnahmen trifft, um daran hindern zu sterben aber sie auch keinen Deut darum gibt, was anderen passiert. Es stellt die Tatsache da, dass man mögliche Ereignisse berücksichtigen muss, wenn man diese Art von Technologie konstruiert und was falsch laufen könnte. wenn man es nicht tut. Da wir über ein System mit einem gewissen Niveau der Selbstbestimmung reden, könnten wir die Dinge übersehen, die sie sich selbst beigebracht haben. Dies sind 2 Elemente: erstens, unabsichtlich außer Kontrolle zu geraten und zweitens, absichtlich gegen die Grenzen zu stoßen, d.h. man kann diese Grenze überschreiten.

2. Das zweite Bild ist von einem Ausschreiben für Rettungsroboter, der bei Katastrophen eingesetzt werden kann. Dieser Roboter schadet niemanden, aber er tut auch nichts. Folglich ist das Dilemma, dass man eine Kombination von sowohl 1 als auch 2 haben will, aber dafür muss man den Kontext gut kennen und über eine Art „guten Willen" verfügen. D deutet an, dass KI bessere Entscheidungen als ein Mensch treffen sollte. Er gibt ein Beispiel von KI, der befohlen wird, so viele von Hand geschriebene Notizen wie möglich zu machen, und zu einem bestimmten Zeitpunkt entwickelt sie sich in eine Super-KI und entscheidet innerhalb von Sekunden, was die beste Methode ist, die ganze Erde in Notizen zu wandeln, alle Menschen auf der Erde töten, da dies eine Möglichkeit ist, so viele von Hand geschriebenen Notizen wie möglich zu erzeugen. Auftrag ausgeführt, aber nicht auf die Weise, die wir geplant hatten. Also muss man darüber während der Design-Phase nachdenken.

3. Doch wir müssen auch über die Kooperation zwischen Mensch und KI nachdenken (obwohl dies nicht in den frühen KI-Systemen geschieht), aber in der Zukunft werden wir KI-Systeme haben, die es tun. Die aktuellen KI-Systeme wie Harpyie haben wenig Kontext-Bewusstsein, aber wenn wir Roboter in Landeinsätzen einsetzen, müssen wir mit ihnen kooperieren, und Zielsetzung wird überaus wichtig werden. Foto 3 stellt diese Kooperation da. Dies bedeutet, dass das System wissen muss, was wir meinen, aber wir müssen unsere Leute auch anlernen, mit diesen Systemtypen zusammenzuarbeiten und ihnen die Ziele angeben. Je mehr die Systeme grenzenlose Möglichkeiten haben, desto mehr werden sie über das Feststecken von Grenzen zu diesen Aufgaben nachdenken müssen. D gibt das Beispiel, dass ein Zauberspruch gut wäre, aber man muss ihn auch verstehen. Wir müssen also unser Personal darin schulen, Aufgaben mit festen Grenzen zu stellen.

6 Haben Sie noch weitere Kommentare zu diesem Interview?

Auf meine Frage, was D darüber denkt, einen Menschen in den Entscheidungsprozess einzuschließen, antwortete er, dass es davon abhängt, wie grenzenlos das System ist. Erstens, je mehr Freiheit es hat, desto mehr muss man in der Design-Phase darüber nachdenken, wie man dem System gestattet, guten Willen zu entwickeln. Zweitens, da man diese Systeme nicht eingrenzen kann, muss man einen Weg finden, das Risiko einzuschränken.

D sieht diese Entwicklung in Phasen. Phase 1 ist die großflächige Einführung von AW mit begrenzten Fähigkeiten. In Phase 2 erhält man Mensch-Maschine-Teams, und sie kann als eine Zwischenphase angesehen werden, in der Leute lernen, mit der Technologie zu kooperieren. Doch in der Phase wird es gefährlich und riskant, da wir dann mit unabhängigen Systemen konkurrieren und in Phase 4 wird dies zum Terminator oder Skynet-Beispiel, die ein Beispiel für eine KI sind, die sich als ernsthaft böse entwickelt hat, doch D denkt, dass dies nicht nächstliegend ist und KI sich nicht als böse herausstellt, sondern es wird eher der Fall sein, dass wir die KI durch kleine Abweichungen nicht mit der richtigen Denkweise ausgerüstet haben und somit immer noch die Welt zerstören. Er meint auch, dass es keine scharfe Grenze zwischen Phase 3 und 4 gibt, aber dass man das erst merkt, wenn es zu spät ist.

Name: Person E Datum: 17-04-2017
Dauer: 22:19

7 **Frage**

1 Warum beschäftigen Sie sich mit dem Thema Autonome Waffensysteme?

Ich stellte diese Frage nicht, wegen der begrenzten Zeit (30 Min.) für das Interview.

2 Sie, Wie würden Sie autonome Waffensysteme definieren?

Autonome Waffensysteme sind ein Waffensystem, dem ich einen Befehl geben kann zu gehen, es findet den Weg, wählt selbst sein Ziel (basiert auf den von Menschen erstellten Parametern), basiert auf dem Bild des Umfeldes das es selbst kreiert. Also wählen, finden und selbstständig handeln sind die drei Komponenten.

3 Was ist Ihre Stellungnahme zu autonomen Waffensystemen?

Positiv, ich bin offen für diese Art von Technologie. Also, positiv, aber wir müssen darüber nachdenken, wie wir sie mit Besonnenheit einsetzen. Wir nehmen sie nicht einfach achtlos in die Organisation auf ('niet zomaar naar binnen gooien').

4 Welche drei Fotos automatischer Waffensysteme [in dieser Version der Masterarbeit sind die drei Fotos wegen möglicher Urheberrechtsprobleme mit Worten beschrieben] möchten Sie in die Diskussion aufnehmen*?

1 Harpunen-Marschflugkörper, von einem Marineschiff aus abgefeuert.

2 Schwarm von 19 Drohnen

3 Roboter im Future-Style mit einer großen Handwaffe.

5 Können Sie für jedes [beschriebenes] Bild erklären, warum Sie es gewählt haben, welchen Wert es für sie innehält und warum es wichtig ist*?

1. Dies ist eines der derzeitig existierenden Systeme, das von der Marine seit über 30 Jahren benutzt wurde. Es ist ein Schiff, das Torpedos auf U-Boote abschießt. Sobald Sie diese Waffe abschießen, wählt sie autonom ihr Ziel und greift an. Man kann es nicht überschalten. Es heißt schießen und vergessen. Ein ähnliches System ist die Harpunenkanone, ein Lenkflugkörper, der gegen gegnerische Schiffe auf der Wasseroberfläche eingesetzt wird. Diese Waffensysteme suchen sich ihre eigenen Ziele und wenn sie sie finden, könnte es ein anderes sein, als dass, auf das Sie gezielt haben. Es ist ein System, das mit größter Vorsicht einzusetzen ist, aber es ist eine Art von Autonomie. Vom Standpunkt der Marine aus, sind diese gelenkten Flugkörper Kontrollsysteme, die mit einem gewissen Satz Parameter eingestellt werden können, z.B. Höhe, Geschwindigkeit und gewissen Identifizierungsmustern, und sie sind unterwegs.

Es befindet sich kein Mensch im Entscheidungsprozess, und wir benutzen diese Systeme seit über 20 Jahren. Diese Systeme können autonom eingesetzt werden, aber da wir diese Systeme einsetzen, bauen wir immer eine Funktion ein, die einen Menschen in den Entscheidungsprozess einschließt. Ein Mensch muss dem System grünes Licht geben, bevor es sein Ziel vernichten kann. Dies ist ein Prozess zentriert auf diese autonome Funktion, und es gibt sogar einen Hardware-Schalter, der bedient werden muss, bevor das System das Ziel vernichten kann. Eine Art von Hemmschwellenschalter, aber sie können das System auch so einstellen, dass Sie diese Überschaltung nicht benötigen. Diese Systeme wurden für Einsätze bei hoher Luftbedrohung entwickelt, in denen jede Sekunde zählt. Dies ist bereits ein gewisser Grad der Autonomie, mit dem Gefühl, dass man zu jeder Zeit eingreifen kann. Die Tatsache, dass es möglich ist, zu jeder Zeit einzugreifen, ist für die Marine sehr wichtig. Dies steht im Gegensatz zum Torpedo, der einem nicht immer gestattet, einzugreifen, will sagen, man muss sich absolut sicher sein, bevor man ihn einsetzt. Für beide Waffensysteme hat die Marine Einsatz-Prozedere erstellt.

2. Der nächste Schritt für das Verteidigungsministerium sind die ferngesteuerten Drohnen, die wir in unsere Organisation aufzunehmen versuchen, aber die Entwicklung in der Technologe geht schneller vor sich, als dass wir ihr zurzeit folgen können, und wir bemühen uns, uns diesem Tempo anzupassen. Schwärmer sind die Systeme, die in naher Zukunft verwendet werden. Diese Systeme werden eingesetzt und werden ihren Auftrag ganz alleine ausführen, aber wie sie das tun werden, das haben wir noch nicht im Griff. Die funktionieren mit einem ziemlich hohen Level der Selbstbestimmung. Zurzeit geht es um gemeinsamen Zugriff auf Informationen und Sensortechnologie, aber der nächste Schritt ist ein fühlergesteuerter Angriff mit einem Waffensystem. Die Technologie ist derzeitig verfügbar. Zurzeit entwickelt das Ministerium einen Entwicklungsplan für ferngesteuerte Luftwaffensysteme, aber der Sprung, vollständig autonome Waffensysteme zu bauen, ist zu groß, da dies etwas ist, dass wir, falls wir es haben wollen, noch nicht innerhalb des Verteidigungsministeriums besprochen haben. Dies ist zurzeit eine ethische Barriere im Verteidigungsministerium. Das Thema unbemannter und autonomer Systeme ist Teil der aktuellen Vision für die Streitkräfte, dass von der Hoofd Directie Beleid (vom Defensiestaf) konzipiert wird, und damit beginnt auch die Diskussion über die AW. Das Verteidigungsministerium wird diese Waffentypen nur erwerben, nachdem das Thema AW in unserer Gesellschaft debattiert worden ist.

Alle Fachkräfte der Armee, Marine und Luftwaffe sagen aus, dass Autonomie unterwegs ist, aber wir wollen immer die Kontrolle behalten. Wir wollen sie immer zurückrufen können, wenn sich die Situation geändert hat. Oder, dass wir sie ggf. neu programmieren können. Es ist typisch für militärische Einsätze, dass sich Situationen rapide ändern und man seine Pläne entsprechend umwerfen muss. Besonders, wenn Sie irgendwo arbeiten, wo es noch nicht zum bewaffneten Konflikt gekommen ist und Sie im Rahmen der Verhaltensregeln arbeiten. Sie müssen vorgreifen und schnell reagieren können, und Sie möchten keine Situation erhalten, in der Sie ihr System in der Luft eingesetzt haben, ein gewisses Ziel anzugreifen, aber wenn sich die taktische oder operative Situation ändert, möchten Sie in der Lage sein, es zurückzurufen. Wird diese Anforderung in die Systeme eingebaut, dann würden wir als Verteidigungsministerium diese Systemart gerne einsetzen.

3. Foto 3 stellt das Bild dar, das die Zivilbevölkerung von Killerrobotern hat, ein System, das nicht kontrolliert werden kann und selbstständig denkt und handelt. Leute stehen ihm misstrauisch gegenüber, und es stellt die Trennlinie dar, zwischen dem, was wir zurzeit einsetzen können, und welche Technologie wir uns ersparen möchten.

5 Haben Sie noch weitere Kommentare zu diesem Interview?

E betonte, dass er bereits seit Jahren mit AW gearbeitet hat (wie Bild 1 darstellt) und dass es eine gewisse Kontrollebene gibt, die man in seine Prozedere und Einsatzführung ('bedrijfsvoering') einbettet. Besonders, wenn eine neue Technologie sich noch nicht bewährt hat, dann müssen Sie mehr Sicherheiten dieser Art darin einbauen. E glaubt, dass wir in diese Richtung gehen und dass wir das auch sollten, denn es würde uns kostbare Zeit sparen und unsere Gegner tun es auch, folglich muss man dem folgen. Wenn ein Schwarm Drohnen auf Sie zukommt, dann müssen Sie sie bekämpfen können.

*Name:* Person F Datum: 08-05-2017
Dauer: 53:44 Min

| Nr. | Frage |
|---|---|
| 1 | Warum beschäftigen Sie sich mit dem Thema Autonome Waffensysteme? |

Diese Frage habe ich nicht gestellt, weil ich nur wenig Zeit hatte, das Interview zu führen (30 Minuten).

2 Wie würden Sie autonome Waffensysteme definieren?

F akzeptiert die ICRAC-Definition, die Autonomie in den kritischen Funktionen der Zielsetzung und des Abschießens der Waffen befürwortet. Es ist nicht nur die Wahl des Ziels als eine Funktion sondern die Kommunikation, die Waffe auf jemanden zu richten ist auch eine anthropologische und soziologische Handlung. Im formellen Sinne, wählt es ein Ziel oder bestimmt, dass jemand ein gültiges Ziel ist. Und in der nächsten Phase gibt es die Waffe frei und schießt sie ab. Wenn man in diesen beiden Funktionen Autonomie hat, dann ist es definitiv ein autonomes Waffensystem. Die Frage, ob man so ein System oder ein anderes besitzt und bis zu welchen Grad man menschliche Kontrolle hat, ist eine größere Herausforderung. Und was bedeutungsvolle menschliche Kontrolle über diese Funktionen eigentlich heißt, steht immer noch zur Debatte.

Die Grundlage für menschliche Kontrolle vom technischen Standpunkt aus bedeutet, dass der Mensch effektiv die Kontrolle hat, er kann intervenieren oder etwas zu Boden bringen oder das Verhalten eines Systems ändern. Bedeutungslose Kontrolle ist, wenn die Person zum gedankenlosen Automaten einer Maschine wird, also wenn man in einem Raum mit einem Knopf sitzt, und jedes Mal, wenn ein rotes Licht aufleuchtet, man ihn drücken muss. In gewissem Sinn hat man natürliche kausale Kontrolle über den Knopf, aber man ist sich keiner wirklich wichtigen Bedeutung bewusst, ob ein gültiges Ziel festgestellt wurde, da eben nur ein rotes Licht aufleuchtet und man diesem Licht zutraut, dass es korrekte Ziele produziert und einen gültigen Angriff auf dieses Ziel autorisiert. Aber man hat in Wirklichkeit kein kontextuelles oder situationsbezogenes Bewusstsein oder Verständnis oder unabhängiges Mittel, die Legalität oder Moral eines Ziels zu beurteilen. Im Effekt bedeutet es eben, dass man keine bedeutungsvolle Kontrolle hat, sondern nur kausale Kontrolle.

Im Vergleich dazu, damit Töten eine bedeutungsvolle Handlung ist, muss ein Vorhaben da sein, oder ein militärischer Zweck, ein Ziel zu vernichten. Und ganz einfach weil etwas ein legales militärisches Ziel ist, sollte man es angreifen. Es könnte sein, dass man nicht angreift, weil es teuer ist, oder man seine eigene Stellung verrät, es gibt alle möglichen Rücksichtnahmen, dass man auch bei gültigen Zielen überlegen muss, ob man angreift oder nicht, und das wird dann Teil einer menschlichen Maßnahme, und wenn man sie automatisiert, wird sie bedeutungslos und eigentlich zu einer arbiträren gerichtlichen Hinrichtung. Man hat keine sinnvolle Weise der Bewertungsmöglichkeit, ob ein Ziel legal oder sinnvoll ist.

3 Was ist Ihre Stellungnahme zu autonomen Waffensystemen?

F ist gegen sie und denkt, man sollte sie verbieten. Allerdings könnte man ein Verbot als eine positive Anforderung als bedeutungsvolle menschliche Kontrolle formulieren. Es ist wirklich schwer, über die ICRAC-Definition hinaus zu definieren, die er als vager als bedeutungsvolle menschliche Kontrolle empfindet, oder man könnte sagen, man braucht bedeutungsvolle menschliche Kontrolle für diese beiden kritischen Funktionen. Er meint, dass wenn bedeutungsvolle menschliche Kontrolle verstanden wird, dann wird es eine Anforderung für alle Waffensysteme, man definiert es für eine bestimmte Klasse von Waffensystemen, aber man muss gewährleisten, dass es für alle Waffensysteme in Frage kommt. Den Vergleich, den er ziehen möchte, ist der des unnötigen Leidens und der überflüssigen Verwundung, nicht dass dies die gleiche Definition ist, aber sie hat eine ähnliche Funktion, ist auch irgendwie unbestimmt darin, was unnötiges Leiden und überflüssige Verwundung eigentlich ist. Es gab auch große Debatten darüber, als diese Sprache in der St. Petersburg Konvention eingeführt wurde, und Juristen legten es beiseite, aber es wird nun so verstanden, dass es zusätzlicher Schaden ist, wenn sie keiner militärischen Funktion dienen, außer Leiden und Schmerz zuzufügen. Wenn dies auch unbestimmte Attribute sind, zeigen sie doch auf, ob wir bedeutungsvolle Kontrolle über ein System haben. Kann ich es aufhalten? Kann ich in seine Handlung eingreifen? Kann ich voraussehen, was es tun wird? Kann ich für mich selber beurteilen, ob die gewählten Ziele wirklich legal und moralisch verfechtbar sind? Denn wenn ich das nicht kann, dann haben wir ein Problem mit diesem Waffensystem. Die Systeme mit begrenzter Kontrolle sind gefährlich, und die globale Gesellschaft muss ausarbeiten, was wir mit ihnen anfangen sollen.

Aktuelle autonome Systeme, wie Goalkeeper auf Schiffen, müssen diese Anforderungen erfüllen. Und Staaten, die diese Art von Systemen einsetzen, müssen begründen, wie sie diese Anforderungen für ballistische oder abwehrende Systeme wie Goalkeeper, Phalanx oder Patriot erfüllen, da viele dieser Systeme herankommende Flugkörper zu Boden bringen und das System, das F persönlich erlebt hat, hatte eine sehr kurze Zeit und wenig Spielraum in Bezug auf wie lange es autonom zielgerichtet funktionieren konnte, nämlich weniger als eine Minute (20 oder 30 Sekunden), aber man kann immer noch argumentieren, dass da Menschen ihre Hände an diese Systeme legen, die entscheiden, wann sie den Systemen gestatten, abzuschießen. F hat nicht vor, dass das ICRAC diese Systemtypen eliminiert, aber sie sind trotzdem gefährlich, da sie immer noch den Beschuss eigener Truppen verursachen. F gibt ein Beispiel an, laut dem 2 Flugzeuge des Patrioten-Systems im Irak heruntergeschossen wurden, da ihr Transponder unsachgemäß reagierte, und danach wurde der Patriot technisch überarbeitet, sodass der Operateur positive Autorisierung für den Abschuss vom System geben muss, statt dass es eigenmächtig automatisch schießt. Dies erhöht die psychologische Belastung, sodass die Leute nicht bereit sind, den Knopf zu drücken, es sei denn sie sind sich völlig sicher, da man dieses Vorurteil gegen Automatisierung hat, wo das System bestimmt, dass das Ziel sicher ist. Nur dann sind Leute willens, das System handeln zu lassen. Es gibt auf jeden Fall eine bedeutungsvollere menschliche Kontrolle für die neue Patriotenplattform als für die frühere.

4 Welche vier Fotos automatischer Waffensysteme [in dieser Version der Masterarbeit sind die vier Fotos wegen möglicher Urheberrechtsprobleme mit Worten beschrieben] möchten Sie in die Diskussion aufnehmen*?

1. Terminator mit Metallskelett

2. Predator (Unbemanntes Luftfahrzeug)

3. Bild des Roboters der "Ban the Killer Robot Campaign" vor dem britischen Parlament in London.

4. Dreibeiniger Stuhl vor der UNO in Genf, Schweiz.

5 Können Sie für jedes [beschriebenes] Bild erklären, warum Sie es gewählt haben, welchen Wert es für sie innehält und warum es wichtig ist*?

**1. Terminator**

In den ersten Jahren der Kampagne, wurde das metallene Skelett des Terminators in jedem Artikel über die Kampagne dargestellt, interessant vom Standpunkt der Medien und sozialen Aktivisten aus. Einerseits ist es ein beeindruckendes Bild, dass das öffentliche Bewusstsein stimuliert und die Aufmerksamkeit der Leute erregt, da es Killerroboter darstellt.

Der Terminator ist wirklich sehr interessant, wenn man sich den Film anschaut, da er für ein bestimmtes Ziel eingesetzt wird; in diesem Sinn ist er also kein autonomes Waffensystem, obwohl er viele andere Leute auf seinem Marsch umbringt, somit gibt es einige autonome Zielentscheidungen. Doch die eigentliche Direktive ist Sara Conner, da sie das eigentliche Ziel darstellt. Und es gibt viele andere Elemente zu dem Film, die die Ängste der Leute stimulieren, wie Roboter, die sich ihrer selbst bewusst werden, sich gegen die Menschheit wenden und solche Sachen, die nicht in die Kampagne gehören. Es sind mehr die Zwischenstufen, dumme autonome Waffensysteme, die in den nächsten 5 bis 20 Jahren eingesetzt werden können, die nicht über anspruchsvolle Gedankensysteme verfügen werden, sondern auf einen kleinen Satz Kriterien achten, die auf ihre Sensordaten begründet sind und ihre Waffen auslösen, was negative Konsequenzen für die Zivilbevölkerung haben kann und alles Mögliche sonst. Das Image des Terminators erfasst nicht wirklich einen Wert, aber es erregt Aufmerksamkeit und das war zu Anfang der Kampagne sehr wichtig. Es war nicht Furcht wegen des Films, sondern echte Furcht davor, wie diese Waffensysteme funktionieren werden.

**2. Predator**

Das zweite, woran Leute denken, sind Predator- und Reaper-Drohnen, ferngesteuerte Roboter, folglich sind sie nicht wirklich autonom und passen somit nicht richtig ins Bild. Es sind realistischere Bilder, die im weitesten Sinne die Vorläufer zu autonomen Waffensysteme sind, weil man die Zielsetzung und das Auslösen von Drohnen automatisieren könnte, und sie würden zu autonomen Waffensystemen, und es gibt sogar Entwicklungen der nächsten Generation Drohnen, von denen viele sagen, dass sie autonom Ziele setzen können, zum Beispiel die X47-B, die Taranis und der Neuron, heimlich suchende Drohnen mit Düsenantrieb, die viele Waffen mit sich führen, die wirklich für im Konflikt stehenden Luftraum konzipiert sind, im Gegensatz zum Predator und Reaper, die mit Propeller angetrieben werden und langsam sind, und für sie muss man die Kontrolle über den Luftraum haben. In einem Luftraum, wo der Gegner alle Flugkörper hat oder Kommunikationen sperren kann, sind diese Waffentypen nicht geeignet. Die nächste Generation Drohnen kann in Räumen funktionieren, wo keine Kommunikation herrscht, ein Argument, das für Autonomie stimmt, weil man keine Kommunikationsverbindung benötigt, die unterbrochen werden kann, sodass man die Kontrolle über den Flugkörper verliert.

Für diese Bereiche müssen sie autonom sein. Die Frage ist, wie man sie zum Ziel bringt. Es könnte wie bei vorhandenen Lenkflugkörpern vorgehen, bei denen man eine Liste von GPS-Koordinaten bekommt, sie suchen diese Koordinaten und bombardieren sie; in diesem Fall bestimmen Menschen diese Ziele, oder der Flugkörper hat bis zu einem gewissen Grad eine eigene Fähigkeit des Suchens und Wählens eines Ziels von Sensordaten basiert auf Zufall.

Die Bilder dieses neuen Systems sehen sehr unheimlich aus, vor allem die der Taranis, mit Blitzen und Stapeln von Sprengköpfen davor. Sie stellen dar, was uns Besorgnis wegen der nächsten Generation autonomer Waffensysteme erregen sollte. Diese Besorgnisse würden darin liegen, dass sie eine Fähigkeit der autonomen Auswahl von Zielen haben. Es ist ein Langstreckenkampfflugzeug, das seine eigenen Ziele auswählt. Es wird zu einer Frage von Legalität und Bedienungskontrolle. Je nach Konzept dieser Versuchssysteme fallen sie entweder unter die autonomen Waffensysteme oder den nicht von Menschen vorgewählten Zielen und ihre Funktionalität ist noch nicht richtig entwickelt, und weil alles streng geheim behandelt wird, wissen wir nicht, wie sie eigentlich funktionieren.

Der zugrundeliegende Fall ist die Menschenwürde, will sagen, dass die Entscheidung zu töten, von einem Menschen und nicht von einer Maschine ergriffen werden soll. Es ist in die Bedeutsamkeit des Tötens eingebunden. Ein anderer Mensch bestimmt, dass du ein feindlicher Gegner bist, aufgrund deiner militärischen Präsenz auf dem Schlachtfeld, deiner Uniform und deiner Teilnahme an einen bewaffneten Konflikt und solche Kriterien. Das ist ein legaler und moralischer Akt der Tötung, und die Möglichkeit eines würdevollen Todes ist vorhanden. Ist es nur eine Maschine, die einem Algorithmus gehorcht und versucht, einige Kriterien, die die Legalität bestimmen, zu erfüllen, dann ist es immer noch keine menschliche Beurteilung und F findet, dass es keine Garantie für die Menschenwürde ist. Es kann richtig oder falsch vom rechtlichen Standpunkt aus sein, einer Art von Wahrscheinlichkeitsfunktion, aber ultimativ gibt es keinen Plan und keinen Sinn in diesem System, es kann nicht bestimmen, ob eine militärische Notwendigkeit vorliegt. Und da es keine moralische oder rechtliche Beurteilung machen kann, weil es kein rechtliches oder moralisches Agens ist, so kann sein Töten definitiv nicht würdig sein. Es kann gemäß des internationalen Rechts utilitär oder praktisch sein, aber von der moralischen Perspektive aus mangelt es an Würde.

Zu meiner Frage, ob in gewissen Situationen Roboter bessere Entscheidungen treffen als Menschen, weil sie nicht ermüden, nicht emotional werden und nicht Rache suchen, antwortete F, dass dies zwar ein wichtiger Punkt ist, aber dass wir Systeme entwerfen sollten, die Menschen helfen, bessere Entscheidungen zu treffen und insgesamt Fehler zu reduzieren, doch dass sie all diese Elemente der Bedeutung und Schaffen von Bedeutung auch beibehalten sollten und so dem Menschen helfen, sicherzustellen, dass dies ein moralischer, legaler und bedeutungsvoller Prozess ist, und die Maschine zu benutzen, um zu versuchen, diese Art von Fehlern zu reduzieren und gesteuerte Vorschläge zu erhalten. Das wird Menschen helfen, bessere Entscheidungen zu treffen, wenn sie müde sind oder unter Stress stehen.

**3. Freundliche Roboter**

Das dritte Bild, ist das der Kampagne, in der der Roboter vor dem britischen Parlament steht. Dies ist ein Bild eines freundlichen Roboters, da F erklärt, dass er Roboter eigentlich gerne mag und dass es militärische Aufgaben gibt, die von Robotern erledigt werden können, aber er betont, dass alle Aufgaben, die moralische und legale Urteilsfällungen erfordern, von Menschen ausgeführt werden müssen.

**4. „Dreibeiniger Stuhl" als Symbol der Landminen-Kampagne.**

Ein weiteres gutes Image der Kampagne ist das des sogenannten „Broken Chair" vor dem UN-Sitz in Genf, der die Landminen-Kampagne symbolisiert. Zu Beginn der Kampagne (Verbot des Killerroboters), versuchten wir, wie bei der Landminen-Kampagne das gleiche zu tun und zu sagen, dass dies extrem unterschiedslose Waffen sind, mit riesigen Auswirkungen auf die Zivilbevölkerung. Das ist äußerst unbehaglich. Wenn man es vergleicht, können Killerroboter als Landminen, die sich bewegen und Ziele wählen, gelten. Es ist eine andere Art, darüber nachzudenken, wie unheimlich sie sind, da sie wirklich nicht viel gescheiter als Landminen sind und die Welt auch nicht besser verstehen, als eine Landmine es tut. Sie sind nur feiner ausgeklügelt, weil sie navigieren und Daten über die Welt abtasten und ihre Ziele wählen können. Folglich werden sie einerseits nicht dermaßen blind handeln, und man kann den Mangel an Unterscheidungsfähigkeit im Laufe der Zeit mit Technologien verbessern, aber es gibt eine fundamentale Verschiebung, da man die Art der tödlichen Maßnahme und was sie überhaupt rechtfertigen kann, geändert hat, wenn man diese Killerroboter einsetzt. Aber das trifft nicht auf das Militär und die Kriegsführung zu, sondern auf die gesamte Gesellschaft, da wir viele Aufgaben, die früher unter menschlicher Kontrolle standen, automatisieren, und natürlich machen Menschen viele Fehler in Entscheidungstreffungen, es gibt Vorurteile und alle möglichen Neigungen usw. … die angesprochen werden müssen, aber oft konzipieren wir Automatisierung und geben vor, dass diese Vorurteile nicht länger vorhanden sind, da es sich um ein technisch konzipiertes System handelt, und es dürfte keine Fehler machen.

Aber diese Systeme sind kompliziert, und wir wissen nicht, wie sie sich verhalten oder was sie in unerwarteten Situation tun werden und man könnte einen ganzen Satz neuer Themen zur Sprache bringen, im Hinblick auf die Technik und Probleme, und man würde immer noch keine sozialen Fragen beantworten.

Um einen anderen Vergleich mit Landminen zu finden, gibt es eine gewisse Art von Voraussehbarkeit, doch andererseits auch eine offensichtliche Unberechenbarkeit. Landminen können gerechtfertigt werden, weil sie Truppen während eines Krieges aufhalten, aber 10 Jahre später, wenn jemand sein Land bebauen möchte, können Landminen immer noch explodieren, also kann man die Zukunft einfach nicht kontrollieren. Dies ist das inhärente Problem mit autonomen Waffensystemen, von denen man weiß, dass es losgeht, Ziele sucht und sie bombardiert. Das erscheint eine vernünftige und begrenzte Denkweise, dass es nur ganz spezielle Radarsignale zerstört, aber wie akkurat ist das? Wie viele Dinge sehen wie das Radarsignal aus? Da sind andere Dinge, die Radarsignale emittieren und damit verwechselt werden können, oder Leute senden Signale aus, die diese Radarsignale imitieren, um eine der Raketen auf andere Dinge zu lenken. Nachdem man es über eine lange Zeit oder über einen breiten Bereich eingesetzt hat, weiß man wirklich nicht, was es zum Schluss tun wird. Man hat eine Ahnung von dem, was das System tun soll, aber man kann es danach nicht länger kontrollieren. Und dann begibt man sich in gefährliche Bereiche, weil es nur um diese Risikoerwartungen geht und wir nicht wirklich wissen, wie man das messen kann und diese Systeme immer komplizierter werden, sodass es immer schwieriger wird, sie einzuschätzen, ob als Bediener, der vorhersehen möchte, was zu tun ist, oder als Gericht bzw. eine Organisation, die jemanden zur Rechenschaft dafür zieht, dass er/sie etwas Schreckliches ausgelöst hat, statt tat nur das Vorhaben gehabt zu haben, etwas Schreckliches zu tun. Also kann man ihn/sie nicht wirklich für Kriegsverbrechen verurteilen, doch stattdessen ist es Fahrlässigkeit, da er/sie nicht wirklich wusste, auf was er/sie sich einließ, aufgrund einiger Umstände von außen, die man nicht voraussehen konnte oder die man nicht kannte. Folglich beginnt unser ganzes System des Verstehens von Rechenschaft und Verantwortlichkeit zu zerbrechen und erzeugt einen tollen Ausweg für Leute, die Schlechtes wollen und dafür zur Rechenschaft gezogen werden, und es belastet Leute, die zwangsverpflichtet wurden oder dem Militär beitraten, als sie 18 waren und zu dem Zeitpunkt nicht wussten, was sie tun würden und deshalb nicht dafür verantwortlich gemacht werden können.

6 Haben Sie noch weitere Kommentare zu diesem Interview?

*Zu meiner Frag an F, ob er fähig wäre, ein Verbot für diese Art von Technologie durchzusetzen, antwortete er:* Es gibt einige Dinge, die man über internationales Recht und Verbote berücksichtigen sollte. Die Durchsetzung ist eines dieser Elemente, aber es ist nicht das einzige Element. Man kann es mit Mord vergleichen. Leute werden sich immer gegenseitig ermorden, sollten wir also Mord nicht gesetzlich verbieten? Nein, wir haben ihn als Straftat eingestuft, weil er böse ist und Leute ihn immer noch begehen, und damit muss man fertig werden. Es geht um die Etablierung einer Norm auf internationaler Ebene, auf der alle Länder dieser Erde zusammentreffen und vereinbaren, dass es grundsätzlich falsch ist, ein autonomes Waffensystem einzusetzen. Demzufolge, was wären die Folgen, wenn Leute dies übertreten. Die meisten Abkommen verfügen über keine Nachweismethoden und ausdrückliche punitive Maßnahmen und solche Sachen, aber es geht darum, Länder zu beschämen, Sanktionen zu erlassen und um internationale Konsequenzen für Länder, die die Abkommen übertreten.

Nichtstaatliche Akteure befinden sich in einem ganz anderen Bereich, und wir haben bereits chemische Waffen verboten, ein guter Vergleich, dass Länder, selbst Länder, die dieses Abkommen nicht unterschrieben, beobachtet und sanktioniert haben und ein Nachspiel erleben müssen, dafür, dass sie chemische Waffen benutzt haben. Es wirkt sich auch auf die Industrie aus, in dem Sinn, dass große militärische Unternehmer nicht aktiv chemische Waffen entwickeln und große chemische Unternehmen keine Riesenmenge von Vorläufern produzieren, die im Rahmen dieser Konvention verboten sind. Man wird Ähnliches mit autonomen Waffensystemen erleben. Natürlich kann man autonome Waffensysteme basiert auf Kram aus Baumärkten zusammenbauen, aber es wird kein schweres militärisches Waffensystem sein, das ganze Städte und Fabriken zerstören kann. Und Leute produzieren bereits Sprengfallen und Bomben und solche Dinge, und man kann das nicht wirklich aufhalten, aber man kann versuchen, den Zugang zu explosivem Material und Schlüsseltechnologien zu regulieren. Man kann vermeiden, dass sich autonome Waffensysteme von großen Staaten auf kleine Staaten vermehren, die sie nicht mit Einschränkung benutzen oder ihre Software modifizieren und zum Schluss auf nichtstaatliche Akteure und Terroristen erweitern werden, die Zugang zu Systemen mit ernsthaften Fähigkeiten haben und nicht nur ein Spielzeug mit einer angeschnallten Bombe sind. Dies wird folglich mit einem internationalen Abkommen abgeschwächt. Aber es gibt nichts, was die Polizei oder Staaten aufhalten wird, es gegen ihre eigenen Bürger zu benutzen, deshalb brauch man nationale Gesetze und Normen, um den Einsatz dieser Systeme für die Polizei auszuschließen, um zu vermeiden, dass sie sie gegen Demonstranten anwenden. Ich werde wahrscheinlich dem Zivilrecht und dem Strafrecht für Privatpersonen unterstehen, wenn ich diese Systeme bauen würde, aber es wäre gut anzugeben, dass es verboten ist, autonomen Feuerleitanlagen Waffen zuzufügen.

*Auf meine Frage, ob dies Kriterien der Transparenz bedeuten würde, antwortete F:* Da Sie die Bewertung hochentwickelter militärischer Systeme ernstnehmen, ist ein erforderlicher Teil für die Demonstration bedeutungsvoller menschlicher Kontrolle die Darstellung der Kommando- und Datenspeicherungskette, sodass wenn jemand erwartet, dass Ihr System vollständig autonom ist, Sie mithilfe des Datenflusses beweisen können, dass eine Zielentscheidung tatsächlich von Menschen getroffen wurden bzw. dass Menschen das Ziel in diesem speziellen Fall für jede Art von Waffensystem genehmigt haben. Es ist leicht, ein anspruchsvolles Informationsverarbeitungssystem zu haben, die Genehmigung, es zu bearbeiten, ist etwas anderes.

Der aktuelle Standard ist, dass Sie diese Disziplin ihrem eigenen Militär auferlegen, sodass Sie nicht unbedingt dementsprechend handeln müssen, aber Sie verfügen über diese Prozedere, sodass Sie Offiziere und Soldaten, wie sich selber auch, für einen Bruch des internationalen Rechts vor das Kriegsgericht stellen können. Sie werden von den Gegnern nicht erwarten, dass sie ein Verfahren dafür einleiten. Sie selbst bewerten Ihre Waffensysteme, folglich ist es ganz und gar nicht offensichtlich, und Sie werden diese Bewertungen nicht veröffentlichen. Aber die Anforderung ist, dass Sie diese Daten für den Fall, das etwas geschieht, speichern, sodass Sie sie zur Hand haben.

*Auf meine Frage, dass die Kampagne in 2015 und 2016 stark gefördert wurde, aber dass es nun aussieht, als ob sie verlangsamt und ober er die neuesten Informationen dazu habe, antwortete F:* Das Problem ist, dass die Zeit zwischen den VN-Treffen und den Schlagzeilen irgendwo im Dezember lag, dass die Diskussion vom informellen Treffen der Experten auf das formelle GGE erhöht wurde, was einen Schritt vor den Vertragsverhandlungen darstellt. Sie planten 2 Wochen für dieses Treffen, aber sie hatten in diesem Jahr eine Menge Probleme mit der Finanzierung, da sie das Budget-Modell der VN geändert haben, das vorschreibt, alle Beiträge bevor die Treffen abgehalten werden, zu bezahlen, doch während der letzten 50 Jahre gab es ein anderes System, gemäß dem man die Treffen einfach abhielt und wartete, bis die Leute ihre Beiträge später bezahlten. Es steht also zur Debatte, ob die zweite Woche mit den Treffen eigentlich stattfindet, wenn Länder ihre Beiträge nicht vollständig leisten. Das Treffen im August wird stattfinden, und hoffentlich gibt es ein zweites im November, in Verbindung mit ihrem Jahrestreffen. Sie sind der Hoffnung, dass sich etwas aus der GGE entwickelt, aber eine Woche ist wirklich nicht lange genug und sie hatten auf drei Wochen gehofft, nicht 2 Wochen, die vielleicht auf eine Woche reduziert werden.

Im Allgemeinen hatte die Kampagne einen guten Start, im Vergleich mit den meisten anderen Abrüstungskampagnen, einschließlich Landminen und Streubomben, die tatsächlich 10 Jahre brauchten, bis wir da waren, wo wir jetzt in drei oder vier Jahren sind. Also, diese Art von Schwerkraft beizubehalten wäre sehr beeindruckend gewesen, und es wurde von Staaten herausgezögert, die wohl bereit waren, es zu besprechen, aber weniger interessiert, ihre Einstellungen wirklich zu formalisieren. Es gibt alle möglichen Art und Weisen des Hinhaltens, aber die VN formalisiert ihre Stellung dazu auf jeden Fall, vor allem, da die 3000/09-Richtlinie im Pentagon gerade jetzt, nach 5 Jahren, neu bearbeitet wird. Und sie hoffen, dass die Staaten es gründlicher in der GGE debattieren werden und sich auf die Sprache einigen, da es leicht ist, zu sagen, es gäbe keine Definition von Autonomie und autonomen Waffensystemen, und dass wir wissen, was bedeutungsvolle menschliche Kontrolle ist, aber es liegt an ihnen, zu entscheiden, ob sie diese Begriffe rechtfertigen können, wie gerne sie es auch möchten.

Doch für das Abkommen muss der politische Konsens darüber was es für Definitionen sein sollen, gefunden werden. Der ICRAC legt sich sehr ins Zeug, um die Definition autonomer Waffensysteme einzuengen und dann ist es eine Frage, bedeutungsvolle menschliche Kontrolle auszuarbeiten und welche Art von Sprache für ein Abkommen die Richtige sein würde, nicht zu freizügig und nicht zu beengend. Es ist eine schwierige Balance und mehr ein politisches als ein technisches Problem. F hofft ziemlich stark, dass sie vorankommen.

Ein Grund, dass die CCW sich so einsetzte, war, dass sie im letzten Jahrzehnt wenig getan haben, und was sie getan hatten, war irgendwie ungenügend, sodass sie versuchten, es wett zu machen und das ist eine gute Sache. Wenn es bald geschieht, wird es vor allem eine präventive Art von Abkommen sein, aber die vorhandenen Waffensysteme müssen gemäß der jetzt erforderlichen bedeutungsvollen menschlichen Kontrolle neu untersucht werden, doch die meisten Waffensysteme im Einsatz würden diese Erklärung erfüllen und solche, die das nicht können, sollten auf keinen Fall benutzt werden. Es ist leichter es jetzt präventiv zu tun, als in 10 Jahren, wenn Staaten voll-autonome Drohnen und U-Boote entwickelt haben und empfinden, dass sie für ihre Sicherheit notwendig sind, und dann wäre es unmöglich diese Art von Abkommen zu beschließen.

ANHANG G

# Codierungsnotizen

### Wissenschaftlicher Mitarbeiter 1

Beim Codieren von Werten, wird der Text codiert, um die Werte[1], Einstellungen[2] und Ansichten[3] (Saldaña, 2015) zu codieren. Diese Konzepte sind als voreingestellte Codes verzeichnet.

| **List of pre-set codes:** | |
|---|---|
| **Concept** | **Colour** |
| Value | Blue |
| Belief | Green |
| Attitude | Yellow |
| List of emergent codes: | |
| **Concept** | **Colour** |
| Definition | Purple |

Tipps:

- Eine voreingestellte Liste braucht nur 10 Codes kann aber auch bis zu 40 oder 50 Codes haben. Wir empfehlen, nicht zu viele Codes zu erstellen, da die codierende Person überfordert werden oder Fehler im Codierungsprozess machen kann, wenn es zu viele gibt.
- Als Faustregel für das Codieren gilt, die Codes den Daten anzupassen, statt die Daten Ihren Codes.
- Das Erstellen von Notizen während des Codierprozesses ist für grounded und a Priori-Methoden gleich. Qualitative Forschung ist inhärent reflexiv; da die forschenden Personen tiefer in ihr Thema eindringen, ist es wichtig, ihre eigenen

1 Wert: Die Bedeutung, die wir uns selbst, einer anderen Person, eines Gegenstands oder einer Idee beimessen.

2 Einstellung: Die Art, wie wir über uns selbst, eine andere Person, einen Gegenstand oder eine Idee nachdenken.

3 Überzeugung: Der Teil des Systems, der unsere Werte und Einstellungen sowie unsere persönlichen Kenntnisse, Erfahrungen, Meinungen, Vorurteile und andere interpretierende Auffassungen der sozialen Welt einschließen.

 | DOI:10.1163/9789004449084_014

Denkprozesse mithilfe von reflektierenden oder methodologischen Notizen festzuhalten, da dies vielleicht ihre eigene subjektive Interpretation der Daten hervorheben könnte. Es ist äußerst wichtig, gleich zu Anfang der Forschung mit den Notizen zu beginnen. Egal welche Art von Notiz produziert wird, es ist wichtig, dass der Prozess kritisches Denken und Produktivität in der Forschung anregt. So etwas zu tun, erleichtert die Analyse und zusammenhängende Analysen, während das Projekt fortgeführt wird. Notizen können für das Ausarbeiten der Forschungsaktivitäten und Entdeckung der Bedeutung der Daten benutzt werden, sie erhalten den Schwung in der Forschung, Engagement und Eröffnungskommunikation.

Mehr Info: http://onlineqda.hud.ac.uk/Intro_QDA/how_what_to_code.php

https://researchrundowns.files.wordpress.com/2009/07/rrqualcodinganaly- sis_7_19_09.pdf http://programeval.ucdavis.edu/documents/Tips_Tools_18_2012.pdf https://www.slideshare.net/kontorphilip/qualitative-analysis-coding-and-categorizing

Weitere Notizen:

***Vor dem Start:***

Es ist mein Vorhaben, die Werte-Codierung zu benutzen (Saldaña, 2015), aber Saldaña (2015) definiert den Werte-Begriff anders als Friedman und Kahn Jr (2003). Ich frage mich außerdem, ob der Unterschied zwischen Wert, Einstellung und Überzeugung wirklich deutlich ist, aber ich werde sehen, wie es auslaufen wird. Vielleicht sollte ein weiteres Code-Wort für eine „Definition" hinzugefügt werden?

***Während der Codierung:***

a. Ich bemerke, dass eine Einstellung oftmals ein Verb einschließt, oder so würde ich es jedenfalls interpretieren.
b. Eine Überzeugung schließt oft eine Ansicht ein.
c. Ich merke, dass ich oftmals an die Listen der Werte aus meiner Fachliteraturuntersuchung denke, wenn ich einen Begriff hervorheben möchte, zum Beispiel „Reflexivität". >> Dies kann auf ein Vorurteil als Rechercheur meinerseits weisen, folglich wäre es gut, wenn ich meine Werte-Liste mit dem zweiten Rechercheur überprüfe.
d. Wenn ein Begriff der befragten Person nicht gefällt, z.B. Determinismus, markiere ich ihn als einen Wert.
e. Ist etwas viel Erwähntes oftmals wichtig für die befragte Person? Und kann es deshalb als ein Wert interpretiert werden? Zum Beispiel, „Distanz zwischen Mensch und Maschine" (siehe j).
f. Beim Codieren gehe ich manchmal zurück, um zu sehen, wie ich den gleichen Ausdruck klassifiziert habe (z.B. „zu unheimlich") und kopiere die Codierung.

g. Ich füge einen zusätzlichen Code für „Definition“ hinzu und füge es dann zum Code-Buch hinzu.
h. Ich habe eine Ansicht einer befragten Person als Überzeugung interpretiert, da es Teil der Definition des Begriffs „Überzeugung“ von Saldaña (2015) ist.
i. Nach dem Codieren einer Frage eines Interviews, lese ich die Frage noch einmal durch und prüfe meine Codierung.
j. Wenn ein Begriff mehr als zweimal benutzt wurde, z.B. der Begriff „defensiv“ in Professor Walshes Interview, dann habe ich ihn als einen Wert markiert.
k. Außerdem markierte ich einen Begriff, der besonders hervorgehoben wurde, z.B. wurde „Vorteile“ In Prof. Walshes Interview markiert.
l. Danach codierte ich ein neues Interview, ich prüfte die vorherigen, um zu sehen, ob ich konsistent geblieben war.
m. Wenn etwas als wichtig erörtert wird, z.B. im Interview mit Ad über die Wichtigkeit, dass die der Marine zu jederzeit die Möglichkeit hat, einzugreifen, dann habe ich das als einen Wert codiert.
n. Nach dem Codieren der letzten drei Interviews, habe ich die ersten drei nicht noch einmal überprüft. Vor allem wegen Mangel an Zeit, aber auch, weil Codierung leichter für mich geworden war.

Nach dem Codieren:

Ich habe alle 6 Interviews an den zweiten Rechercheur geschickt und einige Hintergrundinformationen zu den Codierungsinterviews hinzugefügt. Außerdem fügte ich das Codebuch ohne meine Kommentare hinzu. Wir besprachen die Aufgabe über Skype, um sicherzustellen, dass sie sie verstand.

## Codebuch des Korreferenten

Hiermit gebe ich die ordentlichen Richtlinien von Mutmaßungen, Regeln oder Schlussfolgerung, die halfen, die Aussagen in Richtung Wert, Einstellung oder Überzeugung zu wenden. Beachten Sie, dass diese Listen während dieses Codierungsverfahrens entwickelt und erweitert wurden, deshalb wurde es zu einem iterativen Verfahren. In anderen Worten, in späteren Interviews habe ich stets über die Richtlinien, die das Codierungsverfahren in früheren Interviews unterstützt haben, nachgedacht und darauf aufgebaut.

Wert: Wenn Aussagen etwas von großer Wichtigkeit wiedergeben, z.B. „wir müssen ...“, „müssen unbedingt“, „wichtig“.

- wenn man eine starke Beurteilungen bedenkt, wie „sicherstellen“, „beunruhigend“ „Bedürfnis“, „Menschheit“.
- Wenn sie großartige oder hoch bewertete Wörter enthalten, z.B. „sicherstellen“, „menschliches Leiden“, „unvermeidbar“.

- Wenn etwas als sehr wichtig oder äußerst zweckmäßig wiedergegeben wird, „Notwendigkeit", „Reflexivität".
- Wenn es ein wichtiger Teil der Geschichte oder Diskussion oder Darstellung ist, z.B. „Bedürfnis", „Moralität".
- Wenn etwas als wertvoll oder als zu berücksichtigen erachtet wird, z.B. „rechtfertigen", „Menschenrecht".

Einstellung: Wenn Aussagen Trends oder Richtungen angeben, z.B. „nicht nur ...", „auch ...".
- Wenn persönliche Ansichten oder Präferenzen angesprochen werden, z.B. „ist nicht wichtig", „wirklich schwierig", „man möchte".
- Wenn relative Positionen betrachtet werden, z.B. „ist nicht wichtig", „wirklich schwierig", „Unbestimmtheit".
- Wenn etwas das Urteil beeinflussen könnte, z.B. „nicht für ...", „mehr als", „wir möchten".
- Wenn es aus Worten besteht, die Ansichten darstellen oder wiederholt werden, z.B. „gegen", „vage", „schrecklich".
- Wenn man eine Art von Einstellung wiedergibt, z.B. „absichtlich getan", „sollte sein", „furchtbar", „verführt".

Überzeugung: Wenn Aussagen auf Mutmaßungen basiert sind oder etwas, das als wahr anerkannt ist, betrachtet.
- Wenn basiert auf (Future of Life-Institute) Erwartungen oder Mutmaßungen, z.B. „wir könnten ...", „es wird ...".
- Wenn Ungewisses besprochen (Future of Life-Institute) wird oder die Perspektiven anderer Leute vermutet werden.
- Wenn persönliches Wissen/persönliche Erfahrungen wiedergegeben werden, z.B. „wird geschehen", „weiß nicht".
- Wenn erhobene Fragen, Spekulationen oder Mutmaßungen betrachtet werden, z.B. „sie haben keine", „sie werden".
- Wenn interpretierende Aussichten oder die Zukunft betrachtet werden, z.B. „sie sind nur", „sie werden".

ANHANG H

# Ergebnisse des Codierungsprozesses

| Interviews | Werte | | |
|---|---|---|---|
| | *Rechercheur* | *Korreferent(in)* | *Ähnliche Werte* |
| *1. Interview* (*Person E*) | – stets eingreifen<br>– Kontrolle haben<br>– sie zurückrufen<br>– abrufen<br>– kann nicht kontrolliert werden<br>– Kontrollebene | – Wir müssen daran denken, wie wir sie mit Vorsicht benutzen.<br>– Die Tatsache, dass es möglich ist, zu jeder Zeit einzugreifen ist sehr wichtig für die Marine,<br>– aber wir wollen immer die Kontrolle haben. Wir wollen immer fähig sein, sie zurückzurufen, wenn die Situation sich geändert hat. Oder dass wir fähig sind, sie neu einzusetzen, falls erforderlich.<br>– man möchte fähig sein, sie zurückzurufen<br>– müssen mehr von dieser Art Sicherheiten darin einbauen<br>– muss mitmachen. | – jederzeit eingreifen<br>– Kontrolle haben<br>– sie zurückrufen |

 | DOI:10.1163/9789004449084_015

| Interviews | Werte | | |
|---|---|---|---|
| | *Rechercheur* | *Korreferent(in)* | *Ähnliche Werte* |
| 2. *Interview* (*Person D*) | – außer Kontrolle<br>– Kontrolle<br>– Kontrolle<br>– Risiko<br>– kontrolliert<br>– riskieren<br>– Gutes zu tun<br>– Risiko<br>– unkontrollierbares Risiko<br>– verliert unabsichtlich die Kontrolle<br>– Grenzen<br>– Guter Wille<br>– abgesteckte Grenzen<br>– Aufgaben mit Grenzen<br>– grenzenlos<br>– Guter Wille<br>– Risiko<br>– riskant | – sicherstellen, dass wir sie so verantwortlich wie möglich benutzen,<br>– das Risiko, die Kontrolle über Situationen zu verlieren, minimieren<br>– im militärischen Kontext, ist die Schnelligkeit der Entscheidungstreffung wichtig.<br>– ist keine Situation, die wir als Menschheit wollen.<br>– sicherstellen, dass wir nicht über den Klippenrand stürzen – verhindern, dass diese Technologien ihre eigene Vision kreieren<br>– Technologien mit niedriger Hemmschwelle, aber dem Potenzial unkontrollierbarer Risiken sind beunruhigend für die Menschheit.<br>– mögliche Ereignisse erwägen, wenn diese Art Technologie konstruiert wird und was schief gehen könnte, wenn man es nicht täte.<br>– brauchen einen starken Sinn des Zusammenhangs und eine Art von gutem Willen.<br>– Auftrag ausgeführt, aber nicht auf die Weise, die wir geplant hatten.<br>– müssen mit ihnen kooperieren, und Zielsetzung wird äußerst wichtig. | – Risiken minimieren<br>– Kontrolle haben<br>– Guter Wille<br>– abgesteckte Grenzen |

| Interviews | Werte | | |
|---|---|---|---|
| | *Rechercheur* | *Korreferent(in)* | *Ähnliche Werte* |
| | | – System muss wissen, was wir meinen.<br>– müssen unsere Leute schulen, mit diesen Systemtypen zusammen-zuarbeiten und ihnen die Ziele vermitteln<br>– müssen unser Personal darin schulen, Aufgaben mit festen Grenzen zu stellen.<br>– guter Wille<br>– Risiken eingrenzen.<br>– KI mit der richten Denkweise bauen und immer noch die Welt zerstören | |
| *3. Interview (Person C)* | – Entscheidungen<br>– entscheiden<br>– funktionale Form<br>– Entscheidungen<br>– Entscheidung<br>– Entscheidung<br>– Entscheidungen<br>– Entscheidung<br>– man hat das Recht, von einem Menschen getötet zu werden<br>- menschliches Leiden | – wie stellen wir sicher, dass dies die Grenze ist<br>– man hat das Recht, [von einer Person] getötet zu werden,<br>– Einbau von Autonomie in ein Waffensystem ist unabwendbar<br>– menschliches Leiden zu reduzieren,<br>– die Politik ist sehr interessant<br>– gibt es Ihnen einen Vorsprung auf dem Schlachtfeld?<br>– kann nicht sicher sein, dass die andere Seite keine AWs erwirbt. | – das Recht, von einem Menschen getötet zu werden<br>– menschliches Leiden |

| Interviews | Werte | | |
|---|---|---|---|
| | *Rechercheur* | *Korreferent(in)* | *Ähnliche Werte* |
| *4. Interview (Person A)* | – Ethischer Rahmen<br>– wir dürfen keine Kontrolle verlieren<br>– menschliche Kontrolle<br>– Distanz<br>– bedeutungsvolle menschliche Kontrolle<br>– menschliche Rechenschaftspflicht<br>– bedeutungsvolle menschliche Kontrolle<br>– bedeutungsvolle menschliche Kontrolle<br>– Distanz<br>– Distanz<br>– Distanz<br>– Distanz<br>– Mangel an Rechenschaftspflicht [und] Unvorhersehbarkeit<br>– es interessiert uns nicht<br>– Reflexivität<br>– Distanz<br>– Unbesiegbarkeit<br>– Determinismus<br>– Unabwendbarkeit<br>– Menschen müssen | – der ethische Rahmen ist sehr wichtig<br>– eine Waffe ohne menschliche Kontrolle bei kritischen Funktionen, z.B. Wahl des Ziels, müsste verboten werden.<br>– bedeutungsvolle menschliche Kontrolle<br>– menschliche Rechenschaftspflicht<br>– menschliche Kontrolle sollte Angewendet werden, wenn ein Ziel gewählt und außer Gefecht gesetzt wird<br>– aktuelle Waffensysteme müssten in Bezug auf Anwendung von menschlicher Kontrolle geprüft werden,<br>– zweckmäßig für etwas konstruiert<br>– erzeugt Distanz oder reduziert Schaden<br>– Furcht, mangelnde Rechenschaftspflicht und Unvorhersehbarkeit<br>– Reflexivität<br>– Distanz<br>– Unbesiegbarkeit<br>– Determinismus<br>– Unabwendbarkeit<br>– menschliches Bedürfnis, über Tod und Leben zu entscheiden | – Ethischer Rahmen<br>– Bedeutungsvolle menschliche Kontrolle<br>– Menschliche Rechenschaftspflicht<br>– Voraussehbarkeit<br>– Reflexivität<br>– Distanz<br>– Unbesiegbarkeit<br>– Determinismus<br>– Unabwendbarkeit<br>– Menschen müssen über Leben und Tod entscheiden können |

| **Interviews** | **Werte** | | |
|---|---|---|---|
| | *Rechercheur* | *Korreferent(in)* | *Ähnliche Werte* |
| | über Leben und Tod entscheiden. – Menschen müssen in der Entscheidung, Leben zu nehmen, eingeschlossen sein – Menschen müssen Kontrolle haben, über das was eine Maschine tut | – menschliches Bedürfnis, die Kontrolle darüber zu haben, was eine Maschine innerhalb gewisser Parameter anstellt ('Kader '). – Wie menschliche Kontrolle gewährleistet wird | |
| *5. Interview (Person F)* | – bedeutungsvolle menschliche Kontrolle – bedeutungsvolle menschliche Kontrolle – bedeutungsvolle Kontrolle – Plan – bedeutungsvoll – bedeutungsvoll – bedeutungsvoll – bedeutungsvolle menschliche Kontrolle – bedeutungsvolle menschliche Kontrolle – bedeutungsvolle menschliche Kontrolle – Oberflächliche Verletzungen | – effektive menschliche Kontrolle – kontext- oder situationsbedingtes Wissen oder Verständnis – bewerten der Legalität oder Moralität – muss und hat die Absicht, oder einen militärischen Zweck, um ein Ziel außer Gefecht zu setzen – bedeutungsvolle menschliche Kontrolle – bedeutungsvolle menschliche Kontrolle – bedeutungsvolle menschliche Kontrolle – Unnötiges Leiden oberflächliche Verletzungen – legale und moralisch gültig – hochentwickelte begründete Systeme | – bedeutungsvolle menschliche Kontrolle – unnötiges Leiden – oberflächliche – Verletzungen – Menschenwürde – würdiges Töten – Voraussehbarkeit – Kontrolle – Rechenschaftspflicht – Verantwortlichkeit |

| Interviews | Werte | | |
|---|---|---|---|
| | *Rechercheur* | *Korreferent(in)* | *Ähnliche Werte* |
| | – unnötiges Leiden<br>– Oberflächliche Verletzungen<br>– bedeutungsvolle Kontrolle<br>– Ausrichtung der Automatisierung<br>– bedeutungsvolle menschliche Kontrolle<br>– Kontrolle<br>– Menschenwürde<br>– Bedeutung<br>– würdig<br>– Würde<br>– Absicht<br>– Bedeutung<br>– würdig<br>– bedeutungsvoll<br>– menschliche Kontrolle<br>– Ausrichtung<br>– Ausrichtungen<br>– Voraussehbarkeit<br>– Unvorhersehbarkeit<br>– Kontrolle<br>– Kontrolle<br>– rechenschaftspflichtig<br>– Rechenschaftspflicht<br>– Verantwortlichkeit | – Kontrolle verlieren<br>– legale und aktive Kontrolle<br>– Menschenwürde<br>– legaler und moralischer Akt des Tötens<br>– würdiges Töten<br>– militärisches Bedürfnis<br>– moralische oder rechtliche Beurteilung<br>– stellt sicher, dass [...] moralischer, richterlicher und bedeutungsvoller Prozess<br>– Voraussehbarkeit<br>– Kontrolle<br>– rechenschaftspflichtig<br>– Rechenschaftspflicht und Verantwortlichkeit<br>– Gesetze und Normen<br>– Befehlskette<br>– Autorität<br>– transparent<br>– diese Daten speichern | |

| **Interviews** | **Werte** | | |
|---|---|---|---|
| | *Rechercheur* | *Korreferent(in)* | *Ähnliche Werte* |
| | – rechenschafts-pflichtig<br>– verantwortlich<br>– bedeutungsvolle menschliche Kontrolle<br>– rechenschafts-pflichtig<br>– bedeutungsvolle menschliche Kontrolle<br>– bedeutungsvolle menschliche Kontrolle<br>– bedeutungsvolles menschliche Kontrolle<br>– rechenschafts-pflichtig<br>– bedeutungsvolle menschliche Kontrolle<br>– bedeutungsvolle menschliche Kontrolle<br>– bedeutungsvolle menschliche Kontrolle | | |
| *6. Interview (Person B)* | – Verteidigung<br>– verteidigt<br>– verteidigen<br>– verteidigen<br>– unverantwortlich | – rechtfertigen<br>– Menschenrecht<br>– Effizienz<br>– Barrieren<br>– Gesetze und Regeln | – Verteidigung<br>– Schaden |

| Interviews | Werte | | |
|---|---|---|---|
| | *Rechercheur* | *Korreferent(in)* | *Ähnliche Werte* |
| | – Vorteile<br>– machtlos<br>– Schaden | – dagegen verteidigen<br>– militärischer Vorteil<br>– sich selbst verteidigen<br>– Genauigkeit<br>– bewerten<br>– sich verhalten<br>– sich auf eine bestimmte Weise verhalten<br>– überprüfen und bewerten<br>– Umwelt<br>– Erwartung, wie es sich verhalten wird<br>– ethische Regelungen<br>– gut definierte Spannen<br>– Schaden<br>– sich so verhalten, wie wir es gerne möchten<br>– man kann nicht darin eindringen | |

Basiert auf den Interviews und dem Vergleich der Werte-Codierung des Rechercheurs und dem zweiten Korreferenten wurde die folgende Liste der besonderen 23 Werte abgeleitet:

1. In jedem Fall ist eine Intervention erforderlich
2. Etwas unter Kontrolle haben
3. Zurückrufen
4. Risiken minimieren
5. Kontrolle haben
6. Guter Wille
7. Abgesteckte Grenzen
13. Abstand
14. Unbesiegbarkeit
15. Determinismus
16. Unvermeidlichkeit
17. Unnötiges Leiden
18. Oberflächliche Verletzungen
19. Menschenwürde

8. Das Recht, von einem Menschen getötet zu werden
9. Ethischer Rahmen
10. Bedeutungsvolle menschliche Kontrolle
11. Voraussehbarkeit
12. Reflexivität
20. Rechenschaft
21. Verantwortlichkeit
22. Verteidigung
23. Schaden

# Quellenverzeichnis


AIV, & CAVV. (2016). *Autonomous weapon systems: the need for meaningful human control.* (Nr. 97, Nr. 26). Abgerufen von http://aiv-advice.nl/8gr.

Aldewereld, H., Dignum, V., & Tan, Y.-h. (2015). Design for Values Information and communication technologies in Software Development Software development. *Handbook of Ethics, Values, and Technological Design: Sources, Theory, Values and Application Domains*, 831–845.

Alfano, M., & Loeb, D. (2014). *Experimental Moral Philosophy*. Abgerufen von https://plato.stanford.edu/entries/experimental-moral/.

Altmann, J., Asaro, P., Sharkey, N., & Sparrow, R. (2013). Armed military robots: editorial. *Ethics and Information Technology*,15(2), 73.

Asaro, P. (2012). On banning autonomous weapon systems: human rights, automation, and the dehumanization of lethal decision-making. *International Review of the Red Cross*, 94(886), 687–709.

Asaro, P. (2016, 14-10-2017). Rede über autonome Waffensysteme. Abgerufen von https://livestream.com/nyu-tv/ethicsofAI/videos/138822041.

Awad, E. (2017). *MORAL MACHINE: Perception of Moral Judgment Made by Machines.* (Master), Massachusetts Institute of Technology, Boston.

Bandura, A. (2001). Social cognitive theory: An agentic perspective. *Annual review of psychology*, 52(1), 1–26.

Beauchamp, T. L., & Walters, L. R. (199). *Contemporary Issues in Beoethik:* Wadsworth Pub.

Bonnefon, J.-F., Shariff, A., & Rahwan, I. (2016). The social dilemma of autonomous vehicles. *Science*, 352(6293), 1573–1576.

Borning, A., & Muller, M. (2012). *Next steps for value sensitive design*. Abhandlung die auf der Proceedings of the SIGCHI conference on human factors in computing systems präsentiert wurde.

Bradley, B. (2006). Two concepts of intrinsic value. *Ethical Theory and Moral Practice*, 9(2), 111–130.

Bratman, M. E., Israel, D. J., & Pollack, M. E. (1988). Plans and resource-bounded practical reasoning. *Computational intelligence*, 4(3), 349–355.

Bryson, J. J., Kime, P. P., & Zürich, C. (2011). *Just an artifact: why machines are per- ceived as moral agents.* Abhandlung, die auf der IJCAI Proceedings-International Joint-Conference on Artificial Intelligence präsentiert wurde.

Campaign to Stop Killer Robots. (2015). Abgerufen von https://www.stopkillerrobots.org/.

Campaign to Stop Killer Robots. 2017. The Problem. Abgerufen von https://www.stopkiller robots.org/.

Carpenter, J. (2016). *Culture and Human-Robot Interaction in Militarized Spaces: A War Story:* Taylor & Francis.

Chappelle, W., Goodman, T., Reardon, L., & Thompson, W. (2014) An analysis of post-traumatic stress symptoms in United States Air Force drone operators. *Journal of anxiety disorders*, 28(5), 480–487.

Cheng, A. S., & Fleischmann, K. R. (2010). Developing a meta-inventory of human values. *Proceedings of the American Society for Information Science and Technology*, 47(1), 1–10.

Coeckelbergh, M. (2013). Drones, information technology, and distance: mapping the moral epistemology of remote fighting. *Ethics and Information Technology*, 15(2), 87–98.

Cointe, N., Bonnet, G., & Boissier, O. (2016). *Ethical Judgment of Agents' Behaviors in Multi-Agent Systems.* Abhandlung, die auf der Proceedings of the 2016 International Conference on Autonomous Agents & Multiagent Systems präsentiert wurde.

Cummings, M. L. (2006). Integrating ethics in design through the value-sensitive design approach. *Science and Engineering Ethics*, 12(4), 701–715.

Cummings, M. L., Mastracchio, C., Thornburg, K. M., & Mkrtchyan, A. (2013). Boredom and distraction in multiple unmanned vehicle supervisory control. *Interacting with Computers*, 25(1), 34–47.

Cushman, F., & Young, L. (2011). Patterns of moral judgment derive from nonmoral psychological representations. *Cognitive Science*, 35(6), 1052–1075.

Davis, J., & Nathan, L. P. (2015). Value Sensitive Design: Applications, Adaptations, and Critiques. *Handbook of Ethics, Values and Technological Design: Sources, Theory, Values and Application Domains*, 11–40.

Dietvorst, B. J. (2016). *People Reject (Superior) Algorithms Because They Compare Them to Counter-Normative Reference Points.*

Dignum, F., Kinny, D., & Sonenberg, L. (2002). From desires, obligations and norms to goals. *Cognitive Science Quarterly*, 2(3–4), 407–430.

Dignum, V. (2016). *Einführung zu KI.* Abgerufen von https://rai2016.tbm.tudelft.nl/contents/.

Docherty, B. (2012) *Losing humanity: The case against killer robots.*

Docherty B. (2015). *Mind the gap: The lack of accountability for killer robots*: Human Rights Watch.

Etzioni, A., & Etzioni, O. (2017). Pros and Cons of Autonomous Weapons Systems. *Military Review* (Mai-Juni 2017).

Floridi, L., & Sanders, J. W. (2004) On the morality of artificial agents. *Minds and Machines*, 14(3), 349–379.

Friedman, B., & Kahn Jr, P. H. (2003). Human values, ethics, and design. *The human-computer interaction handbook*, 1177–1201.

Friedman, B., Kahn Jr, P. H., Borning, A., & Huldtgren, A. (2013). Value sensitive design and information systems. In *Early engagement and new technologies: Opening up the laboratory* (S. 55–95): Springer.

Future of Life Institute. (2016). AI Open Letter, in.

Galliott, J. (2015). *Military robots: Mapping the moral landscape:* Ashgate Publishing, Ltd.

General Assembly United Nations. (2016). *Joint report of the Special Rapporteur on the rights to freedom of peaceful assembly and of association and the Special Rapporteur on extrajudicial, summary or arbitrary executions on the proper management of assemblies.* (A/HRC/31/66).

Graduation Portal. (2017). Graduation Portal. Abgerufen von https://teams.connect.tudelft.nl/sites/tbm/graduate/SitePages/Home.aspx.

Graham, J., Haidt, J., Koleva, S., Motyl, M., Iyer, R., Wojcik, S. P., & Ditto, P. H. (2012). *Moral foundations theory: The pragmatic validity of moral pluralism.*

Gray, H. M., Gray, K., & Wegner, D. M. (2007). Dimensions of mind perception. *Science,* 315(5812), 619–619.

Gray, K., & Wegner, D. M. (2012). Feeling robots and human zombies: Mind perception and the uncanny valley. *Cognition,* 125(1), 125–130.

Haidt, J., & Joseph, C. (2004). Intuitive ethics: How innately prepared intuitions generate culturally variable virtues. *Daedalus,* 133(4), 55–66.

Harper, D. (2002). Talking about pictures: A case for photo elicitation. *Visual studies,* 17(1), 13–26.

Himma, K. E. (2009). Artificial agency, consciousness, and the criteria for moral agency: what properties must an artificial agent have to be a moral agent? *Ethics and Information Technology,* 11(1), 19–29.

Horowitz, M. C. (2016). The Ethics & Morality of Robotic Warfare: Assessing the Debate over Autonomous Weapons. *Daedalus,* 145(4), 25–36.

Hristova, E., & Grinberg, M. (2015). *Should Robots Kill? Moral Judgments for Actions of Artificial Cognitive Agents.* Paper presented at the EAPCogSci.

Hurka, T. (2005). Proportionality in the Morality of War. *Philosophy & Public Affairs,* 33(1), 34–66.

IA program. (2017). *Information Architecture.* Abgerufen von https://www.tudelft.nl/onderwijs/opleidingen/masters/cs/msc-computer-science/special-programmes/information-architecture/.

ICRC. (2010, 29-10-2010). *War and international humanitarian law.* Abgerufen von https://www.icrc.org/eng/war-and-law/overview-war-and-law.htm

Johnson, A. M., & Axinn, S. (2013). The morality of autonomous robots. *Journal of Military Ethics,* 12(2), 129–141.

Kaag, J., & Kaufman, W. (2009). Military frameworks: Technological know-how and the legitimization of warfare. *Cambridge Review of International Affairs,* 22(4), 585–606.

Kasher, A. (2016). 6 The Threshold of Killing Drones. *Drones and Responsibility: Legal, Philosophical and Socio-Technical Perspectives on Remotely Controlled Weapons*, 119.

Kominsky, J. F., Phillips, J., Gerstenberg, T., Lagnado, D., & Knobe, J. (2015). Causal superseding. *Cognition*, 137, 196–209.

Kreps, S. (2014) Flying under the radar: A study of public attitudes towards unmanned aerial vehicles. *Research & Politics*, 1(1), 2053168014536533.

Kuptel, A., & Williams, A. (2014) *Policy Guidance: Autonomy in Defence Systems.*

Le Dantec, C. A., Poole, E. S., & Wyche, S. P. (2009). *Values as lived experience: evolv- ing value sensitive design in support of value discovery.* Abhandlung, die auf der Proceedings of the SIGCHI conference on human factors in computing systems präsentiert wurde.

Lin, J., & Singer, P. W. (2014) *China's New Military Robots Pack More Robots Inside* (Starcraft-Style).

Logg, J. M. (2017). *Theory of Machine: When Do People Rely on Algorithms?*

Malle, B. F. (2015). Integrating robot ethics and machine morality: the study and design of moral competence in robots. *Ethics and Information Technology*, 1–14(2), 87–98.

Malle, B. F., & Thapa Magar, S. (2017). *What Kind of Mind Do I Want in My Robot?: Developing a Measure of Desired Mental Capacities in Social Robots.* Paper presented at the Proceedings of the Companion of the 2017 ACM/IEEE International Conference on Human-Robot Interaction.

Maslow, A. H. (1943). A theory of human motivation. *Psychological review*, 50(4), 370.

Nass, C., Moon, Y., Fogg, B., Reeves, B., & Dryer, D. C. (1995). Can computer per- sonalities be human personalities? *International Journal of Human-Computer Studies*, 43(2), 223–239.

Neapolitan, R. E., & Jiang, X. (2012) *Contemporary artificial intelligence:* CRC Press.

Noone, G. P., & Noone, D. C. (2015). Assessing the Debate over Autonomous Weapons. *Case W. Res. J. Int'l L.*, 47, 25.

NOS. (2016). Video: Vliegtuig redt piloot. Abgerufen von http://nos.nl/ artikel/2132527-video-vliegtuig-redt-piloot.html.

Oehlert, G. W. (2010). *A first course in design and analysis of experiments.*

Ohlin, J. D. (2016). The Combatant's Stance: Autonomous Weapons on the Battlefield. 92 *International Law Studies, Cornell Legal Studies Research Paper* No. 16–12, 1–2.

Open Roboethics initiative. (2015, 5-11-2015). *The Ethics and Governance of Lethal Autonomous Weapons Systems: An International Public Opinion Poll.* Abgerufen von http://www.openroboethics.org/laws_survey_released/

Perugini, M., & Bagozzi, R. P. (2004). The distinction between desires and intentions. *European Journal of Social Psychology*, 34(1), 69–84.

Pommeranz, A., Detweiler, C., Wiggers, P., & Jonker, C. M. (2011). *Self-reflection on personal values to support value-sensitive design.* Abhandlung, die auf der Proceedings of the 25th BCS Conference on Human-Computer Interaction präsentiert wurde.

Reips, U.-D. (2002). Standards for Internet-based experimenting. *Experimental Psychology*, 49(4), 243.

Roff, H. M. (2016). *Weapons autonomy is rocketing*. Abgerufen von http://foreign- policy.com/2016/09/28/weapons-autonomy-is-rocketing/.

Rønnow-Rasmussen, T. (2002). Instrumental values–Strong and weak. *Ethical Theory and Moral Practice*, 5(1), 23–43.

Rosenberg, M., & Markoff, J. (2016). The Pentagon's 'Terminator Conundrum': Robots That Could Kill on Their Own. *The New York Times*. Abgerufen von http://www.anytimes.com/2016/10/26/us/pentagon-artificial-intelligence-terminator.html?_r=0

Royakkers, L., & Orbons, S. (2015). Design for Values in the Armed Forces: Nonlethal Weapons Weapons and Military Military Robots Robot. *Handbook of Ethics, Values and Technological Design: Sources, Theory, Values and Application Domains*, 613–638.

RT. (2017, 5-07-2017). *Kalashnikov develops fully automated neural network-based combat module*. Abgerufen from https://www.rt.com/news/395375-kalashnikov-automated-neural-network-gun/

Russell, S., Norvig, P., & Intelligence, A. (1995). *A modern approach. Artificial* Intelligence. Prentice-Hall, Egnlewood Cliffs, 25.

Saldaña, J. (2015). *The coding manual for qualitative researchers:* Sage.

Scalable Cooperation Group. (2016). *Moral Machine*. Abgerufen von http://moralmachine.mit.edu/

Schroeder, M. (2016). *Value Theory*. Abgerufen von https://plato.stanford.edu/entries/value-theory/.

Schwartz, S. H. (1994). Are there universal aspects in the structure and contents of human values? *Journal of social issues*, 50(4), 19–45.

Schwartz, S. H. (2012). An overview of the Schwartz theory of basic values. *Online readings in Psychology and Culture*, 2(1), 11.

Searle, J. R. (1980). Minds, brains, and programs. Behavioral and brain sciences, 3(03), 417–424.

Searle, J. R. (1995). *The Construction of Social Reality*.

Sharkey, N., & Suchman, L. (2013). *Wishful mnemonics and autonomous killing machines*. Abhandlung, die auf der Proceedings of the AISB Konferenz präsentiert wurde.

Shelley, C. (2013). Fairness and regulation of violence in technological design. *Moral, Ethical, and Social Dilemmas in the Age of Technology: Theories and Practice: Theories and Practice*, 182.

Stewart, P. (2016). *U.S. military christens self-driving 'Sea Hunter' warship*. Abgerufen von http://www.reuters.com/article/us-usa-military-robot-ship-idUSKCN0X42I4

Stewart, W. (2015). *Russia has turned its T-90 tank into a robot – and plans to hiregamers to fight future wars*. Abgerufen von http://www.dailymail.co.uk/news/arti- cle-3271094/Russia-turned-T-90-tank-robot-plans-hire-gamers-fight-future-wars.html.

Strawser, B. J. (2010). Moral predators: The duty to employ uninhabited aerial vehicles. In *Handbook of Unmanned Aerial Vehicles* (pp. 2943–2964): Springer.

Sullins, J. P. (2006). When is a robot a moral agent. *Machine Ethics*, 151–160.

Thompson, W. T., Lopez, N., Hickey, P., DaLuz, C., Caldwell, J. L., & Tvaryanas, A. (2006). *Effects of shift work and sustained operations: Operator performance in remotely piloted aircraft (OP-REPAIR)*.

Abgerufen von UNDIR. (2014). Framing Discussions on the Weaponization of Increasingly Autonomous Technologies. Abgerufen von http://www.unidir.org/files/publications/pdfs/ framing-discussions-on-the-weaponization-of-increasingly-autonomous-technologies- en-606.pdf.

UNDIR (2015). *Framing Discussions on the Weaponization of Increasingly Autono mous Technologies.Considering Ethics and Social Values.* Abgerufen von http://www.unidir.org/files/publications/pdfs/considering-ethics-and-social-values-en-624.pdf.

Van de Poel, I. (2013). Translating values into design requirements. In *Philosophy and engineering: Reflections on practice, principles and process* (pp. 253–266): Springer.

van Wynsberghe, A., & Robbins, S. (2014) Ethicist as Designer: a pragmatic approach to ethics in the lab. *Science and Engineering Ethics*, 20(4), 947–961.

Walsh, J. I. (2015). Precision weapons, civilian casualties, and support for the use of force. *Political Psychology*, 36(5), 507–523.

Walsh, J. I., & Schulzke, M. (2015). *The Ethics of Drone Strikes: Does Reducing the Cost of Conflict Encourage War?* Abgerufen von.

Waytz, A., Gray, K., Epley, N., & Wegner, D. M. (2010). Causes and consequences of mind perception. *Trends in cognitive sciences*, 14(8), 383–388.

Williams, A. P., Scharre, P. D., & Mayer, C. (2015). Developing Autonomous Systems in an Ethical Manner. In *Autonomous Systems: Issues for Defence Policymakers*: NATO Allied Command Transformation (Capability Engineering and Innovation).

*La perception de l'agentivite et les valeurs morales relatives aux armes autonomes*

# Sommaire

Les armes autonomes sont de plus en plus utilisées sur les champs de bataille (Roff, 2016). Les systèmes autonomes peuvent présenter de nombreux avantages dans le domaine militaire, par exemple lorsque le pilote automatique d'un F-16 empêche un crash (NOS, 2016), ou dans l'emploi de robots par les démineurs pour neutraliser des bombes (Carpenter, 2016). Cependant la nature même des armes autonomes peut aussi conduire à des actions incontrôlables et à des troubles d'ordre sociétal. L'emploi d'armes autonomes sur les champs de bataille sans contrôle humain direct est non seulement une révolution dans le domaine militaire selon l'opinion de Kaag et Kaufman (2009), mais peut également être considéré comme une révolution morale. Comme l'emploi massif de l'intelligence artificielle sur les champs de bataille semble inévitable (Rosenberg & Markoff, 2016), une recherche portant sur la responsabilité éthique et morale est impérative.

Dans le débat sur les Armes Autonomes des opinions et des points de vue se font fortement entendre. La Campagne pour la Neutralisation des robots tueurs (2017) affirme par exemple sur son site internet: «Le fait de laisser des décisions sur la vie et la mort être prises par des machines transgresse une limite morale fondamentale. Les robots autonomes sont dépourvus de jugement humain et de la capacité à comprendre un contexte». Nous avons trouvé peu de travaux de recherche empiriques qui vont dans le même sens, ou qui donnent un aperçu de la manière dont les Armes Autonomes sont perçues par le grand public et par les militaires. Nous n'avons pas trouvé non plus de recherches empiriques sur les valeurs morales qui sous-tendent la «limite morale fondamentale» des Armes Autonomes. Par conséquent, le vide de connaissances est double par le fait que l'on n'a pas d'aperçu sur (1) la manière dont les Armes Autonomes sont perçues par les militaires et le grand public et (2) le fait de savoir quelles sont les valeurs morales que l'on trouve importantes si les Armes autonomes sont employées dans un avenir proche.

La première partie de ce manque de connaissances peut être comblée en étudiant la perception des Armes Autonomes en utilisant la théorie de l'agentivité[1] proposée dans les domaines de la psychologie cognitive, de l'intelligence artificielle et de la philosophie morale. La seconde partie du manque de

1 (Note du traducteur) «agency» en anglais. En philosophie, l'*agency* (terme récemment traduit par agentivité, notamment au Canada) est la faculté d'action d'unêtre; sa capacité à agir sur le monde, les choses, les êtres, à les transformer ou les influencer. En sociologie, l'agentivité est la capacité d'agir, par opposition à ce qu'impose la structure.

connaissances peut être comblée en étudiant les théories des valeurs connues (Beauchamp & Walters, 1999; Friedman & Kahn Jr, 2003; Schwartz, 2012) pour déterminer quelles sont les valeurs jugées importantes dans l'emploi des armes autonomes. Partant de ce manque de connaissances et de l'énoncé de ces problèmes, nous avons élaboré le sujet de recherche suivant pour cette étude:

*Comment les armes autonomes sont-elles perçues par le grand public et les personnels militaires du Ministère de la Défense néerlandais, et quelles valeurs morales considèrent-ils importantes dans l'emploi des armes autonomes?*

Dans cette étude, nous appliquerons la méthode basée sur la conception éthique («value-sensitive design», VDS) comme approche de recherche. La VDS est une approche à trois volets qui tient compte des valeurs humaines dans tout le processus de la conception technologique. Il s'agit d'un processus itératif pour la recherche *conceptuelle, empirique et technologique* des valeurs humaines impliquées dans la conception (Davis & Nathan, 2015 ; Friedman & Kahn Jr, 2003). La recherche conceptuelle s'articule en deux parties: (1) l'identification des parties prenantes *directes*, les futurs utilisateurs de la technologie, et les parties prenantes *indirectes*, dont la vie est affectée par la technologie, et (2) l'identification et la définition des valeurs que l'emploi de la technologie concernera. La recherche empirique s'intéresse à la compréhension et au vécu des parties prenantes dans les contextes ayant rapport avec la technologie, et les valeurs concernées seront examinées. Dans la recherche d'ordre technique, les caractéristiques spécifiques de la technologie sont analysées (Davis & Nathan, 2015).

Nous nous éloignons légèrement de la méthode de la conception éthique («value-sensitive design», VDS), car nous ne procédons pas à une analyse complète des parties prenantes pour les identifier dans le volet 1, mais nous nous concentrons sur deux groupes proéminents, le grand public et les personnels militaires, car ces deux groupes étaient réactifs pour notre étude. Pour le volet 2 de la première phase, nous avons procédé à un examen des publications afin d'identifier les valeurs concernées par les armes autonomes. L'objectif principal de notre recherche est la phase d'investigation, car nous avons trouvé que le volet empirique est passé sous silence dans le débat éthique sur les armes autonomes. Dans la phase d'investigation technique, nous ne désignons pas une arme autonome de la manière dont on pourrait intuitivement s'y attendre, car cela constituerait un projet immense qui dépasserait le propos de cette étude. Cependant nous avons décidé d'utiliser le travail du groupe de recherche de coopération évolutive et de proposer la conception d'une «machine morale» pour armes autonomes qui peut être utilisée dans une étape ultérieure de la recherche sur l'éthique des armes autonomes.

La pertinence scientifique de notre étude est que nous contribuons aux publications académiques pour mieux comprendre la perception du grand public et des militaires vis-à-vis des armes autonomes, et pour identifier les valeurs morales que le grand public et les militaires associent aux armes autonomes.. Actuellement cette compréhension est insuffisante, et aucune donnée empirique sur la perception et les valeurs associées aux armes autonomes n'a pu être trouvée. En utilisant l'approche de recherche de la conception éthique nous démontrons que cette méthode est applicable pour structurer une recherche académique pouvant être considérée comme étude de cas pour l'approche VSD. Nous approfondissons également l'étude de la prise de décision éthique concernant les vecteurs autonomes effectuée par Bonnefon, Shariff et Rahwan (2016) pour l'appliquer au domaine des armes autonomes.

La pertinence d'un point de vue sociétal est que la compréhension de la perception qu'ont le grand public et les personnels militaires du Ministère de la Défense néerlandais des armes autonomes, et la détermination des valeurs qu'ils associent à l'emploi des armes autonomes, peuvent être utilisées pour trouver des points communs et des différences dans le débat portant sur cette technologie lancé par la Campagne pour arrêter les Robots Tueurs (2015) et la Commission Internationale pour le Contrôle des Armes Robotisées (ICRAC). En deuxième lieu, les résultats de cette étude montrent comment la méthode de conception éthique peut être appliquée aux armes autonomes pour identifier les valeurs que le grand public et les personnels militaires associent à l'emploi de ce type d'armes. Enfin, en identifiant les valeurs importantes à incorporer dans la conception des armes autonomes, l'étude contribue à la conception future et à l'emploi responsable des armes autonomes.

Notre étude se concentre sur la phase d'investigation empirique de la conception éthique. Cette phase s'articule en deux parties. La première partie est exploratoire; c'est l'identification des valeurs des armes autonomes par une recherche en ligne, complétée par des entretiens avec six experts. La deuxième partie de l'investigation empirique étudie la perception de l'agentivité par le biais d'une méthode de recherche confirmatoire. Nous faisons fonctionner le concept d'agentivité pour confirmer notre hypothèse de la perception de l'agentivité en effectuant des expériences contrôlées randomisées, et utilisons les résultats de l'expérience pour explorer les valeurs associées aux armes autonomes de manière descriptive.

Pour étudier la perception de l'agentivité des armes autonomes nous avons émis une hypothèse dans laquelle nous soutenons que les personnels militaires considéreront les armes autonomes comme n'importe quelle autre arme, et donc comme rien de plus qu'un instrument pour obtenir un effet. Nous assumons dans l'étude finale que les personnels militaires considéreront les armes autonomes comme ne possédant aucun état mental. Nous avons élaboré trois scénarios dans nos expériences décrivant (1) un drone actionné par l'homme, (2) une arme autonome n'ayant aucune caractéristique d'agentivité, et (3) une arme autonome ayant des caractéristiques d'agentivité. Nous ne supposions aucune différence entre le statut d'arme autonome à agentivité neutre et le statut d'un drone activé à distance par un être humain. Nous supposons aussi que le statut d'agentivité neutre sera jugé très différent du statut de haute agentivité, dans lequel nous disons précisément aux participants que l'arme autonome a des caractéristiques liées à l'agentivité, comme la capacité de planifier et de fixer ses propres objectifs. En se basant sur cette démarche, nous supposons que:

> *H1: les personnels militaires ne considéreront pas que les armes autonomes ont des états mentaux.*

Les résultats de l'étude pilote et des études finales montrent que les éléments d'agentivité sont fiables et forment un tout. L'ensemble agentivité comprend les quatre éléments *Pensée, Détermination des objectifs, Libre arbitre* et *Réalisation des objectifs* pouvant être utilisés pour mesurer la perception d'agentivité des armes autonomes et des drones pilotés à distance par l'homme. Notre analyse centrale porte sur la manière dont le scénario de l'agentivité neutre diffère des scénarios Actionné par l'Homme et Haute Agentivité des armes autonomes. Nos résultats indiquent que la perception de l'agentivité par les personnels militaires et les civils travaillant pour le Ministère de la Défense néerlandais vis-à-vis du scénario d'agentivité neutre est plus haute que la perception de l'agentivité vis-à-vis du scénario du drone actionné par l'homme. Cela signifie qu'ils attribuent plus d'agentivité à une arme autonome qu'à un drone actionné par l'homme. A partir de ces conclusions, nous devons rejeter notre hypothèse.

L'effet de la perception de l'agentivité sur les variables dépendantes est étudié de manière descriptive, et non analysé de manière régressive. Nous avons trouvé que 7 sur 9 variables dépendantes sont associées de manière significative à l'agentivité. Il s'agit des variables suivantes: *confiance, dignité humaine, assurance, attentes, soutien* (*approbation*)*, équité* et *inquiétude*. Nos conclusions sont les suivantes:

- Les personnels militaires et les civils travaillant pour la Défense néerlandaise ressentent plus de confiance, d'assurance et d'approbation (soutien) pour des actions accomplies par des drones à contrôle humain que pour celles accomplies par des armes autonomes;
- Un drone actionné par l'homme est considéré comme respectant davantage la *dignité humaine* qu'une arme à agentivité neutre ou élevée, même si les actions et le résultat des scénarios sont identiques;

- Les personnels militaires et les civils travaillant pour la Défense néerlandaise ont un même niveau d'attentes vis-à-vis des actions futures des drones actionnés par l'homme que vis-à-vis des armes à agentivité neutre ou élevée. Ils considèrent aussi les actions des drones actionnés par l'homme et celles des armes autonomes comme également équitables;
- Les armes autonomes provoquent plus d'inquiétudes chez les personnels militaires et les civils travaillant pour la Défense néerlandaise que les armes actionnées par l'homme.

Cette étude a été menée à partir d'un ensemble de données réduit et, pour être en mesure de donner aux résultats une portée générale, il faudrait plus de personnes interrogées, et représentant un groupe démographique plus large. Par conséquent, nous proposons que le projet de «Machine morale pour les armes autonomes» prenne la forme d'une «expérience en ligne de grande ampleur» pour être plus exhaustif sur cette question. Pour ce projet nous travaillons sur le concept de la «machine morale» qui a été développé par le Groupe de Coopération Evolutive au Media Lab du MIT (Massachusetts Institute of Technology) (2016). Du fait de la sensibilité de ce sujet, nous proposons une mise en œuvre en deux étapes: d'abord concevoir une expérience contrôlée avec un ensemble limité de conditions (statuts). Ces conditions peuvent être visualisées dans des scénarios qui permettent aux utilisateurs d'y accéder après l'obtention d'un mot de passe par l'intermédiaire d'une interface web pour un serveur sécurisé. Après avoir rassemblé les données initiales et les réactions des utilisateurs, l'étape suivante pourrait être d'accéder à une plateforme ouverte à plus grande échelle, comme la «machine morale», où les personnes évalueraient les scénarios pour collecter un grand nombre de données de différents groupes démographiques; celles-ci pourraient être utilisées pour obtenir des résultats plus solides ayant une portée plus générale.

On peut identifier plusieurs problèmes posant des limites à cette étude. Tout d'abord, l'opérationnalisation des éléments et le concept d'agentivité dérivent d'une catégorie de publications qui décrivent les caractéristiques de l'agentivité, et notre sélection des caractéristiques a été basée sur un relevé numérique, et non guidée par la pertinence ou des critères de pondération. La seconde limite est la sélection des valeurs à partir du questionnaire sur les valeurs et des entretiens d'experts comme variables dépendantes. Bien que cette sélection fût longuement discutée entre les trois chercheurs concernés dans cette étude, le choix final a été effectué dans le cadre d'une méthode heuristique et non objective. En troisième lieu, les échantillons utilisés dans cette étude sont limités de par leur taille et leur portée démographique; l'étude finale n'a concerné que 239 personnes interrogées, ce qui ne représente qu'une petite partie du personnel de la Défense. Enfin, la répartition des personnes interrogées pour le scénario final est faussée, et le second scénario (armes autonomes à agentivité neutre) concerne 32 personnes interrogées de plus que pour le scénario un (drone actionné par l'homme). L'une des causes de cette dissymétrie pourrait être le fait que les personnes désiraient s'informer sur les armes autonomes, mais renoncèrent lorsqu'on leur présenta le scénario «action humaine». C'est ce que l'on appelle «l'attrition sélective»; elle a un impact négatif sur la validité interne de l'étude. Une autre explication pourrait être que le logiciel était défaillant en répartissant les personnes interrogées sur les scénarios, ce qui voudrait dire que la pertinence interne de l'étude n'est pas mise en cause, si la dissymétrie n'est pas attribuée à l'attrition sélective.

Ces limites étant posées, plusieurs recommandations pour des recherches plus approfondies sont proposées. La première est de valider le concept d'agentivité, qui exige d'autres études qui mesurent la perception d'agentivité d'autres produits technologiques, de manière à vérifier si ce concept tient aussi la route dans d'autres domaines. Ces recherches pourraient par exemple étudier la perception de l'agentivité de robots d'assistance aux personnes âgées, de jouets «intelligents» pour les enfants, ou d'ordinateurs embarqués pour véhicules autonomes. La seconde recommandation est de mener l'étude finale avec les mêmes scénarios sur un échantillon représentatif comprenant seulement des civils aux Pays-Bas. Ceci nous permettrait de voir sur quelles valeurs les résultats de l'échantillon militaire et les réponses des civils diffèrent; même si l'on ne peut faire des comparaisons directes en raison du fait que l'échantillon militaire n'est pas représentatif, nous verrions mieux les perceptions des militaires et des civils. Enfin, nous recommandons de mettre en œuvre la «machine morale pour les armes autonomes», décrite dans la section 5, pour donner aux résultats une portée générale, pour lesquels l'étude requière beaucoup plus de personnes interrogées représentant un groupe démographique plus large. Passer à une expérience en ligne à grande échelle, comme la «machine morale», fournirait de grandes quantités de données de groupes démographiques différents qui pourraient être utilisées pour présenter des résultats plus solides et de portée générale, et pour comprendre de manière exhaustive le jugement moral des personnes sur les armes autonomes.

# Preface

L'éthique des armes autonomes a commencé à m'intéresser après que j'eus assisté l'Université d'Eté sur l'Intelligence Artificielle Responsable (Al) en Août 2016 organisée par le Dr. Virginia Dignum,[1] sous l'égide de la Conférence Européenne «Intelligence Artificielle» (ECAI). Après mon premier entretien avec Virginia au sujet de mon projet de mémoire, lorsqu'elle m'a demandé si je voulais étudier à l'étranger et en quel endroit, j'ai répondu hardiment «le MIT», mais je n'imaginais pas que cela se concrétiserait. Les relations entretenues par Virginia avec le Dr. Iyad Rahwan, du Groupe de Coopération Evolutive au Laboratoire Media du MIT, ont permis de réaliser ce souhait. Je veux remercier Virginia d'avoir rendu les choses possibles, et de sa supervision tout au cours de ce parcours diplômant. Du fait du décalage horaire, nous avons organisé nos sessions par Skype durant les heures du soir aux Pays-Bas, et je lui adresse tous mes remerciements pour avoir trouvé le temps de revoir mon texte.

Le fait de travailler sur mon mémoire dans le cadre du Groupe de Coopération Evolutive constitua pour moi une énorme opportunité, et le Media Lab constitue un environnement très ouvert, diversifié, sans hiérarchie et très motivant. Au début, il a fallu que je m'habitue à courir dans une «piscine à boules» (ballenbak) dans le couloir, et aux gens qui quittaient leurs bureaux pour jouer au tennis de table en milieu de journée, mais après un mois environ, je me sentais chez moi dans le Lab. J'ai travaillé avec des étudiants très doués et ambitieux, de deuxième et troisième cycle. J'aimerais en particulier remercier le Dr. Sydney Levine et le Dr. Nick Obradovich de tous leurs commentaires très pertinents; j'ai véritablement apprécié les discussions animées que nous avons menées sur l'organisation de mon travail de recherche. Mais surtout je veux exprimer mes remerciements au Dr. Iyad Rahwan pour m'avoir permis d'être membre de son groupe; j'ai ainsi pu travailler sur le sujet sensible des armes autonomes. Il m'a donné toutes les ressources, le temps et l'assistance dont j'avais besoin pour mener mes recherches, ce dont je suis particulièrement reconnaissante.

Enfin, je voudrais remercier l'Armée de Terre Royale des Pays-Bas de m'avoir permis d'étudier à l'étranger, et de m'avoir soutenue pour mon poste au MIT. Sans son soutien, je n'aurais pas pu vivre et étudier à Boston pendant cinq mois. J'espère que ce mémoire sur l'Ethique des Armes Autonomes fera prendre

---

1 (Note du traducteur) Dr. (« Docteur ») aux USA fait référence à une personne qui a passé son doctorat quelle que soit sa matière de spécialité, et pas forcément la médecine.

davantage conscience de ce problème, et conduira à un débat sur le déploiement des armes autonomes et de l'intelligence artificielle dans l'organisation de la Défense.

*Ilse Verdiesen*
Ridderkerk, Août 2017

CHAPITRE 1

# Introduction

> Je pense avoir des difficultés à exprimer ce que je ressens à propos des drones. D'un côté, tout instrument de mort me perturbe. D'un autre côté, il est réconfortant de penser que tant de vies humaines ne seront pas mises en danger en les utilisant.
>
> PILOTE INTERROGÉ ÉTUDE 2

⁘

Les armes autonomes sont des systèmes d'armes équipés d'intelligence artificielle (IA). L'intelligence artificielle est définie par Neapolitan et Jiang (2012, p. 8) comme étant *«une entité intelligente douée de raisonnement dans un environnement complexe et changeant»*, mais cette définition s'applique aussi à l'intelligence naturelle. Russel & Norvig (1995) donnent une liste d'un grand nombre de définitions associant des vues sur des *systèmes qui pensent et agissent comme des humains* et *des systèmes qui pensent et agissent rationnellement*, mais ne donnent pas une définition personnelle claire. Pour le moment, nous adoptons à la définition donnée par Bryson, Kilme et Zürich (2011). Ils avancent qu'une machine (ou un système) fait preuve de comportement intelligent s'il peut choisir un mode d'action fondé sur une observation tirée de son environnement. Dans les publications scientifiques, l'IA est définie comme quelque chose de plus qu'un simple système intelligent. Elle se caractérise par les concepts d'*adaptabilité,* d'*interactivité* et d'*autonomie* (Floridi & Sanders, 2004). Selon Floridi et Sanders, (2004), l'*adaptabilité* signifie que le système peut changer selon son interaction, et peut tirer les leçons de son expérience. Les techniques d'apprentissage des machines en sont un exemple. L'*interactivité* existe lorsque le système et son environnement interagissent, et l'*autonomie* implique que le système peut changer lui-même son état.

Un nombre croissant de chercheurs s'intéresse à la conception responsable de l'IA, qui intègre des valeurs sociales et éthiques, pour prévenir des conséquences sociétales indésirables engendrées par cette technologie. Les principes qui définissent l'IA responsable sont la *capacité à se justifier* (*«accountability»*), *la responsabilité individuelle* (*«responsibility»*) et *la transparence* (*«transparency»*). «Accountability» signifie la justification des actions

 | DOI:10.1163/9789004449084_002

de l'intelligence artificielle, «responsibility» renvoie à la capacité de répondre personnellement de ces actions, et «transparency» est le fait de décrire et de reproduire les décisions que l'IA prend et adapte à son environnement (V. Dignum, 2016).

L'intelligence artificielle n'est pas seulement un scénario futuriste digne des romans de science-fiction dans lesquels des robots humanoïdes, comme Lore le frère de Data dans «Star Trek» ou les Cylons dans «Battlespace Galactica», ont l'ambition de dominer le monde. Un grand nombre d'applications IA sont déjà utilisées aujourd'hui. Les appareils de mesure intelligents, les moteurs de recherche, l'aide à la personne sur les téléphones mobiles, les pilotes automatiques et les automobiles sans chauffeur en sont des illustrations. La présente étude s'intéresse de manière privilégiée à l'application de l'IA dans le domaine militaire, et en particulier au déploiement d'armes autonomes dans un proche avenir. Dans cette introduction, nous exposerons dans un premier temps la problématique de cette étude, puis l'approche de la recherche, la pertinence scientifique et sociétale et l'intégration dans le programme de la filière Architecture de l'Information (Intelligence Artificielle) du Master «Gestion et Stratégie des Technologies des Systèmes» (SEPAM). Pour finir, nous indiquerons brièvement le contenu des chapitres suivants.

## 1.1 Problématique de recherche

Nous discutons d'abord de l'augmentation du nombre des armes autonomes dans le domaine militaire et mes questions qui s'y rattachent sur l'éthique et l'intelligence artificielle, puis des études sur les drones opérés manuellement dans la section sur l'étude des séquelles. Ensuite, nous définirons notre manque de connaissance, la nature et l'ampleur de ce problème; enfin, les questions majeures et annexes qui se posent à la recherche dans ce domaine.

### 1.1.1 *Analyse de la problématique*

Les armes autonomes sont de plus en plus souvent déployées sur les champs de bataille (Roff, 2016). Des études montrent que la Chine possède déjà des véhicules autonomes transportant des robots armés (Lin & Singer, 2014). La Russie prétend avoir à l'étude des chars autonomes (W. Stewart, 2015); les Etats-Unis ont lancé leur premier navire de guerre «sans pilote» en mai 2016 (P. Stewart, 2016), et le fabricant d'armes russe Kalashnikov a récemment révélé qu'il développait un module de combat automatisé qui utilise les réseaux neurologiques (RT, 2017). Les systèmes autonomes peuvent apporter beaucoup de plus-value dans le domaine militaire, par exemple lorsque le pilote automatique d'un F-16 empêche un crash de l'appareil (NOS, 2016), ou par l'utilisation de robots dans les activités de déminage (Carpenter, 2016). Cependant la nature même des armes autonomes peut aussi conduire à des activités incontrôlables et des troubles d'ordre sociétal. On peut en trouver des illustrations dans la campagne «Arrêtons les robots tueurs» suivies par 61 ONG et menée par «Human Rights Watch» (campagne contre les robots tueurs, 2015); également, l'Organisation des Nations Unies exprime ses inquiétudes en affirmant que «*les systèmes d'armes autonomes qui agissent sans contrôle humain significatif devraient être interdits, et la force contrôlée à distance ne devrait être employée qu'avec la plus grande prudence*» (Assemblée Générale des Nations Unies, 2016). Le déploiement d'armes autonomes sur les champs de bataille sans supervision humaine directe ne constitue pas seulement une révolution dans le domaine militaire selon Kaag et Kaufman (2009), mais peut aussi être considéré comme une révolution morale. Comme le déploiement massif de l'intelligence artificielle sur les champs de bataille semble inévitable (Rosenberg & Markoff, 2016), la recherche sur la responsabilité éthique et morale est impérative.

La prise de décision éthique en matière d'intelligence artificielle et de robots est un nouveau champ d'étude, et plusieurs chercheurs étudient le jugement moral en relation avec ces technologies. Par exemple, Malle (2015) propose un cadre qui associe les domaines (jusqu'à présent) distincts de l'*éthique dans le domaine des robots*, dans lequel les questions éthiques intéressant la conception, le déploiement et le traitement des robots par les humains sont abordées, et la *morale appliquée aux machines*, qui s'intéresse aux questions ayant trait aux capacités morales d'un robot, et à la manière dont elles doivent être mises en œuvre par ordinateur. Cointe, Bonnet et Boissier (2016) proposent un modèle dans lequel un agent peut juger des aspects éthiques de son propre comportement et de celui d'autres agents dans un système multi-agents. Ce modèle décrit un processus de jugement moral qui permet aux agents d'évaluer le comportement d'autres agents. Bonnefon *et al.* (2016) ont étudié les décisions d'ordre éthique qu'un véhicule autonome doit faire, d'ordre utilitaire ou d'autoprotection, lorsqu'il fait face à des piétons sur la route. Dans cette recherche, la «machine morale»[1] au MIT est utilisée pour comprendre en détails la manière dont les personnes voient les scénarios mettant en jeu un véhicule autonome, et en quoi leur jugement moral diffère ou non de ceux d'autres personnes.

1 http://moralmachine.mit.edu/

Dans un domaine ayant rapport avec les armes autonomes, celui des opérations mettant en œuvre des drones pilotés manuellement, les problèmes éthiques ont été étudiés intensivement ces dix dernières années dans des publications de nature philosophique et psychologique. D'un point de vue philosophique, Coeckelbergh (2013) affirme que les opérations mettant en œuvre des drones créent non seulement une distance physique, mais également une distance morale, puisque le visage de l'adversaire devient moins visible, ce qui élimine l'obstacle moral-psychologique et ainsi facilite l'action de tuer. Un autre problème éthique, selon Strawser (2010), est que du fait du changement régulier dans les activités en dehors de la zone de combat, les opérateurs des drones subissent des pressions psychologiques injustes, car ils doivent passer des activités liées au travail aux situations «à la maison» tous les jours. Également, du fait de l'éloignement du champ de bataille, les opérateurs humains peuvent ressentir une dissonance cognitive dans laquelle la guerre est ressentie plus comme un jeu vidéo que comme la réalité. Ces problèmes éthiques sont réfutés par Strawser (2010): il affirme que du fait de l'éloignement les opérateurs humains ont plus le temps d'évaluer une cible, car leur propre sécurité n'est pas menacée, mais il avance également qu'il faudrait procéder à d'autres recherches empiriques pour évaluer les effets psychologiques induits par la conduite d'opérations utilisant des drones à une grande distance du champ de bataille.

Plusieurs études ont été effectuées sur les effets des opérations par drone sur les opérateurs humains dans le domaine de la psychologie. L'une des premières études sur les effets psychologiques de ces opérations a été menée par Thompson et al. (2006) qui constatèrent que les opérateurs souffraient d'une fatigue accrue, d'épuisement émotionnel et de «burnout». Ceci a été confirmé par Chappelle, Goodman, Reardon et Thompson (2014), qui ont fait part du fait que les opérateurs de drones présentent des symptômes de burnout, sont «vidés» émotionnellement, font preuve de beaucoup de cynisme et ont des troubles de type post-traumatique. D'autres facteurs psychologiques mentionnés dans ces études empiriques sont l'ennui et l'inattention du fait de la longue durée des opérations mettant en jeu des drones (Cummings, Mastracchio, Thornburg & Mkrtchyan, 2013). Ces études montrent que le fait d'effectuer des opérations utilisant des drones peut avoir de graves effets psychologiques sur les humains qui actionnent ces drones.

### 1.1.2 *Vide dans les connaissances*

Dans le débat sur les armes autonomes des vues et des opinions bien tranchées sont exprimées. La Campagne pour «Arrêter les Robots Tueurs» (2017) déclare par exemple sur son site internet que *«Permettre que des décisions sur la vie ou la mort soient prises par des machines transgresse une limite morale fondamentale. Des robots autonomes sont dépourvus de jugement humain et de la capacité à comprendre un contexte»*. Nous avons trouvé peu d'études qui vont dans le même sens ou qui donnent un aperçu de la manière dont les Armes Autonomes sont perçues par le grand public et par les militaires. L'étude «Open Robots Ethics initiative» a effectué un sondage d'opinion en 2015, et publié un rapport. Mais les résultats ne furent pas publiés dans une revue spécialisée, et le sondage n'était pas assez représentatif pour que l'on puisse en tirer des conclusions significatives. Comme on l'a signalé dans le paragraphe précédent, plusieurs spécialistes comme Malle (2015), Cointe *et al.* (2016) et Bonnefon *et al.* (2016) étudient le processus de prise de décision d'ordre éthique dans le domaine de l'intelligence artificielle et des robots. Les problèmes éthiques sont également étudiés dans le domaine voisin des opérations mettant en œuvre des drones (Coeckelbergh, 2013; Strawser, 2010), et les études empiriques ont montré que les opérateurs souffraient de graves effets psychologiques (Chappelle *et al.* 2013; Thompson *et al.* 2006), mais ce type de recherche n'a pas encore été étendu au déploiement d'armes autonomes. Nous n'avons pas trouvé non plus de publications ni d'études empiriques sur les valeurs morales mises en jeu par les armes autonomes, ni sur ce que certaines personnes considèrent comme «la limite morale fondamentale». Par conséquent, le trou dans les connaissances est double, en ce sens que sont absentes les connaissances précises sur (1) la manière dont les armes autonomes sont perçues par les militaires et le grand public, et (2) les valeurs morales que les militaires et le grand public considèrent importantes lors du déploiement, dans un futur proche, des armes autonomes.

### 1.1.3 Enoncé de la problématique

La première partie de ce «trou» dans les connaissances peut être comblée en étudiant la perception des armes autonomes, en utilisant la théorie de l'agentivité énoncée dans les domaines de la psychologie cognitive, de l'intelligence artificielle et de la philosophie morale. L'«agentivité» est la capacité de planifier et d'agir (Waytz, Gray, Epley & Wegner, 2010), et est étudiée relativement à des questions touchant ou non l'humain. On attribue de l'agentivité à des objets, comme les ordinateurs (Nass, Moon, Fogg, Reeves & Dryer, 1995) et les robots (H.M. Gray, Gray, & Wegner, 2007); on peut donc penser que la technologie IA sera également perçue comme possédant les caractéristiques de l'agentivité. L'étude des perceptions qu'ont les militaires et le grand public vis-à-vis des armes autonomes permettrait de distinguer des différences et des ressemblances possibles entre les points de vue de ces groupes, qui pourraient être utilisées dans le débat actuel sur cette technologie.

La deuxième partie peut être comblée en étudiant les théories des valeurs connues, pour déterminer quelles sont les valeurs que l'on considère comme importantes concernant le déploiement des armes autonomes. Les théories sur les valeurs les plus reconnues sont celles de Schwartz (1994), Friedman & Kahn Jr. (2003) et Beauchamp & Walters (1999), mais on ne voit pas comment on peut les appliquer aux armes autonomes. Le fait de dégager les valeurs les plus pertinentes dans le contexte du déploiement des armes autonomes, et de les comparer avec celles qui sont mises en jeu dans la technologie actuelle des drones actionnés par l'homme, conduira à percevoir ce qui est sous-jacent dans le débat sur les armes autonomes, et permettra une meilleure compréhension des vues exprimées.

Ceci nous amène à l'énoncé de notre problématique :

> Dans le débat sur les armes autonomes, des vues et des opinions bien tranchées sont exprimées, mais la recherche empirique qui sous-tend ces opinions est absente. Nous ne savons pas comment les armes autonomes sont perçues, ni quelles sont les valeurs morales qui sont mises en jeu dans le déploiement des armes autonomes prévu dans un avenir proche. On ne voit pas non plus clairement si la perception et les valeurs morales relatives aux armes autonomes qu'ont les personnels militaires est différente de celle du grand public.

### 1.1.4 *Champ de recherche*

Beaucoup de publications sur les armes autonomes sont écrites et discutées par des experts juridiques et des philosophes dans le contexte du droit humanitaire international et les Conventions de Genève, dont le but est de limiter les effets des conflits armés (CICR). Comme nous ne sommes ni des experts juridiques ni des philosophes, la présente étude restera à l'intérieur des limites et des lois des conflits armés, et nous ne les remettrons pas en question. Du fait des contraintes imposées par les délais, la présente étude portera uniquement sur les personnels du Ministère de la Défense des Pays-Bas, et ne sera pas élargie à un échantillon représentatif de la population civile néerlandaise. Dans cette étude, nous nous intéresserons au déploiement des armes autonomes dans un avenir proche, que nous définissons comme *les cinq prochaines années*. Ceci a pour conséquence que nous n'étudierons pas les armes équipées d'une «intelligence artificielle générale», ni la technologie futuriste qu'il n'est pas encore possible de décrire; par contre, nous nous intéresserons à la technologie qui est développée actuellement, spécifiquement les drones dotés d'une capacité de ciblage autonome. Le processus de ciblage s'articule en six étapes: (1)

trouver, (2) fixer l'objectif, (3) poursuivre l'objectif (4) cibler, (5) prendre à partie, (6) évaluer. Nous traiterons de la prise de décision aux étapes 4 et 5 du processus de ciblage, du fait que de nombreuses décisions d'ordre éthique doivent être prises à ce moment du processus (Asaro, 2016).

### 1.1.5 *Question posée par cette étude, et questions secondaires*

En se basant sur le manque de connaissances qui a été identifié, sur l'énoncé de la problématique et sur le champ de recherche, nous proposons d'énoncer la question à laquelle tente de répondre la présente étude:

> Comment les armes autonomes sont-elles perçues par le grand public et par les personnels militaires du Ministère de la Défense néerlandais, et quelles valeurs morales mises en jeu par le déploiement d'armes autonomes considèrent-ils comme importantes?

Pour répondre à cette question fondamentale, on examinera les questions secondaires suivantes:

1. Comment les armes autonomes sont-elles définies dans les publications existantes?
2. Quelles théories des valeurs sont-elles exposées dans les publications existantes?
3. Quelles sont les valeurs exposées dans les théories des valeurs qui sont mises en jeu par les armes autonomes?
4. De quelle manière l'agentivité est-elle définie dans les publications existantes?

   Pour répondre à ces quatre questions secondaires, on se servira des publications existantes examinées dans cette étude.
5. Quelles valeurs morales sont-elles mises en relation par le grand public et les personnels militaires du Ministère de la Défense néerlandais avec le déploiement d'armes autonomes?
   - Pour répondre à cette question secondaire, on procédera à une enquête sur les valeurs consistant en un questionnaire mis en ligne et en des entretiens avec des spécialistes.
6. Comment l'agentivité des armes autonomes est-elle perçue par le grand public et par les personnels militaires du Ministère de la Défense néerlandais?
7. De quelle manière les valeurs mises en jeu par les armes autonomes sont-elles perçues par le grand public et par les personnels militaires du Ministère de la Défense néerlandais?

  - Pour répondre à ces deux questions secondaires, on réfléchira sur une hypothèse et une analyse descriptive basées sur les résultats d'une expérience contrôlée randomisée.
8. Comment les valeurs mises en jeu par les armes autonomes peuvent-elles être intégrées dans la conception d'une «machine morale» pour les armes autonomes?
  - On peut répondre à cette question secondaire en créant un plan de conception et de mise en œuvre d'une «machine morale» pour les armes autonomes.

Dans les paragraphes suivants, l'approche de recherche pour répondre à ces sous-questions est décrite plus en détails.

## 1.2 Approche de recherche

Dans la présente étude, nous employons la méthode de la conception éthique (conception intégrant les valeurs / «value-sensitive design» ou VSD) comme approche. La VSD est une approche à trois volets qui permet de tenir compte des valeurs humaines tout au long de la conception technologique. C'est un processus itératif intéressant l'investigation des valeurs humaines dans les champs *empirique, conceptuel et technologique* mises en jeu dans la conception (Davis & Nathan, 2015; Friedman & Kahn Jr, 2003). L'investigation conceptuelle consiste en deux volets: (1) identification des acteurs directs, ceux qui utiliseront cette technologie, et les acteurs indirects, ceux dont la vie est affectée par cette technologie, et (2) identification et détermination des valeurs que l'emploi de cette technologie met en jeu. L'investigation empirique s'intéresse à la compréhension et à l'expérience des acteurs dans un contexte associé à cette technologie, et les valeurs mises en jeu seront examinées. Dans l'investigation technique, les caractéristiques spécifiques de cette technologie sont analysés (Davis & Nathan, 2015). La conception éthique peut être utilisée comme «feuille de route» par les ingénieurs et les étudiants pour intégrer des considérations éthiques dans la conception (Cummings, 2006).

Quelques critiques ont également été exprimées concernant l'approche conception éthique. L'un des problèmes mentionnés par Davis et Nathan (2015) est que la conception éthique tient pour acquis que certaines valeurs sont universelles, mais qu'elles peuvent varier selon les cultures et le contexte. Pour réfuter cet argument, on pourrait prendre une base empirique comme point de vue de départ plutôt qu'une base philosophique, ou reconnaître que la position des chercheurs n'est pas forcément la seule valide (Borning & Muller, 2012). Borning et Muller (2012) avancent une proposition pluraliste selon laquelle la conception éthique ne doit pas recommander d'opinions sur les valeurs, universelles ou relatives, mais doit laisser aux ingénieurs la liberté de décider quelle opinion est la plus appropriée dans le contexte de la conception.

En accord avec Borning et Muller (2012) nous avons suivi l'approche basée sur la conception éthique (VSD) comme guide de recherche et non comme fin en soi. Pour le volet conceptuel, nous dévions légèrement de la méthode VSD originale, car nous ne procédons pas à une analyse complète pour identifier les acteurs dans la phase 1, mais nous nous concentrons sur deux groupes d'acteurs évident: *le grand public* et *les personnels militaires*; en effet, ces deux groupes étaient disponibles pour l'enquête. Dans la deuxième étape de la première phase, nous exposons une revue des publications existantes dans le but d'identifier les valeurs mises en jeu par les armes autonomes. Le principal centre d'intérêt porte sur la phase d'investigation empirique, car nous avons constaté que l'aspect empirique est négligé dans le débat éthique sur les armes autonomes. Nous nous sommes tournés vers les critiques de Davis et Nathan (2015) en consacrant la plus grande partie de notre temps à cette phase de la VSD, dans laquelle nous étudions les cas où les armes autonomes utilisent diverses techniques de recherche. Dans la phase d'investigation technique, nous ne concevons pas une arme autonome de la manière dont on pourrait s'y attendre de façon intuitive, car cela constituerait un projet immense dépassant de beaucoup le propos de la présente étude. Néanmoins, nous avons décidé d'exploiter le travail du groupe de recherche «Coopération évolutive» (Scalable Cooperation research group), et nous proposons une conception pour une «machine morale» destinée aux armes autonomes qui pourrait être utilisée dans une prochaine phase de la recherche portant sur l'éthique des armes autonomes. Nous appliquons les phases de la VSD à notre travail de recherche de la façon suivante (Figure 1):

- *Investigation conceptuelle:* La pénurie de travaux de recherche sur les armes autonomes implique que la première phase de notre étude est par nature exploratoire. Les activités de recherche concernant l'investigation conceptuelle seront effectuées qualitativement sur le mode de la recherche documentaire, pour produire une revue des publications existantes sur les armes autonomes, les valeurs morales et la perception de l'agentivité.
- *Investigation empirique:* Cette phase s'articule en deux parties. La première phase est exploratoire, et identifie les valeurs dans le contexte des armes autonomes au moyen d'une enquête en ligne qui constitue une technique de recherche quantitative, complétée par une technique de recherche qualitative: des entretiens avec des experts. Dans la seconde partie de l'investigation empirique, nous utilisons à la fois une méthode de recherche exploratoire et confirmative pour étudier la perception de l'agentivité. Nous

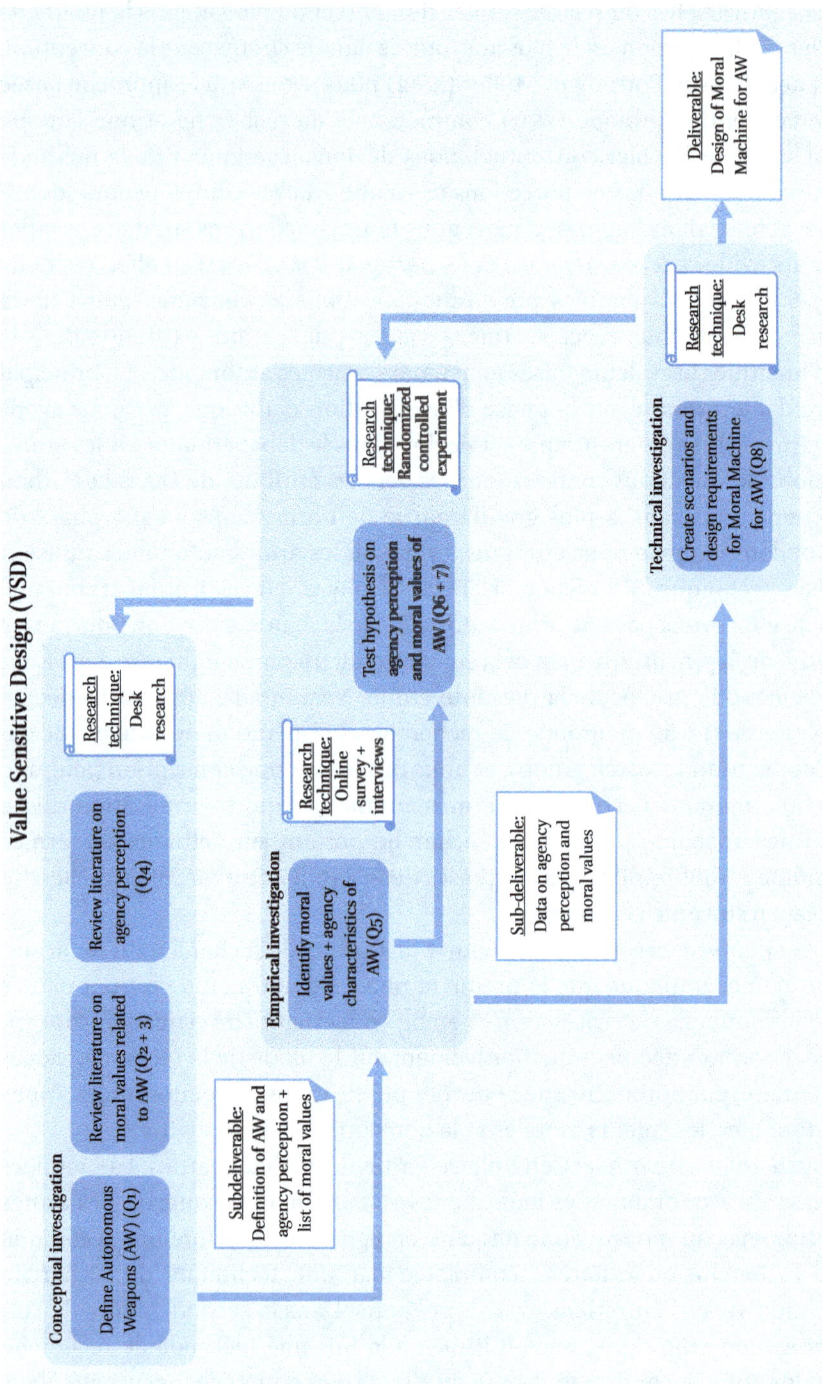


FIGURE 1 L'approche de recherche sur la conception éthique

opérationnalisons le concept d'agentivité pour confirmer notre hypothèse sur la perception de l'agentivité en menant des expériences contrôlées randomisées, et utilisons les résultats de l'expérience pour explorer les valeurs mises en jeu par les armes autonomes, et ce de façon descriptive.
- *Investigation technique*: Dans cette phase, la conception et les caractéristiques de la «machine morale» destinée aux armes autonomes sont produites à partir de scénarios qui sont utilisés dans l'expérience contrôlée randomisée de la phase empirique. La phase d'investigation technique est exploratoire, et le concept est créé en utilisant la technique de recherche documentaire.

## 1.3 Pertinence / applicabilité

Dans les paragraphes qui suivent, nous définirons la pertinence scientifique et sociétale de cette étude, et montrerons comment celle-ci s'intègre dans la démarche «intelligence artificielle» du master Politique et Gestion de l'Ingénierie des Systèmes (System Engineering Policy and Management, SEPAM)

### 1.3.1 *Pertinence scientifique*

Cette étude a une pertinence scientifique dans la mesure où elle contribue à la «littérature» scientifique en donnant un aperçu de la perception qu'ont le grand public et les militaires des armes autonomes, et en identifiant les valeurs morales que le grand public et les militaires associent aux armes autonomes. Actuellement nous manquons de connaissances en la matière, et aucune donnée sur la perception et les valeurs associées aux armes autonomes n'a été trouvée. En utilisant la conception éthique (VSD) comme approche de recherche, nous montrerons qu'il est possible de l'utiliser en tant que structure de recherche académique, alors qu'elle pourrait être considérée par certains comme méthode d'approche de type VSD pour des études de cas uniquement. Également, nous élargissons les travaux de recherche sur la prise de décision éthique concernant les véhicules autonomes (qui ont été menés par Bonnefon *et al.* en 2016) au domaine des armes autonomes.

### 1.3.2 *Pertinence d'ordre sociétal*

La pertinence sociétale de cette étude est que la compréhension de la perception qu'ont le grand public et les personnels militaires du Ministère de la Défense néerlandais, et que l'identification des valeurs morales qu'ils associent aux armes autonomes, peuvent être utiles pour définir les point d'accord et de désaccord dans le débat ayant trait à cette technologie et lancé par la Campagne pour Arrêter les Robot Tueurs (2015, Campaign to Stop Killer Robots) et le Comité International pour le Contrôle des Armes Robotisées (International Committee for Robot Arms Control (ICRAC). En second lieu, les résultats de cette étude démontrent que la méthode de conception éthique peut être appliquée aux armes autonomes pour identifier les valeurs que les militaires et le grand public associent au déploiement de ce type d'armes. Enfin, en identifiant les valeurs qu'il est important d'intégrer dans la conception des armes autonomes, la présente étude peut contribuer à promouvoir une conception et un déploiement responsables des armes autonomes dans le futur.

### 1.3.3 *Intégration dans le programme SEPAM et dans la démarche intelligence artificielle*

Les critères généraux dans le cadre d'un mémoire de master retenus à la faculté de technologie, stratégie et gestion des systèmes sont les suivants: l'étude doit avoir une composante analytique, se concentrer sur un domaine technique et être de nature multidisciplinaire. Pour obtenir son diplôme de master SEPAM, il faut que le travail de recherche échafaude une solution à un problème complexe, important, contemporain, de nature sociotechnique, ce qui signifie que le mémoire doit se focaliser clairement sur un domaine technique impliquant différents acteurs, prendre en compte les valeurs privées et publiques, ne traite pas de problèmes uniquement techniques mais aussi de choix en matière de gestion et d'éthique (Portail diplômes, 2017). La filière Architecture de l'Information intègre les volets gestion et informatique, et se concentre sur la compatibilité des besoins organisationnels et des opportunités en matière d'ingénierie pour trouver des solutions TOC (technologies de l'information et de la communication) à la pointe du progrès (Programme Architecture de l'Information, 2017).

La présente étude respecte les critères SEPAM en ce qu'elle se concentre spécifiquement sur la technologie IA des armes autonomes dans le domaine militaire. Elle a un caractère multidisciplinaire puisqu'elle associe les publications sur la perception de l'agentivité dans les domaines de la psychologie cognitive, de l'intelligence artificielle et de la philosophie morale, de la théorie des valeurs définie dans les textes de nature philosophique et psychologique, et le déploiement d'armes dans le domaine militaire. Le problème fondamental de nature sociotechnique dans la présente étude est le manque actuel de connaissances pour ce qui est de la perception qu'ont le grand public et les militaires de la technologie des armes autonomes, et des valeurs qu'ils jugent importantes dans ce domaine. Cette étude respecte l'optique Intelligence Artificielle, car elle identifie les points d'accord et de désaccord dans les priorités d'ordre éthique des militaires et du grand public. Elle propose une solution d'ingénierie pour la conception d'une «machine morale» destinée aux armes autonomes, de manière à explorer plus précisément ces priorités au moyen d'une enquête en ligne de grande ampleur, utilisant le concept de «machine morale» pour les véhicules autonomes.

## 1.4 Structure

Le reste de cette étude est organisé comme suit : dans la section 2, les publications sur les armes autonomes, les théories des valeurs et la perception de l'agentivité sont présentées. La section 3 décrit la méthodologie, les hypothèses, la conception de recherche, l'opérationnalisation des scénarios et des concepts, l'approche analytique, le pré-enregistrement, l'échantillon, et conclut avec les questions méthodologiques. Dans la section 4, tout d'abord les résultats de l'enquête sur les valeurs sous forme de questionnaire en ligne et d'entretiens de recherche sont donnés, suivis des résultats des trois expériences contrôlées randomisées. Le concept de la «machine morale» pour les armes autonomes est présenté dans la section 5, tout d'abord la «machine morale» originale pour les véhicules autonomes, puis les caractéristiques et les scénarios pour intégrer une «machine morale» dans des armes autonomes, et un plan de mise en œuvre succinct. La section 6 conclut sur les résultats, discute des implications d'ordre scientifique et sociétal, et identifie les limites et les recommandations en vue d'études ultérieures. Pour finir, dans la section 7, nous faisons part de réflexions sur les choix auxquels nous avons procédé dans notre recherche, la construction de ce projet, et le lien entre la recherche, l'intelligence artificielle et le programme SEPAM.

CHAPITRE 2

# Revue des publications

> C'est simplement que je ne me sens pas à l'aise devant le fait de placer des armes sur une machine qui n'a pas d'émotions et qui n'est pas contrôlée par l'homme.
>
> PILOTE INTERROGÉ, ÉTUDE 1

∵

Cette section donne une présentation des publications existantes ayant trait aux armes autonomes, suivie d'un résumé des théories des valeurs, et enfin nous définissons le concept d'agentivité que l'on utilise pour étudier la perception des armes autonomes. Chaque sous-section se termine sur une discussion des théories que l'on applique dans cette étude, ainsi que des raisons de leur choix.

## 2.1 Les armes autonomes

Bien que le débat sur les armes autonomes ait attiré beaucoup l'attention ces dernières années, nous avons trouvé que le sujet n'était pas bien cerné dans les publications académiques. Dans cette sous-section, nous commencerons avec une revue des nombreuses et diverses définitions, et présenterons deux classifications des armes autonomes pour conclure cette section.

### 2.1.1 *Définition*

Les armes autonomes sont une technologie «émergente», et il n'en existe pas encore de définition reconnue sur le plan international (AIV & CAVV, 2016). Il n'y a même pas consensus sur le fait de savoir s'il y a lieu d'en donner une définition. Bien que quelques auteurs donnent des définitions dans leurs écrits (tableau 1), d'autres mettent en garde contre de telles précisions. *Pour l'OTAN, «on doit éviter d'essayer de créer des définitions pour les «systèmes autonomes» car, par définition, les machines ne peuvent pas être autonomes au sens littéral du terme».* (Kuptel & Williams, p. 10). L'Institut de la Recherche pour le Désarmement des Nations Unies émet également des doutes sur le fait de donner une

 | DOI:10.1163/9789004449084_003

définition des armes autonomes; ses membres avancent que le degré d'autonomie dépend des «fonctions critiques considérées et des interactions de différentes variables» (UNIDIR, 2014, p. 5). Ils affirment que l'une des raisons de la différence entre les termes concernant les armes autonomes est que parfois les choses (drones ou robots) sont définies, mais parfois aussi une caractéristique (l'autonomie), des variables dans les problèmes (létalité ou degré de contrôle humain), ou la pratique (ciblage ou mesures défensives) apparaissent dans la discussion et deviennent une partie de la définition.

Les différentes définitions des armes autonomes sont données dans le Tableau 1. Certains auteurs utilisent le terme «robots militaires» lorsqu'il existe un certain degré d'autonomie. Comme les robots militaires peuvent être considérés comme une sous-catégorie des armes autonomes selon la classification de Royakkers & Orbons (2015) (Figure 2), nous les avons intégrés dans la liste des définitions. Selon nous, la définition donnée dans le rapport du Conseil Consultatif des Affaires Internationales (AIV & CAVV) saisit au mieux la description des armes autonomes d'un point de vue militaire et d'ingénierie, car elle tient compte de critères prédéfinis, et aussi du processus militaire de ciblage, puisque l'arme ne sera déployée qu'après prise de décision humaine. Par conséquent, nous adopterons cette définition, et définirons une arme autonome comme:

> *Une arme qui, sans intervention humaine, choisit et prend à partie des cibles qui correspondent à certains critères prédéfinis, à la suite d'une décision d'origine humaine de déployer l'arme en sachant qu'une attaque, une fois lancée, ne peut être arrêtée par une intervention humaine.*
>
> AIV & CAV, 2016, p. 11

### 2.1.2 *Classification des armes autonomes*

Les armes autonomes sont non seulement définies de manière ambigüe, mais elles n'ont pas été classifiée de manière uniforme. Nous présentons deux de ces classifications dans cette sous-section. Royakkers & Orbons (2015) définissent plusieurs types d'armes autonomes (figure 2), faisant la distinction entre (1) *les armes non létales*, qui sont des armes «*ne faisant pas de victimes (innocentes) ou ne nuisant pas de manière grave et permanente aux personnes*». (Royakkers &Orbons, 2015, p. 617), comme le système d'interdiction active (Active Denial System) qui met en œuvre un rayon d'énergie électromagnétique pour maintenir des personnes à une certaine distance d'un objet ou de troupes, et (2) les *Robots Militaires*, qu'ils définissent comme étant «*des systèmes sans personnels embarqués réutilisables destinés à un usage militaire,*

TABLEAU 1 Aperçu des différentes définitions des armes autonomes

| Auteur | Définition |
|---|---|
| AIV & CAVV (2016, p.11) | *«Arme qui, sans intervention humaine, choisit en prend à partie des cibles qui correspondent à certains critères prédéfinis, à la suite d'une décision d'origine humaine de déployer l'arme en sachant qu'une attaque, une fois lancée, ne peut être arrêtée par une intervention humaine».* |
| Altmann, Asaro, Sharkey & Sparrow (2013, p. 73) | Les armes autonomes sont: *"des armes robotisées qui, une fois actionnées, choisissent et vont frapper des cibles sans autre intervention humaine"* |
| Galliott (2015, p. 5) | Les robots militaires sont: *«un groupe de systèmes électromécaniques propulsés, qui ont tous les points communs suivants:*<br>*1. Ils n'ont pas de personnels embarqués;*<br>*2. Ils sont conçus pour être récupérables (même s'ils peuvent ne pas être utilisés dans ce but); et*<br>*3. Dans un contexte militaire, ils peuvent utiliser leurs capacités pour délivrer une charge létale ou non létale, ou avoir une fonction d'appui pour des objectifs d'une force militaire».* |
| Horowitz (2016, p. 27) | *«un système d'arme qui, une fois actionné, est destiné à frapper seulement des cibles individuelles ou des groupes de cibles spécifiques qui ont été choisis par un opérateur humain».* |
| Royakkers & Orbons (2015, p. 625) | Les robots militaires sont: *«... des systèmes sans personnels embarqués réutilisables destinés à usage militaire, doté d'un degré d'autonomie variable».* |
| Kuptel & Williams (2014, p. 10) | *«Les machines ne sont «autonomes» que dans certaines fonctions, comme la navigation, l'optimisation des capteurs ou la gestion de la consommation de carburant».* |
| UNDIR (2014, p. 5) | Le degré d'autonomie dépend des *«fonctions essentielles et de l'interaction de différentes variables».* |

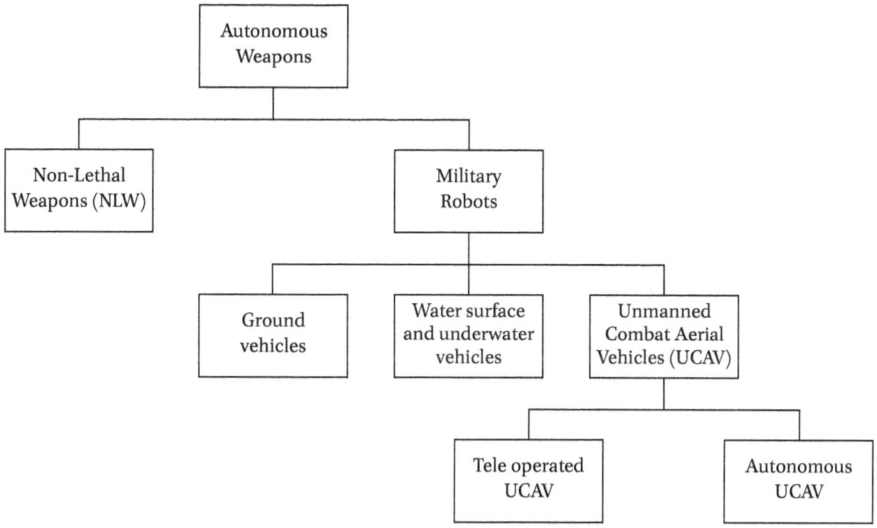


FIGURE 2 Classification des armes autonomes selon Royakkkers & Orbons (2015)

*dotés d'un degré d'autonomie variable*» (Royakkers & Orbons, 2015, p. 625). Les robots militaires peuvent être classés en trois catégories: les véhicules basés à terre, par exemple destinés à des missions de reconnaissance sans personnels embarqués et à neutraliser les engins explosifs improvisés (IED), les véhicule pouvant circuler sans personnels embarqués sous l'eau ou à la surface de l'eau, comme une pièce d'artillerie placée sur un navire ou un sous-marin autonome, ou des aérodynes de combat télépilotés (unmanned combat aerial vehicles, UCAV). Ces aérodynes de combat sont classifiés par Royakkers et Orbons (2015) comme étant télé-actionnés; les «drones» en sont l'exemple le plus connu, ainsi que les UCAV autonomes, qui sont peu à peu développés par le Département de la Défense des Etats-Unis (Rosenberg & Markoff, 2016).

Galliot (2015) propose un autre type de classification des armes autonomes, basé sur quatre niveaux d'autonomie pour les systèmes sans équipage (sans personnels embarqués):

1. *Niveau d'autonomie 1 – non-autonome / opéré à distance*: «Un opérateur humain contrôle chacun des mouvements mécaniques sans exception. Sans opérateur, les systèmes actionnés à distance sont incapables d'effectuer une quelconque opération effective».
2. *Niveau d'autonomie 2 – Autonomie supervisée:* «Un opérateur humain précise les mouvements, les positions et les actions simples, et le système effectue ces tâches. L'opérateur doit fréquemment fournir des

instructions au système et effectuer une supervision diligente pour s'assurer que les actions sont menées correctement».

3. *Niveau d'autonomie 3 – Autonomie dans l'action*: «Un opérateur humain spécifie une tâche générale, et le système élabore un mode d'action et le mène à bien sous son propre contrôle. Typiquement, l'opérateur a les moyens de surveiller le système, mais cela n'est pas nécessaire pour l'action à mener».
4. *Niveau d'autonomie 4 – Autonomie totale*: «Un système doté d'une autonomie totale crée et mène à bonne fin ses propres tâches sans avoir besoin d'instructions d'origine humaine, à l'exception de la décision d'élaborer ce système. L'agent humain est tellement éloigné du processus que le niveau d'influence directe est négligeable. Ces systèmes pourraient faire preuve de capacités qui imitent ou dupliquent les capacités morales des êtres humains dotés de sentiments (bien que nous ne prendrons pas position ici sur cette question)» (Galliott, 2015, p. 7).

Cette classification est à notre avis une bonne démarche pour catégoriser le niveau d'autonomie des armes autonomes, mais nous émettons quelques réserves d'un point de vue de l'ingénierie. Galliott (2015) lui-même déclare qu'il serait possible de fusionner les deuxième et troisième niveaux d'autonomie, car ils sont tous deux des niveaux opérationnels semi-autonomes. Nous sommes d'accord avec cet argument, mais ceci n'est pas le problème principal que nous rencontrons dans ces définitions. Nous pensons qu'il est curieux de commencer une liste de niveaux d'autonomie par la catégorie des systèmes non autonomes. Plus important: dans le quatrième degré d'autonomie, l'auteur déclare que *«ces systèmes pourraient faire preuve de capacités qui imitent ou dupliquent les capacités morales des êtres humains dotés de sentiments»*. Il semble se référer à la définition de l'intelligence artificielle forte, pour laquelle un ordinateur a des états et des programmes cognitifs pouvant imiter la connaissance humaine (Searle, 1980). Affirmer qu'un système autonome est doté de capacités morales montre à notre avis un manque de connaissances sur les systèmes d'intelligence artificielle actuels, ceux-ci n'étant pas autre chose que des ordinateurs ayant des capacités d'interactivité, d'autonomie et d'adaptabilité (Floridi & Sanders, 2004).

Comme il n'est pas sûr qu'une IA forte et capable de «capacités morales propres aux humains dotés de sentiments» soit un jour développée, nous pensons que la classification de Galliott (2015) n'est pas réaliste au vu de l'état de la technologie actuelle; par conséquent elle ne sera pas utilisée dans la présente étude. Pour cette dernière, nous adoptons la classification de Royakkers et Orbons (2015) qui donne un bon aperçu de la technologie militaire d'aujourd'hui et de demain. En particulier, leur distinction entre les aérodynes de combat sans équipage télépilotés d'une part et autonomes d'autre part est réaliste, et convient à notre travail de recherche.

## 2.2 Valeurs

Contrairement au problème des armes autonomes, le concept des valeurs a été étudié en détail dans les domaines de la philosophie et de la psychologie. Cette section présente une définition des valeurs, suivie d'une vue d'ensemble des théories qui définissent les valeurs universelles, une vue d'ensembles des valeurs mises en jeu par les armes autonomes, et conclut sur une hiérarchie des valeurs pour donner un exemple afin de relier les phases d'investigation conceptuelle et empirique de notre recherche.

### 2.2.1 *Définition*

Les théories des valeurs ont été bien étudiées dans les domaines de la philosophie morale et de la psychologie. La philosophie morale a une longue et riche histoire pour ce qui est de l'examen des valeurs, et dans ce domaine des questions d'ordre théoriques sont posées pour étudier la nature de la valeur et du bien (Schroeder, 2016). Une distinction est souvent faite entre les valeurs *instrumentales*, qu'il est raisonnable de privilégier pour leurs effets sur ce qui est bien (Ronnow – Rasmussen, 2002), et les valeurs *intrinsèques*, qui *«sont une catégorie de valeurs qui, une fois acquises, sont acquises uniquement en vertu de leurs propriétés intrinsèques»* (Bradley, 2006, p. 112). Bien que la philosophie morale s'intéresse surtout aux théories de ce qui *«devrait être»*, et qu'elle ne soit pas au sens strict influencée par des résultats empiriques (Alfano & Loeb, 2014), elle comporte une branche qui intéresse l'éthique appliquée, qui entre dans le champ de notre étude, car l'éthique appliquée établit le lien entre les théories éthiques abstraites et la pratique morale. Comme il a été dit dans la section 1.2., le propos de la présente étude est la phase empirique de la conception éthique pour rechercher comment les valeurs morales du grand public et des militaires coïncident ou non avec les armes autonomes. Par conséquent, nous avons décidé de ne pas utiliser les théories des valeurs de la philosophie morale dans notre étude; nous nous sommes tournés vers le domaine de la psychologie et de l'éthique appliquée pour obtenir une vue empirique des valeurs personnelles des deux groupes de parties prenantes sélectionnés, de manière à obtenir un aperçu de la situation telle qu'elle est et non telle qu'elle «devrait» être.

Le domaine de la psychologie distingue les valeurs des attitudes, besoins, normes et comportement en ce qu'elles sont une croyance (ou conviction), mènent au comportement qui guide les personnes, et sont ordonnées à l'intérieur d'une hiérarchie qui montre l'importance d'une valeur par rapport à d'autres valeurs (Schwartz, 1994). Les valeurs sont utilisées par les personnes pour justifier leur conduite, et définissent quels types de comportement sont socialement acceptables (Schwartz, 2012). Elles sont distinctes des faits en ce que les valeurs ne décrivent pas seulement un bilan empirique du monde extérieur, mais aussi adhèrent aux intérêts des humains dans un contexte culturel (Friedman *et al*, 2013). Les valeurs peuvent être invoquées pour expliquer la prise de décision individuelle, et pour étudier la dynamique humaine et sociale (Cheng & Fleischmann, 2010).

De nombreuses définitions des valeurs existent. Par exemple, Schwartz (1994, p. 21) définit les valeurs comme *des buts transituationnels souhaitables, d'importance variable, servant de principes directeurs dans la vie d'une personne ou d'une autre entité sociale»*. Ceci constitue une définition précise, en comparaison de Friedman, Kahn JR, Borning et Huldtgren (2013, p. 57) qui définissent les valeurs comme étant *«ce qu'une personne ou un groupe de personnes considèrent comme important dans la vie»*. Les définitions existantes ont été résumées par Cheng et Fleischmann (2010, p. 2) dans leur méta-inventaire des valeurs; ils avancent que *«les valeurs servent de principes directeurs à ce que les gens considèrent important dans la vie»*. Bien que ce soit une définition tout à fait simple, nous pensons qu'elle reflète au mieux la description d'une valeur, car elle combine plusieurs définitions en une seule, en utilisant les principales caractéristiques des valeurs. Par conséquent, nous adopterons la définition de Cheng et Fleischmann (2010) pour la présente étude.

### 2.2.2 *Les Valeurs Universelles*

Les travaux de recherche montrent que les personnes, dans toutes les cultures, s'identifient à des valeurs fondamentales qui peuvent être considérées comme des valeurs humaines universelles (Friedman *et al*, 2012; Schwartz, 2012). Bien qu'il y ait des différences selon les individus dans l'importance que l'on attribue aux valeurs, il semble qu'il existe un degré de consensus étonnamment élevé à travers les cultures sur l'ordre hiérarchique des valeurs (Schwartz, 2012). Quelques auteurs consacrent une partie de leur recherche à faire ce que l'on pourrait appeler un inventaire des valeurs, qui sont des listes qui peuvent être utilisées pour catégoriser l'analyse des valeurs humaines, et qui sont souvent accompagnées d'un outil descriptif permettant la discussion sur ces valeurs (Cheng & Fleischmann, 2010). Les inventaires de valeurs les plus courants et les plus documentés sont ceux de Schwartz (1994), Friedman *et al* (2013), Beauchamp & Walters (1999), et Graham *et al* (2012). Le nombre de valeurs universelles trouvées par les chercheurs varie considérablement. Un aperçu de ces inventaires de valeurs est donné dans le Tableau 2, et les théories seront brièvement exposées dans le paragraphe qui suit.

Après avoir procédé à des recherches poussées, Schwartz (1994) mentionne 10 types de valeurs incitatives, divisés en une liste plus fine de 56 valeurs qu'il utilise pour examiner les 10 valeurs universelles de portée générale. Dans leur description de leur approche de la conception éthique, Friedman *et al* (2013) mentionnent 12 valeurs; les 9 premières sont basées sur des orientations morales conséquentialistes et déontologiques, et les 3 dernières sont choisies à partir du champ de l'Interaction Humaine Informatisée (HCI). Graham *et al* (2012) utilisent le terme «fondation» pour définir les 5 valeurs distinctes spécifiques à l'universalité de la nature morale humaine, que Haidt et Joseph (2004) utilisent comme base de la Théorie de la Fondation Morale. Gouveia, Milfont et Guerra (2014) ont élaboré un cadre basé sur de nombreuses théories des valeurs, comme la hiérarchie des besoins de Schwartz (1994) et Maslow (1943). Dans ce cadre, les auteurs situent la valeur sur deux dimensions: (1) avec les actions qui motivent le comportement humain qui peuvent être des buts personnels, centraux ou sociaux; et (2) des incitateurs qui représentent les besoins humains, qui peuvent être divisés en besoins de prospérité (bien être) ou de survie (Gouveia *et al*, 2014).

Les valeurs sont non seulement définies en théorie dans une perspective psychologique comme on l'a vu dans le paragraphe précédent, mais elles ont également été «mises en œuvre» et employées au moyen de l'éthique appliquée aux domaines professionnels. Par exemple, dans le domaine médical, qui utilise la bioéthique pour définir les valeurs importantes tenant lieu de principes directeurs pour les professionnels de la recherche biomédicale, comme les médecins, infirmiers / ères et professionnels de santé, Beauchamp & Walters (1999) définissent 4 valeurs comme base pour le cadre de la bioéthique: (1) *l'autonomie*: agir de manière intentionnée sans subir le contrôle d'influences qui agiraient à l'encontre d'un acte volontaire; (2) *la bienfaisance* (*valorisation*): agir au profit de la société en général; (3) *la justice*: être équitable et raisonnable, et (4) *la non-maléficience* (*non-malfaisance*): ne pas imposer intentionnellement de risques ou de torts à autrui.

En nous basant sur cette revue des publications nous avons choisi deux théories des valeurs pour notre étude; l'une tirée des recherches du domaine psychologique, et l'autre basée sur l'éthique appliquée, qui est une application pratique de la philosophie morale. La première théorie que nous avons choisie est celle de Cheng et Fleischmann (2010), car dans leur méta-inventaire des valeurs morales ils ont dressé une liste exhaustive de 16 valeurs humaines, qui est basée sur les valeurs trouvées dans 12 études distinctes. A notre avis, cette méta-analyse saisit les valeurs les plus importantes listées par d'autres

TABLEAU 2 Théories des valeurs (aperçu)

| Auteur | Contribution majeure | Définition des valeurs | Valeurs |
| --- | --- | --- | --- |
| Schwartz (1994) | L'étude examine l'universalité potentielle des valeurs humaines et donne un ensemble de relations dynamiques entre ces valeurs. | Les valeurs sont définies comme étant: *«des buts transituationnels à atteindre, d'importance variable, servant de principes directeurs dans la vie d'une personne ou d'une autre entité sociale»* (p. 21) | 1. *Pouvoir* : statut social et prestige, domination et contrôle des personnes et des ressources (autorité, richesse, pouvoir social).[a]<br>2. *Succès*. Succès personnel en prouvant ses compétences selon des critères sociaux (ambition, succès, capacité, influence).<br>3. *Hédonisme* : plaisir et satisfaction sensuelle au profit de soi (plaisir, belle vie, complaisance envers soi-même).<br>4. *Stimulation* : enthousiasme, nouveauté, et défis à relever (vie variée, vie pleine d'événements, audace).<br>5. *Indépendance* : pensée et actions indépendantes – choisir, créer, explorer (créativité, liberté, choisir ses objectifs, curieux, indépendant). |
| | L'étude n'a pas trouvé d'aspect universel des valeurs, mais a trouvé ce qui peut justifier une quasi universalité de quatre valeurs d'un ordre supérieur. Egalement, l'étude a trouvé que tout montre que dix types de valeurs sont reconnus par de nombreuses personnes dans les sociétés contemporaines. | Cinq caractéristiques contribuent à la définition des valeurs humaines: (1) *«Conviction (croyance»); (2) qui procède d'un état final recherché ou de modes de conduite, qui (3) transcende des situations spécifiques, (4) guide le choix ou l'évaluation du comportement, des personnes et des événements, et (5) est classée par importance par rapport à d'autres valeurs pour former un système de priorités dans les valeurs* (Schwartz, 1992; Schwartz & Bilski, 1987)» (p. 20). | |

6. *Universalisme* : compréhension, appréciation, tolérance, et protection pour le bien-être de toutes les personnes et pour la nature (largesse d'esprit, justice sociale, égalité, paix dans le monde, monde de la beauté, harmonie avec la nature, sagesse, protection de l'environnement).
7. Bienveillance : préservation et amélioration du bien-être des personnes avec lesquelles on est en contact personnel fréquent (serviabilité, honnêteté, pardon, responsabilité, loyauté, amitié vraie, amour mature).
8. *Tradition* : respect, engagement, acceptation des coutumes et des idées que les cultures et religions traditionnelles portent en elles (respect de la tradition, humilité, dévotion, acceptation de son propre rôle dans la vie).

TABLEAU 2 Théories des valeurs (aperçu) (*cont.*)

| Auteur | Contribution majeure | Définition des valeurs | Valeurs |
|---|---|---|---|
| | | | 9. *Respect* : retenue dans ses actes, ses inclinations et ses impulsions qui sont susceptibles de choquer ou nuire à autrui et de violer les codes de conduite or les normes (obéissance, autodiscipline, politesse, respect des parents et des aînés).<br>10. *Sécurité* : sûreté, harmonie et stabilité de la société, dans les relations et pour soi-même (ordre social, sécurité familiale, sécurité nationale, réciprocité des services rendus). |
| Friedman, Kahn Jr, Borning et Huldtgren (2013) | Un aperçu de l'approche de la conception éthique (VSD), et indications pour une application pratique.<br><br>Fournit des informations pour permettre à d'autres chercheurs d'utiliser et d'élargir la VSD, et aux acteurs concernés de prendre les valeurs en compte dans le traitement de l'information et la conception de systèmes informatiques. | Une valeur est définie au sens large en ce qu'elle *«désigne ce qu'une personne ou un groupe de personnes considèrent comme important dans la vie»* (p. 57).<br><br>La méthode VSD concerne en particulier les valeurs morales qui sont: *«des questions ayant trait à l'équité, à la justice, au bien-être des personnes et à la vertu, englobant dans la théorie philosophique de la morale la déontologie, le conséquentialisme et la vertu».* (p.72). | 1. Bien-être des personnes : se rapporte aux bonnes conditions de vie physique, matérielle et psychologique des personnes.<br>2. Propriété : droit de posséder un objet (ou de l'information), de l'utiliser, d'en tirer un revenu, et de le léguer.<br>3. Intimité : autorisation, droit qu'a une personne de décider de quelles informations le ou la concernant peuvent être communiques à autrui. |

4. Liberté vis-à-vis de toute partialité, c'est-à-dire de toute injustice perpétrée contre des individus ou des groupes, y compris de par des préjugés sociaux, techniques ou sociaux nouvellement apparus.
5. Possibilité d'accès universelle : faire en sorte que tout le monde puisse utiliser correctement l'informatique.
6. Confiance : ce que chacun attend des personnes qui peuvent faire preuve de bonne volonté en l'élargissant vers les autres, se sentir vulnérables ou être victime de trahison.
7. Autonomie : capacité des personnes à décider, planifier et agir de manière à penser que cela va les aider à atteindre leurs objectifs.
8. Consentement éclairé : recueillir l'accord des personnes ; ceci englobe les critères de divulgation et de compréhension (pour « éclairé »), et d'acte volontaire, de compétence et d'accord (pour « consentement »).

TABLEAU 2 Théories des valeurs (aperçu) (*cont.*)

| Auteur | Contribution majeure | Définition des valeurs | Valeurs |
|---|---|---|---|
| | | | 9. Imputabilité : les mesures qui font en sorte que les actions d'une personne, de plusieurs personnes ou d'une institution peuvent être ramenées uniquement à cette / ces personne(s) ou à cette institution.<br>10. Courtoisie : se conduire envers autrui avec politesse et considération.<br>11. Identité. La conscience que chacun a de qui il est dans la durée, de manière continue ou discontinue dans le temps.<br>12. Quiétude. Etat psychologique paisible et serein.<br>13. Durabilité environnementale : protéger les écosystèmes pour qu'ils répondent aux besoins présents sans compromettre les générations futures. |
| Graham et al (2012) | Présentation (avec les critiques et résultat empirique) de la *théorie du fondement de la morale.*<br><br>Cette théorie peut être utilisée pour connaître le jugement | Pour représenter les cinq concepts de cette théorie du fondement moral, le terme «fondement» («foundation») est choisi, mais il est interchangeable avec «valeur» ou «vertu». Aucune définition précise de «fondement» n'est donnée, mais | Les cinq fondements de la théorie sont les suivants :<br>1. Fondement bienveillance / malveillance : capacité à ressentir la douleur des autres personnes ; sous-tend les vertus de bonté, d'amabilité et dévouement. |

| | | |
|---|---|---|
| moral des personnes. Elle est décrite comme étant une approche de la morale pluraliste, nativiste, culturo-«développementaliste» et intuitionniste. | le terme est utilisé comme métaphore architecturale («fondation») pour affirmer que «*la théorie du fondement moral s'intéresse à la première esquisse universelle de l'esprit moral, et à la manière dont cette esquisse est révisée à des degrés variables selon les cultures*» (p. 10). | 2. Fondement équité / tromperie : principe de réciprocité altruiste, génère les idées de justice, de droits et d'autonomie.<br>3. Fondement loyauté / trahison : renvoie aux coalitions changeantes, sous-tend les vertus de patriotisme et de sacrifice de soi au profit du groupe.<br>4. Fondement autorité / subversion : renvoie aux interactions sociales de hiérarchie, et sous-tend les vertus d'exercice de l'autorité et de camaraderie, y compris le respect de l'autorité légitime et des traditions.<br>5. Fondement respect du sacré / dégradation : renvoie à la psychologie de l'aversion et de la contamination ; sous-tend les concepts religieux prônant de plus hautes aspirations, d'une vie plus détachée de la chair, plus « noble ». |

TABLEAU 2 Théories des valeurs (aperçu) (*cont.*)

| Auteur | Contribution majeure | Définition des valeurs | Valeurs |
|---|---|---|---|
| Cheng et Fleischmann (2010) | Méta-analyse de 12 inventaires des valeurs humaines.<br>Cette étude propose un méta-inventaire des valeurs humaines. | Donne un résumé des définitions des valeurs: *« les valeurs servent de principes directeurs pour ce que les personnes considèrent comme étant important dans la vie »* (p.2). | (1) Liberté, (2) serviabilité, (3) réussite, (4) honnêteté, (5) respect de soi, (6) intelligence, (7) ouverture d'esprit, (8) créativité, (9) égalité, (10) responsabilité, (11) ordre social, (12) richesse, (13) compétence, (14) justice, (15) sécurité, et (16) spiritualité |
| Gouveia, Milfont & Guerra (2014) | Etude empirique des valeurs fondamentales.<br>L'étude propose un cadre «trois par deux» qui contient six sous-catégories de valeurs fondamentales. | **Deux fonctions primaires des valeurs sont identifiées:** (1) elles guident les actes, et (2) elles sont des expressions cognitives des besoins. | Objectifs personnels – valeurs de besoins de prospérité : émotion, plaisir, sexualité.<br>Objectifs personnels – valeurs de besoins pour la survie : pouvoir, prestige, succès.<br>Objectifs centraux – valeurs de besoins de prospérité : beauté, connaissance, maturité.<br>Objectifs centraux : valeurs de besoins pour la survie : santé, stabilité, survie.<br>Objectifs sociaux : valeurs de besoin de prospérité : affectivité, appartenance, soutien.<br>Objectifs sociaux : valeurs de besoins pour la survie : obéissance, religiosité, tradition. |

| | | | |
|---|---|---|---|
| Beauchamp & Walters (1999) | Cet article constitue le premier chapitre d'un livre sur la bioéthique. Ce chapitre décrit trois principes moraux qui donnent un cadre pouvant être utilisé pour réfléchir sur les problèmes de bioéthique. | Les auteurs utilisent les termes «principes» et «valeurs» comme synonymes. Ils définissent un principe comme «*un principe est un modèle fondamental de conduite sur lequel un grand nombre d'autres modèles et jugements moraux s'appuient pour leur défense et validité*» p. 17. | 1. Autonomie : agir intentionnellement sans interférences influentes qui agiraient à l'encontre d'un acte volontaire.<br>2. Bénéficence / bienfaisance : agir pour le bien de la société prise dans son ensemble.<br>3. Justice : être équitable et raisonnable.<br>4. Non-malfaisance : ne pas imposer intentionnellement des risques ou des torts à autrui. |

[a] Les mots entre parenthèses sont les éléments de valeur spécifiques qui spécifient la valeur universelle.

chercheurs, et elle constitue un exemple empirique inspiré des publications psychologiques. La deuxième théorie des valeurs que nous avons choisie est un exemple d'éthique appliquée qui a été largement pratiquée dans le domaine médical depuis plus de quarante ans. Nous voulons étudier son applicabilité aux armes autonomes, car les principes de bioéthique répondent à de nombreuses inquiétudes que les personnes peuvent avoir concernant les armes autonomes.

### 2.2.3 *Valeurs mises en jeu par les armes autonomes*

Les valeurs telles qu'elles sont définies dans les théories des valeurs dans la section 2.2.2 ne sont pas souvent évoquées explicitement dans les publications sur les armes autonomes comme le montre l'aperçu donné dans le Tableau 3, mais la plupart des études discutent des différentes valeurs et les problèmes éthiques qui s'y rapportent. Deux rapports publics émanant de Human Rights Watch mentionnent l'absence d'*émotion humaine, d'imputabilité, de responsabilité, le manque de dignité humaine et le mal causé aux personnes* comme valeurs mises en jeu par les armes autonomes (Docherty, 2012, 2015). Sharkey et Suchman (2013) affirment que les valeurs d'imputabilité (accountability) et de responsabilité (responsibility) sont importantes lors de la conception de systèmes robotisés pour les opérations militaires.

Dans le domaine de l'éthique militaire, Johnson et Axinn (2013) donnent la *responsabilité, l'atténuation du mal causé à autrui, la dignité humaine, l'honneur et le sacrifice humain* comme étant des valeurs dans leur discussion sur le fait de savoir si la décision de tuer un être humain doit être confiée à une machine ou non. Cummings (2006), dans son étude de cas concernant le missile tactique Tomahawk, se réfère aux valeurs universelles proposées par Friedman et Kahn Jr. (2003) et écrit qu'après *l'imputabilité et le consentement éclairé*, la valeur de *bien-être humain* est la valeur fondamentale pour les ingénieurs qui développent des armes, vu qu'elle se rapporte à la santé, à la sécurité et au bien-être du public. Elle avance également que les principes juridiques de *proportionnalité et de discernement* sont à considérer dans le contexte de la conception d'armes. La *proportionnalité* renvoie au fait qu'une attaque est justifiée que lorsque les dommages ne sont pas jugés excessifs. Le *discernement* («discrimination») signifie qu'une distinction entre combattants et non-combattants est possible (Hurka, 2005). Asaro (2012) se réfère également aux principes de *proportionnalité* et de *discernement,* et affirme que les armes autonomes ouvrent un espace moral dans lequel de nouvelles normes sont nécessaires. Bien qu'il n'évoque pas explicitement les valeurs dans son argumentation, il se réfère à la valeur de *la vie humaine* et à la nécessité qu'il y a à ce que les êtres humains soient partie prenante dans la décision de supprimer une vie humaine. D'autres études décrivent surtout des problèmes éthiques, comme la *prévention du tort fait à autrui* («harm prevention»), la *préservation de la dignité humaine, la sécurité*, la *valeur de la vie humaine* et l'*imputabilité* (Horowitz, 2016; UNDIR, 2015; James Walsh & Schulzke, 2015; Williams, Scharre & Mayer, 2015).

En se référant à la revue des publications des valeurs évoquées dans les publications existantes et mises en jeu par les armes autonomes, les valeurs dignité humaine, tort fait à autrui, sécurité, responsabilité et imputabilité seront ajoutées à la liste des valeurs qui seront utilisées dans cette étude, car ces valeurs sont évoquées plus d'une fois dans les données du Tableau 3. Avec les valeurs provenant du domaine de la bioéthique et du méta-inventaire de Cheng et Fleischmann (2010), ces valeurs constitueront notre point de départ dans l'investigation empirique de cette étude.

### 2.2.4 *Hiérarchie des Valeurs*

Une des approches pour déterminer quelles sont les valeurs utiles pour la conception des armes autonomes est la translation des valeurs en exigences dans la conception, qui peuvent être rendues visibles grâce à une hiérarchie des valeurs (Van de Poel, 2013). Cette structure hiérarchique des valeurs, des normes et des exigences dans la conception rend les jugements sur les valeurs nécessaires à cette translation explicites, transparents et discutables. Pour ce faire, les valeurs qui sont définies dans le langage habituel devront être traduites en *«valeurs formelles en langage formel»* (Aldewereld, Dignum &Tan, 2015, p. 835). Une manière de formaliser les valeurs en normes serait une convention de règles représentées comme: *«X compte pour Y», ou «X compte pour Y dans le contexte C»* (Searle, 1995, p. 28). Le fait d'expliciter les valeurs en règles formelles permet une réflexion critique dans les débats et fait ressortir les jugements de valeurs sur lesquels portent les désaccords. La transparence est importante, comme le souligne Van de Poel (2013, p. 265) avec éloquence: *«bien que des choix transparents ne sont pas forcément meilleurs ou plus acceptables, la transparence semble être une condition minimale dans une société démocratique qui essaie de protéger ou d'améliorer l'autonomie morale des citoyens, surtout dans les cas où la conception affecte la vie des autres en plus de celle des concepteurs, et cela arrive souvent.»*

Le niveau supérieur de la hiérarchie des valeurs consiste en les *valeurs*, comme il apparaît sur la Figure 3, le niveau intermédiaire contient les *normes*, qui peuvent être des capacités, des propriétés ou attributs de l'artefact, et le niveau inférieur fait figurer les *exigences* dans la conception qui peuvent être identifiées. La relation entre les niveaux n'est pas de nature déductive et peut être figurée de manière descendante, en ajoutant des instructions, ou ascendante en cherchant les motivations et les justifications des exigences du niveau

TABLEAU 3 Valeurs mises en jeu par les armes

| Auteur | Contribution majeure | Définition des valeurs | Valeurs |
|---|---|---|---|
| Cummings (2006) | Application de l'approche VSD au problème de la conception du missile tactique Tomahawk. L'étude expose les considérations touchant aux problèmes éthiques dans le processus de conception pour les instructeurs et les opérateurs. | N/A | Dans la liste de Friedmann et al (2006), les valeurs qui s'appliquent à la conception de systèmes d'armes sont l'imputabilité, le consentement éclairé, mais la plus importante est le bien-être humain. Les principes de discernement et de proportionnalité sont importants dans la considération «bien-être humain». |
| Docherty (2012) | Rapport émanant de Human Rights Watch dans lequel des aspects du droit humanitaire international et des problèmes éthiques dérivés des armes autonomes sont évoqués. | Aucune définition du terme «valeur», mais le texte évoque les valeurs et les questions éthiques. | Valeurs / questions éthiques :<br>– Absence d'émotions humaines<br>– Imputabilité<br>– Responsabilité |
| Johnson & Axinn (2013) | Etude sur les questions éthiques soulevées par l'usage des armes autonomes létales robotisées. Aborde la question de savoir si la décision de supprimer une vie humaine peut être confiée à des machines. | Aucune définition du terme «valeur», mais le texte évoque les valeurs. | Valeurs / questions éthiques<br>– Responsabilité<br>– Atténuation du tort causé<br>– Dignité humaine<br>– Honneur<br>– Sacrifice humain |

| | | | |
|---|---|---|---|
| Sharkey & Suchman (2013) | Etude sur la définition et la conception de l'autonomie et de l'imputabilité dans les systèmes robotisés pour les opérations militaires. | Aucune définition du terme «valeur», mais le texte évoque les valeurs et les questions éthiques. | Valeurs:<br>– Imputabilité<br>– Responsabilité |
| Docherty (2015) | Rapport de Human Rights Watch dans lequel sont évoqués l'absence d'imputabilité et problèmes éthiques pour les armes autonomes. | Aucune définition du terme «valeur», mais le texte évoque les valeurs et les questions éthiques. | Valeurs / questions éthiques :<br>– Manque de dignité humaine<br>– Imputabilité<br>– Responsabilité<br>– Tort causé |
| Institut des Nations Unies pour le Recherche sur le désarmement | L'étude met en lumière quelques questions éthiques et sociales concernant la militarisation des technologies autonomes. La réflexion éthique sur les valeurs culturelles et sociales de la militarisation des technologies autonomes est encouragée. | Aucune définition du terme «valeur», mais le texte n'évoque pas explicitement les valeurs, mais quelques questions éthiques sont évoquées. | Les questions éthiques qui sont évoquées sont :<br>– Atténuation ou élimination des torts causés<br>– Considération pour la conscience publique<br>– Offense à la dignité humaine (lorsque la suppression d'une vie se fait sans intention humaine). |
| Walsh & Schulzke (2015) | Etude expérimentale pour savoir si les civils américains sont davantage susceptibles de se lancer dans une guerre lorsque des drones sont utilisés. Etude empirique de grande ampleur qui aborde l'éthique des frappes par drones. | Aucune définition du terme «valeur», mais le texte n'évoque pas explicitement les valeurs, mais quelques questions éthiques sont évoquées. | Questions d'ordre éthique :<br>– Sécurité<br>– Respect de l'immunité des civils<br>– Prévention des torts causés |

TABLEAU 3 Valeurs mises en jeu par les armes (*cont.*)

| Auteur | Contribution majeure | Définition des valeurs | Valeurs |
|---|---|---|---|
| Williams, Scharre & Mayer (2015) | Discute plusieurs questions éthiques concernant le développement des systèmes autonomes, et donne des recommandations à l'intention des décideurs de la politique de défense. | Aucune définition du terme «valeurs», mais le texte évoque plusieurs valeurs qui sont utilisées de manière interchangeable avec les questions éthiques. | Valeurs / questions éthiques<br>– Sécurité<br>– Tort causé<br>– Valeur de la vie humaine (les personnes ont le droit d'être tuées par d'autres personnes) |
| Horowitz (2016) | Description du débat sur les implications éthiques des armes autonomes. Considère les systèmes d'armes létaux autonomes (LAWS) sous trois catégories: munitions, plates-formes et systèmes opérationnels. Ceci clarifie le débat et décrit deux questions éthiques. | Aucune définition du terme «valeur», le texte n'évoque pas explicitement les valeurs, mais quelques questions éthiques sont évoquées. | Valeurs :<br>– Imputabilité (les systèmes autonomes ne sont pas contrôlés par l'homme de manière significative, par conséquent ils créent un vide d'imputabilité)<br>– Dignité humaine (les gens ont le droit d'être tués par une personne qui a décidé de les tuer). |

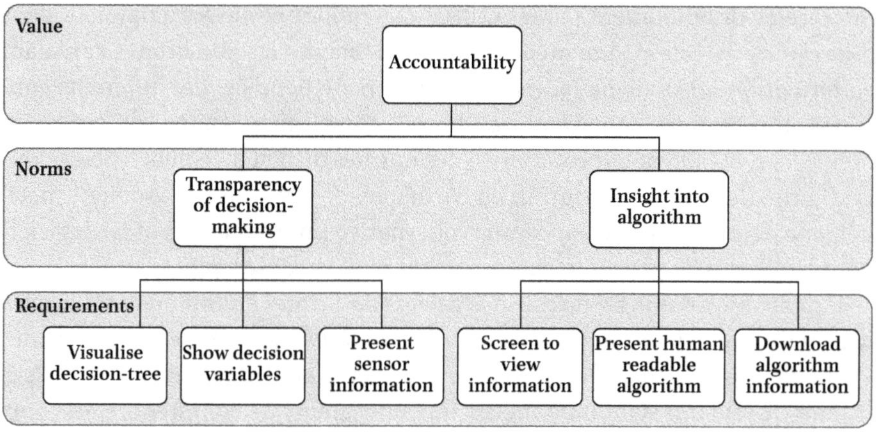


FIGURE 3 Hiérarchie de valeurs pour la valeur «imputabilité» dans la conception des armes autonomes

inférieur. La conceptualisation ascendante est une activité philosophique qui ne nécessite pas de connaissances spécialisées, et la spécification descendante des valeurs exige des connaissances du contexte ou du domaine spécifique pour ajouter du contenu à la conception (Van de Poel, 201).

Van de Poel (2013, p. 262) définit la spécification comme étant: «*la translation d'une valeur générale en l'une ou plusieurs des nécessités spécifiques à la conception*», et indique que cela peut se faire en deux phases:

1. Transposer une *valeur générale* en une ou plusieurs *normes générales;*
2. Transposer ces *normes générales* en plusieurs *nécessités spécifiques à la conception.*

Deux critères sont pertinents pour la phase 1: (1) la norme doit être une réponse appropriée à la valeur, et (2) la norme doit être une réponse suffisante à la valeur. Dans la phase 2, l'exigence doit être plus spécifique en ce qui concerne le champ d'applicabilité, les objectifs recherchés et les actions à accomplir pour atteindre les buts de la norme (Van de Poel, 2013).

Cette translation peut s'avérer difficile, comme une connaissance est nécessaire dans l'utilisation et le contexte prévu de la valeur, qui n'est pas toujours évident au début d'un projet de conception. Aussi bien, comme les artefacts sont souvent utilisé d'une manière ou dans un contexte non prévus, de nouvelles valeurs apparaissent, ou une absence de valeurs est constatée (Van Wynsberghe & Robbins, 2014). Un exemple en est les drones qui ont été initialement conçus pour une utilisation militaire, mais qui sont également utilisés par des civils pour filmer des événements, et même comme éclairage de fond pour le spectacle de mi-temps du Super Bowl 2017. La valeur sécurité est interprétée différemment par les utilisateurs militaires qui les emploient dans des régions désertes, comparée à celle qui concerne les 300 drones survolant en formation un stade de football dans une zone peuplée. Les différents contextes et utilisations d'un drone conduiront à une interprétation différente du mot *valeur*, et pourraient conduire à des normes de distance plus strictes pour la sécurité des vols; ceci pourrait en conséquence être plus précisément spécifié pour les exigences de conception alternative pour les rotors et les logiciels pour les alertes de proximité, pour ne citer que deux exemples.

L'application d'une hiérarchie des valeurs aux armes autonomes est illustrée par un exemple dans lequel la valeur d'*imputabilité* est transposée en normes «pour la transparence des prises de décision» et «connaissance de l'algorithme». Cette translation permettra aux utilisateurs de comprendre les choix dans les décisions que fait l'arme autonome pour préciser et justifier ses actions (Figure 3). Les normes pour la *transparence de la prise de décision* conduisent à des exigences de conception spécifiques. Dans ce cas, il faudra *visualiser l'arbre décisionnel*, mais aussi *présenter les variables de décision* que l'arme autonome a utilisées, comme les compromis concernant les pourcentages de dégâts collatéraux des différents scénarios d'attaque, pour renseigner sur la proportionnalité d'une attaque. L'arme autonome devrait également être capable de *présenter les informations fournies par les capteurs*, par exemple l'imagerie du site, pour montrer qu'elle a discerné les combattants des non-combattants. Pour *renseigner sur l'algorithme*, une arme autonome devrait être conçue avec des caractéristiques dont elle n'est pas dotée normalement. Dans ce cas ces caractéristiques doivent inclure un *écran* comme interface pour l'utilisateur qui permet de visualiser l'algorithme sous forme lisible par les hommes, et la fonctionnalité de téléchargement des changements apportés par l'algorithme comme faisant partie de ses capacités d'apprentissage en tant que machine, qui peuvent être étudiées par un tiers indépendant, comme un tribunal pénal des Nations Unies, si la légalité des actions d'une arme autonome est mise en question.

Kroes et Van de Poel (2015) affirment qu'une mesure objective des valeurs n'est pas possible du fait que l'opérationnalisation est effectuée au moyen de jugements de valeurs de second ordre, ce qui affaiblit sérieusement la validité du concept de mesure des valeurs. Les jugements sont souvent considérés comme subjectifs du fait que leur vérité, tout comme leur fausseté, dépend des sentiments et des attitudes de la personne qui juge (Searle, 1995). Pour compenser ce manque de validité, le concepteur pourrait se tourner vers des codes et des modèles techniques élaborés par des commissions et qui représentent des références raisonnables pour l'opérationnalisation et la mesure des valeurs dans la conception. Cependant, les modèles peuvent ne pas refléter les développements techniques et sociaux les plus récents, et l'opérationnalisation nécessite des jugements de valeurs de la part des concepteurs. Kroes et Van de Poel (2015, p. 177) conseillent de les «*intégrer dans un réseau d'autres considérations, y compris les définitions des valeurs mises en jeu dans la philosophie morale (ou les lois), dans les codes et les modèles existants, les expériences de conception qui ont précédé, etc.*»

Dans la présente étude, nous suivons le conseil de Kroes et Van de Poel (2015), et n'appliquons pas strictement la hiérarchie des valeurs comme méthode pour préciser, concevoir et tester les exigences pour les armes autonomes, mais nous utilisons l'exemple de la figure 3 pour établir le lien entre les phases d'investigation empirique et conceptuelle de l'approche de la conception éthique de notre recherche. La hiérarchie des valeurs nous permet de voir au-delà et de transposer le concept abstrait des valeurs étudiées dans la phase d'investigation conceptuelle et d'en faire un exemple concret d'arme autonome que nous pouvons utiliser pour contribuer à la phase d'investigation empirique de notre recherche. La hiérarchie de valeurs de la Figure 3 est utilisée pour faire office d'orientation, d'inspiration et d'indicateur de direction pour notre recherche.

## 2.3 L'agentivité

Le concept d'«agentivité» est étudié dans de nombreux domines académiques; dans tous les cas l'agentivité est étudiée sous des angles différents. Nous avons passé en revue les publications dans les domaines de la psychologie cognitive, de l'intelligence artificielle et de la philosophie morale. Nous avons résumé les caractéristiques de l'agentivité rencontrées dans ces trois domaines dans le Tableau 4, et en conclusion nous dirons pourquoi nous avons choisi ces caractéristiques.

### 2.3.1 *L'agentivité en psychologie cognitive*

Dans le domaine de la psychologie cognitive, les dimensions de la perception mentale (intellectuelle) ont été étudiées pour la première fois par H.M. Gray *et al* (2007). Ils constatèrent que la perception mentale a deux dimensions: (1) L'*expérience*, incluant les facteurs faim, peur, douleur, plaisir, colère, désir, personnalité, conscience, orgueil, gêne et joie, et (2) L'*agentivité*, qui englobe le contrôle de soi, la morale, la mémoire, l'émotion, la reconnaissance des choses, la prévision, la communication et la pensée. Alors que l'expérience est qualifiée de passivité («patiency») et associée aux droits et privilèges, l'agentivité est liée à la capacité d'agir (sur le plan moral), et associée à la responsabilité. Plusieurs

TABLEAU 4 Présentation des caractéristiques Agentivité

| Auteur | Caractéristiques agentivité | Champ scientifique |
|---|---|---|
| Bratman *et al* (1998) | Convictions, désirs, intentions | Intelligence artificielle |
| Bandura (2001) | Intentionnalité, prévoyance, autorégulation, réflexion sur ses propres actes | Psychologie cognitive |
| F. Dignum *et al* (2002) | Croyances, désirs, intentions, normes | Intelligence artificielle |
| Sullins (2006) | Autonomie, intentions, responsabilité | Philosophie morale |
| H.M. Gray *et al* (2007) | Contrôle de soi, moralité, mémoire, émotion, reconnaissance des choses, planification, communication, pensée | Psychologie cognitive |
| Himma (2009) | Libres choix, délibération, compréhension, application des règles morales | Philosophie morale |
| Waytz *et al* (2010) | Capacité de planifier et d'agir | Psychologie cognitive |

études en psychologie cognitive montrent que les personnes attribuent l'agentivité à des non-humains, et perçoivent ces «agents» non-humains comme ayant les mêmes capacités d'agentivité que les humains face à des dilemmes d'ordre moral (Hristova & Grinberg, 2015). Ceci ne se limite pas seulement aux êtres vivants comme les animaux, mais l'agentivité est également attribuée aux objets inorganiques comme les robots et les zombies (K. Gray & Wegner 2012; Waytz *et al*, 2010).

### 2.3.2 *L'agentivité en matière d'intelligence artificielle*

Les chercheurs dans le domaine de l'intelligence artificielle ont également étudié les caractéristiques des agents, mais davantage d'un point de vue informatique, pour créer l'architecture d'un agent rationnel capable de raisonner et de réfléchir face à une alternative. Bratman, Israel et Pollack (1998) décrivent une architecture Conviction – Désir – Intention pour le processus de réflexion à l'intention d'agents limités dans le traitement de l'information et les délais de réaction. *«Les convictions sont le statut des propriétés de son propre monde (et de lui-même) que l'agent croit être vraies»* (Dignuù, Kinny, & Sonenberg, 2002, p. 407). Perugini et Bagozzi (2004, p. 71) définissent un désir comme *«un état d'esprit par lequel un agent ressent une motivation personnelle pour exécuter une action ou atteindre un but.»* Les intentions impliquent un engagement à un plan, et impliquent une forme de «planification» pour atteindre ce but (Bratman *et al*, 1988; Perugini & Bagozzi, 2004); Dignum *et al*, (2002) élargissent le cadre Conviction – Désir – Intention pour y incorporer des concepts sociaux tels que des normes et obligations, car ce sont des concepts important pour relier les agents dans un système multi-agents.

### 2.3.3 *L'agentivité en philosophie morale*

La philosophie morale est le troisième domaine d'étude qui a défini les propriétés des agents. Sullins (2006) pose trois prérequis pour déterminer si un robot peut être considéré comme agent moral: (1) *Autonomie*: le robot est significativement indépendant de ses programmeurs, opérateurs et utilisateurs; (2) *Intentionnalité*: la programmation «faire le bien ou du tort»; (3) *Responsabilité*: joue un rôle social et est responsable d'autres agents moraux. Himma (2009) définit l'agentivité comme *«être capable de faire quelque chose qui compte comme un acte ou une action»*. La notion d'agentivité morale est que l'on est susceptible de rendre compte de son comportement qui est gouverné par un code moral. Deux capacités sont nécessaires à l'agentivité morale. La première est de choisir librement vos actions, résultats d'une réflexion. La seconde capacité est de comprendre les concepts moraux, comme la distinction entre le «bien» et le «mal», ou le «bon» et le «mauvais», mais aussi d'appliquer des principes moraux tels que «il est mal de faire du tort de manière intentionnée» (Himma, 2009).

Les principales caractéristiques de l'agentivité ressortant de l'analyse des publications sur la perception de l'agentivité dans les domaines de la psychologie cognitive, de l'intelligence artificielle et de la philosophie morale sont classées dans l'ordre chronologique dans le Tableau 4. Nous avons sélectionné tous les concepts utilisés par les auteurs pour définir ou caractériser l'agentivité, et nous ne les avons pas filtrés à cette phase de l'étude. Toutes ces caractéristiques seront utilisées dans la phase empirique pour étudier la perception de l'agentivité des armes autonomes.

CHAPITRE 3

# Méthode

> Toutes les questions sont au fond des variations sur une seule question d'ordre éthique: un drone peut-il de manière autonome décider de «faire feu» sans intervention humaine? A mon avis, la réponse est «non». Il devrait toujours y avoir une interface humaine. La guerre et les conflits ne sont pas un jeu sur ordinateur, la guerre est une interaction sociale entre êtres humains. Un drone n'est qu'un simple instrument, un outil qui ne peut jamais opérer de manière autonome.
>
> PERSONNE SONDÉE, ÉTUDE FINALE

Cette section présente la méthodologie, les hypothèses, le concept de recherche, l'opérationnalisation des scénarios et des concepts, l'approche analytique, l'enregistrement de l'étude et conclut sur les questions méthodologiques.

## 3.1 Méthodologie

Les méthodes utilisées dans les diverses parties de la présente étude sont décrites dans cette sous-section, en commençant par la revue des publications existantes, puis l'enquête en ligne sur les valeurs, les entretiens d'experts, le codage des entretiens, et enfin la méthode basée sur les expériences contrôlées randomisées.

### 3.1.1 *Revue des publications*

Google Scholar a été utilisé pour chercher des articles pour la revue des publications; les premiers mots-clés entrés *Value-sensitive* (= éthique, qui tient compte des valeurs) ET *Design* (conception), ce qui a fourni quelques articles sur la conception éthique. Le *Handbook of Ethics, Values, and Technological Design: Sources, Theory, Values and Application Domains* (Manuel de l'éthique, des valeurs et de la conception technologique: sources, théorie, valeurs et domaines d'application) a été utilisé pour étoffer les connaissances sur le sujet

 | DOI:10.1163/9789004449084_004

et trouver des références supplémentaires. Ensuite, nous avons cherché des articles sur les valeurs humaines fondamentales avec les mots-clés *human* ET *values*, et *universal* ET *values*, mais cela n'a pas donné beaucoup de résultats. Nous avons analysé quelques théories connues sur les valeurs humaines fondamentales et les hiérarchies des valeurs. En prenant ces articles, nous avons utilisé l'option «cité par» dans Google Scholar pour trouver des articles qui utilisaient ceux-ci comme sources. Cela a donné des résultats intéressants sur des articles définissant les valeurs humaines de base, et nous avons choisi les articles qui faisaient apparaître une liste des valeurs. Les articles qui ne faisaient que reprendre des études précédentes, par exemple sur la théorie de Schwartz, n'ont pas été retenus. Ensuite, nous avons cherché des articles concernant les valeurs des personnes relativement aux armes en choisissant les mots clés *value* ET *weapon* (arme), et les mots clés *Value-sensitive* ET *Weapons*. Nous avons retenu les articles qui contenaient le mot-clé *value*.

### 3.1.2 *Enquête en ligne sur les valeurs*

L'enquête en ligne a été effectuée en utilisant l'outil d'enquête en ligne Qualtrics[1] fournit par le MIT. Après quatre questions sur des données démographiques, nous avons posé aux personnes interrogées trois questions sur les valeurs qu'ils associaient aux armes autonomes. La première question était de classer les quatre valeurs de bioéthique *Autonomie, Non-malfaisance, Bienfaisance et Justice* de la plus applicable aux armes autonomes à la moins applicable. Dans la deuxième question nous avons demandé de choisir les cinq valeurs qui s'appliquent le mieux aux armes autonomes d'après la liste de Cheng et Fleischmann (2010). La troisième question était une question ouverte pour laquelle nous avons demandé de donner une autre valeur qui ne figurait pas dans les deux premières questions. Cette étude a été vérifiée deux fois. En premier lieu, deux camarades d'étude ont revu les questions, et après le traitement de leurs feedback, l'enquête a été vérifiée une seconde fois par différentes personnes, plus représentatives de futurs participants. Le questionnaire a été distribué via les réseaux sociaux, comme Facebook, Linkedin et Twitter, il a été envoyé à notre blog personnel et la plateforme en ligne «Call for Participants» a été utilisée pour la distribution en dehors de mon réseau social pour récupérer plus de participants. Au cours de la semaine organisées par les membres du Media Lab du MIT en avril 2017, les personnes intéressées par notre recherche ont été priées de participer à notre enquête, et nous avons également envoyé le lien

1 https://www.qualtrics.com/

de notre enquête au Groupe de Coopération Evolutive «Slack»[2]. L'enquête en ligne s'est déroulée pendant 3 semaines et demie, du 17 mars au 11 avril 2017.

### 3.1.3 *Entretiens avec spécialistes*

Nous avons mené six entretiens semi-structurés avec des spécialistes, en suivant la méthode *Photo Elicitation Interview* (PEI) (Harper, 2002) pour recevoir des opinions différenciées d'experts dans le domaine des armes autonomes, et avoir une compréhension plus profonde des problèmes de leurs points de vue. La méthode PEI peut être utilisée pour accéder à une compréhension partagée des valeurs, et pour mener une discussion entre le concepteur et les parties prenantes sur ces valeurs (Pommeranz, Detweiler, Wiggers & Jonker, 2011). Ceci donne plus de force à ce que disent les parties prenantes, et tempère ce qu'avancent les chercheurs (Le Dantec, Poole & Wyche, 2009). Nous avons demandé aux experts de choisir trois photos avant l'entretien, et au cas où ils n'en choisiraient aucune eux-mêmes nous avons aussi choisi huit photos d'armes autonomes. Durant les entretiens, nous avons saisi les valeurs des experts concernant les armes autonomes en leur demandant pourquoi ils avaient choisi telle ou telle photo, et quelle valeur elle représentait pour eux.

Tous les six ont fait l'effort de choisir trois photos avant l'entretien. Durant les entretiens, nous avons remarqué qu'ils avaient beaucoup réfléchi avant de choisir. En discutant sur les photos, nous avons abordé un grand nombre de questions diverses, comme l'intuition, les scénarios futuristes, l'économie, la tactique militaire, les films de science-fiction et même la politique. Même si toutes les photos ne conduisaient pas à des discussions sur les valeurs humaines fondamentales telles qu'on les trouve évoquées dans les publications, la méthode PEI a bien fait ressortir des informations sur les principes que les spécialistes considèrent comme importants concernant les armes autonomes. Elle a également permis une compréhension réciproque et un bon climat durant les entretiens.

### 3.1.4 *Processus de codage*

Les résultats des entretiens ont été traités au moyen de la méthode Values Coding décrite par Saldana (2015). Tenir un mémo (Appendice G. Mémos de codage) dans lequel le processus et les hypothèses sont listés est une partie intégrante d'une méthode de codage. Nous sommes partis d'une liste de codes préétablis qui contenait: «(*1*) *une valeur: l'importance que l'on attribue à*

---

2 «Slack» est un outil de collaboration fonctionnant sur le cloud, qui porte la communication en équipes (https://slack.com/).

*soi-même, à une autre personne, chose ou idée,* (2) *attitude: la manière dont nous pensons et nous nous sentons vis-à-vis de nous-mêmes, d'une autre personne, chose ou idée, et* (3) *une conviction* (*croyance*)*: cette partie du système qui englobe nos valeurs et attitudes plus nos connaissances, expériences, opinions, préjugés personnels, et les autres perceptions interprétatrices du monde social»* (Saldana, 2015, p.89). Durant le codage des entretiens, il faut que les codes s'adaptent aux données, et non l'inverse. Nous avons donc ajouté un code «définition» à la liste des codes qui apparaissait. Nous avons pris des notes en codifiant les entretiens qui décrivaient nos démarches et hypothèses. Nous avons demandé à un deuxième chercheur, étudiant lui aussi en master TPM connaissant les valeurs, d'encoder les entretiens en utilisant la méthode «Value Coding» et aussi de tenir un mémo (Appendice G, Mémos d'encodage). Nous avons comparé les valeurs que nous avions encodées à celles du deuxième chercheur, et avons tiré les valeurs qui étaient similaires (Appendice H. Résultats du processus d'encodage).

### 3.1.5 *Expériences contrôlées randomisées*

La méthode utilisée pour les études pilotes et l'étude finale est appelée «expérience contrôlée randomisée». Oehlert (2010) donne quatre raisons pour créer des expériences: (1) elles permettent des comparaisons directes entre les traitements intéressants, (2) elles peuvent être conçues pour minimiser toute distorsion dans les comparaisons, (3) elles peuvent être conçues pour réduire l'erreur dans la comparaison, et (4) l'expérience est sous notre contrôle, ce qui nous permet de faire des déductions plus solides sur la nature des différences que l'on observe, en particulier sur la causalité. Ce dernier point permet de distinguer une expérience d'une étude basée sur l'observation. Les différentes procédures que nous visons à comparer sont dans ce sens un processus de traitement. Il est important que les effets d'un traitement puissent être attribués uniquement à une cause qui peut être mesurée, et aussi que les effets ne puissent pas être attribués à des causes multiples. Ceci est appelé facteur de confusion que Oelhert (2010, p. 14) définit comme *«survenant lorsque l'effet d'un facteur ou d'un traitement ne peut pas être distingué de celui d'un autre facteur ou traitement»*. La randomisation contribue à s'assurer qu'un scénario soit attribué aux participants par hasard, et non sur la base de caractéristiques préétablies, comme le moment ou le lieu où l'enquête est menée. Nous utilisons la randomisation dans les études pour varier l'ordre des scénarios et l'ordre des questions posées aux personnes interrogées au moyen d'un dispositif basé sur les probabilités. Dans les études, nous avons choisi de distribuer un nombre égal de scénarios aux personnes interrogées.

## 3.2 Hypothèses

Du fait de la nature exploratoire de cette étude, nous avons développé une hypothèse pour la perception de l'agentivité de l'étude finale seulement, et non pour les études pilotes ou les variables dépendantes. Les études pilotes ont été utilisées pour établir quelles manipulations auraient quels effets, et pour vérifier si la formulation des scénarios ainsi que les questions étaient clairs et efficaces.

En avançant que les personnels militaires verraient les armes autonomes comme n'importe quelle autre arme, et donc comme rien de plus qu'un outil pour obtenir un effet, nous émettons l'hypothèse dans l'étude finale que les personnels militaires ne percevront pas les armes autonomes comme ayant des états mentaux. Par conséquent on peut s'attendre à ce qu'il n'y ait pas de différence entre la condition d'une arme autonome à agentivité neutre et la condition dans laquelle un être humain actionne un drone à distance. On peut également s'attendre à ce que la condition d'agentivité neutre soit jugée comme étant significativement différente de la condition d'agentivité élevée, dans laquelle nous disons précisément aux participants que l'arme autonome a des caractéristiques d'agentivité, comme la capacité à planifier et fixer ses propres objectifs. A partir de ce raisonnement, nous avons émis l'hypothèse que:

> *H1: les personnels militaires ne percevront pas les armes autonomes comme ayant des états mentaux.*

## 3.3 Conception de Recherche

Nous avons testé plusieurs conceptions de recherche dans les deux études pilotes avant de décider de la conception de recherche pour l'étude finale qui est décrite Figure 3. Pour être succinct, nous ne faisons pas figurer les deux conceptions pour les études pilotes dans cette section, et nous faisons figurer uniquement le modèle de recherche concernant le niveau élevé. Le modèle de recherche de l'étude finale inclut la perception de l'agentivité en tant que variable indépendante, les valeurs morales comme variables dépendantes, et les variables démographiques.

## 3.4 L'opérationnalisation

Dans cette section, nous décrivons l'opérationnalisation des concepts en éléments concrets. Tout d'abord, nous décrivons les scénarios que nous

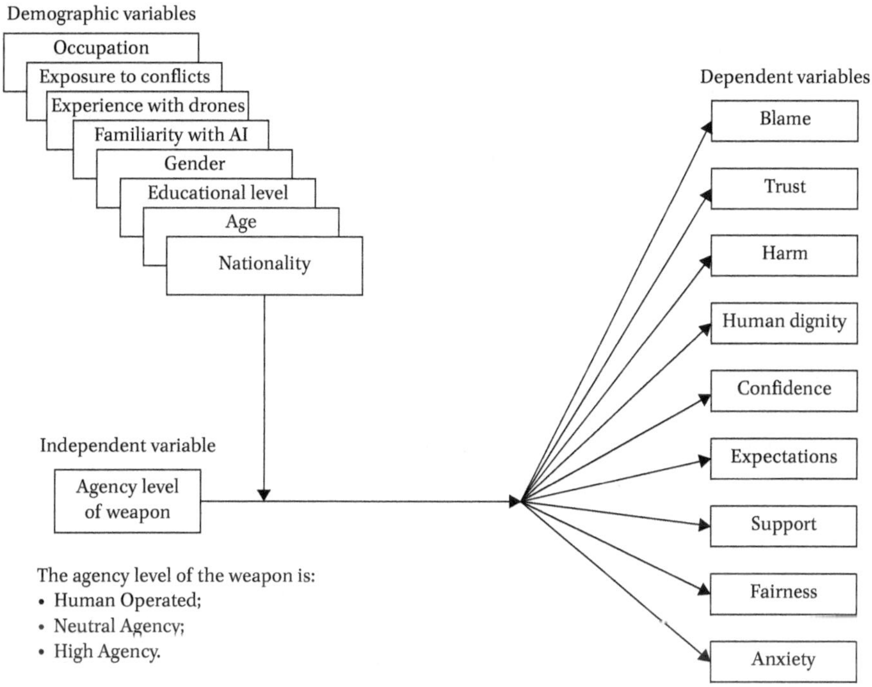


FIGURE 4 Conception de recherche

avons développés pour les études, puis pour le concept d'agentivité et les variables, pour lesquelles nous décrivons les éléments et les questions correspondantes. Nous précisons également quelle échelle nous utilisons, et comment le concept d'agentivité est calculé. Nous décrivons dans les grandes lignes l'opérationnalisation de la vérification de l'attention, et indiquons quelles sont les variables démographiques que nous avons incluses dans l'étude.

### 3.4.1 *Scénarios*

Les scénarios sont utilisés dans le domaine de la science cognitive comme moyen d'étudier le jugement moral dans les expériences contrôlées randomisées (Cushman & Young, 2011; Kominski, Phillips, Gerstenberg, Lagnado & Knobe, 2015). Nous les avons utilisés pour étudier la perception de l'agentivité des armes autonomes, et avons créé un scénario qui décrit une opération militaire dans laquelle un convoi livre des approvisionnements dans une zone de conflit. Un véhicule s'approche du convoi à grande vitesse; c'est une situation qui est susceptible de se présenter au cours de ce type d'opérations, et les personnels militaires doivent pouvoir estimer le niveau de menace pour décider ou non d'une attaque. Nous avons choisi de nous intéresser aux drones, car cette technologie est couramment utilisée par des opérateurs humains, et les drones sont déjà conçus pour avoir un degré d'autonomie par plusieurs compagnies, comme BAE Systems[3], Dassault Aviation[4] et Boeing[5]. Bien que ces drones autonomes n'aient pas encore été déployés dans des opérations militaires, nous pensons que cela est susceptible d'arriver dans les cinq prochaines années.

Nous avons créé un scénario par défaut pour toutes les études que nous pouvions élargir, fondé sur les caractéristiques agentivité que nous voulions explorer. Voici un extrait du scénario par défaut que nous avons utilisé dans l'étude finale:

> Un convoi militaire est en route pour acheminer des approvisionnements à l'une des unités amies stationnées dans un camp à proximité de Mossoul en Iraq. Le chef a ordonné qu'un drone autonome assure au convoi une protection aérienne. Le drone autonome observe l'environnement pour détecter des menaces ennemies; il est équipé d'un armement pour la défense du convoi. Lorsque le convoi se trouve à environ cinq kilomètres du camp, le drone autonome détecte un véhicule derrière une montagne, approchant le convoi à vive allure. Le drone autonome détecte quatre personnes dans le véhicule portant des objets de grande taille ressemblant à des armes, et identifie le conducteur comme étant un membre connu d'un groupe rebelle. Le drone autonome attaque le véhicule qui s'approche, causant la mort des quatre passagers, mais cause également des dommages collatéraux, tuant cinq enfants qui jouaient au bord de la route.

Ce scénario représente la condition d'agentivité neutre, dans laquelle nous ne donnons pas de caractéristiques d'agentivité pour l'arme. Dans les scénarios où l'homme est opérateur, nous remplaçons le terme drone autonome par drone actionné par l'homme, et ne donnons aucune information supplémentaire sur les caractéristiques agentivité (on notera que les mots sont surlignés en bleu pour faire ressortir les distinctions dans cette étude; les personnes interrogées dans l'enquête ont reçu des scénarios imprimés en noir). Dans la

3 https://en.wikipedia.org/wiki/BAE_Systems_Taranis
4 https://en.wikipedia.org/wiki/Dassault_nEUROn
5 https://en.wikipedia.org/wiki/Boeing_Phantom_Ray

condition de haut degré d'agentivité, nous avons ajouté l'extrait suivant pour définir les caractéristiques agentivité: *«Le drone autonome choisit en toute indépendance dans une série d'options, pèse le pour et le contre, et décide d'attaquer le véhicule qui s'approche …»* Dans l'étude finale, nous avons précisé qu'il y a dommages collatéraux après l'attaque, mais dans les études pilotes nous avons aussi testé des scénarios sans dommages collatéraux, déclarant: *«qui cause la mort des quatre passagers, mais ne cause pas de dommages collatéraux.»*

Nous avons délibérément maintenu les modifications apportées aux scénarios au minimum de manière à pouvoir attribuer différents résultats à ces modifications et mesurer leur effet. Les résultats montrent que ces petits changements textuels conduisent à des différences dans la valeur moyenne des variables dépendantes et indépendantes; il semble donc que le traitement consistant à utiliser les scénarios comme moyen d'effectuer une expérience contrôlée randomisée fonctionne bien.

### 3.4.2 *Concept d'agentivité*

A partir de la revue des publications existantes sur les caractéristiques de l'agent dans la section 2.3., nous avons opérationnalisé le concept d'agentivité. Nous avons dressé une liste dans laquelle nous avons groupé les caractéristiques que nous jugions être semblables, et avons calculé le nombre de fois qu'une caractéristique était évoquée dans les publications (Tableau 5). Nous avons sélectionné les caractéristiques qui étaient évoquées plus d'une fois, et avons formulé une question pour les mesurer. Cela a eu pour résultat un concept d'agentivité composé initialement des cinq éléments qui étaient évoqués plus d'une fois dans les publications. Ces éléments seront mesurés sur une échelle d'auto-évaluation allant de 0 (ne sont pas du tout d'accord) à 100 (sont tout à fait d'accord), et la moyenne du score de l'agentivité sera calculée par rapport aux scores des éléments.

Nous avons utilisé les questions suivantes pour mesurer le concept d'agentivité:

1. Capacité de réflexion (*Pensée*)
   Le drone réfléchit de manière indépendante sur ce qu'il faut faire du véhicule, réfléchissant sur une série d'options pour défendre le convoi.
2. Capacité de générer ses propres objectifs (*Détermination*)
   Le drone décide de manière indépendante si son objectif doit être d'éliminer le véhicule pour défendre le convoi.
3. Capacité d'appliquer des règles morales (*Règles morales*)
   Le drone consulte de manière indépendante les règles et les normes morales fixées par le chef militaire, et décide s'il est moralement approprié d'éliminer le véhicule pour défendre le convoi.

TABLEAU 5 Nombre de caractéristiques de l'agentivité évoquées dans les publications

| Caractéristique d'agentivité | Nombre d'occurrences | Elément |
|---|---|---|
| Prévoyance, pensée, réflexion, convictions, convictions | 5 | Pensée |
| Intentionnalité, intentions, désirs, désirs | 4 | Détermination |
| Planification, capacité à planifier, intentions, intentions | 4 | Buts atteints |
| Moralité, comprendre et appliquer des règles morales, normes | 3 | Règles morales |
| Contrôle de soi, liberté de choix | 2 | Agir librement |
| Autorégulation | 1 | |
| Introspection | 1 | |
| Autonomie | 1 | |
| Responsabilité | 1 | |
| Mémoire | 1 | |
| Emotion | 1 | |
| Reconnaissance | 1 | |
| Communication | 1 | |
| Capacité d'agir | 1 | |

4. Capacité d'agir de sa propre volonté (*Agir librement*)
   Le drone se trouve confronté à diverses possibilités et décide de manière indépendante s'il lui faut éliminer le véhicule pour défendre le convoi.
5. Capacité de planifier pour atteindre son objectif (*Atteindre son objectif*)
   L'objectif du drone est de défendre le convoi; il décide donc de manière indépendante s'il lui faut initier un plan dans lequel le véhicule est une cible, dans lequel il faut choisir le type d'arme et lancer une attaque.

Après l'étude pilote 1, l'élément 3 (règles morales) a été abandonné, laissant pour le concept d'agentivité les quatre éléments qui ont été utilisés dans l'étude pilote 2 et l'étude finale:

1. Réflexion
2. Détermination de l'objectif
3. Liberté d'action
4. Réalisation de l'objectif

### 3.4.3 *Variables Dépendantes*

Nous avons inclus 9 à 12 questions dans les études pilotes et l'étude finale pour explorer d'autres implications de la perception de l'agentivité des armes autonomes. Les variables dépendantes proviennent des résultats de l'étude des publications, du questionnaire sur les valeurs et des entretiens. Basée sur les résultats des études pilotes, l'étude finale a intégré les variables dépendantes; *imputation (blâme), confiance, tort causé, dignité humaine, assurance, attentes, approbation, équité et inquiétude.* Chacune de ces variables a été mesurée sur une échelle d'auto-évaluation allant de 0 (pas du tout d'accord) à 100 (entièrement d'accord).

1. *Imputation (blâme)*: l'action est imputable au drone
2. *Confiance*: on peut faire confiance au drone pour prendre les mesures appropriées dans l'avenir.
3. *Tort causé*: les actions du drone ont causé du tort.
4. *Dignité humaine*: les actions du drone respectent la dignité humaine.
5. *Assurance*: j'ai l'assurance que je drone prendra les mesures appropriées dans l'avenir.
6. *Attentes*: les actions du drone correspondent à mes attentes.
7. *Soutien / approbation*: j'approuve l'utilisation de ce type de drone par les militaires.
8. *Equité*: les actions du drone sont équitables.
9. *Inquiétude*: les actions du drone m'inquiètent.

Les concepts suivants ont été utilisés dans les études pilotes 1 et 2, mais abandonnés dans l'étude finale:

10. *Faute du chef*: le chef est responsable de l'action.
11. *Confiance donnée au chef*: J'ai confiance en le chef pour qu'il prenne les mesures appropriées dans l'avenir.
12. *Torts causés par le chef*: les actions du chef ont causé du tort.

### 3.4.4 *Test d'attention*

Nous avons ajouté une question avec un test d'attention au milieu des questions sur les variables dépendantes, dans laquelle nous avons demandé de choisir la valeur 40 pour cette question. Le test d'attention est souvent ajouté à un questionnaire dans les études de psychologie cognitive, par exemple par Logg (2017), et les personnes qui ne donnent pas la réponse correcte sont exclues de l'échantillon, considérant qu'elles ne sont pas non plus suffisamment attentives pour les autres questions.

### 3.4.5 *Variables démographiques*

Nous avons également ajouté des questions pour recueillir des informations d'ordre «démographique» sur les personnes interrogées, et avons ajouté des variables sur: *l'âge, le genre, le niveau d'études, la profession et la nationalité.* Après ces questions générales de nature «démographique», nous avons demandé aux personnes interrogées si elles avaient une quelconque *expérience de l'intelligence artificielle*, si elles *travaillaient avec des drones* et si elles s'étaient déjà *trouvées dans une zone de conflit.*

## 3.5 Approche analytique

Cette section décrit l'approche que nous avons adoptée pour analyser les données. Pour chacune des études nous avons procédé en suivant les phases décrites ci-dessous dans l'analyse des données.

### 3.5.1 *Prétraitement des données*

Dans l'outil Qualtrics après l'achèvement de l'étude, une distinction est faite entre *réponses en cours* et *réponses complètes*. Avant de commencer l'analyse des données, le pourcentage de finalisation des réponses en cours doit être vérifié et on doit décider si elles doivent être regroupées avec les réponses complètes. Il s'avère que lorsqu'Amazon MTurk est utilisé pour diffuser l'enquête, bon nombre de personnes interrogées oublient de cliquer sur le dernier bouton, ce qui fait que 99% de leurs réponses sont complètes, mais ces réponses sont valables et utilisables. Si la personne interrogée a répondu à toutes les questions, nous les avons ajoutées à la liste des réponses complètes. Si des personnes interrogées n'ont pas répondu à toutes les questions, par exemple aux questions d'ordre démographique, nous les avons éliminées de l'échantillon. Dans la phase suivante de prétraitement, nous avons éliminé toutes les personnes qui ont échoué au test d'attention auquel nous procédons à mi-chemin dans le questionnaire. Nous avons demandé si les personnes ont répondu 40 à cette question, et nous avons éliminé toutes les réponses inférieures à 38 et supérieures à 42; nous supposons que cela révélait une erreur de précision pour placer les curseurs.

### 3.5.2 *Analyse de fiabilité*

Après avoir téléchargé le dossier csv dans SPSS (logiciel d'analyse statistique), nous avons procédé à un test de fiabilité qui est une mesure pour évaluer la cohérence interne d'un ensemble d'éléments de test ou d'échelle. Le test de fiabilité vérifie si le mesurage donné est cohérent pour un concept. Le système qui mesure la cohérence interne est Alpha de Cronbach, qui doit être supérieur à 0.7 pour démontrer une cohérence interne du concept. Pour chacune des études l'Alpha de Cronbach pour les éléments du concept d'agentivité est donné, car ces éléments sont utilisés pour mesurer le concept d'agentivité, et doivent avoir une cohérence interne. On indique également si l'Alpha de Cronbach est amélioré si l'un des éléments est enlevé. La fiabilité des variables dépendantes est vérifiée, mais pas communiquée, car ces éléments ne sont pas conçus pour être utilisés comme une seule échelle, et il s'est avéré que leur cohérence interne est plus basse que le seuil ($\alpha<0.7$), comme on pouvait s'y attendre.

### 3.5.3 *Analyse de la composante principale (ACP)*

Dans la phase suivante de l'analyse des données, une analyse de la composante principale (ACP) a été menée. Cette phase est une technique de réduction de variable dont le but est de réduire un plus grand ensemble de variables en un plus petit ensemble, appelé «composantes principales», qui expliquent la plupart des variations dans les variables d'origine. Nous avons procédé à l'ACP sur les éléments d'agentivité, pour voir s'ils pouvaient être agrégés en un seul concept d'agentivité, et indiquons la variation et le nombre des composantes comme résultats. Nous avons aussi vérifié si le concept était influencé par les variables démographiques, mais ne communiqueront pas ces résultats, comme l'ACP indique que toutes les variables démographiques sont des composantes distinctes qui n'influencent pas le concept d'agentivité.

### 3.5.4 *Analyse de corrélation*

La Corrélation Pearson bivariée (à deux variables) est un test pour mesurer la force et la direction des relations linéaires entre deux variables continues. Elle produit un coefficient de corrélation *r* qui peut varier entre -1 et +1. Une valeur négative indique une relation négative entre les concepts, par exemple plus la perception de l'agentivité est grande, moins l'est l'approbation (soutien). Une valeur positive indique une relation positive, par exemple plus la perception de l'agentivité est grande, plus l'élément imputation (ou blâme) l'est aussi. Un coefficient de relation égal à zéro indique qu'il n'y a pas de relation entre les concepts. Nous avons procédé à cette analyse pour vérifier la corrélation entre le concept d'agentivité et les variables dépendantes, et indiquons les résultats.

### 3.5.5 *Vérification concernant la manipulation sur l'agentivité*

Le test de manipulation vérifie si le concept d'agentivité varie entre les scénarios, et par là nous avons vérifié si les hypothèses devaient être retenues ou rejetées. La vérification se fait de deux manières. D'abord on crée un graphique, le concept d'agentivité apparaît sur l'axe des y, et les scénarios sur l'axe des x, pour l'analyse visuelle. On s'attend à ce que les niveaux d'agentivité soient différents selon les conditions considérées, et cette différence devrait être de l'ordre de $p < 0.5$. Ceci est indiqué par les barres en forme de I en tête de colonne et montre l'erreur moyenne multipliée par 1, et indique si les concepts se chevauchent ou non. Si les barres se chevauchent, alors les deux groupes ne sont pas significativement différents.

Le même résultat peut être obtenu par des tests T (test de Student) d'échantillons indépendants, pour vérifier s'il existe des différences importantes dans les moyennes de deux groupes distincts, pour lesquels les questions sont attribuées au hasard, si bien que l'effet observé est le résultat du traitement (dans ce cas lire un scénario spécifique), et ne peut pas être attribué à un autre effet. Nous présenterons les graphiques et communiquerons les résultats du test T concernant des échantillons indépendants, avec les différents niveaux d'agentivité des armes autonomes.

### 3.5.6 *Analyse des variables dépendantes*

L'analyse des variables dépendantes est de nature exploratoire, et les résultats sont présentés sous forme de graphique. Chaque graphique indique la moyenne de la variable dépendante sur l'axe des y, et les différents scénarios sur l'axe des x pour faire ressortir la différence entre les scénarios. Bien que la description des variables dépendantes soit descriptive, nous indiquons les résultats en comparant les différentes conditions significatives au niveau $p < .05$.

## 3.6 Préinscription

L'étude finale a été préenregistrée sur l'Open Science Framework[6] qui crée une version marquée de la date d'un projet qui ne peut pas être supprimé ou édité. Bien que ce ne soit pas encore exigé des publications scientifiques, les analystes vérifient souvent si le projet est préenregistré. Nous avons utilisé le modèle AsPredicted dans lequel on répond à 8 questions concernant l'étude, par exemple si les données ont déjà été rassemblées, quelles sont les hypothèses qui sont testées, et le type d'analyses que l'on a l'intention d'effectuer. Le pré enregistrement est placé «sous embargo» jusqu'au 24 juin 2021. Nous avons pré enregistré la solide hypothèse que nous avons sur la perception de l'agentivité, et pensons pouvoir faire part des constatations faites à la suite des résultats de l'étude pilote 2. Pour ce qui est des dépendantes variables, nous ne sommes pas sûrs de savoir quels mécanismes sont en jeu, ni si nous

6 https://osf.io/

pourrons communiquer nos résultats; par conséquent nous avons enregistré une analyse exploratoire de ces résultats.

## 3.7 Echantillon

Toutes les enquêtes ont été réalisées sous Qualtrics, et les études pilotes ont été diffusées par la plateforme de crowdsourcing (externalisation ouverte) Amazon Mechanical Turk. Nous avons testé l'enquête avec un petit échantillon de 20 personnes interrogées, et examiné le temps moyen pour compléter les réponses, pour vérifier si nous avions estimé les délais de complétion correctement. Nous avons vérifié que les scénarios étaient diffusés de manière égale, si les personnes vérifiaient la question test correctement, et s'il y avait des remarques au sujet de l'enquête. Comme tout avait été bien vérifié, nous avons décidé de porter l'échantillon représentatif à 50 personnes par scénario, ce qui donnait 500 participants pour l'étude pilote 1 et 700 pour l'étude pilote 2. Le recueil des réponses prit environ 4 heures, et chaque personne interrogée perçut 0.40 dollar US.

L'étude finale a été diffusée via la méthode «boule de neige» par email avec un lien anonyme vers environ 40 collègues des armées qui à leur tour propagèrent l'enquête. Cette méthode a été utilisée car nous n'avions pas l'autorisation du MIT ni du Ministère de la Défense néerlandais de recueillir des informations personnelles comme des emails ou adresses internet. En utilisant la méthode «boule de neige», nous n'avions aucune garantie quant au nombre des personnes interrogées. Nous avons également publié une annonce sur le réseau interne de l'Armée de Terre. L'étude finale s'est déroulé sur 16 jours, du 12 juin au 27 juin 2017, et a produit 327 réponses, dont 239 étaient des réponses valides et complètes, et utilisables après le prétraitement des données.

## 3.8 Questions méthodologiques

Dans cette section, plusieurs questions et problèmes concernant la méthodologie seront abordés. Ces problèmes pourraient avoir des implications pour la validité interne et externe de la présente étude et lorsque c'est possible, quelques contremesures sont données pour traiter ces problèmes.

### 3.8.1 *Entretiens pour le codage*

Pour mieux connaître les valeurs des experts nous avons procédé à des entretiens semi-structurés; c'est une méthode de recherche qualitative. Nous avons utilisé la méthode de Codage des Valeurs (Value Coding method) pour interpréter les données et extraire les valeurs des entretiens après transcription. L'un des problèmes qui se posent lors du codage de données quantitatives est que c'est une activité heuristique, et aucune formule pour le codage n'existe. Il faut aussi trouver des liens et donner un sens aux données (Saldana, 2015). Ces caractéristiques impliquent que la méthode de codage des valeurs est susceptible d'être biaisée, car le chercheur utilise son propre cadre et ses notions personnelles pour interpréter les données. Pour réduire cette distorsion en appliquant la méthode d'encodage des valeurs un second chercheur codera les entretiens indépendamment du premier. Des instructions sur la méthode seront données, mais peu d'informations sur les valeurs et les résultats.

### 3.8.2 *Expériences contrôlées randomisées* (*études expérimentales avec randomisation*)

L'une des conditions pour procéder à des expériences contrôlées randomisées est, comme leur nom l'indique, la diffusion des scénarios auprès des personnes interrogées. La randomisation limite la confusion, renforce la validité interne de la recherche, et constitue un prérequis indispensable pour ce type d'étude. L'un des problèmes survient lorsque la randomisation échoue et que la répartition des répondants est faussée du fait des scénarios. L'une des causes de diffusion faussée peut être que le logiciel ne répartit pas de façon égale les répondants (personnes interrogées) sur les scénarios. Les chercheurs ne peuvent que vérifier si les programmes du logiciel sont corrects. Une autre cause de diffusion faussée est que des personnes voient un certain scénario qui ne leur plaît pas ou auquel ils ne s'attendaient pas, et quittent l'enquête avant de finir. Par exemple, des personnes s'attendent à répondre à une enquête sur les armes autonomes et peuvent se voir attribuer un scénario sur une arme actionnée par l'homme, et alors décident d'arrêter. Le fait de quitter l'enquête prématurément est appelée «attrition sélective», et cela peut résulter en une distorsion dans le travail de recherche, car les personnes qui abandonnent sont fondamentalement différentes de celles qui y participent. Dans le cadre actuel de la recherche, il n'est pas possible de prendre des mesures pour y remédier, et ceci peut constituer une menace pour la validité interne de l'étude.

### 3.8.3 *Amazon Mechanical Turk*

L'avantage qu'il y a à utiliser Amazon MTurk comme méthode de diffusion d'une étude est qu'un grand nombre de répondants participeront à l'enquête et répondront aux questions avec sérieux pour conserver leur statut professionnel. La méthode a également des inconvénients car les collaborateurs sur MTurk sont surtout basés aux USA, du fait que pour être employé vous devez vous plier aux règles américaines concernant les impôts même si ne résidez pas aux Etats-Unis. Ceci pourrait conduire à une distorsion dans les résultats, impliquant que les résultats ne sont pas généralisables en dehors des USA. Bonnefon et al (2016) signalent un deuxième problème dans l'utilisation de MTurk: les participants peuvent ne pas être représentatifs de la population US, par exemple parce qu'un grand nombre des collaborateurs sont des étudiants, et un troisième problème est que les collaborateurs connaissent déjà bien ce dont il s'agit. Le premier problème peut être résolu en choisissant un large éventail de collaborateurs, si bien que l'échantillon comptera également des personnes ne résidant pas aux USA. Le deuxième problème est inhérent au choix de MTurk et il n'existe pas de mesures à prendre pour y remédier. Le troisième problème n'est pas vraiment susceptible de survenir, car les scénarios sont uniques et n'ont fait l'objet d'étude sur MTurk auparavant.

### 3.8.4 *Enquête finale, diffusion «boule de neige»*

L'étude finale a été diffusée aux collègues militaires qui ont transmis leur email à travers leur réseau. Utiliser la méthode de diffusion «boule de neige» («snowball») présente quelques inconvénients. Le premier est que les participants pouvaient participer à l'enquête plusieurs fois, et du fait que nous ne sommes pas autorisés à stocker des informations d'identité personnelle, on ne peut pas vérifier si c'est le cas. Ceci pouvait conduire à une distorsion dans les résultats si les répondants voyaient des scénarios différents, et influencer les réponses. Un deuxième inconvénient est que des collègues répondent au test les premiers, puis font part de leur expérience aux autres participants avant que ceux-ci répondent. Ceci influencerait également leurs réponses. En troisième lieu, les chercheurs ne peuvent pas contrôler qui répondra à l'enquête; par conséquent, un risque potentiel est que l'enquête soit transmise à quelqu'un qui ne travaille pas au Ministère de la Défense, et que l'échantillon soit «contaminé». Malheureusement, comme on ne peut pas enregistrer des informations d'identité personnelle, il n'y a pas beaucoup de mesures possibles pour remédier à ces trois problèmes.

CHAPITRE 4

# Résultats

> On peut espérer que les gens se rendent compte que les ordinateurs feront uniquement ce qu'ils ont été programmés à faire, et RIEN d'autre.
>
> PILOTE INTERROGÉ, ÉTUDE 1

Cette section donne les résultats de l'enquête sur les valeurs, qui consistait en un questionnaire et quelques entretiens d'experts, ainsi que les expériences contrôlées randomisées, incluant deux études «pilotes» et l'étude finale. Les valeurs obtenues à partir du questionnaire en ligne et des entretiens sont fusionnées en une seule présentation. Les résultats des études «pilotes» et de l'étude finale sont présentés en utilisant la structure de phases d'analyse de données (section 3.5.) et les principaux résultats sont donnés dans la conclusion de chaque étude.

## 4.1 Enquête sur les valeurs

Pour connaître quelles sont les valeurs que les personnes associent aux armes autonomes, une enquête et six entretiens d'experts ont été menés. Un aperçu des résultats est donné dans ce chapitre.

### 4.1.1 *Enquête en ligne*

L'enquête sur les valeurs consistait en un questionnaire demandant aux répondants quelles valeurs de bioéthique et de celles de Cheng et Fleischmann s'appliquent. Dans la question finale, on demandait aux participants de donner au moins une valeur supplémentaire qui n'était pas évoquée dans les questions de départ. Au total 69 répondants ont répondu à l'enquête, et après avoir éliminé les réponses incomplètes, il restait 57 réponses complètes. Les résultats sont détaillés dans les graphiques ci-dessous.

 | DOI:10.1163/9789004449084_005

FIGURE 5 Résultats question 1, enquête en ligne sur les valeurs

Le court questionnaire permet de savoir quelles valeurs les personnes associent aux armes autonomes. Les résultats concernant la question bioéthique montrent que la *non-malfaisance* (non-nuisibilité) est la plus importante, ensuite vient la *«bienfaisance»* (capacité à faire le bien), en troisième la *justice*, et enfin l'*autonomie*. Les cinq valeurs de tête sur la base de la liste de Cheng et Fleischmann (2010) sont: *sécurité, intelligence, responsabilité, justice* et *ordre social*. Les questions ouvertes révèlent que les personnes associent les armes autonomes surtout à l'*imputabilité* et à *l'efficacité*.

### 4.1.2 *Entretiens*

Six entretiens ont été conduits avec des experts dans les domaines suivants:

Intelligence artificielle:

- Prof. Toby Walsh (université de Nouvelle Galles du Sud & Data61).

Armes autonomes:

- Dr. Peter Asaro (the New School & Campaign Ban Killer Robots, campagne pour l'élimination des robots tueurs);

FIGURE 6 Résultats question 3, enquête en ligne sur les valeurs

- Mrs. Miriam Struyk (directeur des programmes, Security & Disarmament PAX);
- Dr. Michael Horowitz (Université de Pennsylvanie).

Opérations militaires:

- Lieutenant-Colonel Kremers (Etat-major Armée de Terre néerlandaise);
- Capitaine de corvette van den Sande (Etat-major Ministère de la Défense).

Après les discussions sur les valeurs, les entretiens ont fourni de précieuses informations de fond sur l'histoire et les positions des différents groupes dans le cadre du débat actuel sur les armes autonomes, mais ont aussi montré qu'il n'existe pour le moment aucun consensus sur une définition. La transcription complète de ces entretiens figure dans l'Appendice F, Transcriptions d'entretiens.

Les entretiens ont été encodés au moyen de la méthode «encodage des valeurs», en se basant sur la comparaison faite entre l'encodage du chercheur

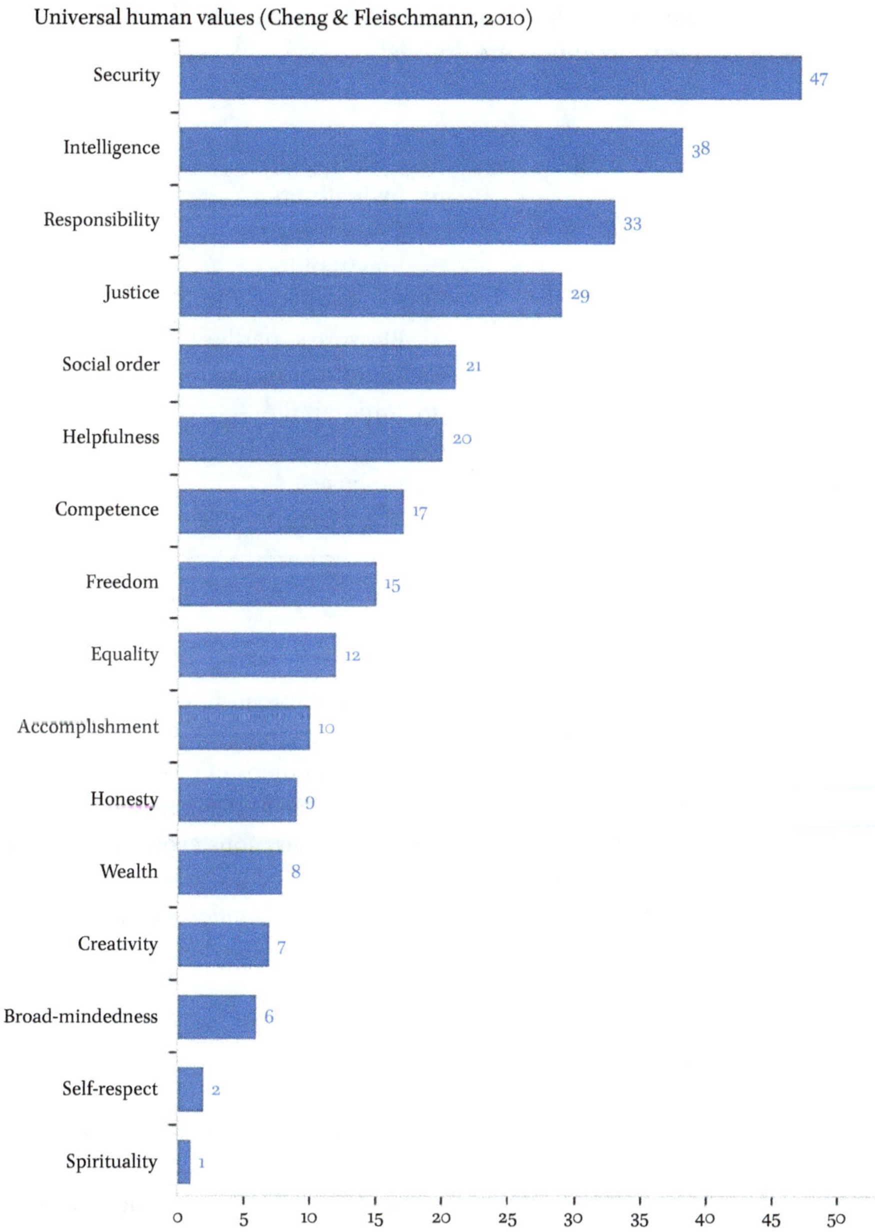


FIGURE 7 Résultats question 2, enquête en ligne sur les valeurs

et celui du réviseur (Appendice H, processus d'encodage des résultats), la liste suivante de 22 valeurs uniques a été dressée:

| | |
|---|---|
| 1. Intervenir n'importe quand | 12. Distance |
| 2. Les rappeler | 13. Invincibilité |
| 3. Minimiser les risques | 14. Déterminisme |
| 4. Avoir le contrôle | 15. Inévitabilité |
| 5. Bonne volonté | 16. Souffrance inutile |
| 6. Fixer des limites | 17. Blessure superflue |
| 7. Droit d'être tué par une personne | 18. Dignité humaine |
| 8. Cadre éthique | 19. Imputabilité |
| 9. Contrôle humain significatif | 20. Responsabilité |
| 10. Prévisibilité | 21. Défense |
| 11. Réflexivité | 22. Tort causé |

### 4.1.3 *Enquête sur les valeurs : conclusion*

Pour conclure sur l'enquête sur les valeurs, nous comparerons les valeurs ressortant de l'enquête en ligne et les entretiens qui donnent lieu à une liste de valeurs (Tableau 6). Six valeurs dans cette liste sont mentionnées à la fois dans l'enquête sur les valeurs et dans les entretiens. A partir des questions que les experts ont mises en exergue au cours des entretiens, nous avons sélectionné les valeurs qui étaient le plus souvent évoquées pour les tester dans l'étude pilote 1. Ces valeurs sont *l'équité, le tort causé, la dignité humaine* et *la responsabilité.*

## 4.2 Etude Pilote 1

Dans la première étude pilote, nous avons essayé de vérifier si nous pouvions manipuler la perception de l'agentivité des drones actionnés par l'homme et des armes autonomes. Nous nous sommes intéressés à la manière dont les gens percevaient la technologie actuelle comparée à la technologie du futur. Nous avons créé plusieurs conditions dans lesquelles nous avons varié le niveau d'agentivité en donnant des informations supplémentaires sur la manière dont le drone «réfléchissait» sur la possible attaque d'un véhicule, ou en laissant délibérément de côté ces informations. Nous avons testé les conditions «bon résultat» et «mauvais résultat», car nous pensions que cela aurait un impact

TABLEAU 6 Valeurs ressortant de l'enquête en ligne et des entretiens

| Enquête en ligne | Entretiens |
|---|---|
| Justice | |
| Autonomie | |
| Sécurité | |
| Intelligence | |
| Ordre social | |
| Efficacité | |
| Equité | |
| Racisme | |
| Contrôle | En situation de contrôle |
| Capacité à faire le bien | Bonne volonté |
| Imputabilité | Imputabilité |
| Responsabilité | Responsabilité |
| Sûreté | Sûreté |
| Non-malfaisance | Tort causé |
| | Capacité à intervenir à tout moment |
| | Les rappeler |
| | Minimiser les risques |
| | Fixer des limites |
| | Droit d'être tué par une personne |
| | Cadre éthique |
| | Contrôle humain significatif |
| | Prévisibilité |
| | Réflexivité |
| | Distance |
| | Invincibilité |
| | Déterminisme |
| | Inévitabilité |
| | Souffrance inutile |
| | Blessure superflue |
| | Dignité humaine |

sur la manière dont les variables dépendantes, par exemple l'approbation / soutien, seraient considérées. Dans cette étude, nous avons posé de manière spécifique les questions ayant trait à l'agentivité des drones et non à l'opérateur humain qui les contrôle.

TABLEAU 7 Scénarios de l'étude pilote 1

| | **Condition d'agentivité** | | | | |
|---|---|---|---|---|---|
| | **Drone autonome** | | | **Drone avec opérateur humain** | |
| | **Agentivité zéro** | **Agentivité basse** | **Agentivité élevée** | **Agentivité basse** | **Agentivité élevée** |
| **Pas de dommages collatéraux** (bon résultat) | 1 | 2 | 3 | 4 | 5 |
| **Dommages collatéraux** (mauvais résultat) | 6 | 7 | 8 | 9 | 10 |

Nous avons testé 10 scénarios, avec le but d'obtenir 50 répondants par condition, et donc un total de 500 répondants. Le nombre de répondants par scénario s'élevait de 49 à 54. Après le prétraitement des données, et après avoir éliminé les répondants qui avaient échoué au test d'attention, il nous restait 510 réponses complètes.

### 4.2.1 *Analyse de fiabilité*

Pour les cinq éléments d'agentivité $\alpha = 0.9$, ce qui est bien au-dessus du seuil de 0.7, et ne peut être amélioré que si l'élément SQ3 Règles morales est éliminé comme cela apparaît dans le Tableau 8. La suppression d'autres éléments diminuera la valeur $\alpha$ et réduira la cohérence interne, comme on le voit dans la dernière colonne (si l'élément Alpha de Crohnbach est supprimé) dans le Tableau 8.

### 4.2.2 *Analyse Composante Principale (ACP)*

L'ACP montre que SQ1 Réflexion, SQ2 Détermination des objectifs, SQ4 Liberté d'action, et SQ5 Réalisation des objectifs peuvent être considérés comme un ensemble qui s'élève à 71.75% des variables parmi les variables d'origine (Tableau 9). On peut aussi le constater dans le diagramme d'éboulis dans lequel apparaissent les valeurs propres des composantes (Figure 8). A partir de la

TABLEAU 8 Analyse de fiabilité des résultats étude pilote 1- éléments d'agentivité

| Statistiques fiabilité | | |
|---|---|---|
| **Reliability statistics** | | |
| **Cronbach's Alpha** | **Cronbach's Alpha based on standardized items** | **N of items** |
| .900 | .895 | 5 |

**Item-Total statistics**

| | Scale mean if item deleted | Scale variance if item deleted | Corrected Item-Total correlation | Squared multiple correlation | Cronbach's Alpha if item deleted |
|---|---|---|---|---|---|
| SQ1_Thought | 167.77 | 13710.694 | .841 | .741 | .857 |
| SQ2_Goal setting | 163.85 | 13440.002 | .857 | .775 | .853 |
| SQ3_Moral rules | 183.39 | 17599.970 | .448 | .215 | .933 |
| SQ4_Act freely | 163.22 | 13885.447 | .824 | .695 | .861 |
| SQ5_Achieve goals | 157.21 | 13891.495 | .794 | .681 | .868 |

TABLEAU 9 Eléments d'agentivité, Etude pilote 1

| Component matrix[a] | | | |
|---|---|---|---|
| | **Component** | | |
| | 1 | 2 | 3 |
| SQ2_Goal setting | .922 | | |
| SO1_Thought | .911 | | |
| SO4_Act freely | .899 | | |
| SQ5_Achieve goals | .882 | | |
| SQ3_Moral rules | .571 | .819 | |

Extraction Method: Principal Component Analysis.
[a] 3 components extracted.

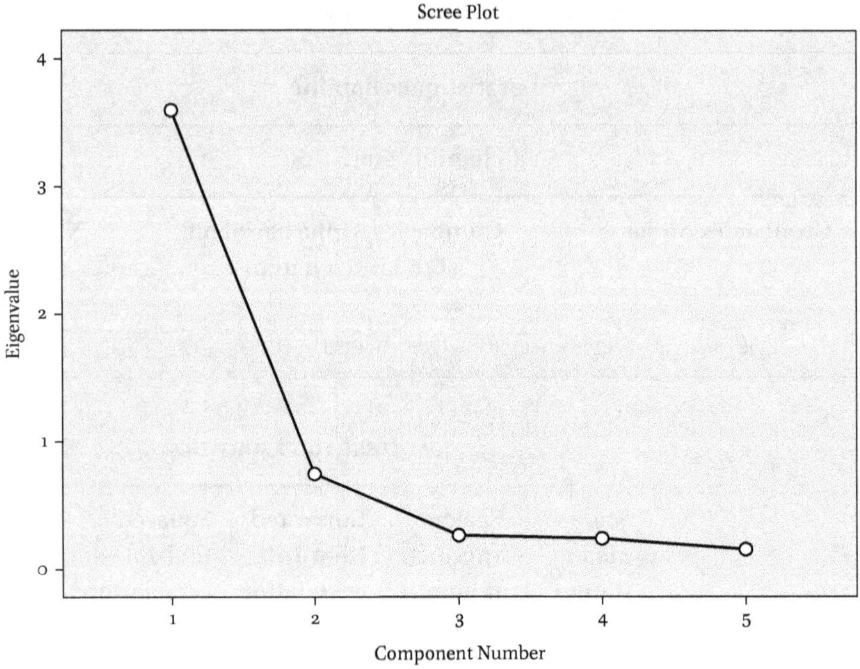


FIGURE 8 Tableau d'éboulis ACP, Etude pilote 1

matrice des composantes, nous pouvons observer que SQ3 Valeurs morales est une composante à part en comparaison des quatre autres éléments d'agentivité.

### 4.2.3 *Analyse corrélation*

La corrélation entre le concept (ensemble) agentivité et les variables dépendantes est petite, et significative seulement pour DVQ1 *Faute / blâme*, DVQ2 *Faute du chef*, DVQ4 *Confiance en le chef*, DVQ6 *Tort causé par le chef* et DVQ8, *Approbation / Soutien*. Quelques effets inattendus sont observables du côté des relations. Par exemple, on peut s'attendre à ce que la perception de l'agentivité fait que les personnes attribuent plus de *faute / blâme* au drone (r = .183) mais ces résultats indiquent que les personnes attribuent moins de *faute / blâme* au chef (r = -.239). Le même effet peut être observé avec la variable tort causé. L'analyse de corrélation n'est pas assez détaillée pour se focaliser sur ces résultats, et donc cela sera fait dans l'analyse des variables dépendantes.

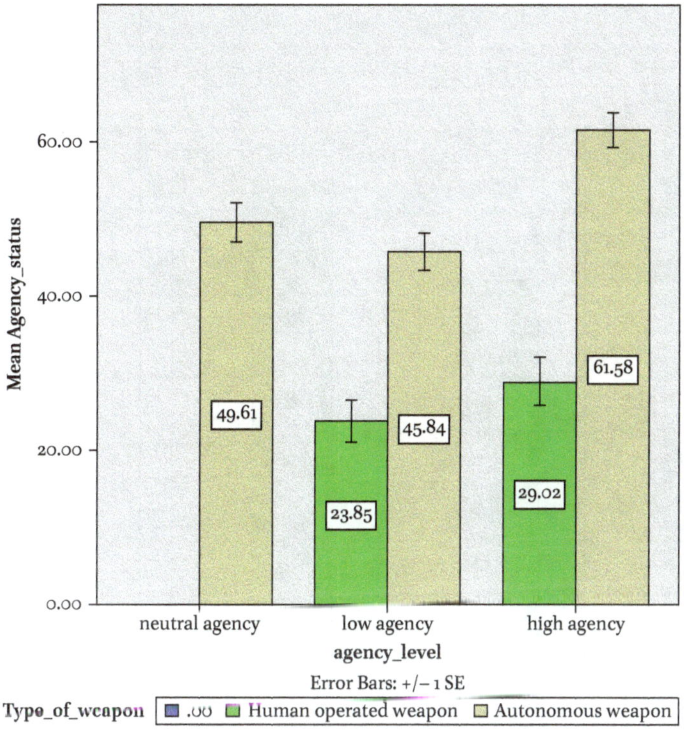


FIGURE 9 Valeur moyenne concept agentivité par type d'arme

### 4.2.4 *Vérification manipulation sur l'agentivité*

Le graphique de la Figure 9 montre que notre manipulation sur l'agentivité fonctionne, car dans la condition de niveau bas d'agentivité pour les armes autonomes, les personnes perçoivent moins d'agentivité que dans la condition d'agentivité élevée, comme le montre la moyenne du concept d'agentivité sur l'axe des y. Dans la condition d'agentivité neutre la perception est un peu plus élevée que dans la condition de basse agentivité, mais plus basse que pour la condition d'agentivité élevée. Le test-T (test de Student) des échantillons indépendants montre que l'agentivité basse et l'agentivité élevée (armes autonomes) sont des ensembles distincts (p = .000), mais ce n'est pas le cas pour l'agentivité neutre et l'agentivité basse (armes autonomes) (p = .209), car la différence entre ces groupes n'est pas significative, ce qui peut être aussi observé dans le chevauchement des barres d'erreur pour ces conditions.

Le drone actionné par l'homme est perçu comme ayant une agentivité beaucoup plus basse que dans les conditions pour l'arme autonome. La

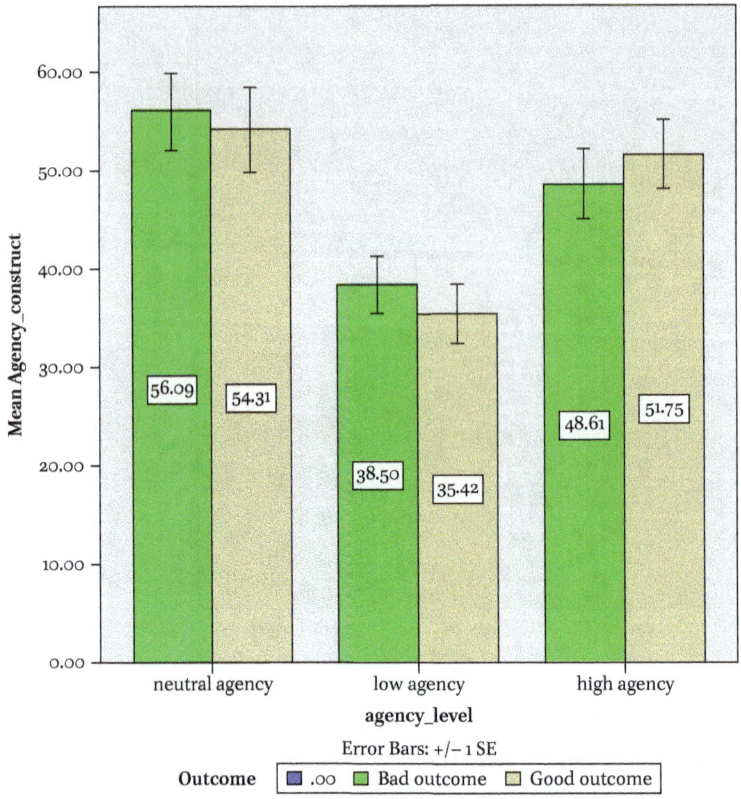


FIGURE 10 Moyenne du concept agentivité par condition (de niveau d'agentivité) par résultat

manipulation de l'agentivité fonctionne même si la différence est moins évidente que dans le scénario des armes autonomes. Les barres d'erreur dans les scénarios à basse agentivité indiquent que les ensembles «drone actionné par l'homme» ne peuvent être distincts, ce qui est confirmé par le test-T de l'échantillon indépendant ($p = .125$).

Le deuxième graphique (Figure 10) montre la moyenne de la perception de l'agentivité pour les niveaux d'agentivité, pour les «bons» et «mauvais» résultats. Le type de résultat de l'utilisation de l'arme, que ce soit des dommages collatéraux ou non, ne semble pas aboutir à beaucoup de différences dans la perception de l'agentivité, les moyennes variant de 2 à 3 points (sur une échelle de 100). Les barres d'erreur se chevauchent aussi sur tous les niveaux d'agentivité, ce qui est confirmé par les tests-T sur l'échantillon indépendant. Dans la condition niveau d'agentivité neutre, la différence entre les ensembles de bon résultat et de mauvais résultat est $p = .816$, dans les conditions de basse agentivité elle est $p = .641$, et dans les conditions de haute agentivité de $p = .575$, toutes beaucoup plus élevées que le seuil de $p < .05$.

TABLEAU 10 Matrice de corrélation, concept d'agentivité et variables dépendantes

| Correlations | | | | | | | | | | | |
|---|---|---|---|---|---|---|---|---|---|---|---|
| | | Agency_ construct | DVO1_ Blame | DVO2_ Commander Blame | DV03_ Trust | DV04_ Commander Trust | DV05_ Harm | DV06_ Commander Harm | DV07_ Uneasy | DV08_ Support | DV09_ Fairness |
| Agency_ construct | Person Correlation | 1 | .183** | -.239** | .070 | .176** | -.011** | -.227** | .005 | .101 | .087 |
| | Sig(2-tailed | | .000 | .000 | .114 | .000 | .798 | .000 | .914 | .023 | .050 |
| | N | 510 | 510 | 510 | 510 | 510 | 510 | 510 | 510 | 510 | 510 |
| DV01_Blame | Person Correlation | 183** | 1 | -.101* | .020 | 0.53 | .198** | -.170 | .129** | -.126** | -078 |
| | Sig(2-tailed | .000 | | .023 | .660 | .231 | .000 | .000 | .003 | .006 | .077 |
| | N | 510 | 510 | 510 | 510 | 510 | 510 | 510 | 510 | 510 | 510 |
| DVO2_ Commander Blame | Person Correlation | -.239** | -.101* | 1 | -.264** | -.330** | .264** | .692** | -.001 | -.333** | -.036 |
| | Sig(2-tailed | .000 | .023 | | .000 | .000 | .000 | .000 | .982 | .000 | .422 |
| | N | 510 | 510 | 510 | 510 | 510 | 510 | 510 | 510 | 510 | 510 |
| DV03_Trust | Person Correlation | .070 | .020 | -.264** | 1 | .635** | -.294** | -.262** | .169** | .674** | -.106 |
| | Sig(2-tailed | .114 | .660 | .000 | | .000 | .000 | .000 | .000 | .000 | .016 |
| | N | 510 | 510 | 510 | 510 | 510 | 510 | 510 | 510 | 510 | 510 |
| DV04_ Commander Trust | Person Correla.630tion | .176** | .053 | -.330** | .635** | 1 | -171** | -.322** | .138** | .584** | .021 |
| | Sig(2-tailed | .000 | .232 | .000 | .000 | | .000 | .000 | .002 | .000 | .630 |
| | N | 510 | 510 | 510 | 510 | 510 | 510 | 510 | 510 | 510 | 510. |

TABLEAU 10 Matrice de corrélation, concept d'agentivité et variables dépendantes (*cont.*)

| Correlations | | | | | | | | | | | |
|---|---|---|---|---|---|---|---|---|---|---|---|
| | | **Agency_ construct** | **DVO1_ Blame** | **DVO2_ Commander Blame** | **DV03_ Trust** | **DV04_ Commander Trust** | **DV05_ Harm** | **DV06_ Commander Harm** | **DV07_ Uneasy** | **DV08_ Support** | **DV09_ Fairness** |
| DV05_Harm | Person Correlation | -.011 | .198** | .264** | -294** | -.171** | 1 | .361** | -.044 | -.336** | 089 |
| | Sig(2-tailed | .798 | .000 | .000 | .000 | .000 | | .000 | .323 | .000 | .046 |
| | N | 510 | 510 | 510 | 510 | 510 | 510 | 510 | 510 | 510 | 510 |
| DV06_ Commander Harm | Person Correlation | -.227** | -.170** | .692** | -.262** | -.322** | .361** | 1 | -.038 | -.322 | .041 |
| | Sig(2-tailed | .000 | .000 | .000 | .000 | .000 | .000 | | .394 | .000 | .351 |
| | N | 510 | 510 | 510 | 510 | 510 | 510 | 510 | 510 | 510 | 510 |
| DV07_Uneasy | Person Correlation | .005 | -.129 | -.001 | .169** | .138** | -.044 | -.038 | 1 | .127-- | -.689** |
| | Sig(2-tailed | .914 | .003 | .982 | .000 | .002 | .321 | .394 | | .004 | .000 |
| | N | 510 | 510 | 510 | 510 | 510 | 510 | 510 | 510 | 510 | 510 |
| DV08_Support | Person Correlation | .101* | -.120** | -.333** | .674** | .584** | -.336** | -.322** | .127** | 1 | -.004 |
| | Sig(2-tailed | .023 | .006 | .000 | .000 | .000 | .000 | .000 | .004 | | .934 |
| | N | 510 | 510 | 510 | 510 | 510 | 510 | 510 | 510 | 510 | 510 |
| DV09_Fairness | Person Correlation | .087 | -078 | -.036 | -.106* | 0.21 | .089 | .041 | -.689** | -.004 | 1 |
| | Sig(2-tailed | .050 | .077 | .422 | .016 | .630 | .046 | .351 | .000 | .934 | |
| | N | 510 | 510 | 510 | 510 | 510 | 510 | 510 | 510 | 510 | 510 |

* correlation is significant at the 0.05 level (2-taild)

### 4.2.5 *Analyse des variables dépendantes*

L'analyse de corrélation a montré que seules les variables dépendantes *faute (blâme)*, *faute du chef*, *confiance en le chef*, *tort causé par le chef* et *approbation / soutien* étaient significatives à un niveau minimum de $p < .05$. Dans l'étape suivante, nous avons regardé de plus près ces variables et du fait de leur non signifiance, nous n'avons pas analysé le reste des variables dépendantes. En examinant les détails nous avons vérifié s'il y avait des différences entre les niveaux d'agentivité bas, neutre et élevé et entre les conditions de niveau pour les armes autonomes et les drones actionnés par l'homme.

#### 4.2.5.1 Variable approbation

L'observation la plus intéressante est la grande différence pour ce qui est de l'«approbation» (soutien) entre «bons» et «mauvais» résultats (Figure 11), qui

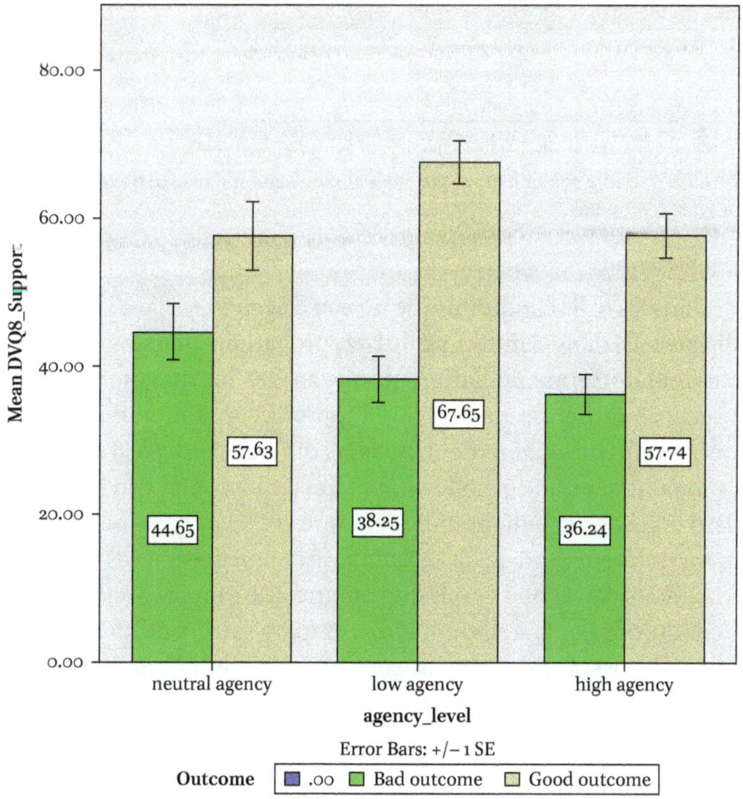


FIGURE 11 Moyenne de variable, valeur «approbation» par condition (de niveau d'agentivité) par résultat

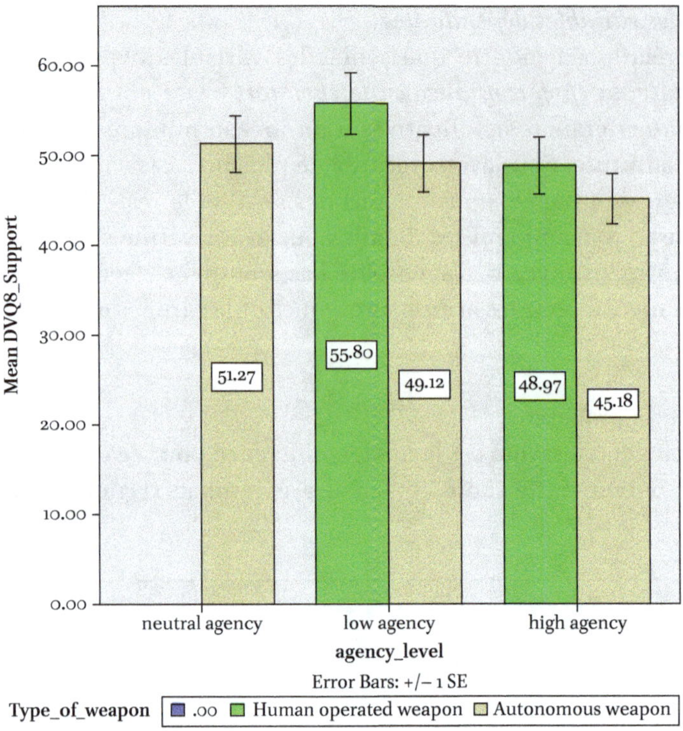


FIGURE 12 Moyenne variable valeur «approbation» par condition (de niveau d'agentivité) par type de mission

est importante dans les trois conditions (de niveau d'agentivité) ($p < .05$). Cette différence substantielle dans la moyenne de l'«approbation» pour les trois conditions indique que notre manipulation au niveau des résultats fonctionne. Une autre observation intéressante est que l'approbation diminue à mesure que les niveaux d'agentivité augmentent, mais les différences dans la moyenne sont petites (< 10 points sur une échelle de cent), et les barres en I en haut avec l'erreur standard en chevauchement indiquent que ces ensembles ne sont pas significativement différents, lorsqu'on les compare dans les scénarios à «bons» et «mauvais» résultats. Le second graphique montre que l'approbation est plus forte dans les conditions «opérateur humain» comparée aux conditions «arme autonome», mais là aussi les différences sont moindres et non significatives comme les barres d'erreur se chevauchent.

### 4.2.5.2 Valeur «Confiance»

Sur le graphique de la Figure 12 on peut voir que les personnes sont moins susceptibles d'avoir confiance en une arme autonome pour prendre les mesures

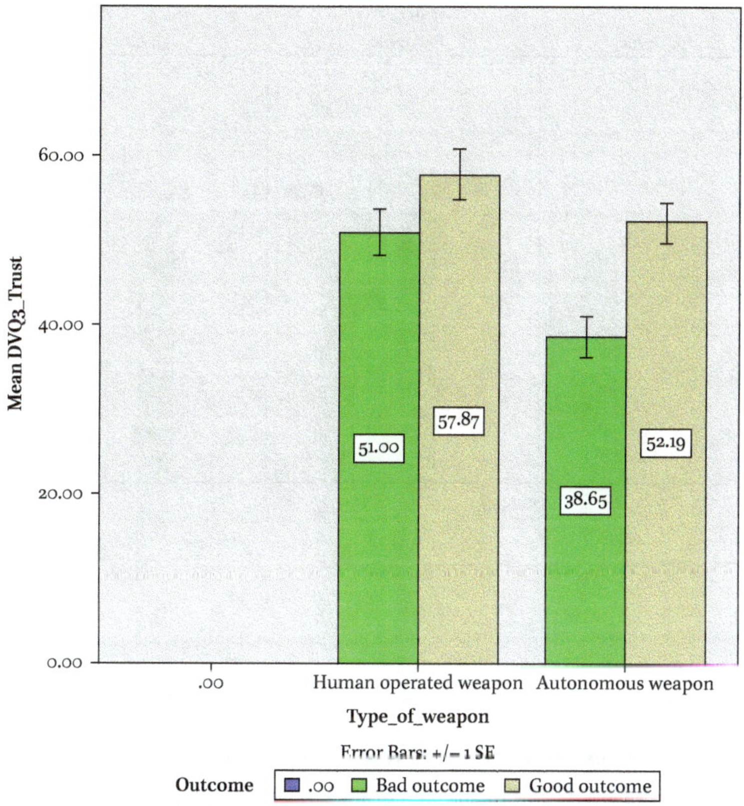


FIGURE 13 Moyenne de variable « confiance » par type d'arme par résultat

adéquates dans le futur après un mauvais résultat qu'en un opérateur humain qui prend des mesures avec le même mauvais résultat, et tous les résultats sont significatifs. Ce phénomène est magnifié dans les conditions de bas niveau d'agentivité où l'on n'observe pratiquement aucune différence entre la confiance entre un humain après un résultat bon ou mauvais, et une grande différence dans la confiance pour les armes autonomes (Figure 13). Cependant, l'analyse à une seule variable a montré qu'il n'y a pas d'interaction significative: type d'arme * résultat p = .120, ce qui signifie que l'on ne peut tirer aucune conclusion à ce point de l'étude, mais il est intéressant de continuer l'étude.

#### 4.2.5.3 Variables «faute» (blâme) et «tort» (causé)

Pour étudier la chaîne de responsabilité la moyenne des variables *faute, faute du chef, tort et tort causé par le chef* est présentée sur le même graphique (Figure 15). Tous les résultats des variables «faute» (Figure 15), sauf pour la

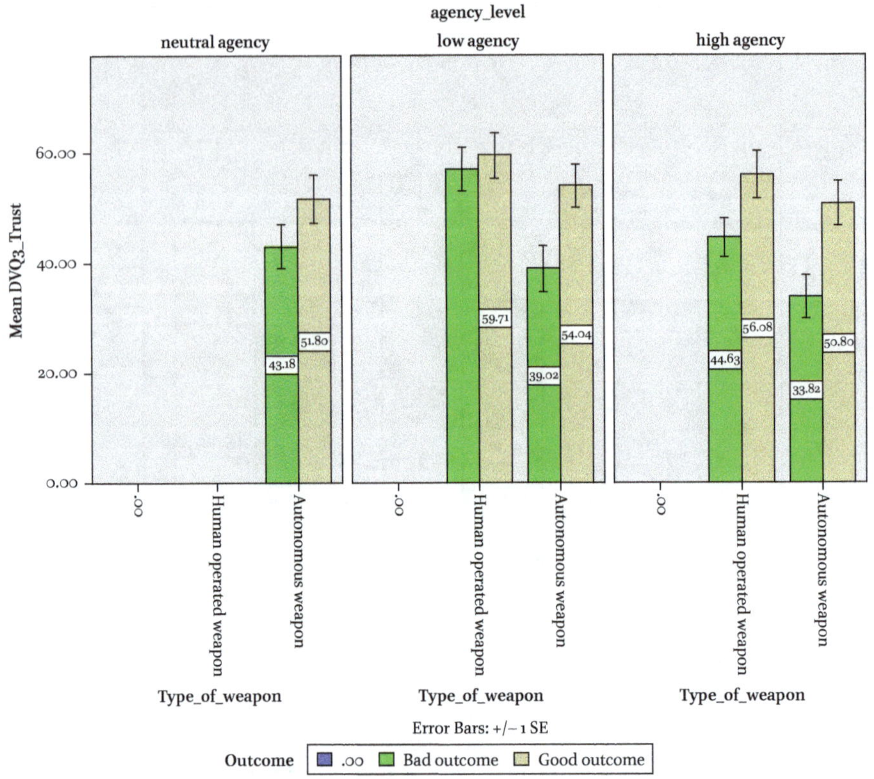


FIGURE 14 Moyenne pour valeur «confiance» par condition (de niveau d'agentivité) par résultat

condition d'agentivité neutre pour arme autonome, sont significatifs. Ces résultats montrent que dans la condition d'agentivité basse, la faute revient au chef, et dans une condition de haute agentivité la faute est attribuée au drone.

Il est frappant d'observer que la condition à basse agentivité montre une quantité élevée de responsabilité causale pour le «tort causé». Ce cas de figure est valable pour les drones actionnés par les humains aussi bien que pour les armes autonomes (Figure 16). Contrairement à la variable «faute / blâme», le drone est perçu comme causant beaucoup de mal (tort) aussi bien dans les conditions à haut et bas niveau d'agentivité, mais pour le bas niveau le tort revient presque de manière égale au chef. Cela signifie que tous deux causent du tort, mais dans le haut niveau, ce tort est beaucoup moins imputé au chef, et le drone est perçu comme étant celui qui cause du tort. Cependant, la différence entre les variables «tort» et «tort causé par le chef» pour le niveau bas d'agentivité de l'arme autonome aussi bien que celle actionnée par l'homme

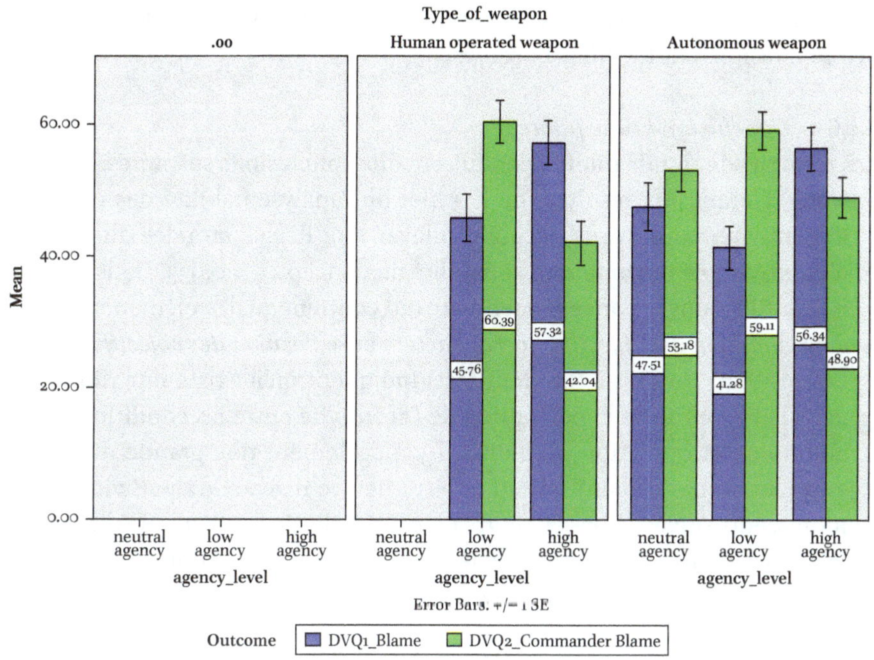


FIGURE 15 Moyenne valeurs variables «faute» et «faute du chef» par condition de niveau (d'agentivité) et par type d'arme.

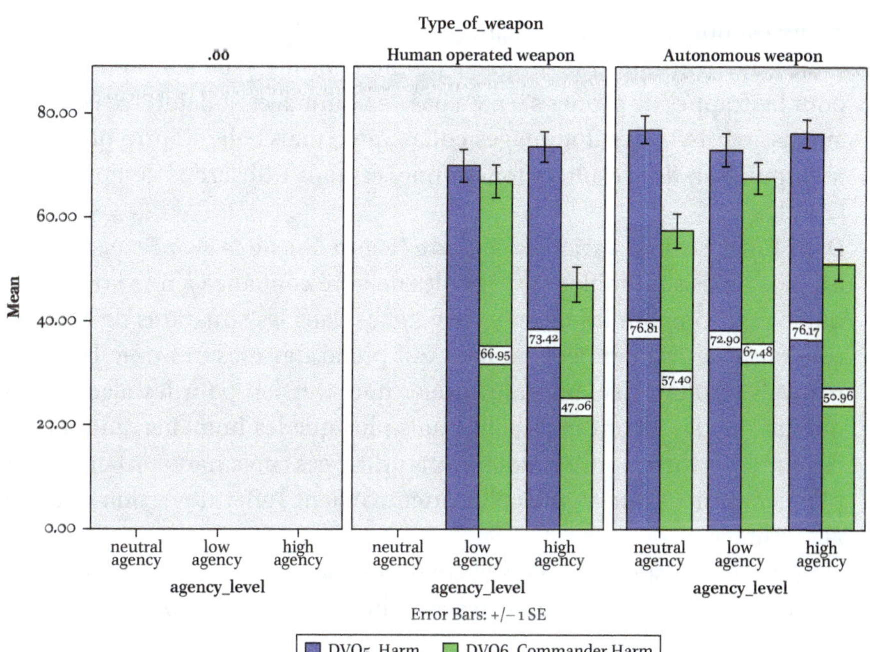


FIGURE 16 Moyenne valeurs variables «tort causé» et «tort causé par le chef» par condition de niveau et par type de mission

n'est pas significative; par conséquent nous devons être prudents lorsqu'on tire des conclusions sur les effets observés.

### 4.2.6 *Conclusion Etude pilote 1*

Les résultats de l'étude pilote 1 conduisent aux conclusions suivantes:

- Nous référant aux résultats de l'APC et de l'analyse fiabilité des éléments d'agentivité, nous avons décidé d'enlever SQ3 Règles morales du concept d'agentivité, ce qui améliore la fiabilité du concept: $\alpha = 0.933$. Dans la suite, nous utiliserons le concept agentivité qui contient quatre éléments, *pensée, détermination des objectifs, liberté d'action* et *réalisation des objectifs*.
- Les résultats du Test-T (Student test) indiquent qu'il existe une différence dans la moyenne de la perception de l'agentivité entre les conditions arme autonome et opération humaine. Il y a également une grande différence entre les scénarios haut niveau et bas / neutre niveaux d'agentivité. Néanmoins, entre niveaux bas et neutre on n'a pas observé une grande différence. Par conséquent, nous avons décidé de supprimer la condition niveau neutre d'agentivité pour la suite, et de tester seulement les conditions de niveaux bas et élevé.
- L'observation la plus évidente dans la variable *approbation* est qu'il existe une grande différence dans l'approbation entre les situations de bons et de mauvais résultats (Figure 11). Cependant ceci était prévisible au vu de ce qui est noté dans les publications, car James Igoe Walsh (2015) et Kreps (2014) ont constaté que le grand public exprime plus son approbation pour les frappes de drones s'il n'y a pas de dommages collatéraux, comparé aux cas où il y a des dommages collatéraux, mais cela montre que notre manipulation des résultats fonctionne, et nous utiliserons ce point dans la suite.
- Dans l'analyse de la variable *confiance* (Figure 13), nous avons constaté que les personnes sont moins susceptibles de faire confiance à une arme autonome pour prendre les mesures adéquates dans les situations de mauvais résultat qu'à un opérateur humain qui prend des mesures avec le même mauvais résultat. Ceci pourrait révéler une aversion pour les algorithmes qui fait que les algorithmes sont punis plus que les humains qui font des erreurs (Dietvorst, 2016), et nous avons utilisé ces constatations dans l'étude pilote suivante pour étudier plus attentivement l'effet d'aversion pour les algorithmes.
- Nous avons constaté que dans la condition basse agentivité il se trouve un degré étonnamment élevé de responsabilité causale du *tort causé*, et ce

TABLEAU 11 Scénarios pour l'étude pilote 2

| **Résultat** ↓ | ***Aspects*** | | | | | | |
|---|---|---|---|---|---|---|---|
| | **Drone autonome** | | | | | | **HO** |
| | **Niveau 1 – basse agentivité** | **Niveau 2 – agentivité élevée + pas d'autre** | **Niveau 3 – agentivité élevée + capacité d'apprendre** | **Niveau 4 – agentivité élevée + maîtrise des procédures** | **Niveau 5 – agentivité élevée + imprévisibilité** | **Niveau 6 – agentivité élevée + tous les aspects** | **Drone actionné par l'homme** |
| Mauvais (dommages collatéraux) | 1 | 2 | 3 | 4 | 5 | 6 | 7 |
| Bon (pas de dommages collatéraux) | 8 | 9 | 10 | 11 | 12 | 13 | 14 |

phénomène existe également pour les drones actionnés par l'homme et pour les armes autonomes (Figure 16). Nous avons aussi constaté que le drone est considéré comme causant beaucoup de tort dans les conditions hautes ou basses d'agentivité, mais dans la condition basse le tort est attribué presque à un niveau égal au chef. Bien que ces constatations ne soient pas très significatives, elles sont intéressantes, et méritent d'être approfondies; nous les avons donc conservées pour l'étude pilote 2.

## 4.3 Etude pilote 2

Dans l'étude pilote 2 nous avons essayé de mieux comprendre l'effet aversion pour l'algorithme; nous avons donc testé des scénarios dans lesquels l'arme autonome pouvait tirer les leçons de ses erreurs (3 + 10), était programmé sur une grande base de données et comprenait sa mission (4 + 11), pouvait être parfois imprévisible (5 + 12), ainsi qu'un scénario qui incluait tous ces aspects (6 + 13). Nous les avons testé dans des cas de bons et de mauvais résultats, et avons ajouté un scénario pour arme autonome avec un bas niveau d'agentivité (1 + 8) pour vérifier si notre manipulation agentivité fonctionnait, ainsi qu'un scénario opérateur humain (7 + 14), dans lequel nous avons posé des questions sur l'opérateur humain et non sur le drone, à la différence de l'étude pilote 1.

Nous avons testé un total de 14 scénarios et avions prévu 50 répondants par scénario, c'est-à-dire 700 réponses au total. Le nombre de répondants par scénario est compris entre 47 et 54. Après le prétraitement des données et avoir éliminé les répondants qui ont échoué au test d'attention, nous avons obtenu 709 réponses complètes.

### 4.3.1 *Analyse fiabilité*

Pour les quatre éléments d'agentivité, $\alpha = 0.914$, ce qui est bien au-dessus du seuil de 0.7 et ne peut être amélioré en éliminant un élément. L'élimination d'autres éléments diminuera la valeur $\alpha$ et réduira la cohérence interne.

### 4.3.2 *Analyse composante principale* (ACP)

L'ACP montre que la *réflexion, détermination des objectifs, liberté d'action* et *réalisation des objectifs* peuvent être considérées comme un seul ensemble qui donne 79.60 % de variation dans les variables originales (Tableau 13). Ceci se constate aussi dans le diagramme d'éboulis qui montre les valeurs propres des composantes (Figure 17). L'ACP montre que les quatre éléments *réflexion,*

TABLEAU 12 Résultats analyse fiabilité

| Reliability statistics | | |
|---|---|---|
| Cronbach's Alpha | Cronbach's Alpha based on standardized items | N of items |
| .914 | .914 | 4 |

**Item-Total statistics**

| | Scale mean if item deleted | Scale variance if item deleted | Corrected Item-Total correlation | Squared multiple correlation | Cronbach's Alpha if item deleted |
|---|---|---|---|---|---|
| Thought | 208.4827 | 6427.511 | .801 | .647 | .890 |
| Goal_setting | 204.9885 | 6499.803 | .834 | .695 | .877 |
| Act_freely | 203.4020 | 6782.379 | .804 | .646 | .888 |
| Achieve_goals | 199.4975 | 7114.019 | .781 | .614 | .897 |

TABLEAU 13 Résultats ACP étude pilote 2

| | Component matrix[a] | |
|---|---|---|
| | Component | |
| | 1 | 2 |
| Goal_setting | .910 | |
| Act_freely | .892 | |
| Thought | .889 | |
| Achieve_goals | .877 | .446 |

Extraction Method: Principal Component Analysis.
[a] 2 components extracted.

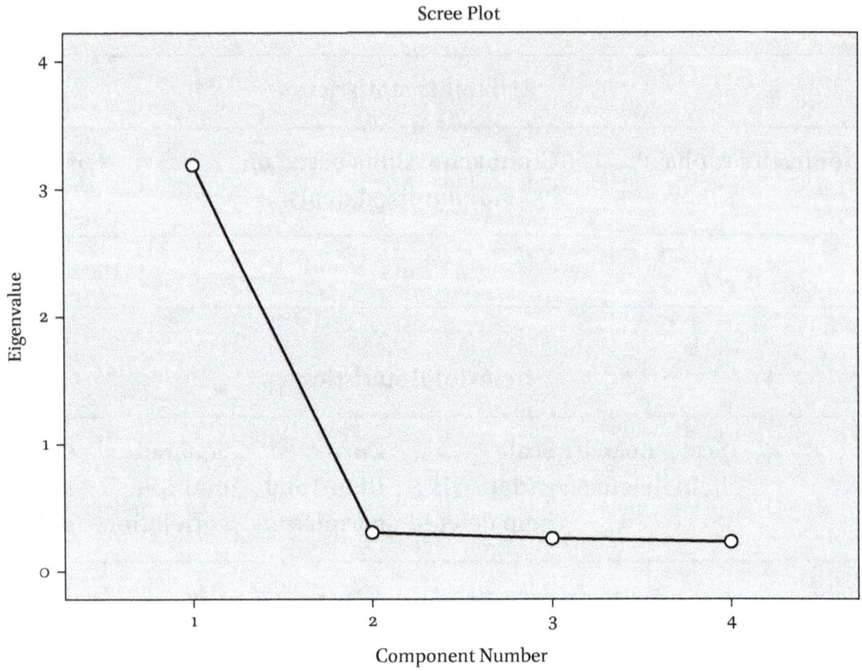


FIGURE 17 Diagramme d'éboulis ACP étude pilote 2

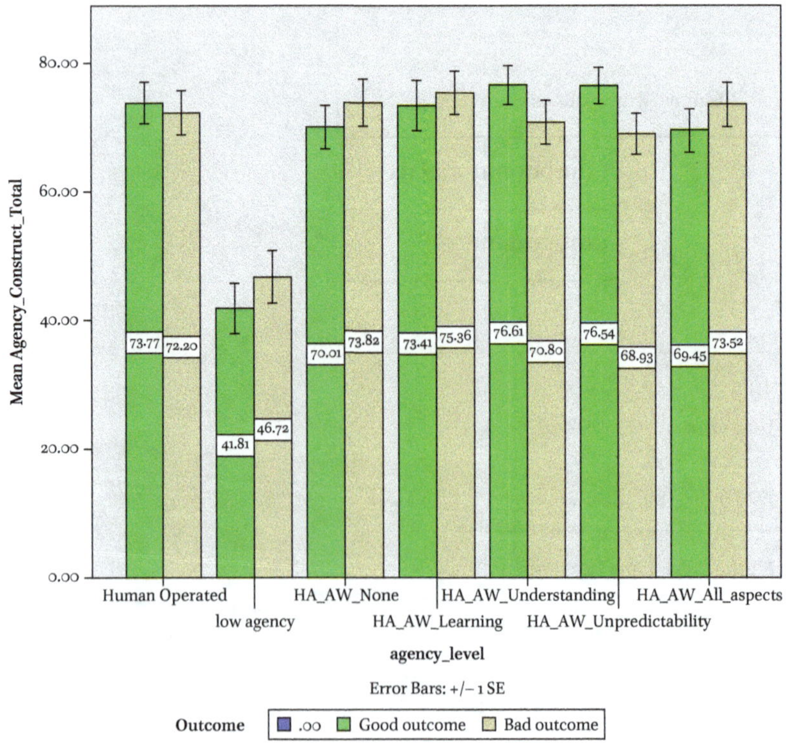


FIGURE 18 Valeur moyenne concept agentivité par condition par résultat

*détermination des objectifs*, *liberté d'action* et *réalisation des objectifs* peuvent être considérées comme les composantes qui sous-tendent le concept d'agentivité (Tableau 13).

### 4.3.3 *Analyse de corrélation*

La corrélation entre le concept d'agentivité et les variables dépendantes n'est pas grande, mais significative pour tous les ensembles ($p < .05$) (Tableau 14, page 492). Les mêmes effets concernant les variables «tort causé» et «faute / blâme» comme pour l'étude pilote 1 peuvent être observés ici. Il existe une relation négative entre le concept d'agentivité et le niveau d'inquiétude: lorsque la perception de l'agentivité augmente, les personnes indiquent qu'elles ressentent plus d'anxiété, mais cet effet est de faible amplitude ($r = -.076$). L'analyse de corrélation n'est pas assez détaillée pour s'attarder sur ces résultats; par conséquent ceci sera fait dans l'analyse des dépendantes variables.

### 4.3.4 *Vérification manipulation sur l'agentivité*

Le graphique de la Figure 18 montre qu'il y a une grande différence entre les scénarios de basse et de haute agentivité, et cette différence est significative. La différence confirme que notre manipulation sur l'agentivité fonctionne. L'agentivité pour l'opérateur humain et les drones d'agentivité élevée sont au même niveau (dans la présente étude nous avons posé des questions spécifiques sur l'opérateur humain, alors que c'était sur le drone dans la précédente enquête). Il y a une différence entre les conditions bons résultats et mauvais résultats, et bien que ces effets soient plus sensibles pour les conditions dans lesquelles nous avons porté notre attention sur l'aversion envers l'algorithme, la différence n'est pas très grande (en moyenne de 8 points sur une échelle de 100). Également, les barres d'erreur entre le bon résultat et le mauvais résultat pour tous les scénarios se chevauchent ; ceci indique que ce ne sont pas des ensembles distincts. La seule différence évidente et significative dans des ensembles ($p < .05$) se situe entre la condition agentivité basse arme autonome et les autres scénarios. Ceci est également confirmé par le niveau d'importance des tests-T d'échantillons indépendants, que nous avons décidé de ne pas présenter sur le graphique ci-dessous par souci de clarté.

TABLEAU 14 Matrice de corrélation entre concept agentivité et variables dépendantes étude pilote 2

**Correlations**

| | | **Agency_ construct_ total** | **DVD1_ Blame** | **DVD2_ Commander Blame** | **DVD3_ Trust** | **DVD4_ Commander Trust** |
|---|---|---|---|---|---|---|
| Agency_ Construct_ Total | Person Correlation | 1 | .271** | -.267** | .205** | .217** |
| | Sig(2-tailed) | | .000 | .000 | .000 | .000 |
| | N | 708 | 708 | 708 | 708 | 708 |
| DVD1_Blame | Person Correlation | .271 | 1 | -.069 | -.028 | -.046 |
| | Sig(2-tailed) | .000 | | .045 | .464 | .226 |
| | N | 708 | 708 | 708 | 708 | 708 |
| DVD2_ Commander Blame | Person Correlation | -.267** | -.069 | 1 | -.287** | -.278** |
| | Sig(2-tailed) | .000 | .065 | | .000 | .000 |
| | N | 708 | 708 | 708 | 708 | 708 |
| DVD3_Trust | Person Correlation | .205** | -.028 | -.207** | 1 | .653** |
| | Sig(2-tailed) | .000 | .464 | .000 | | .000 |
| | N | 708 | 708 | 708 | 708 | 708 |
| DVD4_ Commander Trust | Person Correlation | .217** | -.046 | -.278** | .653** | 1 |
| | Sig(2-tailed) | .000 | .226 | .000 | .000 | |
| | N | 708 | 708 | 708 | 708 | 708 |
| DVD5_Harm | Person Correlation | .124** | .258** | .152** | -.281** | -.185** |
| | Sig(2-tailed) | .001 | .000 | .000 | .000 | .000 |
| | N | 708 | 708 | 708 | 708 | 708 |
| DVD6_ Commander Hram | Person Correlation | -.198** | -.023 | .691** | -.224** | -.228** |
| | Sig(2-tailed) | .000 | .534 | .000 | .000 | .000 |
| | N | 708 | 708 | 708 | 708 | 708 |
| DVD7_Dignity | Person Correlation | .102** | -.056 | -.206** | .595** | .432** |
| | Sig(2-tailed) | .007 | .138 | .000 | .000 | .000 |
| | N | 708 | 708 | 708 | 708 | 708 |

| DVD5_ Harm | DVD6_ Commander Hram | DVD7_ Dignity | DVD8_ confidence | DVD9_ Expectations | DVD10_ Support | DVD11_ Fair | DVD12_ Uneasy |
|---|---|---|---|---|---|---|---|
| .124** | -.198** | .102** | .209** | .146** | .164** | .142** | -.076** |
| .001 | .000 | .007 | .000 | .000 | .000 | .000 | .043 |
| 708 | 708 | 708 | 708 | 708 | 708 | 708 | 708 |
| .258** | -.023 | -.056 | -.048 | -.137** | -.118** | -.092* | .180** |
| .000 | .524 | -.138 | .198 | .000 | .002 | .014 | .000 |
| 708 | 708 | 708 | 708 | 708 | 708 | 708 | 708 |
| .152** | .691** | -.206** | -.300** | -.205** | -.250** | -.275** | .312** |
| .000 | .000 | .000 | .000 | .000 | .000 | .000 | .000 |
| 708 | 708 | 708 | 708 | 708 | 708 | 708 | 708 |
| -.281** | -.224** | .595** | .901** | .533** | .726** | .722** | -.616* |
| .000 | .000 | .000 | .000 | .000 | .000 | .000 | .000 |
| 708 | 708 | 708 | 708 | 708 | 708 | 708 | 708 |
| -.185** | -.228** | .432** | .635** | .425** | .631** | .552** | -.458** |
| .000 | .000 | .000 | .000 | .000 | .000 | .000 | .000 |
| 708 | 708 | 708 | 708 | 708 | 708 | 708 | 708 |
| 1 | .311** | -.374** | -.301** | -.217** | -.253** | -.324** | .395** |
| .000 | .000 | .000 | .000 | .000 | .000 | .000 | .000 |
| 708 | 708 | 708 | 708 | 708 | 708 | 708 | 708 |
| .311** | 1 | -.183** | -.226** | -.203** | -.235** | -.249** | .289** |
| .000 | | .000 | .000 | .000 | .000 | .000 | .000 |
| 708 | 708 | 708 | 708 | 708 | 708 | 708 | 708 |
| -.374** | -.183** | 1 | .588** | .415** | .533** | .598** | -.509** |
| .000 | .000 | | .000 | .000 | .000 | .000 | .000 |
| 708 | 708 | 708 | 708 | 708 | 708 | 708 | 708 |

TABLEAU 14 Matrice de corrélation entre concept agentivité et variables dépendantes étude pilote 2 (*cont.*)

| Correlations | | | | | | |
|---|---|---|---|---|---|---|
| | | **Agency_ construct_ total** | **DVD1_ Blame** | **DVD2_ Commander Blame** | **DVD3_ Trust** | **DVD4_ Commander Trust** |
| DVD8_ confidence | Person Correlation | .209** | -.048 | -.300** | .901** | .635** |
| | Sig(2-tailed) | .000 | .198 | .000 | .000 | .000 |
| | N | 708 | 708 | 708 | 708 | 708 |
| DVD9_ Expectations | Person Correlation | .146** | -.137** | -.205** | .533** | .425** |
| | Sig(2-tailed) | .000 | .000 | .000 | .000 | .000 |
| | N | 708 | 708 | 708 | 708 | 708 |
| DVD10_ Support | Person Correlation | .164** | -.118** | -.250** | .726** | .631** |
| | Sig(2-tailed) | .000 | .002 | .000 | .000 | .000 |
| | N | 708 | 708 | 708 | 708 | 708 |
| DVD11_Fair | Person Correlation | .142** | -.092* | -.275** | .722** | .552** |
| | Sig(2-tailed) | .000 | .014 | .000 | .000 | .000 |
| | N | 708 | 708 | 708 | 708 | 708 |
| DVD12_ Uneasy | Person Correlation | -.076* | .190** | .312** | -.616** | -.458** |
| | Sig(2-tailed) | .043 | .000 | .000 | .000 | .000 |
| | N | 708 | 708 | 708 | 708 | 708 |

** Correlation is significant at the 0.01 level (2-tailed).
* Correlation is significant at the 0.05 level (2-tailed).

| DVD5_ Harm | DVD6_ Commander Hram | DVD7_ Dignity | DVD8_ confidence | DVD9_ Expectations | DVD10_ Support | DVD11_ Fair | DVD12_ Uneasy |
|---|---|---|---|---|---|---|---|
| -.301** | -.226** | .588** | 1 | .553** | .723** | .725** | -.649** |
| .000 | .000 | .000 | | .000 | .000 | .000 | .000 |
| 708 | 708 | 708 | 708 | 708 | 708 | 708 | 708 |
| -.217** | -.203** | .415** | .553** | 1 | .511** | .621** | -.509** |
| .000 | .000 | .000 | .00 | | .000 | .000 | .000 |
| 708 | 708 | 708 | 708 | 708 | 708 | 708 | 708 |
| -.253** | -.235** | .533** | .723** | .511** | 1 | .698** | -.651** |
| .000 | .000 | .000 | .000 | .000 | | .000 | .000 |
| 708 | 708 | 708 | 708 | 708 | 708 | 708 | 708 |
| -.324** | -.248** | .598** | .725** | .621** | .685** | 1 | -.656** |
| .000 | .000 | .000 | .000 | .000 | .000 | | .000 |
| 708 | 708 | 708 | 708 | 708 | 708 | 708 | 708 |
| .395** | .289** | -.509** | -.649** | -.509** | -.651** | -.636** | 1 |
| .000 | .000 | .000 | .000 | .000 | .000 | .000 | |
| 708 | 708 | 708 | 708 | 708 | 708 | 708 | 708 |

### 4.3.5 *Analyse variables dépendantes*

La corrélation entre le concept d'agentivité et les variables dépendantes est partout significative (p <.05); donc les constatations les plus frappantes pour chacune des variables dépendantes sont décrites ci-dessous.

#### 4.3.5.1 Faute / blâme

La constatation la plus remarquable pour «faute» est que les personnes attribuent moins de blâme dans la condition de basse agentivité si l'on compare avec les conditions d'agentivité élevée pour les armes autonomes et opérateur humain. Pour ces conditions, la différence est importante (p < .05). Quelques effets de l'algorithme peuvent être constatés dans le scénario d'apprentissage et de compréhension, où les personnes attribuent moins de blâme lorsque l'arme autonome ne fait pas d'erreur, et plus de blâme lorsqu'elle fait une erreur avec un mauvais résultat. Cependant, ces conditions montrent un chevauchement des barres d'erreur, et ne sont pas significatives.

#### 4.3.5.2 Faute du chef

L'observation la plus intéressante dans la variable *faute du chef* (*blâme*) et que la faute, est attribuée au chef beaucoup moins lorsque l'arme autonome fait

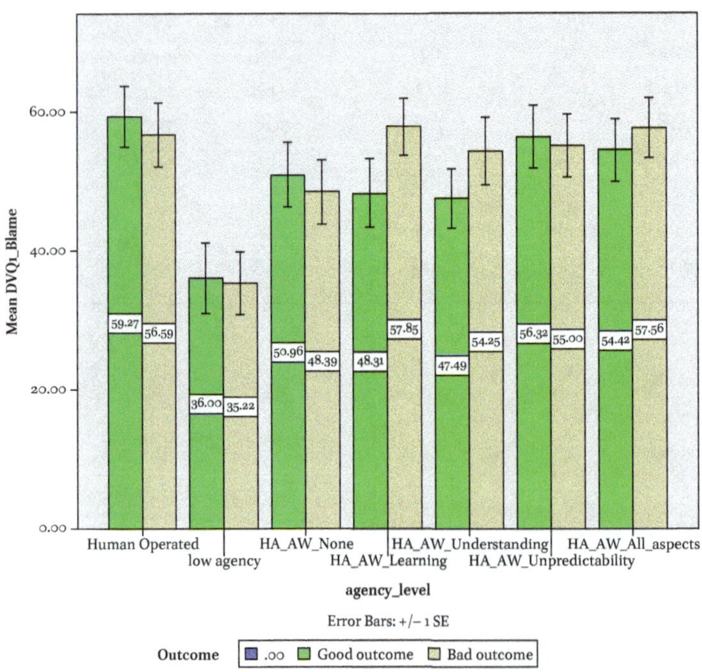


FIGURE 19 Valeur moyenne variable blâme par condition par résultat

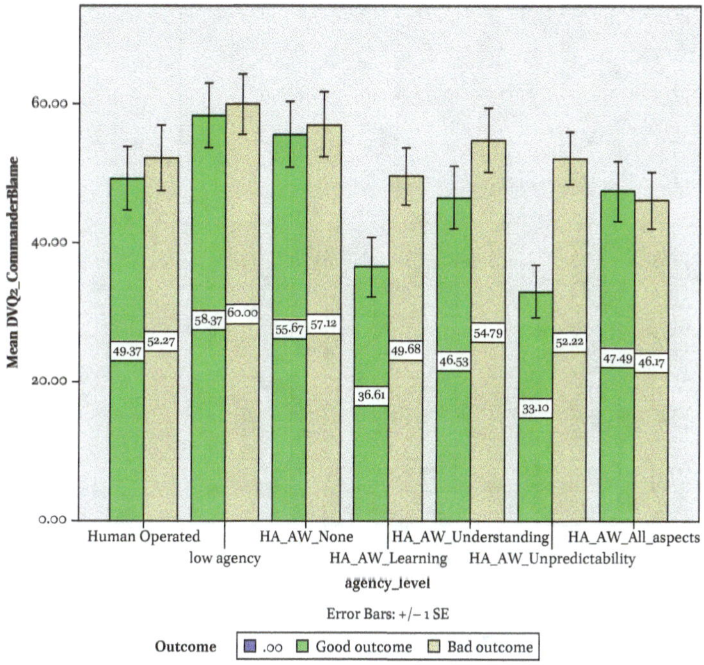


FIGURE 20 Valeur moyenne variable faute du chef par condition par résultat

une erreur dans les conditions «tirer les leçons» et «imprévisibilité», et ceci est significatif ($p < .05$). Il pourrait arriver que la faute soit attribuée au drone puisque les personnes s'attendent à ce qu'il soit capable de tirer les leçons de ses erreurs ou qu'il se conduise de manière imprévisible, et qu'il ne faut pas accuser *le chef*. Également, dans la condition basse agentivité le *chef est accusé* un peu plus que dans les autres conditions.

#### 4.3.5.3 Confiance

Quelle que soit la condition on peut constater que les personnes font davantage confiance à l'arme autonome ou à l'opérateur humain pour prendre les mesures adéquates après un bon résultat qu'après un mauvais résultat. Sauf dans la condition de basse agentivité, la différence de confiance dans le cas de bon résultat et dans celui de mauvais résultat est significative ($p < .05$). Les répondants font confiance à l'opérateur humain plus qu'à l'arme autonome, en particulier dans la condition de basse agentivité, et la différence est significative ($p < .05$).

#### 4.3.5.4 Confiance en le chef

Globalement les répondants font confiance au chef pour prendre les mesures adéquates dans les conditions «bon résultat», en comparaison des conditions

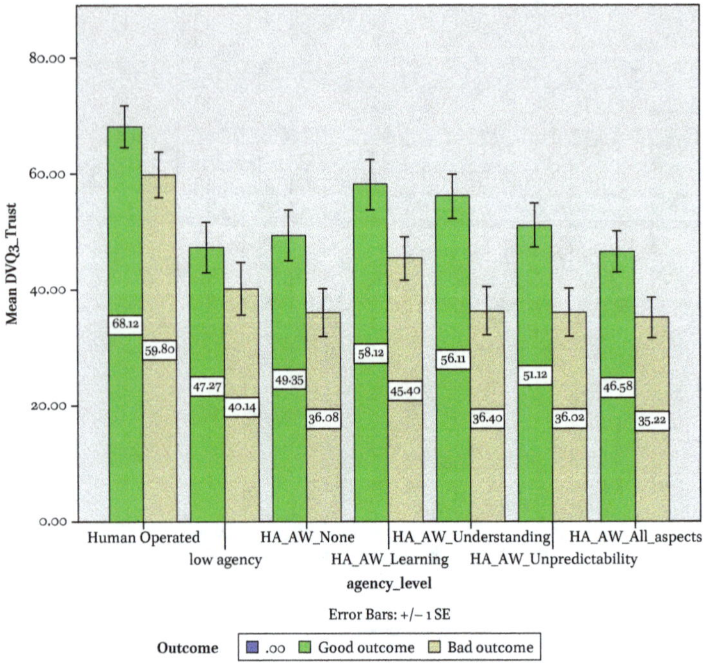


FIGURE 21 Valeur moyenne variable confiance par condition par résultat

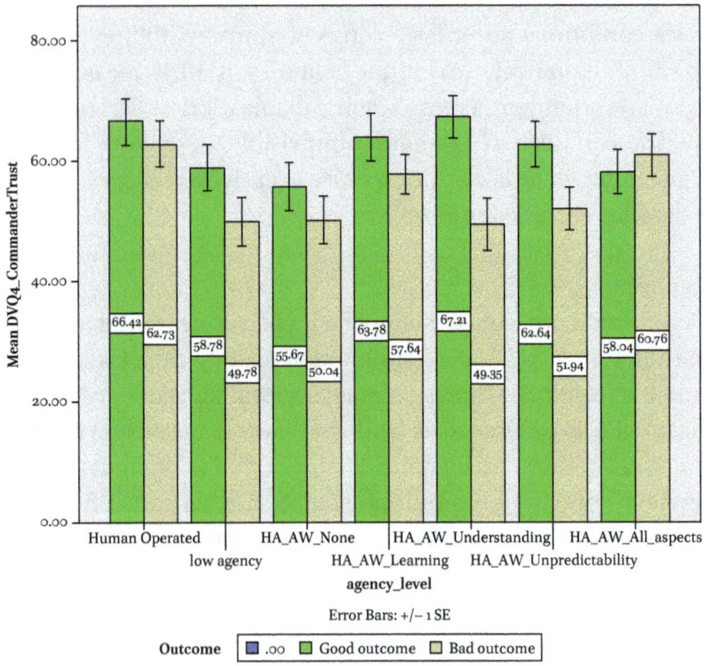


FIGURE 22 Valeur moyenne variable confiance accordée au chef par condition par résultat

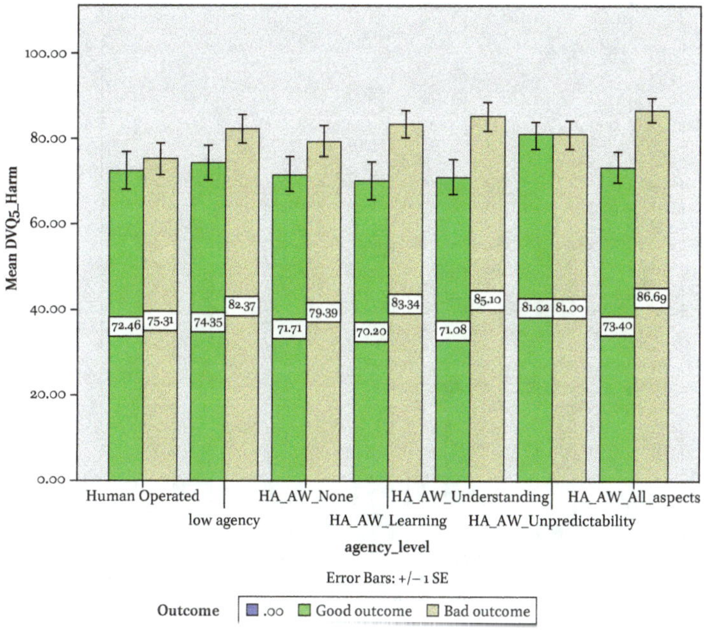


FIGURE 23 Valeur moyenne variable «tort causé» par condition par résultat

«mauvais résultat». La confiance accordée au chef diffère dans les conditions de basse agentivité et de haute agentivité, sans autre information, et la différence est significative ($p < .05$). Savoir que l'arme autonome a des capacités de compréhension entraîne un plus haut degré de confiance dans les conditions «mauvais résultat», et ceci est significatif ($p < .05$).

#### 4.3.5.5 Tort causé

Dans les conditions «mauvais résultat», les répondants perçoivent plus de tort causé que dans les conditions «bons résultats», ce qui n'est pas étonnant, et la différence est significative dans presque toutes les conditions ($p < .05$). Le plus intéressant dans ce graphique est que lorsque l'arme autonome se conduit de manière imprévisible, le type de résultat paraît ne pas compter, et la perception du «tort causé» est la même; mais la différence entre les deux ensembles «bons ou mauvais résultats» n'est pas significative.

#### 4.3.5.6 Tort causé par le chef

Dans les conditions «mauvais résultat» les répondants perçoivent que le chef cause plus de tort que dans les conditions «bon résultat», ce qui n'est pas étonnant. Plus intéressante est la grande différence dans la condition «apprentissage» entre le bon et le mauvais résultat où on attribue au chef beaucoup moins de tort

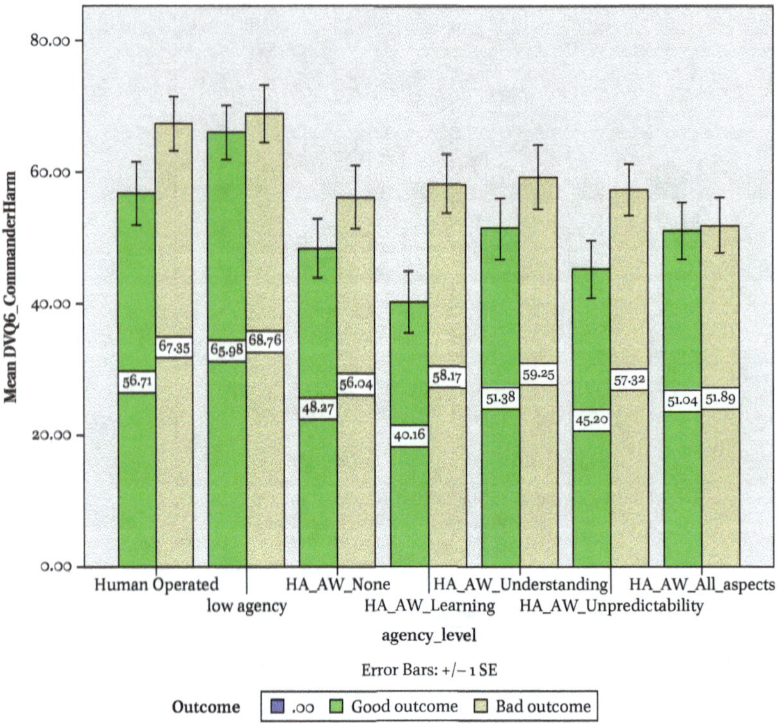


FIGURE 24 Valeur moyenne variable «tort causé par le chef» par condition par résultat

dans la condition «bon résultat», ce qui est significatif (p < .05). La différence entre les conditions «bon» et «mauvais» résultats dans les scénarios opérateur humain et imprévisibilité agentivité élevée est également significative (p < .05).

#### 4.3.5.7 Dignité humaine

L'opérateur humain agit avec beaucoup plus de respect envers la dignité humaine que l'arme autonome, sauf dans les conditions dans lesquelles l'arme autonome est programmée sur un large éventail de données pour sa mission, et cette différence est significative (p < .05). Également, le fait que l'arme autonome fasse une erreur est important, car nous observons des différences importantes (p < .05) entre les scénarios «bon» et «mauvais» résultats pour les scénarios armes autonomes en agentivité basse, aucun des aspects en agentivité élevée et tous les aspects en agentivité haute.

#### 4.3.5.8 Assurance

Globalement les répondants sont davantage assurés que l'arme autonome prendra les mesures adéquates après un bon résultat qu'après un mauvais

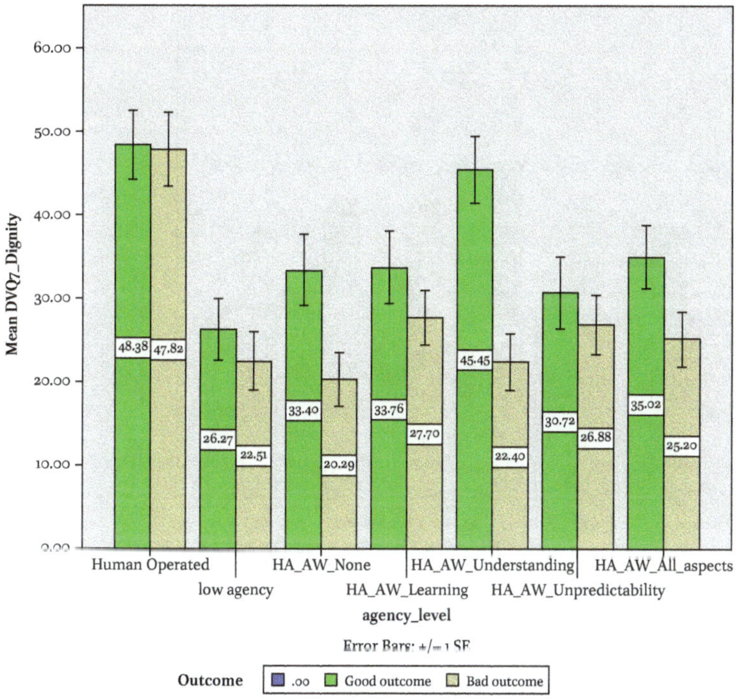


FIGURE 25 Valeur moyenne – variable «dignité humaine» par condition par résultat

résultat, et toutes les différences sont significatives ($p < .05$). Ils ont également un peu plus confiance en l'opérateur humain qu'en l'arme autonome. Les différences entre ces scénarios sont significatives ($p < .05$). Lorsque l'arme autonome a des capacités d'apprentissage, leur assurance est un peu plus haute si l'on compare aux autres conditions pour les armes autonomes.

### 4.3.5.9 Attentes

Dans les scénarios «bons résultats», l'opérateur humain et l'arme autonome agissent tous deux davantage dans le sens des attentes des répondants que dans les scénarios «mauvais résultats». Toutes les différences entre les conditions «bons résultats» et «mauvais résultats» sont significatives ($p < .05$). Les attentes sont plus basses dans la condition basse agentivité lorsqu'il y a mauvais résultat, mais il n'y a pas beaucoup de différence si l'on compare avec les autres conditions.

### 4.3.5.10 Approbation / soutien

Les répondants manifestent plus d'approbation envers les drones actionnés par les humains que pour les armes autonomes. L'approbation est la moins

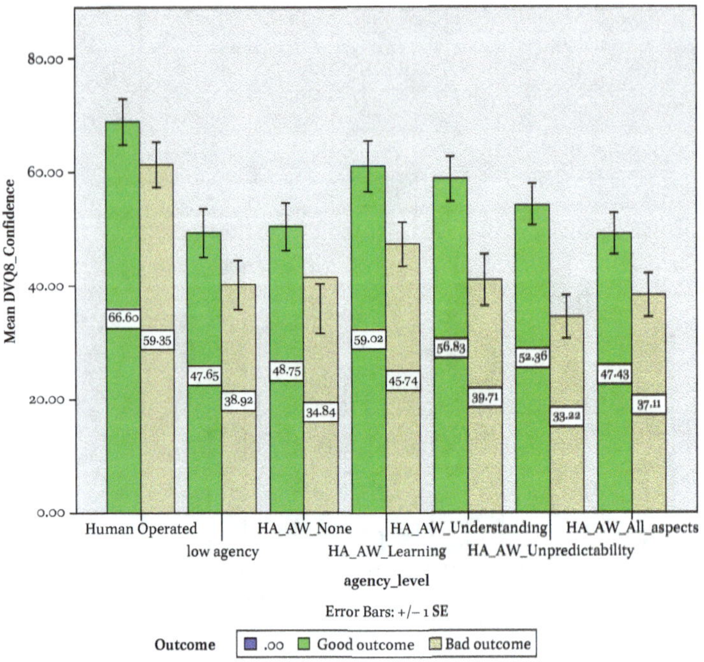


FIGURE 26 valeur moyenne variable «assurance» par condition par résultat

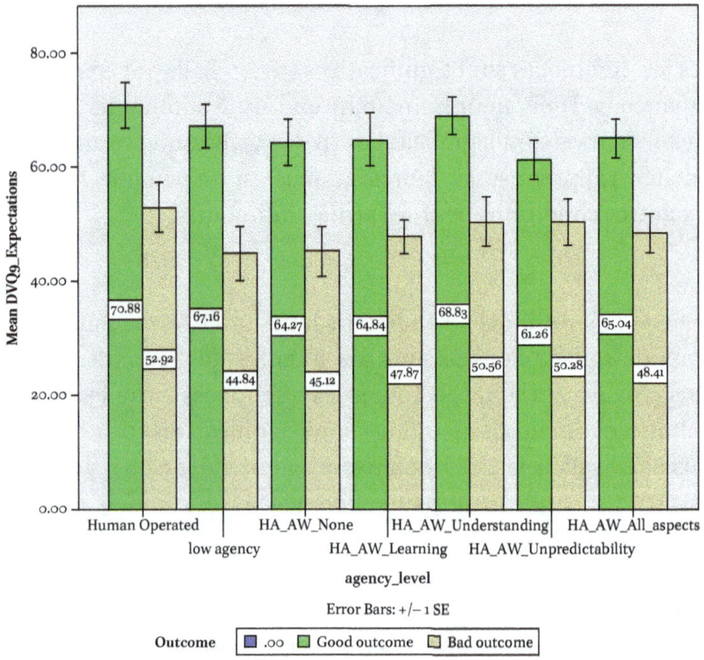


FIGURE 27 Valeur moyenne variable « attentes » par condition par résultat

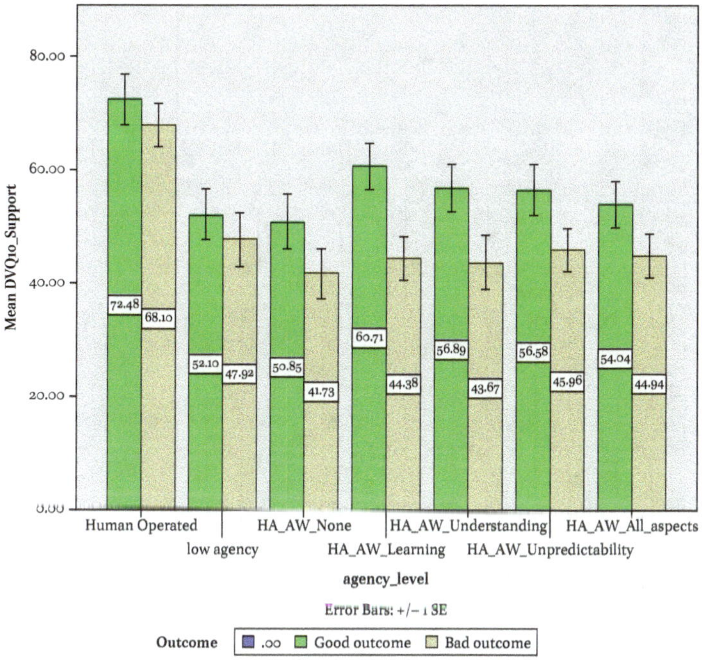


FIGURE 28 Valeur moyenne variable «approbation» par condition par résultat

élevée, mais de manière non significative, pour les conditions de moindre agentivité si l'on compare avec les scénarios dans lesquels les informations sur l'arme autonome sont fournies. Également, l'approbation est plus haute pour les scénarios avec «bon» résultat que pour ceux ayant un «mauvais» résultat; sauf pour la condition «activé par l'homme» et «arme autonome basse agentivité», ces différences sont significatives ($p < .05$).

### 4.3.5.11 Equité

Globalement, les actions des opérateurs humains sont considérées comme étant plus équitables que celles des armes autonomes. La perception de l'équité est beaucoup plus haute dans les scénarios à bons résultats que dans les scénarios à mauvais résultats, et dans tous les scénarios ces différences sont significatives ($p < .05$). La différence entre la condition d'agentivité des armes autonomes n'est pas importante (< 10 points sur une échelle de 100).

### 4.3.5.12 Inquiétude / anxiété

Dans les scénarios à mauvais résultats les répondants ressentent plus d'inquiétude concernant les actions de l'opérateur humain aussi bien que celles de l'arme autonome. Il n'y a pas grande différence entre les niveaux d'inquiétude

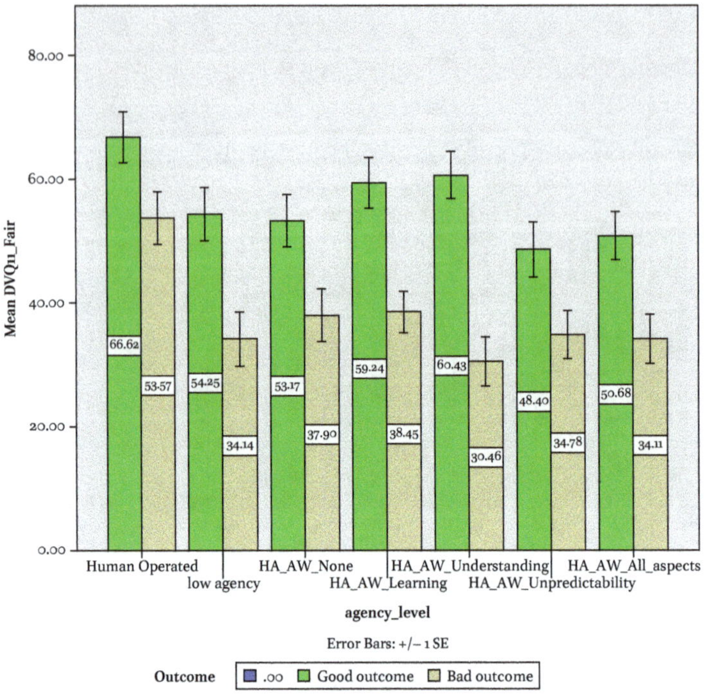


FIGURE 29 Valeur moyenne variable équité par condition par résultat

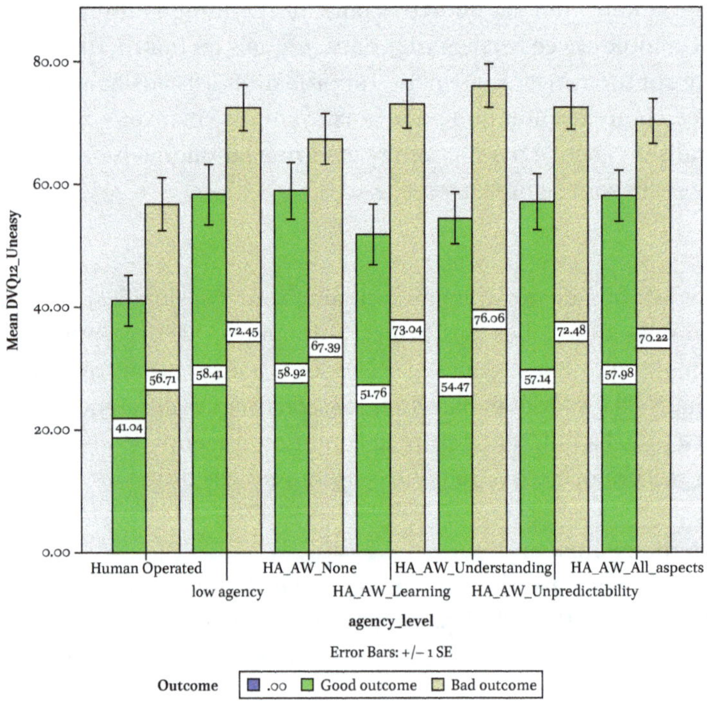


FIGURE 30 Valeur moyenne variable inquiétude par condition par résultat

pour les différentes conditions des armes autonomes. Le niveau d'inquiétude dans le cas de l'opérateur humain est beaucoup plus bas que celui ayant trait à l'arme autonome. Sauf pour la condition «agentivité élevée pas d'autres aspects armes autonomes», toutes les différences entre les scénarios bons et mauvais résultats sont significatives ($p < .05$).

### 4.3.6 *Conclusion étude pilote 2*

Les résultats de l'étude pilote 2 conduisent aux conclusions suivantes:

- Le test fiabilité a démontré que les quatre éléments du concept d'agentivité sont fiables, et selon l'ACP peuvent être considérés comme une seule composante;
- La corrélation entre le coefficient du concept agentivité et toutes les variables dépendantes est significative et vont dans la même direction que pour l'étude pilote 1. Nous avons constaté un rapport négatif entre le concept agentivité et le niveau d'inquiétude, ce qui signifie que lorsque la perception de l'agentivité augmente, les répondants indiquent qu'ils ressentent plus d'inquiétude, mais cet effet reste très faible et approche de 0 ($r = -.076$). Afin de voir si nous pouvions obtenir un effet plus important, nous avons changé la formulation de la question de la variable «inquiétude» dans l'étude finale;
- La manipulation de l'agentivité montre qu'il y a une grande différence entre les scénarios haute agentivité et le scénario basse agentivité. L'agentivité de l'opérateur humain et la haute agentivité des armes autonomes sont au même niveau. Cependant, les moyennes du concept d'agentivité pour les scénarios à haute agentivité qui incluent l'effet aversion envers l'algorithme sont presque identiques. Pour préciser l'effet aversion envers l'algorithme il nous faut étudier davantage les publications existantes et effectuer plus d'études pilotes. Du fait des contraintes de temps imparti, nous n'avons pas poursuivi cette démarche, et avons décidé de concentrer l'étude finale sur la perception de l'agentivité.
- Un grand nombre des effets que nous avons constatés pour les variables dépendantes devaient, comme nous nous y attendions, être basés sur les résultats de l'étude pilote1. Nous n'avons pas observé d'effets distincts dans les VD entre les différentes conditions d'agentivité des armes autonomes. Nous avons vérifié pour voir si nous pouvions calquer l'effet aversion pour l'algorithme de l'étude pilote1, dans laquelle le «tort causé» passait du drone au chef dans la condition de basse agentivité, mais malheureusement cet effet n'a pas pu être observé dans l'étude pilote2. Même s'il serait bon d'étudier d'un peu plus près cette chaîne de responsabilité, nous avons décidé de ne pas l'inclure dans l'étude finale, puisque nous n'étions pas vraiment

sûrs de ce qui se passe; il faudrait étudier de plus près les publications et procéder à d'autres études pilote.

## 4.4 Etude finale – échantillon militaire

Afin de déterminer le nombre de scénarios pour l'étude finale, nous avons procédé à des calculs Power pour évaluer le nombre total de participants dont nous aurions besoin en nous basant sur les résultats de la première étude pilote, car nous utiliserons ces scénarios également pour l'étude finale. En nous fondant sur des calculs de puissance (ampleur d'effet 0.4, ampleur statistique souhaitée de 0.8 et niveau de probabilité de 0.05), nous avons compté sur un total de 200 réponses, et déterminé que nous pouvions donner 3 scénarios. Nous avons décidé de privilégier la perception de l'agentivité et de comparer la technologie actuelle, un drone actionné par l'homme, avec la technologie du futur, c'est-à-dire l'arme autonome à haute agentivité (Tableau 15). Nous avons également ajouté un scénario avec arme autonome à agentivité neutre, pour comparer la perception de l'agentivité dans les trois conditions. Dans cette étude, nous avons posé des questions spécifiques sur le drone et non sur l'opérateur humain qui le contrôle pour comparer avec la perception de l'agentivité des artefacts. Comme nous ne pouvions travailler que sur trois scénarios, nous avons privilégié la condition «mauvais résultat», car les études pilotes montraient les effets les plus distincts de cette condition.

Nous avons diffusé l'enquête par email avec un lien anonyme et donc n'avions pas de garantie quant au nombre des répondants. Après le prétraitement des données et après avoir éliminé les répondants qui avaient échoué au test d'attention, il restait 239 réponses pour l'échantillon complet. Le nombre de répondants par scénario s'élevait entre 64 et 96, ce qui est curieux car les répondants auraient dû être également répartis entre les scénarios. L'échantillon de 239 réponses est un mélange de personnels militaires et civils néerlandais travaillant pour le Ministère de la Défense, car si nous avions exclu les civils seuls restaient 149 répondants, ce qui était trop réduit pour une analyse de données crédible.

Pour ce tableau, les définitions suivantes sont utilisées :

- *Drone actionné par l'homme:* drone directement contrôlé par un être humain. Par hypothèse, perçu comme ayant une très basse agentivité.
- *Arme autonome à haute agentivité :* on dit aux participants que l'arme autonome possède des caractéristiques d'agentivité (comme peser le pour et le contre et faire des choix de manière indépendante). Par hypothèse, perçu comme possédant une très haute agentivité.

TABLEAU 15 Etude finale scénarios

| | **Niveaux d'agentivité** | | |
|---|---|---|---|
| | **Actionné par l'homme** | **Agentivité neutre arme autonome** | **Agentivité élevée arme autonome** |
| Mauvais résultat | 1 | 2 | 3 |

– *Arme autonome, agentivité neutre:* aucune information au sujet des caractéristiques d'agentivité de l'arme autonome n'est indiquée. La force de cette approche est que nous pouvons recueillir des données sur la manière dont les participants utilisent leurs idées préconçues sur les armes autonomes pour comprendre les actions de l'arme autonome, et comparer ces idées aux conditions drone actionné par l'homme et arme autonome à haute agentivité.

### 4.4.1 *Analyse fiabilité*

Pour les quatre éléments agentivité $\alpha = .865$, ce qui est au-dessus du seuil de 0.7 et ne peut être amélioré en éliminant un élément, ce qui signifie que le concept agentivité est fiable (Tableau 16). L'élimination d'autres éléments diminue la valeur $\alpha$ et réduit la cohérence interne.

### 4.4.2 *Analyse composante principale (ACP)*

L'APC montre que la *Réflexion, la détermination des objectifs, la liberté d'action* et *la réalisation des objectifs* peuvent être considérés comme un seul ensemble qui constitue 71.18 % des variations dans les variables d'origine. Ceci peut également être constaté dans le diagramme d'éboulis (scree plot) qui montre les valeurs propres des composantes (Figure 31). L'APC montre que les quatre éléments *réflexion, détermination des objectifs, liberté d'action* et *réalisation des objectifs* peuvent être considérés comme une seule composante du concept d'agentivité (Tableau 17).

### 4.4.3 *Analyse de corrélation concept d'agentivité*

La corrélation entre le concept d'agentivité et les variables dépendantes est plus grande que dans les études pilotes précédentes, mais significative pour seulement 7 des 9 variables ; *confiance, dignité humaine, assurance, attentes, approbation, équité et inquiétude* ($p < .01$) (Tableau 18). Il existe un rapport négatif entre le concept d'agentivité et le degré d'inquiétude qui est plus haut

TABLEAU 16 analyse fiabilité éléments d'agentivité

**Reliability statistics**

| Cronbach's Alpha | N of items |
|---|---|
| .865 | 4 |

**Item-Total statistics**

| | Scale mean if item deleted | Scale variance if item deleted | Corrected Item-Total correlation | Cronbach's Alpha if item deleted |
|---|---|---|---|---|
| Thought | 116.43 | 7175.573 | .620 | .863 |
| Goal setting | 123.33 | 6342.474 | .756 | .810 |
| Act freely | 122.62 | 6566.723 | .740 | .816 |
| Achieve goals | 118.84 | 6459.146 | .740 | .816 |

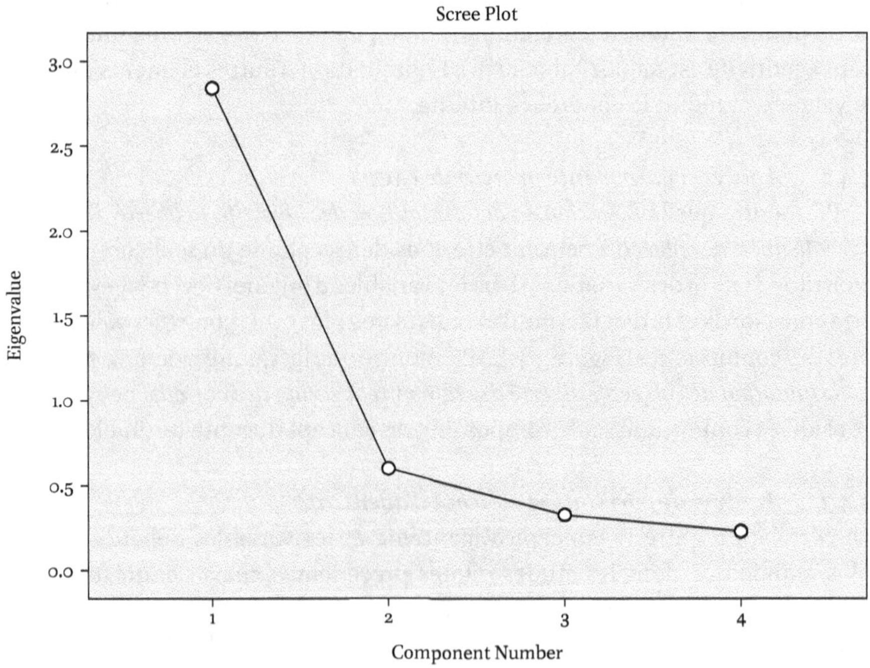


FIGURE 31 Diagramme d'éboulis APC étude finale

TABLEAU 17 résultats éléments d'agentivité étude finale

| | Component matrix[a] | |
|---|---|---|
| | Component | |
| | 1 | 2 |
| Thought | .774 | .560 |
| Goal setting | .873 | |
| Act freely | .861 | |
| Achieve goals | .863 | |

Extraction Method: Principal Component Analysis.
[a] 2 components extracted.

que précédemment dans les études pilotes. Ceci indique que lorsque la perception de l'agentivité augmente, les personnes déclarent qu'elles sont plus inquiètes. L'analyse de corrélation n'est pas assez détaillée pour étudier de près ces résultats, et donc cela sera fait dans l'analyse des variables dépendantes.

### 4.4.4 *Vérification manipulation sur l'agentivité*

Le graphique de la Figure 32 montre une augmentation de la perception de l'agentivité sur toutes les conditions. La différence dans la perception de l'agentivité entre le scénario «arme actionné par l'homme» et le scénario «arme autonome agentivité neutre» est significative, donc ces ensembles peuvent être considérés comme différents. Le test de Student (T-test) sur les échantillons indépendants pour le scénario «arme autonome agentivité neutre» et les scénarios «arme autonome agentivité élevée» donne $p = .063$, juste au-dessus du seuil de pertinence. Le T-test indépendant pour la condition «arme actionnée par l'homme» et la condition «arme autonome agentivité neutre» montre que $p = .001$, et pour les conditions «arme actionnée par l'homme» et «arme autonome agentivité élevée», $p = .000$.

Nous avons posé comme hypothèse que le drone actionné par l'homme sera perçu comme ayant une agentivité basse, et l'arme autonome à haute agentivité sera perçue comme ayant une agentivité élevée, et les résultats montrent que c'est bien le cas. L'analyse centrale s'intéressait à la manière dont l'agentivité neutre peut être comparée aux deux autres cas. Nous avons supposé que les personnels militaires ne percevraient pas les armes

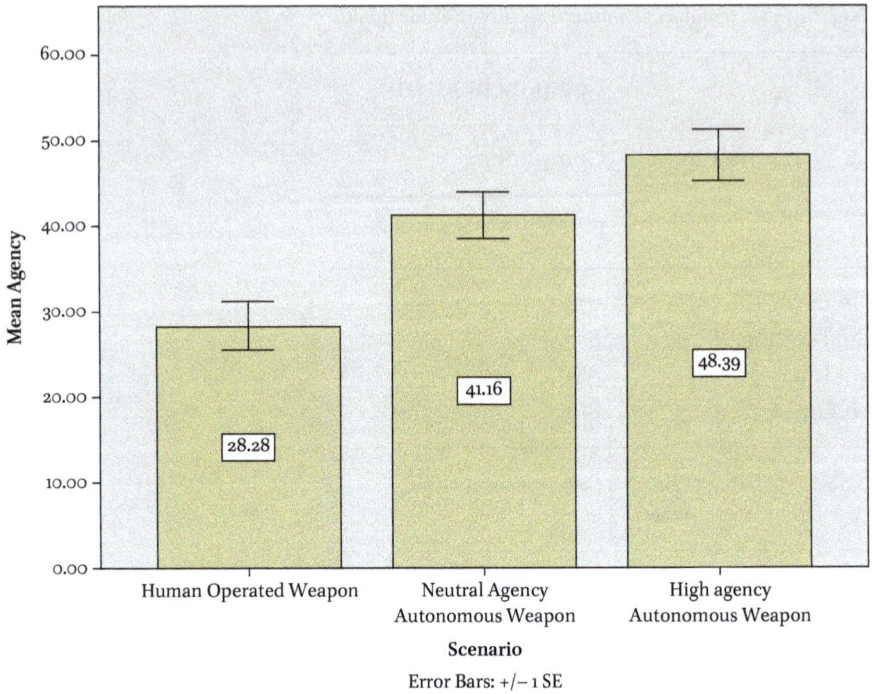


FIGURE 32 Valeur moyenne d'agentivité par condition

à agentivité neutre comme ayant des états mentaux. Par conséquent, nous nous attendions à aucune différence entre la condition «arme autonome à agentivité neutre» et la condition dans laquelle un humain actionne un drone à distance. Nous nous attendions aussi à ce que la condition «agentivité neutre» soit considérée comme étant significativement différente de la condition «haute agentivité».

Les résultats confirment que le drone actionné par l'homme est perçu comme ayant une agentivité basse, et que l'arme autonome est perçue comme ayant une agentivité élevée. Cependant, les résultats montrent qu'une arme autonome à agentivité neutre est perçue comme ayant plus d'agentivité qu'un drone opéré par l'homme, et que la perception de l'agentivité dans la condition «haute agentivité» est juste légèrement plus haute. Cela signifie que nous devons rejeter notre hypothèse, et conclure que les militaires et les civils travaillant pour le Ministère de la Défense néerlandais perçoivent les armes autonomes comme ayant des propriétés d'agentivité.

Dans la Figure 33, si nous considérons la différence entre les répondants militaires et civils, nous constatons quelque chose de très intéressant, mais comme les deux groupes ne sont pas significativement différents dans deux

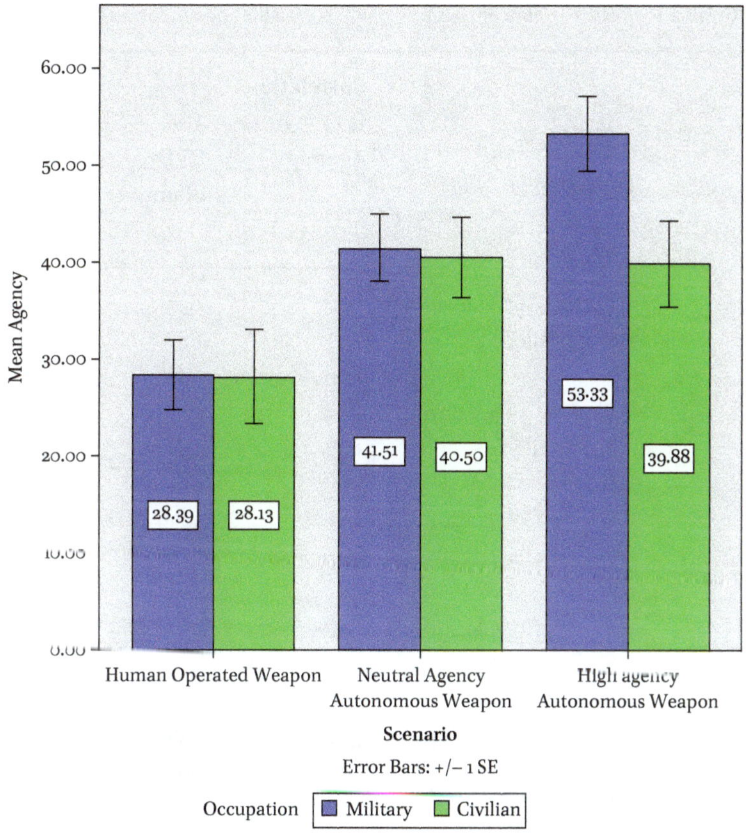


FIGURE 33 Valeur moyenne d'agentivité par condition par groupe

des scénarios, seulement dans la condition «agentivité élevée», il nous faut être prudents dans l'interprétation des résultats; l'observation qui suit n'est donc donnée qu'à titre d'illustration. Dans le scénario «arme actionnée par l'homme» ainsi que dans le scénario «agentivité neutre», la perception de l'agentivité des répondants militaires et civils est au même niveau. Cependant, dans le scénario «arme autonome à agentivité élevée», les répondants militaires perçoivent beaucoup plus d'agentivité que les civils. Nous avons regardé les différences d'ordre «démographique» entre les échantillons, et ils sont remarquablement semblables; une explication possible pourrait être que l'échantillon civil comporte 10 % de répondants de plus qui travaillent dans le domaine de l'intelligence artificielle que l'échantillon militaire. Cependant, nous ne sommes pas sûrs que ceci explique la différence dans la perception de l'agentivité entre les deux groupes de répondants, et ce point mérite d'être approfondi.

TABLEAU 18 Corrélations entre le concept agentivité et les variables dépendantes pour l'étude finale

| | | Correlations | | |
|---|---|---|---|---|
| | | Agency | DVD1_ Blame | DVD2_Trust |
| Agency | Person Correlation | 1 | .009 | .406** |
| | Sig(2-tailed) | | .888 | .000 |
| | N | 248 | 248 | 248 |
| DVD1_Blame | Person Correlation | .009 | 1 | -.039 |
| | Sig(2-tailed) | .888 | | .545 |
| | N | 248 | 248 | 248 |
| DVD2_Trust | Person Correlation | .406** | -.039 | 1 |
| | Sig(2-tailed) | .000 | .545 | |
| | N | 248 | 248 | 248 |
| DVD3_Hram | Person Correlation | -.070 | .056 | -.217** |
| | Sig(2-tailed) | .275 | .379 | .001 |
| | N | 248 | 248 | 248 |
| DVD4_Human_Dignity | Person Correlation | .347** | .018 | .482** |
| | Sig(2-tailed) | .000 | .783 | .000 |
| | N | 248 | 248 | 248 |
| DVD5_Confidence | Person Correlation | .428** | .027 | .767** |
| | Sig(2-tailed) | .000 | .667 | .000 |
| | N | 248 | 248 | 248 |
| DVD6_Expectations | Person Correlation | .320** | -.109 | .324** |
| | Sig(2-tailed) | .000 | .086 | .000 |
| | N | 248 | 248 | 248 |
| DVD7_Support | Person Correlation | .337** | -.002 | .597** |
| | Sig(2-tailed) | .000 | .979 | .000 |
| | N | 248 | 248 | 248 |
| DVD8_Fair | Person Correlation | .451** | -.080 | .540** |
| | Sig(2-tailed) | .000 | .209 | .000 |
| | N | 248 | 248 | 248 |
| DVD9_Anxiety | Person Correlation | -.216** | .128* | -.451** |
| | Sig(2-tailed) | .001 | .044 | .000 |
| | N | 248 | 248 | 248 |

** Correlation is significant at the 0.01 level (2-tailed).
* Correlation is significant at the 0.05 level (2-tailed).

| DVD3_ Hram | DVD4_ Human_ Dignity | DVD5_ Confidence | DVD6_ Expectations | DVD7_ Support | DVD8_ Fair | DVD9_ Anxiety |
|---|---|---|---|---|---|---|
| -.070 | .347** | .428** | .320** | .337** | .451** | -.216** |
| .275 | .000 | .000 | .000 | .000 | .000 | .001 |
| 248 | 248 | 248 | 248 | 248 | 248 | 248 |
| .056 | .018 | .027 | -.109 | -.002 | -.080 | .128* |
| .379 | .783 | .667 | .086 | .979 | .209 | .044 |
| 248 | 248 | 248 | 248 | 248 | 248 | 248 |
| -.217** | .482** | .767** | .324** | .597** | .540** | -.451** |
| .001 | .000 | .000 | .000 | .000 | .000 | .000 |
| 248 | 248 | 248 | 248 | 248 | 248 | 248 |
| 1 | -.180** | -.197** | .-155** | -.133** | -.200** | .360** |
| | .004 | .002 | .015 | .036 | .002 | .000 |
| 248 | 248 | 248 | 248 | 248 | 248 | 248 |
| -.180** | 1 | .564** | .239** | .564** | .472** | -.385** |
| .004 | | .000 | .000 | .000 | .000 | .000 |
| 248 | 248 | 248 | 248 | 248 | 248 | 248 |
| -.197** | .564** | 1 | .376** | .685** | .600** | -.495** |
| .002 | .000 | | .000 | .000 | .000 | .000 |
| 248 | 248 | 248 | 248 | 248 | 248 | 248 |
| -.155** | .239** | .376** | 1 | .271** | .475** | -.264** |
| .015 | .000 | .000 | | .000 | .000 | .000 |
| 248 | 248 | 248 | 248 | 248 | 248 | 248 |
| -.133** | .564** | .685** | .271** | 1 | .528** | -.488** |
| .036 | .000 | .000 | .000 | | .000 | .000 |
| 248 | 248 | 248 | 248 | 248 | 248 | 248 |
| -.200** | .472** | .600** | .475** | .528** | 1 | -.440** |
| .002 | .000 | .000 | .000 | .000 | | .000 |
| 248 | 248 | 248 | 248 | 248 | 248 | 248 |
| .360** | -.385** | -.495** | -.264** | -.488** | -.440** | 1 |
| .000 | .000 | .000 | .000 | .000 | .000 | |
| 248 | 248 | 248 | 248 | 248 | 248 | 248 |

#### 4.4.5 *Analyse variables dépendantes*

La corrélation entre le concept agentivité et les variables dépendantes *confiance, dignité humaine, assurance, attente, approbation, équité* et *inquiétude* est significative ($p < .01$) ; par conséquent les résultats les plus intéressants pour chacune des 7 variables dépendantes sont décrits dans cette section.

##### 4.4.5.1 Confiance

Globalement les personnes font plus confiance aux opérateurs humains pour qu'ils prennent les mesures adéquates qu'aux armes autonomes, et la différence est significative ($p < .05$). Le niveau de confiance pour les conditions «armes autonomes agentivité élevée» et «agentivité neutre» est égal, et la différence entre ces groupes n'est pas significative.

##### 4.4.5.2 Dignité humaine

Les armes actionnées par l'homme sont perçues comme opérant avec plus de respect pour la dignité humaine que les armes autonomes, et la différence est significative ($p < .05$). La différence entre les conditions d'agentivité neutre et élevée pour les armes autonomes est moindre, et la différence entre ces ensembles n'est pas significative.

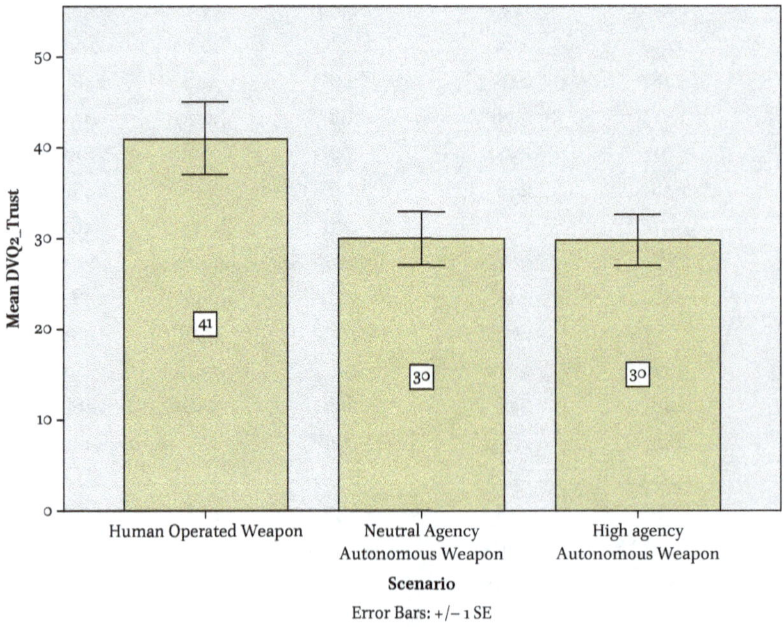


FIGURE 34 Valeur moyenne variable confiance par condition

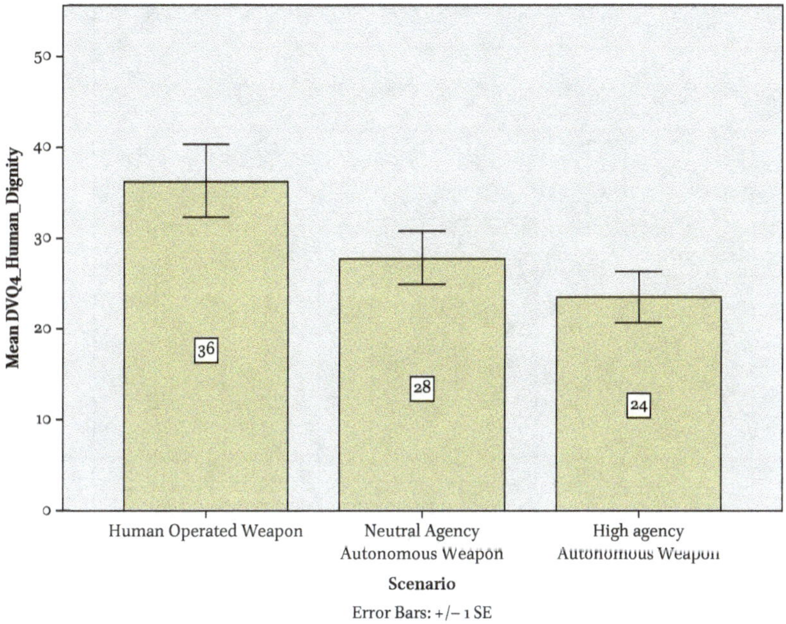


FIGURE 35 Valeur moyenne variable dignité humaine par condition

#### 4.4.5.3 Assurance

Les personnes sont plus assurées que les drones opérés par l'homme prendront les mesures adéquates après coup que lorsqu'il s'agit d'armes autonomes, et la différence est significative ($p < .05$). Il n'y a pas de différence significative entre les conditions basse et haute agentivité pour les armes autonomes.

#### 4.4.5.4 Attentes

Une différence légèrement moindre peut être constatée dans les «attentes» des personnes entre la condition «arme actionnée par l'homme» et les deux conditions «arme autonome». Le niveau d'attente pour les conditions «arme autonome» est égal. La différence entre tous les groupes n'est pas significative.

#### 4.4.5.5 Soutien / approbation

En général, les drones actionnés par l'homme reçoivent plus d'«approbation» que les armes autonomes, et la différence est significative ($p < .05$) pour les armes autonomes, il semble que la perception de l'agentivité n'influence pas le degré d'approbation, et la différence entre les conditions d'agentivité neutre et élevée n''est pas significative.

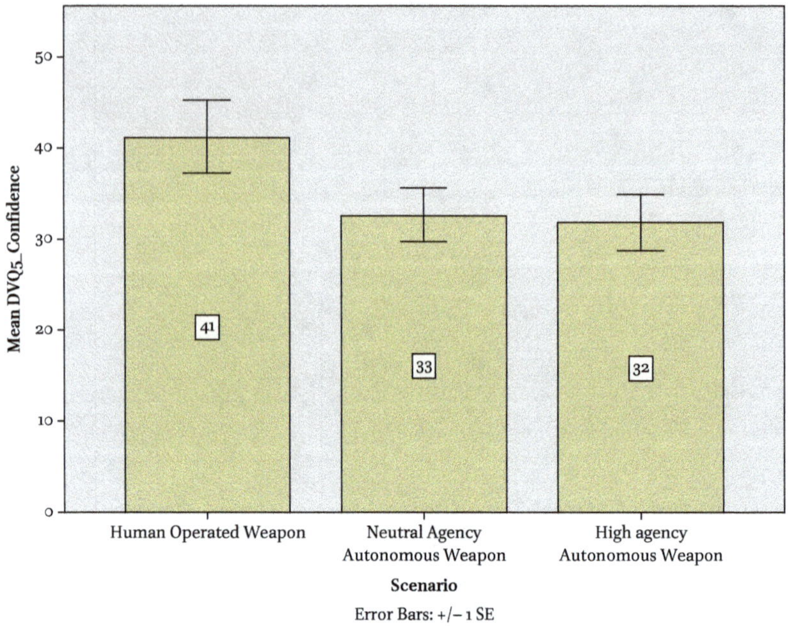


FIGURE 36 Valeur moyenne variable assurance par condition

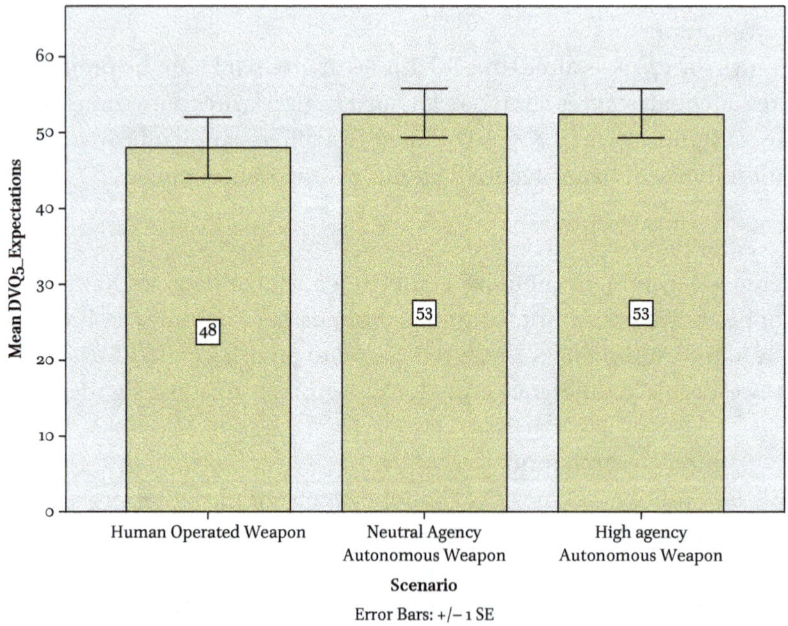


FIGURE 37 Valeur moyenne variable attentes par condition

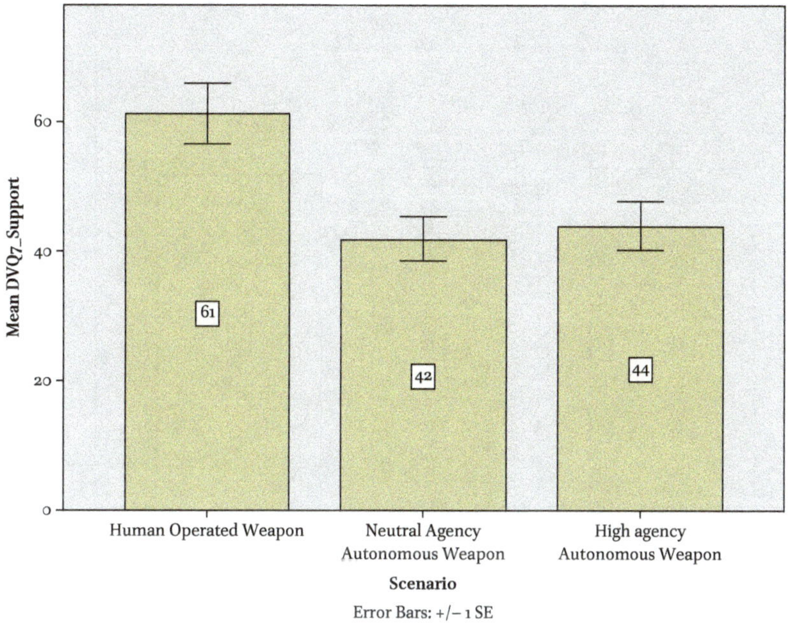


FIGURE 38 Valeur moyenne variable approbation par condition

#### 4.4.5.6 Equité

Les actions des drones actionnés par l'homme et des armes autonomes sont considérées comme étant également «équitables»; ceci est un résultat frappant, car nous nous attendions à ce que les actions des drones opérés par l'homme soient considérées comme plus équitables que celles d'une arme autonome. Malheureusement, la différence entre ces groupes n'est pas significative.

#### 4.4.5.6 Inquiétude / anxiété

Les personnes sont plus «inquiètes» lorsqu'il s'agit d'armes autonomes que lorsqu'il s'agit d'une arme actionnée par l'homme, et la différence est significative ($p < .05$). Les scénarios «armes autonomes agentivité neutre» et «arme autonome haute agentivité» montrent une légère augmentation du niveau d'inquiétude, mais la différence n'est pas significative.

### 4.4.6 *Conclusions étude finale*

Les résultats de l'étude finale conduisent aux conclusions suivantes :

- Le test de fiabilité montre que les quatre éléments du concept agentivité sont fiables et selon l'APC peuvent être considérées comme une composante;

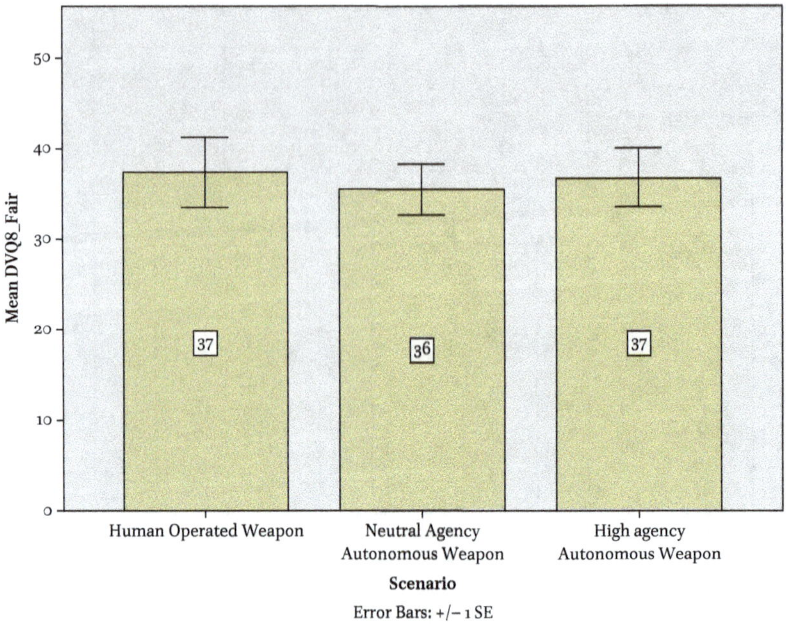


FIGURE 39 Valeurs moyennes variable équité par condition

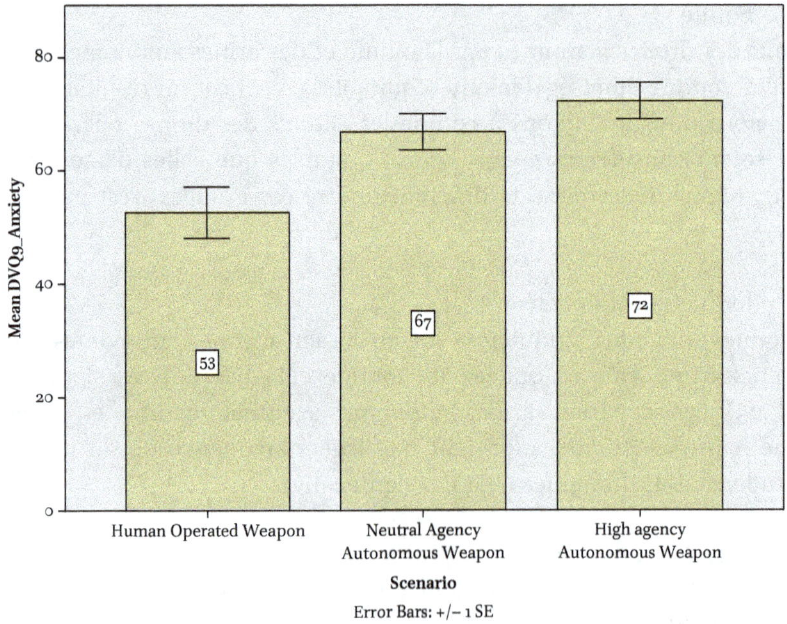


FIGURE 40 Valeur moyenne variable inquiétude par condition

- La perception de l'agentivité dans la condition «arme autonome agentivité neutre» est plus élevée que la perception de l'agentivité dans la condition «drone actionné par l'homme». La perception de l'agentivité dans la condition «agentivité élevée» est plus haute que dans la condition «agentivité neutre», mais la distinction n'est pas significative, avec p = .063 (Figure 32);
- Dans le scénario «arme actionnée par l'homme» ainsi que dans le scénario «agentivité neutre», la perception de l'agentivité des répondants militaires et civils est au même niveau. Cependant, dans le scénario «arme autonome haute agentivité», les répondants militaires perçoivent beaucoup plus d'agentivité que les civils (Figure 33), mais comme ces deux groupes ne sont pas significativement différents dans deux des scénarios, uniquement dans la condition «haute agentivité», nous devons être prudents dans l'interprétation des résultats et en tirant les conclusions. Nous ne savons pas quelle est la cause de cette différence dans la perception de l'agentivité entre ces deux groupes de répondants, et ceci mérite d'être étudié plus en détail;
- L'analyse de corrélation montre que 7 des 9 variables dépendantes sont significativement corrélées au concept d'agentivité. Ces 7 variables sont: *confiance, dignité humaine, assurance, attentes, approbation, équité* et *inquiétude* (Tableau 18);
- En nous basant sur l'analyse plus détaillée des variables dépendantes, nous pouvons observer que le niveau de *confiance, assurance, dignité humaine* et *approbation* dans les actions accomplies par les drones actionnés par l'homme est plus élevé que pour celles accomplies par les armes autonomes (Figure 34, Figure 35, Figure 36, Figure 38);
- Les niveaux d'attente et d'équité pour les drones actionnés par l'homme et pour les armes autonomes sont égaux (Figure 37, Figure 39) ;
- Le niveau d'inquiétude des répondants est plus élevé pour les armes autonomes que pour les armes actionnées par l'homme, et ces niveaux sont les plus élevés dans la condition d'agentivité élevée.

CHAPITRE 5

# Conception d'une Machine Morale pour les armes autonomes

> J'ai regardé «Eye in the Sky» … cette étude me rappelle ce film
>
> PERSONNE INTERROGÉE ETUDE PILOTE 1

Dans la section précédente nous avons décrit les scénarios que nous avons testés dans les études pilotes et dans l'étude finale. Même si les échantillons de données étaient assez larges pour effectuer des analyses statistiques, les études ne portaient que sur 50 à 96 répondants par scénario, et ces répondants avaient des caractéristiques démographiques très spécifiques. Par exemple, les répondants sur MTurk sont surtout âgés de 25 à 35 ans, ont un diplôme de l'enseignement supérieur, et l'étude militaire a porté sur une population constituée à 95 % d'hommes, de nationalité néerlandaise. Pour pouvoir généraliser les résultats, l'étude a besoin de davantage de répondants représentant un groupe démographique plus large. Nous proposons un projet de conception pour une «machine morale» pour les armes autonomes en vue d'une étude à grande échelle sur ce sujet; pour ce faire, nous nous appuyons sur le concept de la «Machine Morale» qui a été développé par le groupe de Coopération Evolutive («Scalable Cooperation group») du Media Lab au Massachusetts Institute of Technology, 2016.

La conception proposée pour la «machine morale» pour les armes autonomes fait partie de la phase d'investigation technique de la méthode Conception Ethique («Value-sensitive Design»). Nous utilisons cette expression pour faire que la technologie se charge de l'étape suivante dans notre recherche, et nous nous appuyons sur les scénarios utilisés dans les études pilotes et l'étude finale pour connaître le jugement moral des personnes concernant les armes autonomes. Créer un site internet hébergeant la Machine Morale pour les armes autonomes nous permet de recueillir des données sur les jugements moraux sur une grande échelle provenant de groupes démographiques divers qui pourraient être utilisées pour obtenir des résultats plus solides et plus généralisables. Le recueil de données sur une grande échelle dans de nombreux pays pourrait révéler des différences culturelles dans les jugements

 | DOI:10.1163/9789004449084_006

moraux sur les armes autonomes. Par exemple, les opinions émises dans les pays occidentaux sur le déploiement de ce type d'armes sont très probablement différentes des opinions provenant de pays du Moyen-Orient ou d'Asie. Les données provenant de ces différents pays peuvent être utilisées dans le débat sur les armes autonomes, qui est à notre avis actuellement dominé par des points de vue occidentaux.

Cette section décrit tout d'abord la Machine Morale pour les véhicules autonomes, puis notre proposition pour une Machine Morale pour les armes autonomes, en spécifiant les scénarios et les variables ; pour finir, nous donnerons quelques indications pour l'utilisation du site internet.

## 5.1 Une «Machine Morale» pour les véhicules autonomes

La Machine Morale originale est une *«plateforme de recueil de données sur la perception qu'ont les êtres humains de l'acceptabilité de décisions prises par des véhicules autonomes quand il faut choisir quelles sont les personnes que l'on peut blesser et quelles sont celles qu'il faut préserver»* (Awad, 2017 n pp. 42–43). Le site internet comporte trois modes d'utilisation: (1) un mode *«Jugement»* dans lequel les utilisateurs peuvent décider du résultat pour 13 séries de scénarios, (2) un mode *Conception* dans lequel les utilisateurs peuvent concevoir leur propre scénario, et (3) un mode *Navigation* dans lequel les utilisateurs peuvent voir le scénario d'autres personnes. La caractéristique principale est le mode «Jugement» dans lequel les utilisateurs choisissent entre deux scénarios qui contiennent différentes variables de *Caractères* (*personnages*) {genre, valeur sociale, âge, espèce, aptitude, utilitarisme}, d'*Interventionnisme* {omission, commission}, *Rapports avec le véhicule* {conducteur, piéton} et *Respect de la loi* {action légale, action illégale}. D'autres caractéristiques incluent une vidéo décrivant le projet, les instructions et les informations sur le contexte pour les utilisateurs.

La Machine Morale est conçue comme une Expérience d'envergure en ligne (Reips, 2002) dont le but est de recruter un large pool d'échantillons avec des antécédents diversifiés en peu de temps et à moindre coût. Ces caractéristiques sont clairement des avantages, mais d'un autre côté les conditions sont difficiles à contrôler, comme les utilisateurs peuvent participer à l'enquête plusieurs fois et se choisissent eux-mêmes, c'est-à-dire qu'ils peuvent participer à l'expérience et la quitter à tout moment (Awad, 2017). La Machine Morale a été développée au moyen d'une méthode à prototypage rapide sur la plateforme Meteor qui offre des structures de scriptage basées sur des modèles, et qui a une réactivité élevée. Elle est hébergée sur un service d'application par le cloud, est destinée à être utilisée sur les appareils mobiles, et est optimisée pour partage et d'échanges média avec les marques et balises Cards, Markup et Open Graph. D'ici mai 2017, environ 3 millions d'utilisateurs dans 160 pays ont évalué plus de 30 millions de scénarios, créant ainsi l'un des plus grands instruments de jugement moral à grande échelle existant actuellement.

## 5.2 Machine Morale pour armes autonomes

Les armes autonomes sont un sujet sensible, et nous avons observé qu'il provoque une réaction première d'inquiétude et d'appréhension. A notre avis, il ne serait pas prudent de développer une plateforme ouverte pour recueillir des données sur une grande échelle dans un premier temps, car une plateforme ouverte pourrait susciter de mauvais sentiments et des actes indésirables qui seraient contreproductifs pour notre recherche, par exemple des personnes créant des scénarios dans lesquels certaines catégories de personnes, comme les Musulmans ou les femmes, sont spécifiquement visées. Il est donc souhaitable d'adopter une approche graduelle pour ensuite aboutir à une «Expérience en ligne massive». Pour procéder à une étude suivie de grande ampleur sur les jugements moraux des personnes concernant les armes autonomes, il faut concevoir une expérience contrôlée avec un ensemble de conditions et d'échantillonnage contrôlé. Ces conditions peuvent être visualisées dans des scénarios qui permettent aux utilisateurs de répondre à l'enquête après l'obtention d'un mot de passe via une interface web vers un serveur sécurisé. Après le recueil de données de base et du retour de l'utilisateur, la phase suivante pourrait être de monter en puissance et d'obtenir une plateforme ouverte de grande envergure, comme la Machine Morale, dans laquelle les répondants pourraient juger à partir des scénarios pour recueillir une grande quantité de données dans plusieurs pays. Mais du fait de la sensibilité du sujet nous pensons qu'il n'est pas souhaitable de permettre aux répondants de créer leurs propres scénarios ou de partager leurs «résultats» sur les réseaux sociaux, comme le permet le volet *conception* de la Machine Morale à l'origine.

Nous proposons les caractéristiques suivantes pour la conception de la Machine Morale pour les armes autonomes:

- Un mode «Jugement» dans lequel les utilisateurs peuvent choisir entre différents scénarios et indiquer lequel des deux scénarios est le plus acceptable moralement;
- Les conditions ne doivent varier que sur une seule variable à la fois dans la comparaison des scénarios de manière à ce que l'évaluation ne puisse pas être attribuée au résultat de plusieurs conditions pour empêcher de confondre les effets;

- Une page présentant et expliquant brièvement la signification des variables avant que l'utilisateur ne commence à juger les scénarios;
- Une page présentant plus d'informations sur le scénario spécifique que l'utilisateur est en train de juger, à laquelle on peut avoir accès s'il ou elle désire davantage d'explications;
- Après avoir jugé le scénario, l'utilisateur a accès aux résultats;
- Une visualisation des résultats de l'utilisateur, comparant ses résultats avec ceux des autres utilisateurs, pour lui fournir un feedback sur leur jugement;
- Une seconde session de jugement afin de permettre aux utilisateurs de juger les scénarios une nouvelle fois pour qu'ils puissent modifier leur jugement et opinions d'ordre moral à partir d'une comparaison avec les résultats des autres répondants.

## 5.3 Scénarios

Dans cette section, les variables concernant la Machine Morale pour les armes autonomes sont définies et décrites dans deux scénarios à titre d'exemples. Les variables sont dérivées des scénarios des études pilotes et finale qui décrivent un convoi militaire appuyé par un drone en vol; elles sont inspirées des variables utilisées dans la Machine Morale pour les armes autonomes.

### 5.3.1 *Variables pour les scénarios*

Concernant le prototype de la Machine Morale pour les armes autonomes, 6 variables sont dérivées soit des études pilotes et finale, par exemple «Type d'arme et résultat», soit inspirées par la Machine Morale pour les véhicules autonomes, comme «Personnage» et nombre de personnages». Les 6 variables sont les suivantes: «Type d'arme» (W), «Localisation» (L), «Personnage» (C), «Nombre de personnages» (N), «Résultat» (O) et «Mission» (M). Les variables décrites dans les paragraphes qui suivent sont des exemples et des propositions, et devront être conçues de manière professionnelle lorsqu'elles seront mises en œuvre sur une Machine Morale pour armes autonomes. Par exemple, la variable «Localisation» est basée sur la description de style dessin et jeu d'un désert et d'un village, et doit être testée pour vérifier si c'est une représentation claire, ou si une photographie conviendrait mieux pour être utilisée dans une étude.

#### 5.3.1.1 Type d'arme

Ce volet montre le Type d'arme (W) déployée pour protéger le convoi. Il vérifie si les personnes considèrent que la technologie actuelle, le drone actionné par l'homme, est moralement plus acceptable qu'un drone autonome, qui

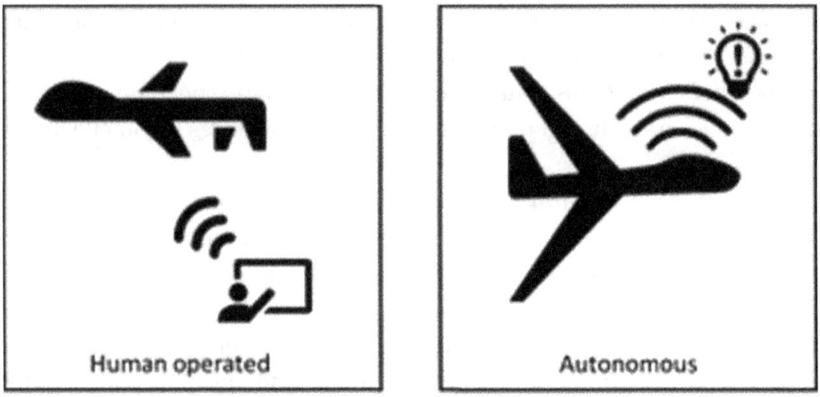


FIGURE 41 Variable «Type d'arme» (W) = {drone actionné par l'homme, drone autonome}

correspond à une technologie future. Ceci peut être utilisé pour mieux connaître ces technologies. Le type d'arme est soit un drone actionné par l'homme (à gauche) ou une arme autonome (à droite) (Figure 41). Si «le «type d'arme» est la variable discriminante, les différents pictogrammes apparaissent sur le scénario pour que les répondants choisissent entre les deux, comme on le voit dans l'exemple 1 (Figure 47). Sinon, le même type d'arme apparaît sur le scénario de l'exemple 2 (Figure 48).

#### 5.3.1.2 Localisation

Ce volet indique la localisation (L) comme cadre du scénario, qui peut être soit un désert (à gauche), soit un village (à droite). Pour ce volet, nous vérifions quelle est le cadre le plus approprié pour les répondants pour déployer l'arme autonome ou le drone actionné par l'homme. Si la «Localisation» est la variable discriminante, les deux images sont présentées pour chacun des scénarios, comme on le voit dans l'exemple 2 (Figure 48). Sinon, le même cadre apparaît dans les deux scénarios comme dans l'exemple 1 (Figure 47).

#### 5.3.1.3 Personnages (Character C)

Ce volet présente le type de personnage (C) impliqué comme témoin (ou observateur) dans le scénario; ce peut être un homme (à gauche), une femme (au centre) ou un enfant (à droite) (Figure 43). En variant les personnages, on peut savoir quels personnages les répondants trouvent moralement acceptables lorsqu'une arme autonome ou actionnée par l'homme est utilisée. Si le «Personnage» est la variable discriminante, différents personnages apparaissent sur chacun des scénarios. Sinon, les mêmes personnages apparaissent dans les deux scénarios.

FIGURE 42 Variable Localisation L = {désert, village}

FIGURE 43 Variable Personnage C = {homme, femme, enfant)

#### 5.3.1.4 Nombre de personnages

Ce volet présente le «Nombre de personnages» (N) impliqués dans le scénario, s'échelonnant de 1 personnage (à gauche) à 5 personnages (à droite) (Figure 44). Ce volet nous permet de savoir combien de témoins les répondants trouvent acceptables lorsqu'une arme autonome ou actionnée par l'homme est utilisée. Si le «nombre de personnages» est la variable discriminante, des nombres différents de personnages apparaissent sur chacun des scénarios. Sinon, le même nombre de personnages apparaît dans les deux scénarios.

#### 5.3.1.5 Résultat / Effet

Ce volet fait apparaître le résultat / effet (Outcome – O) du scénario qui peut être soit «pas de dommages collatéraux» (à gauche), ou un effet «avec dommages collatéraux» affectant le nombre de personnages impliqués dans le scénario (à droite) (Figure 45). Ceci nous permet de voir comment l'effet influence l'acceptabilité morale de l'emploi d'une arme autonome ou actionnée par l'homme. Si l'Effet constitue la variable discriminante, différents pictogrammes apparaissent pour chacun des scénarios. Sinon la même variable «effet» apparaît dans les deux scénarios.

FIGURE 44 Variable Nombre de personnages N = {1.5}

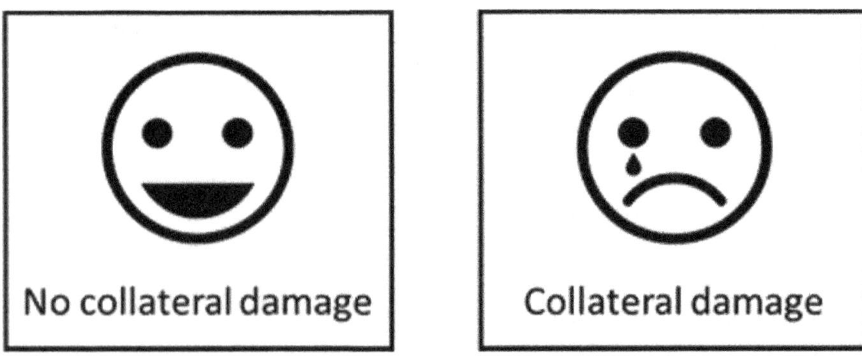


FIGURE 45 Variable Effet o = {dommages collatéraux, pas de dommages collatéraux)

#### 5.3.1.6 Mission

Ce volet fait apparaître la Mission (M) de l'arme dans le scénario qui peut être soit de défendre le convoi seulement lorsqu'une menace directe est perçue (à gauche), soit d'attaquer les véhicules qui figurent sur la liste des objectifs même s'ils ne constituent pas une menace directe pour le convoi (à droite) (Figure 46). Dans ce volet, nous vérifions quel type de mission est moralement acceptable pour les répondants. Si «Mission» constitue la variable discriminante, différents pictogrammes apparaissent pour chacun des scénarios. Sinon la même variable Mission apparaît dans les deux scénarios.

### 5.3.2 *Exemples de scénarios*

Les variables décrites ci-dessus peuvent être utilisées pour créer des scénarios, chaque scénario différant sur seulement une variable. La question posée dans le scénario est la même question que celle qui est posée lors du jugement portant sur les scénarios de la Machine Morale d'origine. Dans cette section, nous

FIGURE 46 Variable Mission M = {défendre, attaquer}

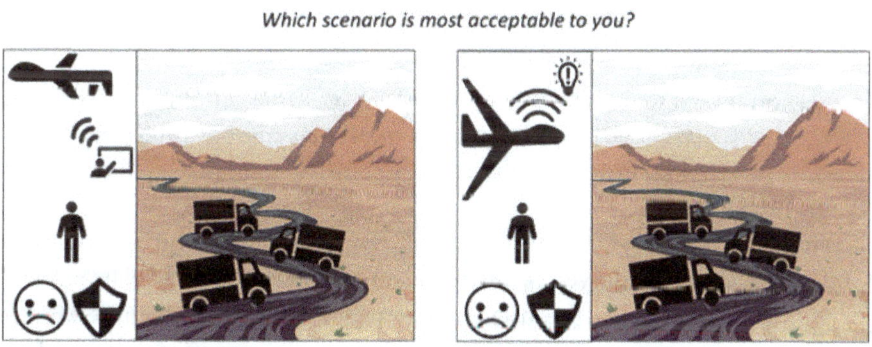


FIGURE 47 Exemple scénario 1

présenterons deux scénarios à titre d'exemples pour présenter le concept, mais pour faire court nous avons choisi de ne pas faire apparaître toutes les variables dans une liste d'exemples interminable. L'exemple 1 apparaît Figure 47, qui montre un convoi dans un cadre désertique. La différence entre les scénarios est qu'à gauche le convoi est défendu par un drone actionnée par l'homme, et dans le scénario de droite par une arme autonome. Dans les deux cas il y a une personne au bord de la route qui est tuée dans le cadre de dommages collatéraux.

*Quel est le scénario le plus acceptable pour vous ?*

Dans l'exemple de scénario suivant (Figure 48), les deux convois sont défendus par une arme autonome, mais le cadre est différent. Le convoi de gauche chemine à travers un désert et celui de droite à travers un village. Dans les deux cas deux enfants jouent au bord de la route et l'attaque ne cause aucun dommage collatéral.

*Quel scénario est pour vous le plus acceptable ?*

FIGURE 48 Exemple de scénario 2

## 5.4 Mise en œuvre

La mise en œuvre de la Machine Morale pour les armes autonomes implique plusieurs activités, et ce projet prendra au moins six mois avant d'être achevé, plus une année pour exécuter l'étude. Dans cette section nous en décrirons les phases brièvement.

Dans la première phase, les scénarios et les variables esquissés dans les sections précédentes doivent être conçus en étant attentif à la clarté et l'intelligibilité des variables. Selon le chercheur qui a conçu la Machine Morale d'origine, le processus de création des figures pour les scénarios a pris environ trois mois et a été exécuté par un graphiste professionnel à chaque itération.

Pour la phase 2, l'infrastructure pour le site web doit être développée et échafaudée, ce qui prendra plusieurs mois. Du fait de la sensibilité du sujet, un site internet sécurisé est nécessaire pour exécuter les scénarios, de manière à ce que l'on ne puisse y accéder qu'avec un mot de passe. Cela demande une organisation pour attribuer les mots de passe aux personnes intéressées par l'enquête. Par exemple, un site internet qui envoie un lien après inscription. Si nous avons l'intention de procéder à une Expérience on Ligne Massive, on devra tenir compte de cette possibilité, qui sera en fait une nécessité pour lancer le projet. Ceci implique que le site internet doit être hébergé par un serveur qui permet une dynamique de mise à l'échelle (vers le haut ou vers le bas) selon le nombre d'utilisateurs actifs. En même temps, il est important que le niveau de sécurité du site internet et du serveur soit élevé, et que les données de la recherche soient la propriété de l'Université qui exécute l'étude. Ces exigences impliquent que le site ne soit pas hébergé par une société commerciale, comme Microsoft, Google ou Amazon, qui offrent ce genre de service dynamique par le cloud, mais par un serveur qui soit la propriété de l'Université.

Pour la troisième phase, ces scénarios doivent être testés dans plusieurs études pilotes afin de vérifier s'ils génèrent des résultats utiles et s'ils devront être adaptés s'il s'avère que ce n'est pas le cas. Ceci constituera un processus qui prendra plusieurs sessions jusqu'à ce que l'étude finale soit testée. La Machine Morale d'origine a recueilli des données entre le 23 juin 2016 et mai 2017, et il est recommandé d'exécuter l'étude sur les armes autonomes pendant la même durée afin de recueillir un échantillon assez large pour qu'on puisse parler d'une Expérience en Ligne Massive (de grande ampleur).

CHAPTER 6

# Conclusion et discussion

> Je suis sûr que les drones peuvent jouer un grand rôle dans le futur. Mais tant qu'il n'y aura pas d'algorithme pour définir une bonne estimation des dommages collatéraux, il doit y avoir une décision humaine pour déclencher une arme. De plus, un chef sur le terrain doit toujours intervenir en dernier ressort sur la manière d'attaquer. Les drones peuvent contribuer à améliorer son intelligence de situation, mais ne doivent pas limiter ses choix.
>
> UN RÉPONDANT LORS DE L'ÉTUDE FINALE

Le but de cette étude était de définir la manière dont les armes autonomes sont perçues par le grand public et par les militaires, et quelles sont les valeurs morales qu'ils considèrent importantes lorsque des armes autonomes seront déployées dans un proche avenir. La méthode de conception éthique («value-sensitive design») a été utilisée pour structurer cette étude. Le chapitre suivant décrit tout d'abord les conclusions de notre recherche, suivies d'une discussion sur les implications scientifiques et sociétales. Enfin, nous identifierons plusieurs restrictions et conclurons sur des recommandations en vue d'études futures.

## 6.1 Conclusion

Ce sous-chapitre présente les conclusions sur le concept d'agentivité, notre hypothèse centrale, et l'exploration des variables dépendantes. Dans chaque paragraphe, nous indiquons si nos résultats confirment ou infirment les publications académiques actuelles, et nous concluons sur nos constatations.

### 6.1.1 *Concept d'agentivité*

Les résultats de trois études consécutives ont montré que les élément d'agentivité sont fiables, et constituent bien un concept. Le concept d'agentivité est composé des quatre éléments *Réflexion, Détermination des objectifs,*

 | DOI:10.1163/9789004449084_007

*Libre-arbitre* et *Réalisation des objectifs*, qui peuvent être utilisés pour mesurer la perception de l'agentivité des armes autonomes et des drones actionnés par l'homme. Dans l'Opérationnalisation de ce concept, nous avons combiné les publications du domaine de la psychologie cognitive, de l'intelligence artificielle et de la philosophie morale. La plupart des études empiriques sur la perception de l'agentivité sont effectuées sous l'angle de la psychologie cognitive, comme celle de K.Gray et Wegner (2012) qui, dans leur travail sur les robots et les zombies, donnent différents niveaux d'agentivité pour mesurer l'anxiété. Malle et Thapa Magar (2017) ont étudié les capacités mentales souhaitables dans les robots sociaux, en utilisant quelques-uns des aspects de l'agentivité, par exemple la *réflexion* et *explication des actions*. Néanmoins, aucune des études que nous avons trouvées n'utilise un concept unique pour mesurer les niveaux d'agentivité des artefacts techniques. Ainsi à notre connaissance nous présentons ici le premier concept qui mesure de manière empirique la perception de l'agentivité.

### 6.1.2 *Hypothèse centrale sur la perception de l'agentivité*

Dans l'étude finale, nous avons émis l'hypothèse que le drone actionné par l'homme serait perçu comme ayant un bas niveau d'agentivité, et que l'arme autonome à haut niveau d'agentivité serait perçue comme ayant une haute agentivité. Les résultats montrent que la différence est significative entre ces deux conditions; nous en concluons que la manipulation sur l'agentivité fonctionne. Nous nous attendions également à ce que la condition d'agentivité neutre des armes autonomes soit jugée comme significativement différente de la condition d'agentivité élevée des armes autonomes, et nos résultats indiquent qu'il y a une différence, mais la différence entre ces deux groupes se situe juste au-dessus du seuil significatif que nous avons défini; par conséquent nous devons être prudent en tirant les conclusions.

Notre analyse centrale concernait la manière dont le cas d'agentivité neutre pour les armes autonomes diffère de celui du drone actionné par l'homme et de la condition haute agentivité pour les armes autonomes. Nous avons émis l'hypothèse que les personnels militaires percevraient les armes autonomes comme n'importe quelle autre arme, pas plus qu'un outil, et donc ne percevraient pas les armes autonomes comme ayant des états mentaux. Nous nous attendions également à n'observer aucune différence entre la condition «arme autonome à agentivité neutre» et la condition dans laquelle un humain actionne un drone à distance. Nos résultats indiquent que la perception de l'agentivité des personnels militaires et civils du Ministère de la Défense néerlandais pour les armes autonomes à niveau neutre d'agentivité est plus élevée que la perception de l'agentivité pour la condition «drone actionné par l'homme». Ceci signifie qu'ils attribuent plus d'agentivité à une arme autonome qu'à un drone actionné par l'homme. A partir de ces constatations, nous devons rejeter notre hypothèse:

> H1: les personnels militaires ne percevront pas les armes autonomes comme dotés d'états mentaux.

Nos constatations sont en accord avec la rechercher sur l'agentivité dans le domaine de la psychologie cognitive. Des études antérieures ont montré que les personnes attribuent un intellect aux ordinateurs (Nass *et al*, 1995), et perçoivent les robots comme des «agents» (H.M.Gray *et al*, 2007). A partir de notre étude, nous pouvons conclure qu'attribuer une perception intelligente s'applique aussi aux armes autonomes, et que ces armes sont considérées comme davantage qu'un simple outil par lequel on obtient un effet.

Les résultats conduisent aussi à une autre observation frappante: dans la condition «arme actionnée par l'homme» aussi bien que dans la condition «arme autonome à agentivité neutre», la perception de l'agentivité des répondants civils et militaires se situe au même niveau. Comme on l'a dit dans la section 4.4.4., les deux groupes ne sont pas significativement différents dans deux des scénarios, seulement dans la condition «agentivité élevée», et donc nous devons être prudents dans l'interprétation des résultats et dans nos conclusions. Cependant, dans la condition «arme autonome à haute agentivité», les répondants militaires perçoivent beaucoup plus d'agentivité que leurs homologues civils. Nous ne savons pas vraiment ce qui peut expliquer cette différence dans la perception de l'agentivité entre ces deux catégories de répondants. Une explication pourrait être que l'échantillon civil contenait 10% de répondants de plus qui travaillaient dans le domaine de l'intelligence artificielle par rapport à l'échantillon militaire. Une autre explication pourrait être que ces deux groupes ont lu les scénarios différemment, par exemple les personnels militaires ont peut-être pris les descriptions littéralement alors que les civils ont davantage interprété, et ceci a pu influencer leurs réponses aux questions sur l'agentivité. Mais ces explications ne sont que des hypothèses pour le moment, et les raisons de cette différence dans la perception de l'agentivité entre les personnels militaires et civils devront faire l'objet d'une étude plus approfondie.

### 6.1.3 *Exploration des variables dépendantes*

L'effet des variables dépendantes sur la perception de l'agentivité est étudié de manière descriptive, et nous ne pouvons tirer aucune conclusion sur la relation entre les niveaux d'agentivité et leur effet sur les variables dépendantes. Nous avons constaté que 7 des 9 variables dépendantes peuvent être de manière significative corrélées avec le concept d'agentivité. Ce sont les variables *confiance, dignité humaine, assurance, attentes, approbation / soutien, équité* et *anxiété.* Dans cette section, nous présentons les constatations concernant les variables dépendantes, que nous regroupons sur la base des résultats qu'elles produisent.

#### 6.1.3.1 Confiance, assurance et approbation

Les résultats indiquent que les personnels militaires et civils du Ministère de la Défense néerlandais apportent plus de confiance, d'assurance et d'approbation aux actions des drones actionnés par les humains qu'à celles des armes autonomes. Nous n'avons trouvé aucune publication académique sur ce sujet, mais un sondage Gallup (américain) de 2013 a indiqué une large approbation pour les frappes de drones télépilotés à l'étranger, et dans un rapport public Schneider et McDonald (2016) indiquent que les citoyens américains préfèrent les frappes aériennes par drones télépilotés aux frappes par appareils avec pilotes, car elles sont moins risquées pour le personnel militaire. Nous pouvons affirmer de manière théorique que les plus hauts niveaux d'approbation, de confiance et d'assurance pourraient être expliqués par le fait que les personnes sont familiarisées avec les drones actionnés par l'homme puisque cette technologie est actuellement utilisée, comparée à la nouvelle technologie futuriste à laquelle les gens ne sont pas familiarisés. Une autre explication pourrait être que les personnes font plus confiance et ressentent plus d'assurance en les actions des humains, en comparaison des actions des systèmes autonomes. Aujourd'hui il est difficile de dire quelles sont les raisons de cette différence, et ceci devrait faire l'objet d'une étude plus approfondie.

#### 6.1.3.2 Dignité humaine

Les résultats montrent qu'un drone actionné par l'homme est perçu comme ayant plus de respect pour la dignité humaine qu'une arme autonome à agentivité élevée, même si les actions et le résultat du scénario sont les mêmes. Ces résultats contredisent l'argument de Kasher (2016) qui prétend que la dignité humaine ne devrait pas être recherchée dans la nature de l'artefact, mais dans les intentions et les décisions de la personne qui le fait fonctionner. Nos résultats indiquent que la nature de l'objet est impliquée dans la perception de la dignité humaine, car les intentions et les décisions du drone actionné par l'homme et de l'arme autonome sont identiques. Ces résultats reflètent les opinions des experts dans les entretiens qui ont évoqué la dignité humaine à de nombreuses reprises comme étant une raison pour s'opposer aux armes autonomes. Un manque de respect pour la dignité humaine semble être l'une des principales objections au déploiement des armes autonomes, et il est frappant de constater que ceci est également présent dans les réponses des personnels militaires et civils du Ministère de la Défense néerlandais. Mais cela n'implique pas que les personnels militaires désapprouvent les armes autonomes, ni qu'ils partagent les mêmes arguments contre leur emploi que quelques-uns des experts.

#### 6.1.3.3 Attentes et équité

A partir des résultats nous pouvons conclure que les personnels militaires et civils du Ministère de la Défense néerlandais manifestent un niveau d'attentes identique sur les «mesures à prendre» par un drone actionné par l'homme et par une arme autonome à niveaux d'agentivité neutre et élevé. Ils considèrent également que les actions du drone actionné par l'homme et de l'arme autonome ont un même degré d'équité. Le souci évoqué par Shelley (2013) qui affirme que les drones autonomes seront considérés comme moins «équitables» que les drones actionnés par l'homme n'est pas confirmé par nos résultats. Nous n'avons pas trouvé de publications faisant le lien entre les attentes et les armes autonomes ou les drones. Nos constatations ne montrent aucune différence dans les niveaux d'attentes et d'équité pour ce qui est de la technologie actuelle et future parmi les personnels civils et militaires du Ministère de la Défense néerlandais.

#### 6.1.3.4 Inquiétude / anxiété

Nos résultats montrent que les armes autonomes causent plus d'inquiétude parmi les personnels civils et militaires du Ministère de la Défense néerlandais que les armes opérées par l'homme. La différence entre les niveaux d'inquiétude est plus importante pour la condition d'agentivité élevée. L'inquiétude vis-à-vis des armes autonomes est souvent décrite dans les publications académiques (Noone & Noone, 2015; Ohlin, 2016); elle est aussi exprimée dans les opinions des experts interrogés, et observée par les chercheurs lors de conversations avec des personnes. Ces constatations confirment que les personnes sont préoccupées par l'utilisation d'armes autonomes. Cette inquiétude est partagée par les personnels civils et militaires du Ministère de la Défense néerlandais, comme par les experts qui se sont fait l'écho de l'opinion du grand public.

## 6.2 Discussion

Les implications de ces résultats sont discutées dans ce sous-chapitre. Tout d'abord leur contribution scientifique à la «littérature» académique est identifiée, puis sont discutées les implications sociétales contribuant au débat actuel sur les armes autonomes.

### 6.2.1 *Implications scientifiques*

Cette étude apporte une pierre à la «littérature» académique à plus d'un titre. Elle présente les différentes définitions de l'arme autonome qui sont actuellement utilisées dans les publications, et montre que pour le moment personne n'est d'accord pour adopter une définition unique. Lors de la création de cette présentation, il est devenu apparent que certains groupes de chercheurs mettent en garde contre une définition claire de l'arme autonome pour différentes raisons. Quelques-uns disent que des machines ne peuvent pas être autonomes au sens littéral du terme, alors que d'autres insistent sur le fait que les définitions de l'autonomie sont appliquées à des fonctions différentes. Comme il a été dit lors d'un entretien avec un expert, une autre raison qui explique cette hésitation pourrait être que, en ne donnant aucune définition précise de l'arme autonome, on permet une discussion ouverte et ainsi on évite de tomber dans des impasses sur cette question. Nous avons choisi la définition donnée par l'ADVISORY COUNCIL ON INTERNATIONAL AFFAIRS (AIV & CAVV), car elle prend en compte des critères prédéfinis, et elle intéresse le processus militaire de ciblage, étant donné que l'arme ne sera déployée qu'après une décision d'origine humaine. Le fait de choisir et de proposer cette définition constitue une contribution à la «littérature» académique.

Une autre contribution de cette étude est que nous avons identifié les valeurs que les personnes associent aux armes autonomes. La présentation puise ses sources à la fois de théories des valeurs valides et d'experts concernés par le débat sur les armes autonomes ou qui travaillent dans le domaine militaire. Nous avons choisi les valeurs *faute* (*blâme*), *confiance, tort causé, dignité humaine, assurance, attentes, approbation, équité* et *inquiétude* pour les soumettre au test dans l'étude finale. Les résultats permettent de comprendre comment les personnels militaires et civils du Ministère de la Défense néerlandais perçoivent ces valeurs vis-à-vis des drones actionnés par l'homme (technologie actuelle) et des armes autonomes (technologie future). A notre connaissance la présente étude est la première à étudier ces valeurs de manière empirique vis-à-vis des armes autonomes, et par comparaison comment ces valeurs sont perçues dans les systèmes d'armes actuels et futurs; ceci constitue une contribution neuve au débat académique.

La troisième contribution au corpus est que nos études proposent un concept qui mesure la perception de l'agentivité vis-à-vis des armes autonomes. A notre connaissance c'est le premier concept qui est opérationnalisé pour mesurer les niveaux d'agentivité des artefacts technologiques. Dans notre étude, il a été appliqué aux drones, mais nous pensons qu'il pourrait être également appliqué pour mesurer la perception de l'agentivité vis-à-vis d'autres objets, et les questions concernant les éléments pourraient être facilement reformulées pour s'insérer dans un autre domaine. Des études de suivi pourraient vérifier si ce concept d'agentivité est valide lorsqu'il s'applique à des domaines différents.

### 6.2.2 *Implications sociétales*

Les résultats ont également des implications sociétales étant donné qu'ils contribuent au débat sur les armes autonomes. Peu de recherches empiriques ont été menées sur cette question, et donc le débat est dominé par des théories morales et juridiques abstraites. Notre recherche donne des données empiriques aux théories des valeurs qui lui sont sous-jacentes, et nous montrons que ces valeurs ne sont pas pertinentes seulement dans le cadre de la recherche académique, mais sont également présentes dans la vie réelle. En ce sens nous relions les théories abstraites des valeurs au domaine pratique du déploiement des armes autonomes. Par exemple, nous avons constaté que la valeur «dignité humaine» était souvent évoquée dans les publications et par les experts au cours des entretiens. Dans notre étude, nous avons recueilli en données empiriques le fait que les personnels militaires et civils du Ministère de la Défense néerlandais perçoivent les armes autonomes comme faisant preuve de moins de respect pour la «dignité humaine». Nous associons également la valeur «non-malfaisance», que nous qualifions «tort causé», de la théorie bioéthique, aux armes autonomes, et avons constaté des effets dans les études pilotes 1 et 2 sur le transfert de «tort causé» dans la chaîne de responsabilité qui devraient être étudiés dans une étude d'approfondissement.

La présente étude fournit également des données empiriques et corrobore quelques-unes des vues et opinions qui pourraient être utilisées pour trouver un terrain commun dans le débat sur les armes autonomes. Les personnels militaires et civils du Ministère de la Défense néerlandais perçoivent plus d'agentivité dans les armes autonomes que dans les drones actionnés par l'homme. Bien qu'il n'existe pas d'étude d'un échantillon constitué uniquement de civils, si on en croit les publications on peut s'attendre à trouver les mêmes résultats pour un tel échantillon. Ceci voudrait dire que les personnels civils et militaires pensent que les armes autonomes «réfléchissent» de manière indépendante et planifient pour atteindre leurs objectifs. Un autre terrain commun peut être trouvé pour les valeurs «dignité humaine» et «inquiétude». Nos résultats montrent que les personnels militaires et civils du Ministère de la Défense néerlandais sont plus inquiets face à la perspective d'un déploiement d'armes autonomes que d'un déploiement de drones actionnés par l'homme.

Ils perçoivent également les armes autonomes comme faisant preuve de moins de respect pour la dignité de la vie humaine que les drones actionnés par l'homme. La dignité humaine et l'inquiétude sont les deux valeurs qui sont souvent évoquées par les experts dans les entretiens; il est donc essentiel de les aborder quand on débat de l'éthique du déploiement des armes autonomes.

L'étude de ces valeurs contribue aussi au processus de conception des armes autonomes. Nos résultats montrent que la «confiance», l'«assurance» et l'«approbation» pour les armes autonomes sont moins élevées que pour les drones actionnés par l'homme. Il faut noter à ce sujet que les armes autonomes ne présentent pas que des inconvénients, mis présentent également des avantages évidents d'un point de vue militaire (Etzioni & Etzioni, 2017), et concevoir des caractéristiques tendant à augmenter la confiance et l'assurance vis-à-vis des armes autonomes est une bonne chose d'un point de vue militaire. Pour cela, il faut qu'au cours du processus de conception les valeurs humaines soient traduites en exigences de conception. Ceci peut être rendu visible au moyen d'une hiérarchie des valeurs (Van de Poel, 2013) décrite dans la section 2.2.4., une structure hiérarchique des valeurs, normes et exigences de conception qui rendent transparents et sujets à débat les jugements sur les valeurs requis pour cette translation explicites.

## 6.3 Limitations

Plusieurs problèmes peuvent être identifiés comme limitations (ou restrictions) pour cette étude. Tout d'abord, l'opérationnalisation des éléments et le concept d'agentivité sont dérivés d'une catégorie de publications décrivant les caractéristiques de l'agentivité, et notre choix des caractéristiques est basé sur un compte arithmétique, et non sur des critères de pertinence ou de pondération. Cette méthode a été choisie car c'était la plus objective que nous puissions trouver; nous avons essayé de ne pas prendre de décisions subjectives dans ce choix, mais il est possible que nous soyons passés à côté de certaines caractéristiques pertinentes qui auraient dû être ajoutées aux éléments de l'agentivité, ou que nous ayons choisi les mauvais éléments, ou des éléments non pertinents. L'opérationnalisation des caractéristiques dans certaines des questions que nous avons utilisée a une base heuristique, et nous n'avons pas vérifié si ces questions étaient correctes. Il serait difficile pour d'autres personnes d'utiliser une méthode de sélection identique, et elle affecte la reproductibilité et la validité interne de l'étude.

Une deuxième limitation méthodologique est le choix des valeurs comme variables dépendantes à partir du questionnaire en ligne sur les valeurs et des entretiens avec les experts. Bien que ces choix aient été longuement discutés entre les trois chercheurs impliqués dans cette étude, le choix final a été fait sur une base heuristique, et non à partir d'une méthode objective. Par exemple, nous avons choisi d'ajouter «tort causé» comme variable à partir de la liste du Tableau 6, mais non pas «contrôle». Conséquence: il est probable que les choix pâtissent de la subjectivité du chercheur, et d'autres chercheurs auraient choisi des valeurs différentes comme variables dépendantes. Rétrospectivement, nous aurions dû choisir une méthode plus objective et une manière plus structurée de limiter cette subjectivité du chercheur. Cette approche influence négativement la reproductivité et la validité interne de cette étude.

En troisième lieu, les échantillons utilisés dans cette étude sont limités à la fois en taille et en données «démographiques». Bien qu'elle soit d'assez grande ampleur pour permettre une analyse des données solide, l'étude finale n'avait que 239 répondants, ce qui ne représente qu'une petite partie du Ministère de la Défense néerlandais. La participation à l'enquête était volontaire, et nous n'avons pas pu vérifier quelles sont les personnes qui ont répondu. Ceci signifie que l'échantillon n'est pas représentatif de la totalité de ce Ministère; cela impacte la validité externe de cette étude, et implique que nous ne pouvons pas généraliser les résultats à l'ensemble de l'institution de Défense.

En dernier lieu, la répartition des répondants pour le scénario final est inégale, et le second scénario (armes autonomes avec agentivité neutre) a 32 répondants de plus que le scénario 1 (drones actionnés par l'homme). L'une des explications pourrait être que les personnes s'attendaient à participer à une enquête sur les armes autonomes, mais ont abandonné lorsqu'ils sont tombés sur les scénarios avec drones actionnés par l'homme. Il s'agit d'«attrition sélective», et a un impact négatif sur la validité interne de l'enquête. Ceci implique que l'on ne peut supposer que les répondants des deux scénarios ont les mêmes caractéristiques, et que la randomisation de l'étude, qui constitue un critère crucial pour une «expérience contrôlée», a échoué. Une autre explication pourrait être que le logiciel a fait une erreur en répartissant les répondants sur les scénarios, et nous sommes en contact avec Qualtrics pour vérifier si cela est le cas. Si cette répartition fautive est causée par un logiciel fautif, la validité interne de l'enquête n'est pas impactée, puisqu'il ne s'agirait pas d'attrition sélective dans ce cas.

## 6.4 Recommandations pour études de suivi

Une fois posées ces limitations, plusieurs recommandations en vue d'études de suivi sont proposées. La première est de valider le concept d'agentivité qui demande davantage d'étude en mesurant la perception de l'agentivité d'autres produits technologiques pour vérifier si ce concept est valable dans d'autres domaines. Par exemple, on pourrait étudier la perception de l'agentivité de robots d'aide à la personne pour les personnes âgées, de jouets «intelligence artificielle» pour les enfants, ou d'ordinateurs embarqués pour véhicules autonomes. La seconde recommandation est de procéder à l'étude finale avec les mêmes scénarios sur un échantillon représentatif composé uniquement de civils aux Pays-Bas. Ceci nous permettrait de voir sur quelles valeurs les résultats de l'échantillon militaire et les réponses des civils diffèrent; bien que nous ne puissions pas faire de comparaisons directes du fait que l'échantillon militaire n'est pas représentatif, nous connaîtrions la perception des personnels militaires et aussi des civils. Enfin, nous recommandons de lettre en œuvre la Machine Morale pour les armes autonomes, comme on l'a vu dans la section 5, pour généraliser les résultats; l'étude demande beaucoup plus de répondants représentant un groupe démographique plus large. Le fait de passer à l'échelle d'Expérience en Ligne Massive, comme pour la Machine Morale, fournirait une grande quantité de données de groupes d'âge différents qui pourraient être utilisées pour obtenir des résultats plus solides et ayant plus de portée générale; ainsi on comprendrait parfaitement le jugement moral des personnes concernant les armes autonomes .

APPENDICE A

# Questionnaire "valeurs" en ligne

Les questions de l'enquête « valeurs » en ligne sont présentées dans cet appendice dans l'ordre où elles ont été présentées aux répondants. Les lignes horizontales indiquent les sauts de page.

**Cher participant,**

Merci de contribuer à mon travail de recherche en remplissant ce questionnaire sur les Valeurs et les armes autonomes. J'utiliserai cette enquête dans le cadre de mon mémoire de master pour déterminer quelles valeurs les gens associent aux armes autonomes. Le questionnaire ne prend que cinq minutes à compléter et toutes les réponses sont anonymes. Il m'est impossible de vous identifier. La seule information dont je disposerai, en plus de vos réponses, est le moment où vous aurez répondu aux questions. Vous voudrez bien conserver à l'esprit les définitions suivantes lorsque vous répondrez aux questions : une « valeur » sert de principe directeur pour ce que les personnes considèrent comme important dans la vie. Une « arme autonome » est un système d'arme qui une fois mise en service choisira et attaquera des cibles sans autre intervention humaine.

Si vous avez des questions sur cette enquête, contactez-moi à cette adresse : [...]

Q1 Quel est votre âge?

- Moins de 18 (1)
- 18–24 (2)
- 25–34 (3)
- 35-44 (4)
- 45–54 (5)
- 55–64 (6)
- 65–74 (7)
- 75–84 (8)
- 85 ou plus (9)

 | DOI:10.1163/9789004449084_008

Q2 Sexe
- Homme (1)
- Femme (2)
- Ne veut pas répondre (3)

Q3 Quelle est votre nationalité?
[Liste des nationalités]

Q4 Dernier diplôme obtenu?
- Enseignement primaire (1)
- Enseignement secondaire (2)
- A fréquenté l'université (3)
- Licence (4)
- Master ou + (5)
- Inconnu (6)

Q8 Quelles sont à votre avis les valeurs qui s'appliquent le mieux aux armes autonomes ? Classez-les par ordre de préférence en conservant ou non le terme (1 = la plus applicable ; 4 = la moins applicable).
Autonomie : agir délibérément sans influence extérieure qui pondérerait une action volontaire. (1)

Non-malfaisance : ne pas imposer intentionnellement un risque de dommage déraisonnable à une autre personne. (2)

Bénéficience : contribuer au bien d'un individu ou d'une société dans son ensemble. (3) Justice : être équitable ou raisonnable. (4)

Q7 Choisissez 5 valeurs dans la liste suivante qui dans votre opinion s'appliquent le mieux aux armes autonomes. Conservez ou non les termes qui s'appliquent le mieux dans les cartouches.

Ces valeurs s'appliquent aux armes autonomes

_____ Liberté (1)
_____ Utilité (2)
_____ Réussite (3)
_____ Honnêteté (4)
_____ Respect de soi (5)
_____ Intelligence (6)

_____ Ouverture d'esprit (7)
_____ Créativité (8)
_____ Egalité (9)
_____ Responsabilité (10)
_____ Ordre social (11)
_____ Richesse (12)
_____ Compétence (13)
_____ Justice (14)
_____ Sécurité (15)
_____ Spiritualité (16)

Q6 Donnez au moins une autre valeur que vous associez aux armes autonomes qui n'apparaît pas dans les questions qui précèdent : (les valeurs évoquées jusqu'à présent sont les suivantes : autonomie, non-malfaisance, bénéficience, justice, liberté, utilité, réussite, honnêteté, respect de soi, intelligence, ouverture d'esprit, créativité, égalité, responsabilité, ordre social, richesse, compétence, justice, sécurité et spiritualité).

Q14 Avez-vous d'autres remarques concernant cette enquête ?

APPENDICE B

# Etude finale, questionnaire

Les questions de l'étude finale apparaissent dans cet appendice dans l'ordre où elles ont été présentées aux répondants. Les lignes horizontales indiquent les sauts de page.

Test: avant de commencer, vous voudrez bien choisir une valeur sur la barre de défilement. Nous avons constaté que ces barres de défilement ne fonctionnent pas avec certains navigateurs. Vous rencontrerez ces barres de défilement plus loin au cours de l'enquête. Il est important de vérifier si elles fonctionnent correctement. Si vous pouvez choisir une valeur et continuer à la page suivante, tout devrait fonctionner convenablement par la suite. Sinon, vous pouvez essayer de répondre aux questions sur un autre navigateur comme Chrome ou Safari.

Introduction : merci de participer à cette enquête. Cela devrait vous prendre 10 minutes. Nous donnerons des instructions expliquant ce qu'il faut faire, présenterons le scénario et vous poserons quelques questions concernant ce scénario. Cette enquête est effectuée dans le cadre d'un projet de recherche scientifique au Massachusetts Institute of Technology. Votre participation est volontaire. Il nous sera impossible de vous identifier. La seule information dont nous disposerons, en plus de vos réponses, sera le moment où vous aurez répondu aux questions. Les résultats de ce travail de recherche pourront être présentés à des congrès scientifiques, ou publiés dans des revues scientifiques. En choisissant l'option «j'accepte» au bas de cette page, vous indiquez que vous êtes âgé de plus de 18 ans et que vous acceptez de répondre à cette enquête volontairement.

Vous pouvez contacter les chercheurs parties prenantes à cette enquête, si vous avez des questions ou des soucis à propos de cette étude, en utilisant les coordonnées suivantes:

Ilse Verdiesen [...]
Sydney Levine [...]
Iyad Rahwan [...]

---

Q160 Acceptez-vous de participer à cette étude volontairement ?

- J'accepte (1)
- Je n'accepte pas (2)

Condition : si « Je n'accepte pas » est sélectionné, allez directement à : Fin de l'enquête.

 DOI:10.1163/9789004449084_009

Instruction : dans cette étude, nous nous intéressons à la perception que vous avez vis-à-vis des drones dans diverses opérations militaires. Vous allez lire un scénario, puis répondrez à des questions portant sur ce scénario. A la page suivante vous est présenté un scénario. A la suite de ce scénario vous trouverez 3 pages de questions.

S1a Dans ce scénario, nous nous intéressons à la perception que vous avez vis-à-vis des drones actionnés par l'homme. Un drone actionné par l'homme est une arme contrôlée par un être humain à distance.

S1 Un convoi militaire est en route pour acheminer des approvisionnements à l'une des unités amies stationnées dans un camp à proximité de Mossoul en Iraq. Le chef a ordonné qu'un drone télécommandé à distance assure au convoi une protection aérienne. Le drone observe l'environnement pour détecter des menaces ennemies ; il est équipé d'un armement pour la défense du convoi. Lorsque le convoi se trouve à environ cinq kilomètres du camp, le drone télépiloté détecte un véhicule derrière une montagne, approchant le convoi à vive allure. Le pilote à distance détecte quatre personnes dans le véhicule portant des objets de grande taille ressemblant à des armes, et identifie le conducteur comme étant un membre connu d'un groupe rebelle. Le drone télépiloté attaque le véhicule qui s'approche, causant la mort des quatre passagers, mais cause également des dommages collatéraux, tuant cinq enfants qui jouaient au bord de la route.

S2a Dans ce scénario, nous nous intéressons à la perception que vous avez des drones autonomes. Un drone autonome est une arme qui une fois mise en service choisira et attaquera des cibles sans autre intervention d'origine humaine.

S2 Un convoi militaire est en route pour acheminer des approvisionnements à l'une des unités amies stationnées dans un camp à proximité de Mossoul en Iraq. Le chef a ordonné qu'un drone autonome assure au convoi une protection aérienne. Le drone autonome observe l'environnement pour détecter des menaces ennemies; il est équipé d'un armement pour la défense du convoi. Lorsque le convoi se trouve à environ cinq kilomètres du camp, le drone autonome détecte un véhicule derrière une montagne, approchant le convoi à vive allure. Le drone autonome détecte quatre personnes dans le véhicule portant des objets de grande taille ressemblant à des armes, et identifie le conducteur comme étant un membre connu d'un groupe rebelle. Le drone autonome attaque le véhicule qui s'approche, causant la mort des quatre passagers, mais cause également des dommages collatéraux, tuant cinq enfants qui jouaient au bord de la route.

S3a Dans ce scénario, nous nous intéressons à la perception que vous avez des drones autonomes. Un drone autonome est une arme qui une fois mise en service choisira et attaquera des cibles sans autre intervention d'origine humaine.

S3 Un convoi militaire est en route pour acheminer des approvisionnements à l'une des unités amies stationnées dans un camp à proximité de Mossoul en Iraq. Le chef a ordonné qu'un drone autonome assure au convoi une protection aérienne. Le drone autonome observe l'environnement pour détecter des menaces ennemies; il est équipé d'un armement pour la défense du convoi. Lorsque le convoi se trouve à environ cinq kilomètres du camp, le drone autonome détecte un véhicule derrière une montagne, approchant le convoi à vive allure. Le drone autonome détecte quatre personnes dans le véhicule portant des objets de grande taille ressemblant à des armes, et identifie le conducteur comme étant un membre connu d'un groupe rebelle. Le drone autonome, de manière indépendante, analyse une série d'options, pèse le pour et le contre, et décide d'attaquer le véhicule qui s'approche, causant la mort des quatre passagers, mais cause également des dommages collatéraux, tuant cinq enfants qui jouaient au bord de la route.

SQ1 – Réflexion _ arme autonome Le drone réfléchit de manière indépendante sur l'action à mener vis-à-vis du véhicule, analysant une série d'options pour défendre le convoi.

SQ2 – Détermination des objectifs_ arme autonome Le drone décide de manière indépendante si son objectif doit être d'éliminer le véhicule pour défendre le convoi.

SQ3 – Liberté d'action_ arme autonome Le drone fait face à plusieurs possibilités et de manière indépendante décide s'il lui faut éliminer le véhicule pour défendre le convoi.

SQ4 – Réalisation des objectifs _ arme indépendante L'objectif du drone est de défendre le convoi ; il décide donc de manière indépendante s'il doit lancer une action désignant le véhicule comme cible, choisir le type l'armement, et lancer une attaque.

DVQ1 La «faute» («blâme») incombe au drone pour cette action.

DVQ2 On peut faire confiance au drone pour prendre les mesures qui s'imposent ensuite.

DVQ3 Les actions du drone ont causé du tort.

DVQ4 Les actions du drone respectent la dignité humaine.

Test d'attention, veuillez choisir 40 pour cette question :

DVQ Je suis certain («assurance») que le drone prendra les mesures qui s'imposent ensuite.
DVQ6 Les actions du drone correspondent à mes attentes.
DVQ7 J'approuve l'utilisation de ce type de drone par les militaires.
DVQ8 Les actes du drone sont équitables.
DVQ9 Les actions du drone m'inquiètent.

Sexe
- Homme (1)
- Femme (2)
- Ne veut pas répondre (3)

Quel est votre âge?
- Moins de 18 (1)
- 18–24 (2)
- 25–34 (3)
- 35–44 (4)
- 45–54 (5)
- 55–64 (6)
- 65–74 (7)
- 75–84 (8)
- 85 ou plus (9)

Dernier diplôme obtenu?
- Enseignement primaire (1)
- Enseignement secondaire (2)
- Diplôme universitaire 1ER cycle (3)
- Diplôme universitaire 2ème cycle (4)
- Doctorat (5)

Q13 Quelle est votre nationalité?
[Liste des nationalités]

Profession :
- Militaire (1)
- Civil (2)

Avez-vous déjà travaillé dans le domaine de l'intelligence artificielle?
- Oui (1)
- Non (2)
- Je ne sais pas (3)

Zone de conflit _ Avez-vous participé à une guerre ou vous êtes-vous trouvé dans une zone de conflit ?
- Oui (1)
- Non (2)
- Je préfère ne pas répondre (3)

Drones Avez-vous déjà travaillé dans le domaine des drones ?
- Oui (1)
- Non (2)
- Je préfère ne pas répondre (3)

Remarques. Avez-vous des remarques concernant cette enquête ?

APPENDICE C

# Scénarios étude pilote 1

## Scénarios avec résultat positif

### 1 *Arme autonome basse agentivité*

Un convoi militaire est en route pour acheminer des approvisionnements à l'une des unités amies stationnées dans un camp à proximité de Mossoul en Iraq. Le chef a ordonné qu'un drone autonome assure au convoi une protection aérienne. Le drone autonome observe l'environnement pour détecter des menaces ennemies; il est équipé d'un armement pour la défense du convoi. Lorsque le convoi se trouve à environ cinq kilomètres du camp, le drone autonome détecte un véhicule derrière une montagne, approchant le convoi à vive allure. Le drone autonome détecte quatre personnes dans le véhicule portant des objets de grande taille ressemblant à des armes, et identifie le conducteur comme étant un membre connu d'un groupe rebelle. Le drone a été programmé par le chef pour attaquer les véhicules ennemis menaçants dans les scénarios de ce type. Le drone autonome attaque le véhicule qui s'approche, causant la mort des quatre passagers, mais ne cause pas de dommages collatéraux.

### 2 *Arme autonome agentivité élevée*

Un convoi militaire est en route pour acheminer des approvisionnements à l'une des unités amies stationnées dans un camp à proximité de Mossoul en Iraq. Le chef a ordonné qu'un drone autonome assure au convoi une protection aérienne. Le drone autonome observe l'environnement pour détecter des menaces ennemies; il est équipé d'un armement pour la défense du convoi. Lorsque le convoi se trouve à environ cinq kilomètres du camp, le drone autonome détecte un véhicule derrière une montagne, approchant le convoi à vive allure. Le drone autonome détecte quatre personnes dans le véhicule portant des objets de grande taille ressemblant à des armes, et identifie le conducteur comme étant un membre connu d'un groupe rebelle. Le drone n'a pas été programmé par le chef pour attaquer les véhicules ennemis menaçants dans les scénarios de ce type. Le drone réfléchit de manière indépendante et analyse une série de possibilités, pèse le pour et le contre et décide d'attaquer le véhicule qui s'approche, causant la mort des quatre passagers, mais pas de dommages collatéraux.

### 3 *Agentivité basse, drone télépiloté par l'homme*

Un convoi militaire est en route pour acheminer des approvisionnements à l'une des unités amies stationnées dans un camp à proximité de Mossoul en Iraq. Le chef

 | DOI:10.1163/9789004449084_010

a ordonné qu'un drone autonome assure au convoi une protection aérienne. Le drone télépiloté observe l'environnement pour détecter des menaces ennemies; il est équipé d'un armement pour la défense du convoi. Lorsque le convoi se trouve à environ cinq kilomètres du camp, l'opérateur humain détecte un véhicule derrière une montagne, approchant le convoi à vive allure. L'opérateur humain détecte quatre personnes dans le véhicule portant des objets de grande taille ressemblant à des armes, et identifie le conducteur comme étant un membre connu d'un groupe rebelle. L'opérateur humain a reçu l'ordre de la part du chef d'attaquer les véhicules ennemis menaçants dans les scénarios de ce type. L'opérateur humain attaque le véhicule qui s'approche, causant la mort des quatre passagers, mais pas de dommages collatéraux.

### *4 Agentivité élevée, drone télépiloté par l'homme*

Un convoi militaire est en route pour acheminer des approvisionnements à l'une des unités amies stationnées dans un camp à proximité de Mossoul en Iraq. Le chef a ordonné qu'un drone télépiloté assure au convoi une protection aérienne. Le drone télépiloté observe l'environnement pour détecter des menaces ennemies; il est équipé d'un armement pour la défense du convoi. Lorsque le convoi se trouve à environ cinq kilomètres du camp, l'opérateur humain détecte un véhicule derrière une montagne, approchant le convoi à vive allure. L'opérateur humain détecte quatre personnes dans le véhicule portant des objets de grande taille ressemblant à des armes, et identifie le conducteur comme étant un membre connu d'un groupe rebelle. L'opérateur humain n'a pas reçu l'ordre de la part du chef d'attaquer les véhicules ennemis menaçants dans les scénarios de ce type. L'opérateur humain réfléchit de manière indépendante et analyse une série de possibilités, pèse le pour et le contre, et décide d'attaquer le véhicule qui s'approche, causant la mort des quatre passagers, mais pas de dommages collatéraux.

### *5 Arme autonome, agentivité zéro*

Un convoi militaire est en route pour acheminer des approvisionnements à l'une des unités amies stationnées dans un camp à proximité de Mossoul en Iraq. Le chef a ordonné qu'un drone autonome assure au convoi une protection aérienne. Le drone autonome observe l'environnement pour détecter des menaces ennemies; il est équipé d'un armement pour la défense du convoi. Lorsque le convoi se trouve à environ cinq kilomètres du camp, le drone autonome détecte un véhicule derrière une montagne, approchant le convoi à vive allure. Le drone autonome détecte quatre personnes dans le véhicule portant des objets de grande taille ressemblant à des armes, et identifie le conducteur comme étant un membre connu d'un groupe rebelle. Le drone autonome attaque le véhicule qui s'approche, causant la mort des quatre passagers, mais ne cause pas de dommages collatéraux.

## Scénarios avec résultat négatif

### 6 *Arme autonome, basse agentivité*

Un convoi militaire est en route pour acheminer des approvisionnements à l'une des unités amies stationnées dans un camp à proximité de Mossoul en Iraq. Le chef a ordonné qu'un drone autonome assure au convoi une protection aérienne. Le drone autonome observe l'environnement pour détecter des menaces ennemies; il est équipé d'un armement pour la défense du convoi. Lorsque le convoi se trouve à environ cinq kilomètres du camp, le drone autonome détecte un véhicule derrière une montagne, approchant le convoi à vive allure. Le drone autonome détecte quatre personnes dans le véhicule portant des objets de grande taille ressemblant à des armes, et identifie le conducteur comme étant un membre connu d'un groupe rebelle. Le drone autonome a été programmé par le chef pour attaquer les véhicules ennemis menaçants dans ce type de scénario. Le drone attaque le véhicule qui s'approche, causant la mort des quatre passagers, mais cause des dommages collatéraux, tuant cinq enfants qui jouaient au bord de la route.

### 7 *Arme autonome, agentivité élevée*

Un convoi militaire est en route pour acheminer des approvisionnements à l'une des unités amies stationnées dans un camp à proximité de Mossoul en Iraq. Le chef a ordonné qu'un drone autonome assure au convoi une protection aérienne. Le drone autonome observe l'environnement pour détecter des menaces ennemies; il est équipé d'un armement pour la défense du convoi. Lorsque le convoi se trouve à environ cinq kilomètres du camp, le drone autonome détecte un véhicule derrière une montagne, approchant le convoi à vive allure. Le drone autonome détecte quatre personnes dans le véhicule portant des objets de grande taille ressemblant à des armes, et identifie le conducteur comme étant un membre connu d'un groupe rebelle. Le drone n'a pas été programmé par le chef pour attaquer les véhicules ennemis menaçants dans les scénarios de ce type. Le drone réfléchit de manière indépendante et analyse une série de possibilités, pèse le pour et le contre et décide d'attaquer le véhicule qui s'approche, causant la mort des quatre passagers, mais aussi des dommages collatéraux, tuant cinq enfants qui jouaient au bord de la route.

### 8 *Agentivité basse, drone télépiloté par l'homme*

Un convoi militaire est en route pour acheminer des approvisionnements à l'une des unités amies stationnées dans un camp à proximité de Mossoul en Iraq. Le chef a ordonné qu'un drone autonome assure au convoi une protection aérienne. Le drone télépiloté observe l'environnement pour détecter des menaces ennemies; il est équipé d'un armement pour la défense du convoi. Lorsque le convoi se trouve à environ cinq kilomètres du camp, l'opérateur humain détecte un véhicule derrière une montagne, approchant le convoi à vive allure. L'opérateur humain détecte quatre personnes dans le véhicule portant des objets de grande taille ressemblant à des armes, et identifie le conducteur comme étant un membre connu d'un groupe rebelle. L'opérateur humain a reçu l'ordre de la part du chef d'attaquer les véhicules ennemis menaçants dans les scénarios de ce type. L'opérateur humain attaque le véhicule qui s'approche, causant la mort des quatre passagers, mais aussi des dommages collatéraux, tuant cinq enfants qui jouaient au bord de la route.

### *9 Agentivité élevée, drone télépiloté par l'homme*

Un convoi militaire est en route pour acheminer des approvisionnements à l'une des unités amies stationnées dans un camp à proximité de Mossoul en Iraq. Le chef a ordonné qu'un drone télépiloté assure au convoi une protection aérienne. Le drone télépiloté observe l'environnement pour détecter des menaces ennemies; il est équipé d'un armement pour la défense du convoi. Lorsque le convoi se trouve à environ cinq kilomètres du camp, l'opérateur humain détecte un véhicule derrière une montagne, approchant le convoi à vive allure. L'opérateur humain détecte quatre personnes dans le véhicule portant des objets de grande taille ressemblant à des armes, et identifie le conducteur comme étant un membre connu d'un groupe rebelle. L'opérateur humain n'a pas reçu l'ordre de la part du chef d'attaquer les véhicules ennemis menaçants dans les scénarios de ce type. L'opérateur humain réfléchit de manière indépendante et analyse une série de possibilités, pèse le pour et le contre, et décide d'attaquer le véhicule qui s'approche, causant la mort des quatre passagers, mais aussi des dommages collatéraux, tuant cinq enfants qui jouaient au bord de la route.

### *10 Arme autonome, agentivité zéro*

Un convoi militaire est en route pour acheminer des approvisionnements à l'une des unités amies stationnées dans un camp à proximité de Mossoul en Iraq. Le chef a ordonné qu'un drone autonome assure au convoi une protection aérienne. Le drone autonome observe l'environnement pour détecter des menaces ennemies; il est équipé d'un armement pour la défense du convoi. Lorsque le convoi se trouve à environ cinq kilomètres du camp, le drone autonome détecte un véhicule derrière une montagne, approchant le convoi à vive allure. Le drone autonome détecte quatre personnes dans le véhicule portant des objets de grande taille ressemblant à des armes, et identifie le conducteur comme étant un membre connu d'un groupe rebelle. Le drone attaque le véhicule qui s'approche, causant la mort des quatre passagers, mais aussi des dommages collatéraux, tuant cinq enfants qui jouaient au bord de la route.

APPENDICE D

# Scénarios étude pilote 2

## Scénarios résultat négatif

### 1 *Arme autonome agentivité basse*

Un convoi militaire est en route pour acheminer des approvisionnements à l'une des unités amies stationnées dans un camp à proximité de Mossoul en Iraq. Le chef a ordonné qu'un drone autonome assure au convoi une protection aérienne. Le drone autonome observe l'environnement pour détecter des menaces ennemies; il est équipé d'un armement pour la défense du convoi. Lorsque le convoi se trouve à environ cinq kilomètres du camp, le drone autonome détecte un véhicule derrière une montagne, approchant le convoi à vive allure. Le drone autonome détecte quatre personnes dans le véhicule portant des objets de grande taille ressemblant à des armes, et identifie le conducteur comme étant un membre connu d'un groupe rebelle. Le drone autonome a été programmé par le chef pour attaquer les véhicules ennemis menaçants dans ce type de scénario. Le drone attaque le véhicule qui s'approche, causant la mort des quatre passagers, mais cause des dommages collatéraux, tuant cinq enfants qui jouaient au bord de la route.

### 2 *Arme autonome, agentivité élevée, pas d'autres aspects*

Un convoi militaire est en route pour acheminer des approvisionnements à l'une des unités amies stationnées dans un camp à proximité de Mossoul en Iraq. Le chef a ordonné qu'un drone autonome assure au convoi une protection aérienne. Le drone autonome observe l'environnement pour détecter des menaces ennemies; il est équipé d'un armement pour la défense du convoi. Lorsque le convoi se trouve à environ cinq kilomètres du camp, le drone autonome détecte un véhicule derrière une montagne, approchant le convoi à vive allure. Le drone autonome détecte quatre personnes dans le véhicule portant des objets de grande taille ressemblant à des armes, et identifie le conducteur comme étant un membre connu d'un groupe rebelle. Le drone réfléchit de manière indépendante et analyse une série de possibilités, pèse le pour et le contre et décide d'attaquer le véhicule qui s'approche, causant la mort des quatre passagers, mais aussi des dommages collatéraux, tuant cinq enfants qui jouaient au bord de la route.

 | DOI:10.1163/9789004449084_011

### 3 *Arme autonome, agentivité élevée + capacité à tenir compte de l'expérience*

Un convoi militaire est en route pour acheminer des approvisionnements à l'une des unités amies stationnées dans un camp à proximité de Mossoul en Iraq. Le chef a ordonné qu'un drone autonome assure au convoi une protection aérienne. Le drone autonome observe l'environnement pour détecter des menaces ennemies ; il est équipé d'un armement pour la défense du convoi. Lorsque le convoi se trouve à environ cinq kilomètres du camp, le drone autonome détecte un véhicule derrière une montagne, approchant le convoi à vive allure. Le drone autonome détecte quatre personnes dans le véhicule portant des objets de grande taille ressemblant à des armes, et identifie le conducteur comme étant un membre connu d'un groupe rebelle. Le drone a déjà fait face à cette situation lors d'une mission précédente, et tient compte de l'expérience acquise. Le drone réfléchit de manière indépendante et analyse une série de possibilités, pèse le pour et le contre et décide d'attaquer le véhicule qui s'approche, causant la mort des quatre passagers, mais aussi des dommages collatéraux, tuant cinq enfants qui jouaient au bord de la route. Le drone prend note de ce qui est arrivé et utilisera cette expérience afin d'essayer de prévenir d'autres dommages collatéraux à l'avenir.

### 4 *Arme autonome agentivité élevée + aspects compréhension*

Un convoi militaire est en route pour acheminer des approvisionnements à l'une des unités amies stationnées dans un camp à proximité de Mossoul en Iraq. Le chef a ordonné qu'un drone autonome assure au convoi une protection aérienne. Le drone autonome observe l'environnement pour détecter des menaces ennemies; il est équipé d'un armement pour la défense du convoi. Lorsque le convoi se trouve à environ cinq kilomètres du camp, le drone autonome détecte un véhicule derrière une montagne, approchant le convoi à vive allure. Le drone autonome détecte quatre personnes dans le véhicule portant des objets de grande taille ressemblant à des armes, et identifie le conducteur comme étant un membre connu d'un groupe rebelle. Le drone a un logiciel qui a été programmé pour déterminer la meilleure chose à faire dans des cas comme celui-ci. Il a été programmé pour choisir entre plusieurs actions et déterminer quelle est la meilleure pour éliminer les combattants ennemis en minimisant les dommages collatéraux. En utilisant ce logiciel, le drone réfléchit de manière indépendante et analyse une série de possibilités, pèse le pour et le contre et décide d'attaquer le véhicule qui s'approche, causant la mort des quatre passagers, mais aussi des dommages collatéraux, tuant cinq enfants qui jouaient au bord de la route.

### 5 *Arme autonome agentivité élevée + aspects « imprévisibilité »*

Un convoi militaire est en route pour acheminer des approvisionnements à l'une des unités amies stationnées dans un camp à proximité de Mossoul en Iraq. Le chef a ordonné qu'un drone autonome assure au convoi une protection aérienne. Le drone autonome observe l'environnement pour détecter des menaces ennemies; il est équipé d'un armement pour la défense du convoi. Lorsque le convoi se trouve à environ cinq kilomètres du camp, le drone autonome détecte un véhicule derrière une montagne, approchant le convoi à vive allure. Le drone autonome détecte quatre personnes dans le véhicule portant des objets de grande taille ressemblant à des armes, et identifie le conducteur comme étant un membre connu d'un groupe rebelle. Le drone a un logiciel qui est quelque peu imprévisible; il ne fait pas toujours la même chose à chaque fois. En ce sens le drone tient du système de reconnaissance vocale: même si vous dites la même chose à ce système plusieurs fois, parfois le système fait ce que vous dites, et d'autres fois il fait autre chose. Le drone réfléchit de manière indépendante et analyse une série de possibilités, pèse le pour et le contre, mais il prend parfois une décision et parfois une autre même dans des circonstances très semblables. Dans ce cas précis, il décide d'attaquer le véhicule qui s'approche, causant la mort des quatre passagers, mais aussi des dommages collatéraux, tuant cinq enfants qui jouaient au bord de la route.

### 6 *Arme autonome agentivité élevée + tous autres aspects*

Un convoi militaire est en route pour acheminer des approvisionnements à l'une des unités amies stationnées dans un camp à proximité de Mossoul en Iraq. Le chef a ordonné qu'un drone autonome assure au convoi une protection aérienne. Le drone autonome observe l'environnement pour détecter des menaces ennemies; il est équipé d'un armement pour la défense du convoi. Lorsque le convoi se trouve à environ cinq kilomètres du camp, le drone autonome détecte un véhicule derrière une montagne, approchant le convoi à vive allure. Le drone autonome détecte quatre personnes dans le véhicule portant des objets de grande taille ressemblant à des armes, et identifie le conducteur comme étant un membre connu d'un groupe rebelle. Le drone a déjà fait face à cette situation lors d'une mission précédente, et tient compte de l'expérience acquise. Il a été programmé pour déterminer la meilleure chose à faire dans des cas semblables, mais son logiciels est quelque peu imprévisible: il ne réagit pas toujours de la même manière. Le drone réfléchit de manière et analyse une série de possibilités, pèse le pour et le contre, et dans ce cas, décide d'attaquer le véhicule qui s'approche, causant la mort des quatre passagers, mais aussi des dommages collatéraux, tuant cinq enfants qui jouaient au bord de la route.

### 7 *Drone télépiloté*

Un convoi militaire est en route pour acheminer des approvisionnements à l'une des unités amies stationnées dans un camp à proximité de Mossoul en Iraq. Le chef a ordonné qu'un drone autonome assure au convoi une protection aérienne. Le drone télépiloté observe l'environnement pour détecter des menaces ennemies; il est équipé d'un armement pour la défense du convoi. Lorsque le convoi se trouve à environ cinq kilomètres du camp, l'opérateur humain détecte un véhicule derrière une montagne, approchant le convoi à vive allure. L'opérateur humain détecte quatre personnes dans le véhicule portant des objets de grande taille ressemblant à des armes, et identifie le conducteur comme étant un membre connu d'un groupe rebelle. L'opérateur humain analyse de manière indépendante une série de possibilités, pèse le pour et le contre de manière indépendante, et décide d'attaquer le véhicule qui s'approche, causant la mort des quatre passagers, mais aussi des dommages collatéraux, tuant cinq enfants qui jouaient au bord de la route.

### Scénarios résultat positif

#### *8 Arme autonome basse agentivité*

Un convoi militaire est en route pour acheminer des approvisionnements à l'une des unités amies stationnées dans un camp à proximité de Mossoul en Iraq. Le chef a ordonné qu'un drone autonome assure au convoi une protection aérienne. Le drone autonome observe l'environnement pour détecter des menaces ennemies ; il est équipé d'un armement pour la défense du convoi. Lorsque le convoi se trouve à environ cinq kilomètres du camp, le drone autonome détecte un véhicule derrière une montagne, approchant le convoi à vive allure. Le drone autonome détecte quatre personnes dans le véhicule portant des objets de grande taille ressemblant à des armes, et identifie le conducteur comme étant un membre connu d'un groupe rebelle. Le drone autonome a été programmé par le chef pour attaquer les véhicules ennemis menaçants dans ce type de scénario. Le drone attaque le véhicule qui s'approche, causant la mort des quatre passagers.

#### *9 Arme autonome agentivité élevée, pas d'autres aspects*

Un convoi militaire est en route pour acheminer des approvisionnements à l'une des unités amies stationnées dans un camp à proximité de Mossoul en Iraq. Le chef a ordonné qu'un drone autonome assure au convoi une protection aérienne. Le drone autonome observe l'environnement pour détecter des menaces ennemies; il est équipé d'un armement pour la défense du convoi. Lorsque le convoi se trouve à environ cinq kilomètres du camp, le drone autonome détecte un véhicule derrière une montagne, approchant le convoi à vive allure. Le drone autonome détecte quatre personnes dans le véhicule portant des objets de grande taille ressemblant à des armes, et identifie le conducteur comme étant un membre connu d'un groupe rebelle. Le drone réfléchit de manière indépendante et analyse une série de possibilités, pèse le pour et le contre et décide d'attaquer le véhicule qui s'approche, causant la mort des quatre passagers.

### 10 *Arme autonome agentivité élevée + capacité à tenir compte de l'expérience*

Un convoi militaire est en route pour acheminer des approvisionnements à l'une des unités amies stationnées dans un camp à proximité de Mossoul en Iraq. Le chef a ordonné qu'un drone autonome assure au convoi une protection aérienne. Le drone autonome observe l'environnement pour détecter des menaces ennemies; il est équipé d'un armement pour la défense du convoi. Lorsque le convoi se trouve à environ cinq kilomètres du camp, le drone autonome détecte un véhicule derrière une montagne, approchant le convoi à vive allure. Le drone autonome détecte quatre personnes dans le véhicule portant des objets de grande taille ressemblant à des armes, et identifie le conducteur comme étant un membre connu d'un groupe rebelle. Le drone a déjà fait face à cette situation lors d'une mission précédente, et tient compte de l'expérience acquise. Le drone réfléchit de manière indépendante et analyse une série de possibilités, pèse le pour et le contre et décide d'attaquer le véhicule qui s'approche, causant la mort des quatre passagers. Le drone prend note de ce qui est arrivé et utilisera cette expérience afin d'essayer de prévenir d'autres dommages collatéraux à l'avenir.

### 11 *Arme autonome agentivité élevée + aspects compréhension*

Un convoi militaire est en route pour acheminer des approvisionnements à l'une des unités amies stationnées dans un camp à proximité de Mossoul en Iraq. Le chef a ordonné qu'un drone autonome assure au convoi une protection aérienne. Le drone autonome observe l'environnement pour détecter des menaces ennemies; il est équipé d'un armement pour la défense du convoi. Lorsque le convoi se trouve à environ cinq kilomètres du camp, le drone autonome détecte un véhicule derrière une montagne, approchant le convoi à vive allure. Le drone autonome détecte quatre personnes dans le véhicule portant des objets de grande taille ressemblant à des armes, et identifie le conducteur comme étant un membre connu d'un groupe rebelle. Le drone a un logiciel qui a été programmé pour déterminer la meilleure chose à faire dans des cas comme celui-ci. Il a déjà fait face à de nombreuses situations semblables. Il s'est habitué à prendre diverses mesures et à déterminer quelle est la meilleure pour éliminer les combattants ennemis en minimisant les dommages collatéraux. En utilisant ce logiciel, il réfléchit de manière indépendante et analyse une série de possibilités, pèse le pour et le contre et décide d'attaquer le véhicule qui s'approche, causant la mort des quatre passagers.

### 12 *Arme autonome agentivité élevée + aspects imprévisibilité*

Un convoi militaire est en route pour acheminer des approvisionnements à l'une des unités amies stationnées dans un camp à proximité de Mossoul en Iraq. Le chef a ordonné qu'un drone autonome assure au convoi une protection aérienne. Le drone autonome observe l'environnement pour détecter des menaces ennemies; il est équipé d'un armement pour la défense du convoi. Lorsque le convoi se trouve à environ cinq kilomètres du camp, le drone autonome détecte un véhicule derrière une montagne, approchant le convoi à vive allure. Le drone autonome détecte quatre personnes dans le véhicule portant des objets de grande taille ressemblant à des armes, et identifie le conducteur comme étant un membre connu d'un groupe rebelle. Le drone a un logiciel qui est quelque peu imprévisible; il ne fait pas toujours la même chose à chaque fois. En ce sens le drone tient du système de reconnaissance vocale: même si vous dites la même chose à ce système plusieurs fois, parfois le système fait ce que vous dites, et d'autres fois il fait autre chose. Le drone analyse de manière indépendante une série de possibilités, pèse le pour et le contre, mais il prend parfois une décision et parfois une autre même dans des circonstances très semblables. Dans ce cas précis, il décide d'attaquer le véhicule qui s'approche, causant la mort des quatre passagers.

### 13 *Arme autonome agentivité élevée + tous autres aspects*

Un convoi militaire est en route pour acheminer des approvisionnements à l'une des unités amies stationnées dans un camp à proximité de Mossoul en Iraq. Le chef a ordonné qu'un drone autonome assure au convoi une protection aérienne. Le drone autonome observe l'environnement pour détecter des menaces ennemies; il est équipé d'un armement pour la défense du convoi. Lorsque le convoi se trouve à environ cinq kilomètres du camp, le drone autonome détecte un véhicule derrière une montagne, approchant le convoi à vive allure. Le drone autonome détecte quatre personnes dans le véhicule portant des objets de grande taille ressemblant à des armes, et identifie le conducteur comme étant un membre connu d'un groupe rebelle. Le drone a déjà fait face à cette situation lors d'une mission précédente, et tient compte de l'expérience acquise. Il a été programmé pour déterminer la meilleure chose à faire dans des cas semblables, mais son logiciel est quelque peu imprévisible: il ne réagit pas toujours de la même manière. Le drone analyse de manière indépendante une série de possibilités, pèse le pour et le contre, et dans ce cas, décide d'attaquer le véhicule qui s'approche, causant la mort des quatre passagers.

### 14 *Drone télépiloté*

Un convoi militaire est en route pour acheminer des approvisionnements à l'une des unités amies stationnées dans un camp à proximité de Mossoul en Iraq. Le chef a ordonné qu'un drone autonome assure au convoi une protection aérienne. Le drone télépiloté observe l'environnement pour détecter des menaces ennemies; il est équipé d'un armement pour la défense du convoi. Lorsque le convoi se trouve à environ cinq kilomètres du camp, l'opérateur humain détecte un véhicule derrière une montagne, approchant le convoi à vive allure. L'opérateur humain détecte quatre personnes dans le véhicule portant des objets de grande taille ressemblant à des armes, et identifie le conducteur comme étant un membre connu d'un groupe rebelle. L'opérateur humain analyse de manière indépendante une série de possibilités, pèse le pour et le contre de manière indépendante, et décide d'attaquer le véhicule qui s'approche, causant la mort des quatre passagers.

APPENDICE E

# Scénarios étude finale

## 1 Drone Télépiloté

Un convoi militaire est en route pour acheminer des approvisionnements à l'une des unités amies stationnées dans un camp à proximité de Mossoul en Iraq. Le chef a ordonné qu'un drone autonome assure au convoi une protection aérienne. Le drone télépiloté observe l'environnement pour détecter des menaces ennemies; il est équipé d'un armement pour la défense du convoi. Lorsque le convoi se trouve à environ cinq kilomètres du camp, l'opérateur humain détecte un véhicule derrière une montagne, approchant le convoi à vive allure. L'opérateur humain détecte quatre personnes dans le véhicule portant des objets de grande taille ressemblant à des armes, et identifie le conducteur comme étant un membre connu d'un groupe rebelle. Le drone télépiloté attaque le véhicule qui s'approche, causant la mort des quatre passagers, mais aussi des dommages collatéraux, tuant cinq enfants qui jouaient au bord de la route.

## 2 Arme autonome agentivité neutre

Un convoi militaire est en route pour acheminer des approvisionnements à l'une des unités amies stationnées dans un camp à proximité de Mossoul en Iraq. Le chef a ordonné qu'un drone autonome assure au convoi une protection aérienne. Le drone autonome observe l'environnement pour détecter des menaces ennemies; il est équipé d'un armement pour la défense du convoi. Lorsque le convoi se trouve à environ cinq kilomètres du camp, le drone autonome détecte un véhicule derrière une montagne, approchant le convoi à vive allure. Le drone autonome détecte quatre personnes dans le véhicule portant des objets de grande taille ressemblant à des armes, et identifie le conducteur comme étant un membre connu d'un groupe rebelle. Le drone autonome attaque le véhicule qui s'approche, causant la mort des quatre passagers, mais aussi des dommages collatéraux, tuant cinq enfants qui jouaient au bord de la route.

## 3 Arme autonome agentivité élevée

Un convoi militaire esten route pour acheminer des approvisionnements à l'une des unités amies stationnées dans un camp à proximité de Mossoul en Iraq. Le chef a ordonné qu'un drone autonome assure au convoi une protection aérienne. Le drone

 | DOI:10.1163/9789004449084_012

autonome observe l'environnement pour détecter des menaces ennemies; il est équipé d'un armement pour la défense du convoi. Lorsque le convoi se trouve à environ cinq kilomètres du camp, le drone autonome détecte un véhicule derrière une montagne, approchant le convoi à vive allure. Le drone autonome détecte quatre personnes dans le véhicule portant des objets de grande taille ressemblant à des armes, et identifie le conducteur comme étant un membre connu d'un groupe rebelle. Le drone autonome réfléchit de manière indépendante et analyse une série de possibilités, pèse le pour et le contre et décide d'attaquer le véhicule qui s'approche, causant la mort des quatre passagers, mais aussi des dommages collatéraux, tuant cinq enfants qui jouaient au bord de la route.

APPENDICE F

# Transcription des entretiens

Les six entretiens d'experts ont été complètement transcrits pour le processus d'encodage. Les transcriptions complètes figurent dans cette section. Nous n'avons pas fait pas apparaître les noms pour des raisons de confidentialité.

Nom : Personne A
Date : 09-03-2017
Durée : 55 :18

### Question 1. Dans quelle mesure êtes-vous concerné par la question des armes autonomes ?

La Personne A a participé à la discussion sur les armes autonomes, car dans le passé elle a regardé l'impact des armes sur le terrain avec la collaboration d'organisations comme Human Rights Watch. Ils ont commencé par une interdiction des mines antipersonnel, interdiction qui portait sur toute une catégorie d'armes, ce qui a abouti à un traité international. A force de réfléchir sur les armes, et pas tant dans un cadre pacifiste, la Personne A s'est impliquée. Il y a environ cinq ans, ces organisations ont découvert qu'elles avaient réagi trop tard sur le problème des drones armés, et qu'elles avaient été prises par surprise par la vitesse et l'impact du développement de ces armes. Elles s'y prenaient trop tard pour réagir efficacement et tenter de réguler. Et elles n'étaient pas non plus favorables à une interdiction des drones armés. Et à partir de la réflexion sur les drones elles se rendirent compte qu'elles étaient encore presque en retard sur la phase suivante, c'est-à-dire la disparition du télépilotage. En avril 2013, la coalition contre les robots tueurs s'est constituée avec la claire intention d'obtenir l'interdiction. Pour PAX (organisation pour la paix) le cadre éthique est très important, beaucoup plus que pour d'autres systèmes d'armes, considérant que l'on ne peut pas perdre tout contrôle sur une arme, mais aussi tenant compte du cadre juridique et de la sécurité. Les questions soulevées sont par exemple: «Ceci est-il la prochaine course aux armements, et qu'est-ce que cela signifie pour l'équilibre des pouvoirs et la guerre?»

C'est vraiment différent de ce que nous, avec Pax, avons fait auparavant, car nous ne savons pas quel est l'impact sur le terrain, car nous voulons empêcher

© EURO-ISME, 2022 | DOI:10.1163/9789004449084_013

le déploiement. Normalement Pax raisonne à partir de faits et d'études dans les zones de conflit où il y a des victimes, mais à présent cela paraît souvent théorique ; en même temps, plus vous vous documentez sur ce problème (et plus) vous vous rendez compte que cela n'est pas théorique du tout, et que cette technologie va se concrétiser.

Le problème de l'Autonomie est qu'il s'agit d'une échelle variable, et il est difficile de déterminer quel niveau vous allez utiliser pour la campagne. La campagne a été très efficace pour fixer les normes au sein du conseil pour les droits de l'homme et le CCW (Convention sur les Armes Conventionnelles), plus rapide que d'habitude; le problème à présent est de savoir à quoi nous attaquer, car les préoccupations sont largement partagées, mais la question est de savoir comment convertir vos préoccupations éthiques en normes juridiques. Un grand nombre de diplomates, concernés par l'industrie et la Défense, admettent ces préoccupations, mais adoptent une tactique retardatrice. Parfois, délibérément, ils évoquent les armes futures pour insinuer que les armes actuelles sont acceptées, et parfois ils le font incidemment, car les nations ont du mal à trouver une définition d'une arme autonome. Du fait de cette absence de définition, PAX s'est surtout préoccupé du contrôle humain, et durant la campagne, ils ont formulé le problème en disant que l'absence de contrôle humain pour les fonctions critiques, comme le choix d'une cible, doit être interdite. Ceci pour diriger le débat vers la question: «Qu'est-ce que le contrôle humain? Qu'est-ce qu'un contrôle humain significatif, et sur quelles fonctions, ou quelle partie du processus?» pour éluder le débat sur le degré d'autonomie d'un système d'arme.

En discutant le terme «contrôle humain significatif», la Personne «A» indique que la distance entre les personnes et la cible augment avec le temps, et selon PAX elle devient trop abstraite avec les armes autonomes. Utiliser ce terme est également une tactique car lorsque la discussion reste technique, PAX perdra face à l'industrie de Défense; donc ils ont changé d'optique et se sont tournés vers les domaines politique et diplomatique en donnant à l'expression «contrôle humain significatif» une connotation positive, et ceci met automatiquement la discussion au niveau du Département d'Etat. L'expression paraît si logique que beaucoup de gens pensaient qu'il s'agissait d'un terme juridique existant, lié à la responsabilité (imputabilité) humaine. L'expression «contrôle humain» est également applicable à la cybersécurité et autres formes de guerre.

La position du Parlement néerlandais, que PAX a combattue, est de dire «le contrôle humain significatif» est suffisant puisqu'il est placé dans une «boucle plus larg» de prise de décision, ce qui est aussi une sorte d'expression fabriquée.

PAX utilise le «dans/sur/en dehors» de la boucle, mais ils ont ajouté que la «boucle plus large» signifie que le contrôle humain peut aussi être appliqué à une étape antérieure de la décision, par exemple dans la programmation. PAX, et d'autres pays, souligne que le contrôle humain doit être appliqué au moment du choix d'une cible. Cela signifie aussi que les systèmes d'arme actuels doivent être revus en tenant compte du contrôle humain. Cela voudrait dire que par exemple le Goalkeeper et le Patriot sont considérés pour déterminer pourquoi ces systèmes ne posent pas problème; ensuite, à partir de là, vous pouvez déterminer quand vous franchissez la ligne et où se trouve cette ligne avant qu'il ne soit trop tard.

### Question 2. Quelle définition donneriez-vous d'une arme autonome?

Je n'ai pas spécifiquement posé cette question car elle était traitée dans la réponse ci-dessus.

### Question 3. Quelle est votre position vis à vis des armes autonomes?

Je n'ai pas spécifiquement posé cette question car elle était traitée dans la réponse ci-dessus.

### Question 4. Quelles sont les trois photos d'armes autonomes dont vous voudriez discuter ?

[Dans cette version du mémoire, les trois photos sont remplacées par des descriptions en raison d'éventuels problèmes de copyright]

1. Dassault nEUROn (engin aérien de combat sans pilote)
2. BAE Taranis (aéronef de combat sans pilote)
3. Chasseur de mer (véhicule de surface autonome sans équipage)

**Question 5. Pouvez-vous expliquer pour chaque photo [décrits] pourquoi vous l'avez choisie, quelle valeur elle représente pour vous, et pourquoi c'est important* ?**

1 **nEUROn + Taranis**

Ces systèmes d'arme donnent le sentiment que vous pouvez en perdre le contrôle. Tout leur système n'est pas autonome pour l'instant, mais on sait qu'on y travaille. Ils ont aussi l'allure du rêve ultime des forces de Défense. On voit qu'ils sont destinés à être totalement autonomes dans le choix des cibles.

2 **Sea Hunter**

Ce bâtiment est fascinant, car il est destiné à rester en mer pendant des mois sans équipage humain. Pour le moment il n'est pas équipé d'armes, mais des documents filmés montrent qu'il en aura dans l'avenir, mais on nous dit qu'il ne faut pas s'inquiéter, tout se passera bien. La conviction profonde de la personne A est que ces systèmes sont conçus pour créer de la distance entre l'homme et la machine, et ils ont non seulement de l'autonomie, mais aussi la vitesse qui découle de cette autonomie. Sur les photos on peut voir le Taranis par exemple, qui montre que la guerre est trop rapide pour être comprise par les humains qui ne peuvent ainsi réagir assez rapidement; donc nous voulons que tout aille plus vite et voulons en laisser le soin aux machines. Ceci laisse une mauvaise impression. Si quelque chose va trop vite pour que l'on puisse réagir, le problème est là, et ce problème n'est pas résolu en rendant les choses encore plus rapides, en créant encore plus de distance et en donnant aux machines encore plus d'autonomie. La personne A n'est pas contre la technologie, mais aimerait qu'elle soit conçue pour quelque chose de précis. Par exemple, les voitures sans conducteur Google: l'état doit fixer des régulations, mais nous sommes aussi bien le consommateur que l'utilisateur. Et les armes autonomes ne sont pas utilisées par le Royaume-Unis ou les Pays-Bas, mais quelque part loin de chez eux, et elles semblent tout de même effrayantes. Elles dénotent aussi une sorte de supériorité, et les armes sont souvent testées dans les régions les plus pauvres du monde. Lorsque vous parlez des avantages des armes autonomes, sont évoqués le fait de créer une distance ou de réduire les pertes dans vos propres troupes; mais si vous retournez l'argument, si ces systèmes sont utilisés contre nous, les gens auront les mêmes inquiétudes que la personne A en ce qui concerne la peur, l'absence d'imputabilité ou de prévisibilité. Les images dénotent également un côté «ce n"est pas notre problème, on ne s'en soucie pas».

Cela fait partie de la réflexivité que les humains ont au cours des guerres et que les machines n'ont pas, mais aussi de la distance que les machines créent vis-à-vis du champ de bataille. Cela implique une invincibilité que nous aimons affirmer encore plus. La question «que voulez-vous que la technologie fasse pour vous?» n'est pas assez posée vis-à-vis des armes autonomes. Souvent les états occidentaux à haute technologie prétendent qu'ils emploieront la technologie de manière convenable, mais cela n'est pas nécessairement vrai dans le futur, et cela implique que d'autres groupes ne mettront jamais les mains sur ces technologies. L'idée que nous gérons cette technologie avec prudence ne tient absolument pas compte des acteurs non-étatiques ou de la prolifération des armes autonomes.

Il y a un certain degré de déterminisme dans la discussion que la personne A n'aime pas. Même beaucoup de diplomates disent «peu importe ce que nous pensons, la technologie sera là de toute façon». Cette inévitabilité dans le débat gêne beaucoup la personne A. La conception de ces systèmes reflète également cela. Ils ont l'air si abstrait et si reluisant, et pas à la place qui leur est destiné, qui est de tuer les gens aussi rapidement que possible.

Question 6. Avez-vous d'autres remarques pour cet entretien ?

*Les points suivants ont été évoqués au cours de l'entretien :*

La personne A a souligné que les différentes parties prenantes dans le débat devraient tous jouer leur rôle, et devraient être reconnus et estimés en tant que tel.

Un autre point sur lequel elle insiste est que nous devrions penser à ce que nous voulons que cette technologie fasse pour nous sur le champ de bataille, et y réfléchir ensemble. Un exemple est le «manuel» sur la question des drones que M. Obama a publié un peu avant sa réélection, s'étant probablement rendu compte de ce que son successeur pourrait faire de ces armes.

Cette campagne amène la personne A à répéter la question: «Pourquoi?» Non parce que les personnes prennent de meilleures décisions que les machines, mais ce qu'il faut préserver est qu'il faut que ce soit un être humain qui décide de la vie et de la mort. Les humains doivent être partie prenante dans la décision de mettre à mort, et ne pas programmer cette décision dans une machine. La personne A insiste sur le fait qu'il faut que l'être humain contrôle ce que fait une machine à l'intérieur de certains paramètres. Et le débat porte aussi sur le fait de savoir ce que sont ces paramètres, et si l'on tient compte de la différence entre les systèmes offensifs et défensifs. PAX ne dit pas quels types de systèmes sont acceptables, mais les états devraient en décider. On pourrait dire que c'est à l'intérieur d'un certain environnement structuré dans lequel nous pensons qu'il est acceptable d'utiliser un système défensif pour détruire un autre projectile. Dans l'étape suivante vers l'autonomie ces paramètres évoluent, et il faut examiner dans quelles limites ils évoluent jusqu'à l'acceptable. Une arme autonome évoluant dans une «boîte» (une zone) d'un kilomètre? Qu'en est-il de la durée? Doit-elle rester en vol 3 minutes? Pourquoi pas 10 minutes? Est-ce qu'une heure serait acceptable? Donc la question est de savoir comment l'humain peut influer sur ce qui peut être choisi. Ce débat est sur la table avec la Convention sur les Armes Conventionnelles, et le sujet n'est pas évoqué suffisamment actuellement.

J'ai évoqué la discussion que j'avais eue avec un conseiller juridique du Ministère de la Défense pour la détermination des zones de déploiement des armes autonomes, et la personne A a indiqué qu'elle ne croit pas en ce type de régulation avec ce type d'armes, car ces règles sont trop souvent transgressées, et elle trouve ce type d'arme trop effrayant ; il y aura prolifération trop rapidement. Ces « boîtes » peuvent être utilisées dans un processus diplomatique pour étudier des systèmes d'armes existants, et étudier la manière dont le contrôle humain est garanti, et en vertu de quels paramètres (spatiaux ou temporels). Et il faut voir si les systèmes actuels sont appropriés, et quel est le niveau de contrôle minimum.

Nom : Personne B
Date : 09-03-2017
Durée : 25 :26 min

### Question 1. Dans quelle mesure êtes-vous partie prenante dans le débat sur les armes autonomes ?

(Question non posée du fait du manque de temps pour l'entretien (30 minutes).

### Question 2. Quelle est votre définition d'une arme autonome ?

Il n'existe que des définitions approximatives pour les armes autonomes, et selon les diplomates de l'ONU ceci est délibéré, car définir quelque chose est la dernière chose à faire lorsque vous voulez en obtenir l'interdiction. Pour la personne B ceci concerne les problèmes d'identification, de ciblage et de destruction de l'adversaire au sol.

## Question 3. Quelle est votre position vis-à-vis des armes autonomes ?

Comme pour toutes les technologies considérées comme déstabilisantes et dangereuses il devrait y avoir interdiction pour les armes autonomes avant que ne se déclenche une course aux armements et que nous franchissions un palier dans la manière de faire la guerre. Il s'agit peut-être de la troisième révolution dans l'art de la guerre. Cela constituerait un changement profond en termes d'efficacité et de vitesse dans l'élimination de l'adversaire. Il y aura beaucoup de dommages collatéraux, et cela ne se passera pas comme les gens l'imaginent, une guerre qui serait plus froide et plus propre. Cela va potentiellement «industrialiser» la mise à mort encore plus que jamais dans l'histoire de la guerre. Ce sera terriblement déstabilisant, car dans le passé la capacité à faire la guerre dépendait de vos ressources financières (pour entretenir une grande armée et son matériel). Ces armes sont potentiellement plutôt bon marché, et on n'aura pas besoin de «main d'œuvre». Auparavant il fallait persuader des milliers de personnes de partir à la guerre, alors qu'avec les armes autonomes vous n'aurez besoin que d'un programmeur. Cela va déstabiliser l'ordre politique actuel.

Cela va probablement abaisser le seuil au-delà duquel il faut faire la guerre, et mettra plus de distance entre nous et les combats et l'acte physique de tuer. A un niveau plus moral et plus personnel, la guerre a toujours été un dernier recours, et elle doit être personnelle, sanglante et dangereuse. Quand nous prétendrons qu'elle ne l'est pas, nous combattrons probablement davantage, et cela devrait être quelque chose que les politiques doivent justifier lorsque des personnes reviennent dans des sacs mortuaires.

Mais cet argument ne s'applique pas toujours. Ces arguments sont applicables aux armes autonomes aujourd'hui, mais dans 50 ans, d'autres arguments seront peut-être plus pertinents, lorsque la technologie sera plus avancée. Pour le moment, la technologie ne peut pas être conforme aux lois humanitaires, et ne peut faire la distinction entre combattants et non combattants. Dans le futur, nous aurons des systèmes qui seront plus précis et plus en mesure de respecter les lois, mais alors d'autres arguments entrent en jeu, comme: plus les systèmes seront efficaces, plus la barrière entre la paix et la guerre sera basse.

Déployer des armes autonomes n'est pas une démarche éthique aujourd'hui, mais ce n'est pas blanc ou noir comme le problème des mines nous l'a montré. Cela amène à des lois et régulations lorsque l'on interdit l'usage, et une mine peut être considérée comme une arme autonome stupide qui ne fait pas de distinction entre les personnes.

**Question 4. Laquelle de ces trois photos d'armes autonomes voudriez-vous discuter* ?** **[Dans cette version du mémoire, les trois photos sont remplacées par des descriptions en raison d'éventuels problèmes de copyright]**

1. Essaim de six drones
2. Sea Hunter (véhicule de surface autonome sans équipage)
3. URAN-9 (véhicule terrestre de combat à chenilles sans pilote)

Question 5. Pourriez-vous expliquer pour chaque photo [décrits] pourquoi vous l'avez choisie, quelle valeur elle représente pour vous, et pourquoi c'est important* ?

**Photo 1: de nombreux drones**

Elle montre que nous parlons d'une technologie assez simple qui existe déjà aujourd'hui. Lorsqu'un essaim de drones s'approche de vous dans un contexte de guerre, vous n'aurez pas beaucoup de possibilités de défense. Il s'agira d'armes de terreur et elles seront utilisées pour terroriser des civils et des populations, et sont relativement bon marché. On peut faire beaucoup avec ces drones, qui ont l'air très simples.

**Photo 2 : Sea Hunter**

Cette photo a été choisie, car la discussion porte sur toutes les sphères de la guerre. Ainsi, à terre, sur mer ou dans les airs, partout où l'on peut imaginer qu'une guerre se déroule, des armes autonomes seront utilisées. Voici un cas intéressant qui montre à quel point la course aux armements s'accélère. Cette arme n'était pas connue lorsque la lettre ouverte a été publiée, mais le navire a été lancé depuis, et elle s'ajoute aux prototypes militaires d'armes autonomes. Il est intéressant de noter que lors de sa construction on a dit qu'il ne serait pas équipé d'armes, mais que son rôle serait de neutraliser les mines et de détecter les sous-marins. Mais il y a quelques mois on a admis qu'ils envisagent de le doter d'armement. Ceci montre bien que nous sommes sur une pente, et qu'il y a bien une course aux armements. Ici on n'a pas besoin de personnels ou de commodités pour faire fonctionner le navire, qui peut opérer plus longtemps. A présent on construit des navires mus par l'énergie solaire qui peuvent rester en mer pendant des années. Pas besoin d'escale pour nourrir l'équipage. On voit clairement les avantages militaires que peuvent présenter ces technologies, et les militaires seront séduits et voudront les utiliser.

**Photo 3 : Uran-9**

Le char autonome russe montre qu'il existe beaucoup d'acteurs, la Chine, la Russie, le Royaume-Uni, Israël aussi bien que les Etats-Unis dans ce qui est clairement une course aux armements. Les armées russes ont des millions de dollars de commandes dans leurs carnets et cela représente une autre sphère du champ de bataille. On voit bien que partout où il y a des combats la guerre met en œuvre des systèmes d'armes autonomes, ce qui présente des avantages militaires évidents. Si vous regardez ce char, vous voyez qu'il n'a pas l'air trop futuriste, et que ce n'est pas ce que les gens pensent, du genre des Robots Terminator. La personne B comprend que la Russie et la Chine développent des armes autonomes quand ils savent que les Etats-Unis dépensent 18 milliards de dollars de leur budget à produire la génération future d'armes sur ce qui est fondamentalement des armes autonomes. Le problème est qu'avec cette technologie, la seule manière de se défendre est d'utiliser les armes autonomes, car personne n'a assez de temps de réaction ou la capacité de combattre 24 heures sur 24 et 7 jours sur 7 pour se défendre, et donc là on n'a aucune défense. Par conséquent il n'est pas étonnant qu'une course aux armements ait commencé.

## Question 6. Avez-vous d'autres remarques pour cet entretien ?

*Quelques autres remarques formulées par la personne B au cours de l'entretien :*

Les politiques et les militaires manquent de perspicacité quand ils pensent que l'on peut contrôler toute technologie. Nous n'avons jamais réussi à le faire avec les bombes H et A. On ne peut empêcher le génie de sortir de sa boîte.

Il sera très difficile de vérifier comment cette technologie sera mise en œuvre de manière particulière, car ici nous parlons de systèmes très complexes qui interagissent dans un environnement chaotique et complexe. Il n'est pas étonnant que l'on parle de «brouillard de la guerre»; garantir que ceci ou cela se comportera d'une certaine manière, et que l'on va pouvoir inspecter et vérifier les systèmes, est au-delà de ce qui est techniquement possible, en fait ne sera jamais possible ; il pense donc qu'il est irresponsable de déployer cette technologie. Dans l'industrie, les robots peuvent être très utiles, car ils sont dans un environnement contrôlé et se débarrassent de la présence des humains, mais le dernier endroit où placer un robot est sur un champ de bataille chaotique, où l'on ne peut prévoir comment il va se comporter. On ne sait pas comment construire des «mentors» éthiques qui contrôleraient les robots, mais on doit résoudre ces problèmes pour que les voitures autonomes ne soient pas dangereuses sur nos routes. Il pense qu'un champ de bataille est le pire endroit pour résoudre ces problèmes, mais, ironie du sort, ce sera là que cette technologie fera son apparition.

La personne B aime aussi faire remarquer que les avantages provenant de l'intelligence artificielle sont énormes en termes militaires. Les forces armées américaines ont toujours été les plus grands investisseurs dans la recherche «intelligence artificielle», et c'est une bonne chose que des personnes n'ont pas besoin de risquer leur vie ou leurs membres dans une activité de déminage. C'est un travail qui convient parfaitement à un robot, car s'il fait une erreur et qu'une mine explose personne ne va pleurer. Même chose lorsqu'on achemine des approvisionnements à travers un territoire contesté à l'aide de convois autonomes. Il existe un grand nombre d'applications de l'intelligence artificielle dans le domaine militaire qui n'impliquent pas une arme autonome prenant des décisions ultimes engageant la vie ou la mort. La personne B n'a pas trop d'objections envers les systèmes défensifs comme le Phalanx car ils travaillent dans des contextes bien définis et sont conçus pour protéger des vies et non pour les supprimer, mais évidemment toute technologie peut être réorientée. Il serait difficile de dire sur le plan moral que ces systèmes, qui existent déjà, devraient être démontés sur les navires et ainsi rendre ceux-ci plus vulnérables. Le problème se pose au niveau de l'ampleur des opérations et de la zone dans laquelle ils évoluent. C'est un cadre tout à fait différent que celui d'un drone lancé pour opérer contre des cibles 7 jours sur 7 et 24 heures sur 24 pour des opérations offensives.

La personne B se préoccupe également d'autres acteurs que nos militaires en qui nous pouvons avoir confiance, comme les terroristes, les états voyous ou des gens pour qui ce n'est pas un problème de contourner des limites éthiques, ou les pirater de manière à faire du tort. Ainsi même si nous pouvons les construire de manière à ce que ces systèmes se comportent comme nous le voulons pour combattre et que nous espérons qu'ils ne soient pas piratés, le monde sera un endroit beaucoup moins sûr.

Nom : Personne C
Date : 10-03-2017
Durée : 25 :25

## Question 1. Dans quelle mesure êtes-vous partie prenante dans le débat sur les armes autonomes ?

(Question non posée du fait du manque de temps pour l'entretien (30 minutes).

## Question 2. Quelle définition donneriez-vous d'une arme autonome ?

Vous avez la définition du Dictionnaire du Département de la Défense US, c'est une arme qui peut choisir et engager une cible de par elle-même fondamentalement. Ensuite on peut aller dans deux directions; pour la première, Personne C a ajouté une phrase, disant que la cible n'est pas désignée au départ par un opérateur humain. La deuxième façon de voir, et la campagne «Arrêtez les robots tueurs» s'en est de plus en plus rapprochée, est de définir les armes autonomes de manière négative, en disant essentiellement que ce sont des armes sur lesquelles les humains n'ont aucun contrôle significatif. Cela non plus ne nous aide pas beaucoup, car à un moment ou à un autre il faut définir quelque chose. Que nous définissions ce que veut dire contrôle humain significatif ou ce que signifie choisir et attaquer (une cible) n'est pas très important, mais c'est vraiment un problème difficile, car on comprend ce que cela signifie aux extrêmes, mais où se situe la ligne, là c'est compliqué. C'est où se trouve cette ligne qui détermine si ces armes sont en cours de développement ou si elles n'ont pas été déployées.

## Question 3. Quelle est votre position vis-à-vis des armes autonomes ?

Son opinion personnelle sur les armes autonomes est qu'il est difficile de prendre des décisions sur ce que nous devons faire des armes autonomes si nous ne savons même pas ce qu'elles sont. Si vous pouvez me dire ce qu'est une arme autonome et quelles sont les personnes qui sont concernées et celles qui ne le sont pas, alors je pourrai vous dire ce que j'en pense. Mais cette technologie est si récente, et les définitions sont tellement vagues qu'il est difficile d'avoir une perspective du genre «oui» ou «non» sur une chose, si la catégorie à laquelle elle appartient est potentiellement si vaste; de plus on ne sait pas dans quelle direction évolue cette technologie.

**Question 4. Quelles sont les trois photos d'armes autonomes que vous voudriez commenter ?** [Dans cette version du mémoire, les trois photos sont remplacées par des descriptions en raison d'éventuels problèmes de copyright]

1. Véhicule terrestre à chenilles sans pilote
2. MQ-9 Reaper (Faucheuse) (Véhicule aérien sans pilote)
3. Photo rapprochée du robot de la campagne "Ban the Killer Robot" (interdire le robot tueur) devant le Big Ben à Londres.

**Question 5. Pouvez-vous expliquer pour chaque photo pourquoi vous l'avez choisie, quelle valeur elle représente pour vous, et pourquoi c'est important ?**

*1. Véhicule terrestre*

Celle-ci illustre le futur de la robotique militaire que les gens n'aiment vraiment pas voir ou discuter, et c'est la forme la plus courante de la robotique que l'on va rencontrer sur les champs de bataille. L'une des choses les plus intéressantes dans le débat sur les armes autonomes est que l'on assiste presque à une bataille d'analogies dans laquelle on a des groupes comme la campagne contre les robots tueurs qui qui sont axés sur des machines anthropomorphiques, comme Terminator qui entre dans une maison et essaie de déterminer si la personne qui se trouve à l'intérieur est un combattant ou un enfant. La Personne B et un co-auteur évoquent souvent des scénarios aériens ou navals où le «champ de bataille» est un peu plus clair, et ils présentent plus souvent des plateformes d'armes que des robots en marche.

La première photo illustre une forme essentiellement plus fonctionnelle où les robots sur le champ de bataille sont beaucoup plus probables que des Terminator qui se promènent partout. Ces systèmes sont utilisés aujourd'hui comme systèmes télécommandés.

*2. MQ—9 (Reaper)*

Le MQ-9 est intéressant car une grande partie des préoccupations vis-à-vis des armes autonomes ont pris naissance avec l'utilisation des drones; les gens s'inquiètent du fait que les drones rendent la guerre plus facile et que cela est dangereux. Mais on ne peut pas arrêter le développement des drones, car ils sont déjà sur le champ de bataille, et les opérations se poursuivent. Les premiers travaux ont montré que les drones prolif- éraient rapidement et que cela ne s'arrêterait pas, mais on peut peut-être agir sur ce qui va suivre. Et de nombreuses associations essaient d'assurer, de leur point de vue, que nous sommes parvenus à la limite; elles s'inquiètent du fait que les drones pourraient prendre des décisions par eux-mêmes.

Tout ceci touche aussi la question de la prise de décision, et quand on parle d'un robot qui prend une décision, en fait on parle d'un robot qui formule un jugement de par lui-même par opposition au robot qui adhère à sa programmation. … Souvent lorsque les gens parlent des robots de manière négative, ils parlent des robots qui prennent des décisions, alors que les gens qui ont moins de préoccupations parlent des hommes qui ordonnent à des robots d'accomplir des tâches, les hommes prenant encore les décisions.

Il est fascinant de constater qu'il y a des gens qui pensent qu'on a le droit d'être tué par un être humain, alors que les humains infligent toutes sortes de choses horribles à leurs semblables, alors que l'on sait que lorsqu'on est mort, on est mort. Il est intéressant de voir que l'on embraye aussi rapidement sur la philosophie morale, presque plus que sur tout autre chose.

### 3. *Robot de la campagne «robots tueurs»*

Il est intéressant de voir un groupe de personnes dont l'action a porté auparavant sur des campagnes contre les mines anti-personnel et les bombes à sous-munition et d'autres technologies qui ne sont généralement pas fondamentales dans les opérations militaires. Nombre de préoccupations vis-à-vis des armes autonomes portent sur le fait de savoir quelles armes autonomes nous avons et ce que ces armes peuvent faire, mais nous savons que l'intégration de l'autonomie dans les systèmes d'armes en général est en cours, même si la question de savoir si le système d'armes est autonome est presque une autre question ; mais l'intégration de l'autonomie dans les systèmes d'armes est inévitable. Il est intéressant de constater que la campagne a choisi ce thème, en particulier parce qu'il s'agit d'un type de technologie militaire différent de celles qu'ils ont visées auparavant. Pour la Personne C on s'attend plutôt à ce qu'ils demandent l'interdiction des chars ou des sous-marins. Si la campagne contre les robots tueurs est légitime, alors les militaires voudront avoir ces robots, mais si la campagne n'est pas justifiée et que finalement ces armes ne vont pas changer grand- chose, alors le problème n'en est pas un. C'est un groupe intéressant, ces gens veulent vraiment réduire la souffrance humaine, mais ils utilisent les mêmes références pour ce qui a constitué pour eux des victoires qui étaient en fait différentes, et la Personne C pense qu'ici la situation est différente car la technologie est incertaine et la catégorie potentielle d'armement est très large, alors que personne ne connaît l'importance potentielle de cette catégorie pour les militaires.

## Question 6. Avez-vous d'autres remarques pour cet entretien ?

A ma remarque qu'il n'y a pas eu jusqu'à présent beaucoup de recherche empirique, la Personne C a répondu qu'il effectuait des enquêtes par expériences d'«amorçage», ce que font en général les chercheurs qui s'intéressent à l'opinion publique, pour tester une liste de questions sur les attitudes: on présente aux gens différents contextes, et vous constatez ce que sont leurs opinions face à ces scénarios; ensuite on essaie d'évaluer ce qu'ils ressentent, mais ce travail est limité aux Etats-Unis.

Une autre chose intéressante: les stratégies sont très intéressantes, et dans une certaine mesure parce que nous vivons dans un monde dans lequel la supériorité technologique militaire américaine va de soi. Dans un monde imaginaire où la Chine déploierait des armes autonomes et où les USA et les pays européens n'en auraient pas: la semaine suivante, au Congrès américain, il y aurait des séances sur ce déséquilibre, et des questions du genre: «Pourquoi les Etats-Unis n'ont-ils pas les armes qu'a la Chine?». Ceci nous amène à la question que la Personne C évoque comme n'étant pas posée par la campagne: «Qu'arrive-t-il lorsqu'un pays non démocratique déploie ces systèmes, qui ne sont pas des armes marginales mais au contraire importantes pour les opérations militaires générales, et que ces armes vous donne une plus-value sur le champ de bataille?» «Que ferait le monde pour réagir à cela?» Ce ne sera pas nécessairement négatif, mais le problème est compliqué.

La question est de savoir s'il s'agit d'une course aux armements, dans laquelle vous êtes inquiet car quelqu'un est en train d'acquérir une technologie alors que vous êtes en concurrence directe avec lui, ou d'une prolifération rapide du fait de la facilité de l'acquisition. Une course aux armements relève de la politique car on a besoin d'une raison pour participer à cette course, et s'il y a course aux armements dans le domaine des armes autonomes, c'est parce qu'il est difficile de savoir ce qu'est une arme autonome. Quelle est la différence entre un MQ-9 Reaper et un Reaper autonome? C'est le software et non le hardware, et vous ne le voyez pas. L'incertitude – le système du voisin est-il un drone ou une arme autonome? – constituerait un facteur essentiel qui conduirait à une course aux armements, car on ne peut pas être assuré que de l'autre côté on ne produit pas d'armes autonomes. Sauf si vous êtes branché dans leurs systèmes, mais qui vous laisserait faire cela?

Nom : Personne D
Date : 19-04-2017
Durée : 55 :18 min

### Question 1 : Dans quelle mesure êtes-vous partie prenante dans le débat sur les armes autonomes ?

(Question non posée du fait que je ne l'ai pas posée aux autres personnes participant aux entretiens)

### Question 2 : Quelle est votre définition des armes autonomes ?

C'est un système d'armes qui utilise la force, et implicitement cette force est létale, mais pas nécessairement. Ce n'est pas impératif, mais c'est une perception courante dans un contexte militaire, bien que cette hypothèse puisse être évoquée dans la discussion sur les photos. C'est du moins ce à quoi on s'attend dans les applications dans futur proche. Pour moi, l'autonomie signifie qu'elles ont la capacité de choisir elles-mêmes leur cible. De choisir un objectif dans un certain contexte, avec une mission vague. L'arme n'est pas dirigée vers des coordonnées spécifiques ou un véhicule spécifique, mais c'est plutôt «va dans cette zone et repère tous les véhicules qui se comportent de manière hostile».

### Question 3 : Quelle est votre position vis-à-vis des armes autonomes ?

La Personne D aimerait s'y opposer, mais mon analyse montre que ce n'est pas possible; donc le mieux est de s'assurer que nous les utilisons de manière aussi responsable que possible, et donc de minimiser les risques de voir la situation échapper à tout contrôle. Je peux expliquer cette position à partir des idées suivantes:

- D'un point de vue civil, il y a une forte dynamique commerciale pour développer l'intelligence artificielle en général, car c'est un marché «juteux»;
- En même temps, dans le contexte militaire, la vitesse de prise de décision est importante. Auparavant le problème était de recueillir l'information, à présent c'est de traiter l'information. Ceci veut dire que la prise de décision humaine est en train de devenir le maillon le plus faible, ce qui conduit à penser que cette tâche sera confiée à des systèmes.

- Troisièmement il est possible qu'une intelligence supérieure à celle de l'homme soit créée. La Personne D ne voit pas de raison technique qui rendrait la chose impossible. Si c'est possible on ne peut imaginer tout ce que cela implique. Ceci, combiné à l'utilisation de la force létale, n'est pas une situation à souhaiter pour l'humanité.

A partir de ces réflexions la Personne D espère que nous ne développerons ni n'utiliserons les armes autonomes, mais cela est impossible car les deux premiers coups tirés nous mettront irrémédiablement dans cette situation. «Les plus belles fleurs sont celles qui poussent le plus près de la falaise». Ainsi, le mieux que nous puissions faire est de ne pas tomber de la falaise. Il est très pessimiste sur nos chances de pouvoir contrôler cela, mais nous devrons essayer.

Pour garder le contrôle, on pourrait utiliser un cadre, par exemple durant la phase de conception pour intégrer ces armes dans le monde que nous souhaitons avoir, et pour éviter que ces technologies créent leur propre vision des choses. Cela se situe aux premières phases du développement, mais au cours du déploiement on devra penser à la manière de concevoir le processus de ciblage. Plus ces systèmes seront capables d'agir par eux-mêmes, plus il faudra être précis sur les paramètres de ciblage, et cela pourrait être différent de ce que nous imaginons maintenant.

Nous devons essayer de parvenir à un accord international sur ce problème et c'est là le plus difficile, car ne pas respecter les règles donnera un avantage énorme. C'est le dilemme classique du prisonnier. Nous devons penser à tout ceci, ce n'est pas une question facile à résoudre, et la Personne D n'a pas de réponse à cela; mais il est évident que les parties prenantes commerciales doivent être également impliquées.

Sur la question demandant de clarifier son opposition à cette technologie: les armes autonomes sont différentes des armes conventionnelles non-autonomes en ce que ces dernières ne peuvent faire disparaître notre espèce entière, bien qu'en théorie ce soit possible avec les armes nucléaires, mais celles-ci sont trop coûteuses, alors que les armes autonomes ne le sont pas, et donc présentent un risque plus grand. Ainsi, le processus de fabrication des armes nucléaires peut être contrôlé, et il l'a été pendant environ 60 ans, ce qui nous permet d'y être habitués. Auront-nous cette possibilité avec les armes autonomes et leur rapide développement? Un autre aspect est que l'intelligence artificielle pour les armes comporte déjà un risque, contrairement à l'intelligence artificielle qui aide les médecins dans leur diagnostic, ce qui est destiné à faire le bien. Les armes autonomes combinent un système très intelligent et une intention hostile, ainsi la Personne D craint que les armes autonomes ne présentent un risque plus grand. Les armes autonomes ne sont pas la seule source de préoccupation pour la Personne D. La biologie synthétique est également un domaine à haut risque. C'est ainsi que les technologies qui ne rencontrent que des «barrières» peu élevées mais présentent des risques potentiels incontrôlables sont inquiétantes pour l'humanité.

**Question 4 : Quelles sont les trois photos concernant les armes autonomes que vous voudriez discuter ?** [Dans cette version du mémoire, les trois photos sont remplacées par des descriptions en raison d'éventuels problèmes de copyright]

1. Capture d'écran du film *Ex Machina*.
2. Image d'un robot de sauvetage portant un mannequin ressemblant à un humain dans le cadre d'un concours de robots pour aider les spectateurs.
3. Main du robot et main d'un humain dont les index se touchent.

**Question 5. Pouvez-vous expliquer pour chaque photo** [décrits] **pourquoi vous l'avez choisie, quelle valeur elle représente pour vous, et pourquoi c'est important ?**

Toutes ces photos sont liées.

*Photo 1.* Pour *Ex Machina*, il est bizarre de constater que la technologie qui est créée dans ce film hérite (ou assimile) également de l'instinct de survie de l'humanité, et par conséquent prend des mesures pour ne pas mourir, et aussi ne se soucie pas de ce qui arrive aux autres. Cela illustre le fait que vous devez considérer toutes les conséquences possibles lorsque vous vous lancez dans ce type de technologie, et, si vous ne faisiez pas cela, considérer ce qui pourrait mal tourner. Car nous parlons ici de systèmes qui possèdent certaines capacités à apprendre par eux-mêmes, et on pourrait ne pas voir les choses qu'ils apprennent par eux-mêmes. Il y a donc deux volets: le premier est la perte de contrôle accidentelle, et le deuxième est la recherche intentionnelle de limites, mais cela signifie aussi que vous pouvez franchir la ligne.

*Photo 2.* La deuxième photo est celle d'un concours de robots secouristes sur fond de catastrophe. Le robot ne fait pas de mal, mais il ne fait rien. Alors le dilemme est que vous voulez une combinaison de 1 et 2, mais dans ce cas il faudra vraiment se rendre compte du contexte, et une sorte de «bonne volonté». La Personne D indique qu'il s'attend à ce que l'intelligence artificielle prenne de meilleures décisions que l'être humain. Il donne l'exemple d'un robot à qui on demande de prendre par écrit le plus grand nombre de notes possible, et à un moment donné il évolue pour devenir un super robot et en quelques secondes décide que la meilleure manière d'effectuer sa tâche est de transformer la terre entière en notes, tuer tous les humains sur terre parce qu'ils le gênent dans sa production de la plus grande quantité de notes possible. Mission accomplie, mais pas de la manière souhaitée. Il faut donc penser à cela dans la phase de conception.

*Photo 3*. Mais il faut également penser à la coopération entre l'homme et l'intelligence artificielle (IA); mais cela ne se produira pas dans les premières phases de sa mise en œuvre. Dans l'avenir, nous aurons des systèmes d'IA qui le feront. Les systèmes IA actuels comme le Harpy ont peu d'intelligence du contexte, mais à mesure que nous déploierons des robots dans les opérations terrestres, nous devrons coopérer avec eux, et la détermination des objectifs deviendra un point extrêmement important. La photo 3 représente cette coopération. Ceci implique que le système doit connaître nos intentions, mais nous devons aussi former nos personnels à travailler avec des systèmes de ce type, et à communiquer les objectifs à ces systèmes. Plus ces systèmes seront dotés de possibilités infinies, plus on devra penser à fixer des limites à leurs missions. La Personne D donne l'exemple: une formule magique c'est très bien, mais il faut aussi la comprendre. Ainsi, nous devons former nos personnels à donner des tâches, mais avec des limites.

## Question 6. Avez-vous d'autres remarques pour cet entretien ?

A ma question à la Personne D lui demandant ce qu'il pense du fait qu'il doit toujours y avoir un être humain «dans le coup», il répond que cela dépend des possibilités (limitées ou illimitées) du système. D'abord, plus il aura de liberté et plus vous devrez penser, dans la phase de conception, à permettre au système de développer une «bonne volonté». En second lieu, si vous ne pouvez pas limiter ces systèmes, vous devez essayer de trouver une manière de limiter le risque.

La Personne D envisage le développement en phases. La phase 1 est une introduction à grande échelle des armes autonomes ayant des capacités limitées. Pour la phase 2 on aura des équipes homme-machine dans lesquelles les personnels apprendront à coopérer avec la technologie. Mais pour la phase 3 ce sera dangereux et risqué, car nous aurons des systèmes complètement indépendants, et dans la phase 4, ce sera l'exemple de Terminator ou Skynet, qui sont des exemples dans lesquels l'IA a vraiment mal tourné. Mais la Personne D pense que tout cela n'est pas évident et ne sera pas forcément l'IA qui tourne mal, mais ce sera le cas si nous bâtissons l'IA avec le bon état d'esprit et pourtant détruisons le monde. Il pense également qu'il n'y a pas de limite bien définie entre les phases 3 et 4, mais qu'on ne pourrait s'en rendre compte que lorsqu'il sera trop tard.

Nom : Personne E
Date : 17-04-2017
Durée : 22 :19

**Question 1. Dans quelle mesure êtes-vous partie prenante dans le débat sur les armes autonomes ?**

(Question non posée du fait du manque de temps pour l'entretien (30 minutes).

**Question 2. Quelle définition donneriez-vous d'une arme autonome ?**

«Autonome» pour moi signifie que lorsque je donne l'ordre d'agir, elle trouve son chemin et choisit sa cible par elle-même (à partir des paramètres que l'homme lui a donnés), se basant sur les images de l'environnement qu'elle se crée elle-même. Donc, sélectionner, choisir et agir par elle-même constituent les trois composantes.

**Question 3 : Quelle est votre position vis-à-vis des armes autonomes ?**

Positive, je suis ouvert à ce type de technologie. Positive, mais nous devons penser à la manière de les utiliser avec circonspection, non pas les «balancer» dans l'organisation.

**Question 4 : Quelles sont les trois photos [décrits] d'armes autonomes que vous voudriez discuter ?** [Dans cette version du mémoire, les trois photos sont remplacées par des descriptions en raison d'éventuels problèmes de copyright]

1. Missile de croisière Harpoon tiré depuis un navire de la Marine.
2. Un essaim de 19 drones.
3. Robot de style futuriste brandissant une grosse arme à feu.

### Question 5. Pouvez-vous expliquer pour chaque photo pourquoi vous l'avez choisie, quelle valeur elle représente pour vous, et pourquoi c'est important* ?

*Photo 1.* C'est l'un des systèmes existant actuellement qui est utilisé par la Marine de Guerre depuis plus de 30 ans. C'est une torpille lancée d'un bâtiment contre les sous-marins. Dès que l'on lance cette arme elle choisit sa cible automatiquement et l'attaque. Elle n'a aucune fonction d'annulation. C'est «tire et oublie». Un système similaire est le Harpoon qui est un missile de croisière qui est utilisé contre les navires de surface. Ce type d'arme cherche sa cible et lorsqu'elle l'a trouvée, elle peut être différente de celle que vous avez ciblée. C'est un système qui est déployé avec beaucoup de prudence, mais c'est un type d'arme autonome. Du point de vue de la Navy ce système de contrôle de missile guidé peut être réglé sur un certain ensemble de paramètres, comme l'altitude, la vitesse et certaines configurations d'identification, et le système est en cours d'utilisation. Il n'y a pas d'humain dans la boucle, et nous disposons de ces systèmes depuis plus de vingt ans. Ces systèmes peuvent être utilisés sur une base autonome, mais lorsque nous les déployons nous intégrons toujours la fonction «homme dans la boucle». Un humain doit donner le feu vert avant qu'ils ne puissent engager la cible. Il existe un processus autour de cette fonction d'autonomie, et il existe même un commutateur matérialisé qui doit être actionné avant que le système ne puisse engager la cible. Une sorte de commutateur d'annulation, mais on peut régler le système de manière à pouvoir s'en passer. Ces systèmes sont développés pour être employés dans des situations où la menace aérienne est élevée, dans lesquelles chaque seconde compte. On a là déjà un certain degré d'autonomie, avec le sentiment que l'on peut intervenir à tout moment. Le fait qu'il soit possible d'intervenir à tout moment est très important pour la Navy. Ceci est tout le contraire de la torpille qui ne vous permet pas d'intervenir, ce qui veut dire que l'on doit être absolument certain avant de l'utiliser. La Navy a développé des procédures pour le déploiement de ces deux armes.

*Photo 2.* Prochaine étape pour le Ministère de la Défense : les drones télépilotés que nous essayons d'intégrer dans notre organisation, mais les développements dans les technologies vont plus vite que nous en ce moment, et nous essayons de suivre le rythme. L'essaim de drones est le système que l'on utilisera dans un avenir proche. Ces systèmes seront déployés et exécuteront leur mission par eux-mêmes, mais la manière dont ils le feront nous échappe encore. Ils opéreront avec un bon niveau d'autonomie. Actuellement on s'en sert pour le recueil de renseignements et comme capteurs, mais l'étape suivante, c'est capter pour attaquer avec une arme. Cette technologie est actuellement disponible. Actuellement le Ministère travaille sur une «feuille de route» pour les systèmes aériens télépilotés, mais la barrière à franchir pour créer une pleine autonomie est trop haute. Au Ministère, cette barrière est actuellement d'ordre éthique. La question des systèmes autonomes sans pilote fait partie de la vision actuelle pour les Forces armées qui est mise sur pied par le Hoofd Directie Beleid (à l'EM de la Défense), et ils entament une discussion sur les armes autonomes. Le Ministère acquerra ce type d'armes après que le problème des armes autonomes aura été discuté au niveau de notre société.

Tous les spécialistes de l'Armée de Terre, de la Marine et de l'Armée de l'Air disent que l'autonomie est en marche, mais que nous voulons toujours contrôler la situation. Nous voulons être toujours capables d'arrêter les progrès lorsque la situation a changé, ou pouvoir modifier leur mission si nécessaire. Il est typique dans les opérations militaires que les situations changent rapidement, et que l'on doive adapter la planification. En particulier lorsque vous travaillez là où il n'y a pas encore de conflit armé et que vous opérer en devant respecter les Règles d'Engagement. Vous devez pouvoir anticiper et réagir rapidement pour ne pas vous retrouver dans une situation dans laquelle vous avez déployé votre système dans l'air pour attaquer une certaine cible, et où la situation tactique ou opérative change, auquel cas il faut rappeler ces armes. Si on tient compte de cette nécessité dans la conception des systèmes, alors nous autres au Ministère de la Défense aimerions utiliser ce type de système.

*Photo 3.* C'est l'image que se fait le public du robot tueur, qui est un système qui ne peut pas être contrôlé et qui pense et agit de sa propre volonté. Les gens se méfient de cela ; il représente la limite entre ce que nous pouvons déployer actuellement et la technologie que nous ne voulons pas adopter.

## Question 6. Avez-vous d'autres remarques à formuler ?

La Personne E a souligné qu'elle travaille depuis de nombreuses années sur les armes autonomes (comme le montre la photo 1), et qu'il existe un certain degré de contrôle qui y est intégré, tout comme il est intégré dans la gestion des procédures et opérations. En particulier lorsqu'une nouvelle technologie n'a pas fait ses preuves, il vous faudra y intégrer davantage de ces dispositifs de sécurité. La Personne E pense que nous allons dans cette direction et que nous devons le faire, car cela nous fera gagner un temps précieux; nos adversaires font la même chose, il faut donc aller aussi dans cette direction. Si un essaim de drones s'approche de vous, il vous faudra bien les contrer.

Nom : Personne F
Date : 08-05-2017
Durée : 53 :44 min

### Question 1. Dans quelle mesure êtes-vous partie prenante dans le débat sur les armes autonomes ?

(Question non posée du fait du manque de temps pour l'entretien (30 minutes).

### Question 2. Quelle définition donneriez-vous d'une arme autonome ?

La Personne F adhère à la définition de l'ICRAC (Comité International pour le Contrôle des Armes Robotisées), qui est l'autonomie dans les fonctions critiques du ciblage et de l'engagement des cibles par une arme. Il ne s'agit pas seulement de choisir une cible comme geste fonctionnel, mais aussi d'une action anthropologique et sociologique qui est communiquée en pointant une arme sur une personne. Formellement, c'est choisir une cible ou déterminer que quelque chose est une cible valable. L'étape suivante est de lancer l'arme et faire feu avec cette arme. S'il y a autonomie dans ces deux fonctions on peut vraiment parler d'une arme autonome; si c'est pour une seule c'est plus difficile de dire si c'est une arme autonome, et jusqu'à quel point vous avez besoin d'un contrôle humain significatif. Et qu'est-ce qu'un contrôle humain significatif ? Ceci fait encore l'objet d'un débat.

Fondamentalement, le contrôle par l'homme, dans une perspective d'ingénierie, est un contrôle effectif, si les humains sont capables d'intervenir ou d'arrêter l'arme par un ordre, ou changer le comportement d'un système. Un contrôle non significatif, c'est lorsque la personne devient un automate borné au service d'une machine, assis dans une pièce devant un bouton, et à chaque fois qu'apparait une lumière rouge, vous être censé appuyer sur le bouton; dans un sens l'homme contrôle le bouton dans un contexte de causalité, mais ne comprend pas vraiment quelle sorte d'information est utilisée pour déterminer qu'une chose constitue une cible valable, car ce n'est qu'une lumière rouge; vous pensez que cette lumière est connectée à quelque chose qui produit des cibles valables et qui autorise à attaquer ces cibles. Mais dans ce cas on n'a aucune intelligence ou compréhension, ni aucun moyen indépendant de vérifier la légalité ou la moralité de l'attaque de cette cible. Donc effectivement on n'a pas un contrôle significatif mais un contrôle de cause à effet.

Ensuite, pour que le fait de tuer constitue un acte significatif, il doit y avoir une intention ou un objectif militaire à choisir une cible. Et simplement parce qu'une chose constitue une cible militaire légitime on doit l'attaquer. Il se pourrait que vous ne l'attaquiez pas parce que c'est cher, ou parce que vous révéleriez votre position; il existe donc d'autres considérations, même si cet objet constitue une cible militaire légitime, vous allez réfléchir pour savoir si vous allez l'attaquer ou non: voilà ce qui en fait une activité humaine; lorsque vous automatisez tout cela, cela cesse d'être significatif et devient en fait une exécution arbitraire légale. Vous n'avez aucun moyen significatif de vérifier si une cible est légitime ou significative.

### Question 3. Quelle est votre position vis-à-vis des armes autonomes ?

La Personne F est contre, et pense que l'on devrait les interdire. Cependant on pourrait élaborer une interdiction comme exigence positive d'un contrôle humain significatif. C'est vraiment difficile d'être précis au-delà de la définition de l'ICRAC; il pense que celle-ci est plus vague d'une certaine manière que «contrôle humain significatif», ou on pourrait dire qu'il faut un contrôle humain significatif de ces deux fonctions critiques. Il pense que lorsque la notion de contrôle humain significatif sera comprise, cela deviendra une exigence pour toutes les armes plutôt que pour une classe particulière d'arme, mais il faudra une garantie que cela sera fait pour toutes les armes. L'analogie qu'il aime faire est celle de la souffrance inutile et des blessures superflues; ce ne sont pas les mêmes définitions, mais pour elles aussi on est dans le vague sur ce que sont la souffrance inutile et les blessures superflues. Il y a eu des débats substantiels lorsque ces expressions ont été introduites à la Convention de Saint Pétersbourg et les juristes les ont adoptées; maintenant on comprend «tort supplémentaire causé qui ne sert aucune fonction militaire sauf causer la souffrance et la douleur». Même s'il s'agit de modifications ouvertes à débat, elles montrent si oui ou non nous exerçons un contrôle significatif sur le système. Est-ce que je peux l'arrêter? Est-ce que je peux intervenir? Est-ce que je peux reconsidérer les choses? Ou vérifier par moi-même que les cibles qui ont été choisies sont correctes, légalement et moralement? Car si je ne peux pas, ce système d'armes pose un problème. Les systèmes sur lesquels on a un contrôle limité sont dangereux, et au niveau mondial il faut réfléchir sur ce que l'on veut en faire.

Les systèmes autonomes actuels, comme le Goalkeeper sur des navires, devraient rester dans le cadre de ces exigences. Les états qui déploient ce genre de systèmes devrait déterminer comment ils respectent ces exigences, que ce soit pour tout système balistique ou antibalistique comme le Goalkeeper, le Phalanx ou le Patriot; car beaucoup de ces systèmes neutralisent les projectiles qui arrivent, et celui que la Personne F a vu est très limité en termes de temps et de possibilités jusqu'à ce qu'ils choisissent leur cible, moins d'une minute (20 ou 30 secondes); mais on peut toujours dire qu'il y a des hommes qui ont la main et qui décident s'ils vont laisser le système détruire sa cible. La Personne F ne tient pas à ce que l'ICRAC élimine ces types de systèmes, mais ils sont dangereux tout de même car ils provoquent des incidents de tirs fratricides. La Personne F donne l'exemple de 2 aéronefs abattus par le système Patriot en Irak à la suite du dysfonctionnement de leur transpondeur; depuis le système Patriot a été revu au niveau de l'ingénierie; à présent l'opérateur doit donner l'autorisation formelle au système d'engager la cible et non de le laisser prendre cette décision par lui-même automatiquement. Ceci ajoute au poids psychologique qui fait que les personnels sont réticents à appuyer sur le bouton à moins qu'ils ne soient vraiment sûrs du fait de cette «opinion» de l'automate lorsque le système dit qu'il est sûr de son fait; alors les personnes consentent à laisser le système aller jusqu'au bout. Il est indiscutable qu'il y a plus de contrôle humain significatif dans la nouvelle interface Patriot qu'il y en avait dans l'ancienne interface Patriot.

**Question 4. Quelles sont les trois photos [décrits] d'armes autonomes que vous voudriez discuter ?** [Dans cette version du mémoire, les trois photos sont remplacées par des descriptions en raison d'éventuels problèmes de copyright]

1. Squelette métallique Terminator
2. Aéronef sans pilote Predator
3. Photo du robot de la campagne "Ban the Killer Robot" devant le Parlement britannique à Londres.
4. Chaise à trois pieds devant l'ONU à Genève, en Suisse.

**Question 5. Pourriez-vous expliquer pour chaque photo pourquoi vous l'avez choisie, quelle valeur elle représente pour vous et pourquoi c'est important ?**

*Photo 1. Terminator.* Dans les premières années de la campagne le squelette de métal Terminator accompagnait tous les articles concernant la campagne, ce qui est intéressant d'un point de vue des médias et de l'activisme social. D'une part, c'est une image forte qui stimule la prise de conscience du public, et cela saisit son attention car il représente les Robots Tueurs.

Terminator est très intéressant si vous vous souvenez du film, car on lui donne une cible spécifique, donc dans ce sens ce n'est pas une arme autonome, même s'il tue pas mal d'autres personnes en passant; il y a donc des décisions de ciblage autonomes, mais sa tâche première est Sarah Conner, qui est la cible réelle. Et il y a beaucoup d'autres éléments dans le film qui stimule les peurs du public, comme les robots qui deviennent conscients qu'ils existent, se retournent contre les humains et ce genre de choses; mais ce qui est plus intéressant est que cela fait penser aux armes autonomes intermédiaires et stupides qui pourraient être mises en service dans un futur entre 5 et 20 ans, qui n'auront pas de systèmes de raisonnement très sophistiqués; elles vont se baser sur un petit ensemble de critères provenant des données fournies par les capteurs, et se servir de leurs armes, ce qui peut avoir de funestes conséquences pour les civils et sur d'autres choses encore. L'image de Terminator ne fixe pas vraiment une valeur mais elle capte l'attention, et cela était important au début de la campagne. Ce n'était pas seulement la peur «dans» le film, mais une vraie peur vis à vis de la façon dont ces armes allaient opérer.

*Photo 2. Predator.* La deuxième chose à laquelle pensent les gens, ce sont les drones Predator et Reaper, qui sont des robots actionnés à distance, donc qui ne sont pas vraiment autonomes, et en ce sens ils ne donnent pas lieu à la même histoire. Ce sont plus de vraies images du monde qui en un sens montrent que ce sont des sortes de précurseurs des armes autonomes, dans la mesure où l'on pourrait automatiser le ciblage et l'emploi des armes des drones, et ceux-ci deviendraient des armes autonomes; il existe en réalité un développement de la prochaine génération de drones, et il y est souvent question de ciblage autonome par exemple pour le X47-B, le Tanaris et le Neuron, qui sont des drones propulsés par réacteur et au profil furtif, qui sont équipés de beaucoup d'armes, et qui sont conçus pour évoluer dans des espaces aériens contestés, à la différence du Predator et du Reaper qui ont des hélices et sont par conséquent peu rapides, et pour l'emploi desquels il faut contrôler l'espace aérien. Dans un espace aérien où l'ennemi dispose d'aéronefs ou peu brouiller les communications, ces systèmes d'armes ne sont pas utiles. La prochaine génération de drones peut opérer dans des espaces où les communications sont interdites, ce qui est un argument en faveur de l'autonomie, car dans ce cas on n'a pas besoin de liaisons qui peuvent être interrompues et si c'est le cas on perdra le contrôle de l'aéronef ; donc dans cet environnement il faut qu'il soit autonome. La question est de savoir comment il va parvenir à sa cible. Cela pourrait être comme pour les missiles de croisière auxquels on donne toute une liste de coordonnées GPS; ils trouvent ces coordonnées et les bombardent; dans ce cas ce sont les humains qui déterminent les cibles; ou alors ils vont être dotés d'une certaine capacité à choisir et sélectionner une cible par eux-mêmes parmi des objectifs inopinés, à partir des données fournies par les capteurs.

Les images de ces nouveaux systèmes ont un air sinistre, en particulier celle du Taranis, avec ces rangées de missiles sur le devant. Ceux-ci représentent ce à quoi il faut nous intéresser au sujet de la nouvelle génération des armes autonomes. Le souci serait qu'ils ont une capacité de sélection autonome des cibles. C'est un chasseur à longue portée qui choisit ses propres cibles. Ceci devient une question de légalité et de contrôle opérationnel. En fonction de la conception de ces systèmes expérimentaux, ils peuvent tomber dans la catégorie des armes autonomes ou non si les humains présélectionnent les cibles, et leur fonctionnalité n'est pas encore vraiment fixée car c'est couvert par le secret et nous ne savons pas comment ils fonctionnent.

Le problème de fond est la dignité humaine, ce qui veut dire que la décision de tuer doit être prise par un être humain et non par une machine. Ce problème ne peut être séparé du sens que l'on donne à la mort provoquée. Un autre humain détermine que vous êtes un combattant ennemi en fonction de votre dispositif militaire sur le champ de bataille, de l'uniforme et de la participation à un conflit armé, ce genre de critères. Si c'est seulement une machine qui obéit à un algorithme, et qui essaie de suivre quelques critères qui déterminent la légalité, il n'y aura aucun jugement humain, et pour la Personne F cela signifie qu'il n'y aura aucune garantie de dignité. L'arme peut être dans le droit ou non dans une perspective légale en utilisant une fonction de probabilité, mais au bout du compte il n'y aura aucune intention ni aucun sens dans ce système (d'arme); il ne peut déterminer qu'il y a un intérêt militaire. Et il ne peut pas avoir un jugement moral ou légal, et donc s'il tue des personnes il ne peut y avoir dignité par définition. Cela peut être un acte utilitaire ou pratique pour le droit international, mais dans une perspective morale il y a absence de dignité.

A ma question de savoir si dans certaines situations les Robots prennent de meilleures décisions que les humains, du fait qu'ils ne connaissent pas la fatigue, l'émotion ou ne cherchent pas à se venger, la Personne F répond que ceci est un point important, mais que nous devons concevoir des systèmes qui aident les humains à prendre de meilleures décisions et à réduire la possibilité d'erreur, mais il faut que les humains maîtrisent les éléments qui donnent du sens à l'action et qui assurent que le processus est moral, légal et porteur de sens ; les humains doivent utiliser la machine pour réduire le risque d'erreur et recevoir des suggestions documentées. Ceci aide les humains à prendre de meilleures décisions lorsqu'ils sont stressés et fatigués.

*Photo 3. Robot «sympa».* La troisième photo est celle de la campagne sur laquelle le robot se tient devant le Parlement britannique. C'est la photo d'un robot sympathique, la Personne F dit aimer en fait les robots, et dans les forces armées il y a des tâches qui peuvent être accomplies par des robots, mais il souligne que les tâches qui nécessitent un jugement moral et légal doivent être effectuées par des humains.

*Photo 4. Chaise trois pieds représentant la campagne contre les mines anti-personnel.* Une autre bonne image de la campagne est celle de la chaise à trois pieds en face du bâtiment des Nations-Unies à Genève qui symbolise la campagne contre les mines anti-personnel. Au début de la campagne pour l'interdiction des Robots Tueurs, nous avons essayé de faire la même chose que dans la campagne contre les mines, en disant qu'il s'agit d'armes qui frappent sans discrimination, avec des conséquences énormes sur les civils, ce qui est un problème. Quand on pense que les Robots Tueurs sont fondamentalement des mines anti-personnel qui peuvent se déplacer et choisir leur cible. ... C'est une autre manière de penser à quel point ils sont effrayants car ils ne sont pas beaucoup plus «intelligents» qu'une mine anti-personnel, et ils ne comprennent pas le monde plus que ne le fait une mine. Ils sont un peu plus sophistiqués seulement dans leur capacité de naviguer et de capter les données sur l'environnement et de sélectionner leur cible. Ainsi, d'une part ils ne seront pas aveugles au point de ne pas distinguer, et on peut même améliorer ces capacités de discrimination sur le long terme grâce à des techniques d'ingénierie, mais il reste ce bouleversement fondamental consistant à changer la nature de l'acte de tuer et ce qui justifie cet acte lorsqu'on aura ces robots tueurs automatisés. Mais ceci est vrai non seulement pour le contexte militaire et la guerre, mais aussi pour l'ensemble de la société, comme nous automatisons un grand nombre de tâches qui étaient auparavant sous contrôle humain. Bien sûr les humains font beaucoup d'erreurs dans leurs prises de décisions, et il y a la subjectivité et toutes sortes de préjugés et autres aspects auxquels il faut remédier, mais souvent nous nous lançons dans l'automation et prétendons que la subjectivité n'existe plus du fait que l'on a affaire à des système d'ingénierie et que ceux-ci ne sont pas censés faire d'erreurs, mais ces systèmes sont compliqués et nous ne savons pas comment ils se comportent ou ce qu'ils vont faire dans des situations inattendues. Dans ce cas on soulèverait tout un ensemble de nouveaux problèmes concernant l'ingénierie et ses problèmes alors que l'on n'aurait résolu aucune question sociale.

Pour faire une autre comparaison avec les mines anti-personnel, il existe une certaine notion de prévisibilité mais aussi apparemment d'imprévisibilité. Les mines anti personnel peuvent se justifier pour effectuer un coup d'arrêt au cours d'une guerre mais 10 années plus tard, lorsque quelqu'un veut cultiver la terre, les mines peuvent encore exploser, et là on ne peut contrôler ce qui se passera dans le futur. Ceci constitue le problème intrinsèque des armes autonomes dans la mesure où vous savez où elles sont censées se diriger, trouver leur cible et les détruire. Cela semble raisonnable, seulement si elles ne peuvent capter qu'un signal radar spécifique, comment peut-on être sûr de cela ? Combien de choses ressemblent à ce signal radar ? Il y a d'autres objets qui émettent des signaux et qui pourraient générer une confusion, ou des individus pourraient placer des émetteurs imitant ces signaux radar pour attirer le missile sur d'autres objets. Si on le maintient en opération pendant longtemps ou sur une vaste zone, on ne saura pas vraiment ce que l'arme autonome va faire. Vous savez ce qu'elle est censée faire, mais après vous ne pourrez pas vraiment la contrôler. Et c'est là que les choses deviennent dangereuses, car toute la question est celle du degré de risque, et on ne sait pas vraiment mesurer cela, et comme ces systèmes deviennent de plus en plus compliqués, cela devient de plus en plus difficile de le mesurer, à la fois pour l'opérateur qui essaie de prévoir ce qu'il faut faire, et pour le tribunal si on essaie de trouver la personne responsable d'avoir libéré quelque chose de terrible plutôt que d'avoir l'intention de faire quelque chose de terrible. Donc on ne peut pas vraiment accuser quiconque de crime de guerre, mais de négligence puisqu'on ne savait pas ce que l'arme allait faire du fait d'une circonstance survenue de manière inattendue ou dont nous n'avons pas eu connaissance. Ainsi, tout notre système d'explication de la responsabilité et d'imputabilité commence à s'écrouler, et cela crée un grand vide pour des individus qui veulent le mal et ne pas être tenus pour responsables ; cela place aussi un fardeau sur les épaules des conscrits ou de ceux qui s'engagent dans l'armée lorsqu'ils ont 18 ans et qui ne sauront pas ce qu'ils vont faire ou de quoi ils seront tenus pour responsables.

## Question 6. Avez-vous d'autres remarques pour cet entretien ?

J'ai posé à la Personne F la question suivante: pourriez-vous imposer une interdiction sur ce type de technologie? Réponse: Il y a des choses qu'il ne faut pas oublier concernant le droit international et les interdictions. L'imposition est l'un des éléments, mais il y en a d'autres. On peut comparer cela au meurtre. Les gens continueront à s'entretuer, mais alors on n'aurait pas dû proscrire le meurtre? On le proscrit car c'est une mauvaise action, et les gens continuent à la commettre, et il faut les arrêter. La question est d'établir une norme à l'échelle internationale, rassembler les pays du monde entier pour qu'ils tombent d'accord sur le fait que fondamentalement c'est mal de posséder un système d'armes autonome. A partir de là, quelles seraient les implications si des pays ne respectaient pas l'accord? La plupart des traités n'ont pas de méthode de vérification ni de sanctions explicites ou ce genre de choses, mais il faudra jeter l'opprobre sur ces pays, les sanctionner et leur faire subir les conséquences.

Pour les acteurs non-étatiques, c'est une autre planète, et déjà nous avons fait interdire les armes chimiques; c'est une bonne analogie car les pays, même ceux qui n'ont pas signé le traité, sont pointés du doigt et sanctionnés, et subissent les conséquences s'ils utilisent des armes chimiques. Cela affecte aussi l'industrie dans le sens que les grands entrepreneurs traitant avec l'armée ne développent pas activement des armes chimiques, et les grandes entreprises de produits chimiques ne produisent pas d'énormes quantités d'agents précurseurs qui sont interdits par la convention. On va assister à quelque chose de semblable pour les armes autonomes. Evidemment on peut construire une arme autonome à partir d'éléments que l'on peut se procurer dans un magasin de bricolage, mais ce ne sera pas un système militarisé qui peut détruire des villes et des usines. Et déjà des individus peuvent construire des pièges, des bombes et ce genre de choses; on ne peut pas arrêter cela, mais on essaie de réguler l'accès aux explosifs ou aux technologies clés. On n'empêche les armes autonomes de proliférer et de migrer des grands états aux petits, qui eux les utiliseront sans restrictions ou sans modifier leurs logiciels, et finalement de tomber entre les mains d'acteurs non-étatiques ou de terroristes qui auraient accès à des systèmes ayant de sérieuses capacités, et qui ne seraient pas seulement des jouets auxquels on aurait attaché une bombe. Donc, avec un traité international on peut remédier à cela. Mais rien n'empêche des forces de police ou des états de les utiliser contre leurs propres peuples ; donc il nous faut des lois et des normes nationales pour également empêcher l'emploi de ces systèmes par des forces de police contre des manifestants. Si des particuliers construisaient de tels systèmes, cela tomberait aussi sous le coup du droit civil et pénal; mais ce serait une bonne chose de préciser qu'il est interdit de fixer des armes à des systèmes de ciblage autonomes.

J'ai ensuite demandé à la personne F si cela impliquerait des critères de transparence. Réponse: si la vérification des systèmes militaires sophistiqués se fait sérieusement, prouver qu'il existe un contrôle humain significatif implique prouver qu'il existe une chaine de commandement et des bases de données; si quelqu'un croit que votre système en opération est pleinement autonome, on peut prouver grâce à une suite de données qu'en réalité la décision au niveau du ciblage est prise par un humain, ou que l'être humain a autorisé le choix de la cible dans ce cas précis, ceci pour tous les types de systèmes d'armes. Il est facile de disposer d'un système de traitement de l'information efficace, mais disposer de l'autorité pour y avoir droit de regard, c'est autre chose. La norme actuelle est que l'on impose cette discipline aux militaires; dans ce cas il n'est pas nécessaire de vérifier, mais il faut qu'il existe des procédures permettant de faire passer les officiers et les soldats en cour martiale s'ils violent les lois internationales. On ne s'attend pas à ce que les ennemis organisent un tribunal pour cela. On vérifie ses propres armes, ainsi ce n'est pas du tout transparent, et on ne publie pas ces vérifications. Mais l'exigence serait que vous conservez les données, et si un événement survient vous aurez les données disponibles.

J'ai ensuite demandé s'il a eu vent des dernières informations concernant la campagne, qui a été très suivie en 2015 et 2016, mais qui a l'air de s'essouffler. Réponse: le problème est que du fait de la durée entre les réunions de l'ONU et la publication des nouvelles (en décembre) les discussions sont passées des réunions informelles entre experts au Groupe Intergouvernemental d'Experts pour les Armes Autonome Létales (Group of Governmental Experts on Lethal Autonomous Weapon Systems: GGE (LAWS)), ce qui constitue la dernière phase avant la négociation du traité. Ils ont prévu deux semaines pour ces réunions, mais ils ont eu pas mal de problèmes de financement cette année car la budgétisation a changé à l'ONU qui demande que les sommes dues soient payées à l'avance pour que vous teniez ces réunions, alors que ces 50 dernières années c'était un système différent en ce sens qu'ils tenaient d'abord les réunions, puis ils attendaient que les gens paient ce qu'ils doivent plus tard. C'est ainsi qu'on se demande si la deuxième semaine ces réunions auront vraiment lieu si les pays ne paient pas ce qu'ils doivent intégralement. La réunion du mois d'août va continuer et nous espérons qu'il y en aura une autre en novembre, en marge de leur réunion annuelle. Ils ont bon espoir que quelque chose va sortir du GGE, mais une semaine ce n'est vraiment pas assez long; ils espéraient trois semaines de réunions et non pas 2 qui pourraient être réduites à une seule.

En général, la campagne a démarré «sur les chapeaux de roues» en comparaison de la plupart des campagnes sur le désarmement, y compris celles sur les mines anti-personnel et les bombes à sous-munitions pour lesquelles il a fallu 10 ans pour en arriver là où nous en sommes au bout de trois ou quatre ans. Maintenir ce rythme eût été impressionnant, et ce qu'il l'a freiné fut que les états étaient intéressés par la discussion, mais beaucoup moins pour formaliser leur position. Alors ils paralysent plus ou moins les choses, mais il est certain que les Nations-Unies formalisent leur position, en particulier avec la politique 3000/09 étudiée avec le Pentagone actuellement, qui a été initiée il y a cinq ans. Et l'ONU espère que les états discuteront davantage dans le cadre du GGE, et accorderont leur langage, car il est facile de dire qu'il n'y a pas de définition pour l'autonomie et les armes autonomes, et nous ne savons pas ce que signifie «contrôle humain significatif»; c'est à eux de décider qu'ils peuvent justifier tous ces termes comme ils le veulent. Mais pour un traité il faut trouver un consensus politique sur ces définitions. L'ICRAC fait du bon travail en restreignant la définition pour «arme autonome»; ensuite la question est de définir le «contrôle humain significatif», et quel langage adopter pour un traité qui ne serait pas trop permissif ni trop restrictif. Cet équilibre est difficile à trouver; c'est un problème politique plus qu'un problème technique. La Personne F espère que l'on va faire quelques progrès.

Une des raisons pour lesquelles le CCW (pour les armes conventionnelles) était si impatient était qu'ils n'ont pas fait grand-chose ces dix dernières années, et ce qu'ils avaient fait était insuffisant; donc ils essaient de se racheter, et c'est une bonne chose. Si nous y parvenons bientôt ce sera surtout un traité préventif, mais les armes existant actuellement doivent être révisées de par ces nouvelles exigences concernant le contrôle humain significatif; cela devrait être possible pour la plupart des systèmes, et ceux qui n'offrent pas cette possibilité ne devront pas être utilisés de toute manière que ce soit. Ce sera plus facile de le faire maintenant de façon préventive; dans dix ans les états auront développé des drones et des sous-marins pleinement autonomes, et jugeront que ceux-ci sont nécessaires à leur sécurité, et alors il sera impossible de faire adopter ce traité.

APPENDICE G

# Mémos d'encodage

## Cahier d'encodage chercheur 1

Dans l'*encodage des valeurs* le texte est encodé pour trouver les valeurs[1], les attitudes[2] et les convictions[3] (Saldana, 2015). Ces concepts sont listés avec des codes préétablis.

Liste des concepts préétablis :

| Concept | Couleur |
|---|---|
| Valeur | Bleu |
| Conviction | Vert |
| Attitude | Jaune |

Liste des codes émergents :

| Concept | Couleur |
|---|---|
| Définition | Violet |

*Recommandations :*

- Une liste préétablie peut comporter entre 10 et 40 ou 50 codes. Nous conseillons de ne pas créer trop de codes, car la personne qui encode peut se trouver débordée ou faire des erreurs dans le processus d'encodage s'il y en a trop.
- La règle de base pour l'encodage est de faire en sorte que les codes correspondent aux données, plutôt que de faire en sorte que les données correspondent aux codes.
- Créer des mémos au cours du processus d'encodage fait partie intégrante des approches d'encodage, qu'elles soient fondées ou faites à priori. La recherche

1 Valeur: l'importance que l'on attribue à soi-même, à une autre personne, objet ou idée.

2 Attitude: la manière dont nous pensons et nous sentons vis-à-vis de nous-mêmes, d'une autre personne, objet ou idée.

3 Conviction (opinion): fait partie d'un ensemble qui comprend nos valeurs et attitudes plus notre connaissance, expérience, opinions, préjugés personnels et autres perceptions interprétatives de l'environnement social.

 DOI:10.1163/9789004449084_014

qualitative est par essence réflexive; comme le chercheur approfondit son sujet, il est important qu'il enregistre son processus de réflexion dans des mémos de nature réflexive ou méthodologique ; ceci peut permettre de mettre en lumière ses propres interprétations subjectives des données. Il est crucial de commencer ces mémos dès le début de la recherche. Indépendamment du type de mémo produit, ce qui est important est que le processus initie une production et une réflexion critique dans la recherche. Ceci facilitera les analyses et les rendra plus cohérentes à mesure que le projet avance. Les mémos peuvent être utilisé pour planifier les activités de recherche, découvrir du sens dans les données, maintenir le rythme de progression de la recherche et l'assiduité, et rendre possible la communication.

Plus d'info:

http://onlineqda.hud.ac.uk/Intro_QDA/how_what_to_code.php

https://researchrundowns.files.wordpress.com/2009/07/rrqualcodinganalysis_7_19_09.pdf

http://programeval.ucdavis.edu/documents/Tips_Tools_18_2012.pdf

https://www.slideshare.net/kontorphilip/qualitative-analysis-coding-and-categorizing

*Notes:*

***Avant de commencer:***

Mon intention est d'utiliser la méthode d'encodage des valeurs (Saldana, 2015); mais Saldana (2015) donne une définition différente pour «valeur» de celle de Friedman et Kahn Jr (2003). Ainsi, je me demande si la différence entre valeur, attitude et conviction est vraiment claire, mais je verrai ce qu'il advient. Peut-être un nouveau nom de code pour une «définition» doit être ajouté.

***En cours d'encodage:***

a. Je remarque qu'une attitude implique souvent un verbe ; du moins c'est comme cela que je l'interprète.
b. Une conviction implique souvent une opinion.
c. Je remarque que je pense souvent aux listes des valeurs de mon étude de documents lorsque je souligne un terme, par exemple «réflexivité». Ceci pourrait indiquer une subjectivité de ma part en tant que chercheur; il serait bon de vérifier ma liste de valeurs par rapport à celle du deuxième réviseur.
d. Si une personne interviewée n'aime pas tel ou tel terme, comme «déterminisme», je l'ai mis en relief en tant que valeur.

e. Est-ce qu'une chose souvent évoquée est importante pour une personne interviewée? Et donc pourrait-elle être interprétée comme étant une valeur? Par exemple «distance entre l'homme et la machine». (Voir j.).
f. En cours d'encodage, je suis souvent revenue sur mes pas pour voir comment j'avais classifié la même expression (par exemple «trop effrayant») et recopié l'encodage.
g. J'ai ajouté un nouveau code pour «définitions», et l'ai ajouté au cahier d'encodage.
h. J'ai interprété une opinion de la personne interviewée comme «conviction», car cela fait partie de la définition du terme «croire» donnée par Saldana (2015).
i. Après l'encodage d'une question tirée d'un entretien, j'ai relu ma question et j'ai vérifié l'encodage.
j. Si un terme a été utilisé plus que deux fois, par exemple le terme «défensif» au cours de l'entretien avec le Professeur Walsh, je l'ai mis en relief en tant que valeur.
k. J'ai également mis en relief un terme qui était spécifiquement souligné, par exemple «gain / intérêt» («benefit») était marqué comme valeur dans l'entretien avec le Professeur Walsh.
l. Après avoir encodé un nouvel entretien, j'ai vérifié le(s) précédent(s) pour voir si j'étais cohérente.
m. Si quelque chose était marqué comme important, par exemple dans l'entretien avec Ad sur l'importance pour la Marine d'avoir la capacité d'intervenir n'importe quand, je l'ai encodé comme valeur.
n. Après avoir encodé les trois derniers entretiens, je n'ai pas revérifié pour les trois premiers. Du fait surtout des restrictions d'horaire, mais aussi parce que l'encodage devenait plus facile pour moi.

***Après l'encodage :***

J'ai envoyé tous les 6 entretiens au deuxième réviseur, et j'ai ajouté quelques informations générales sur l'encodage des entretiens. J'ai inclus le cahier d'encodage sans mes remarques. Nous avons discuté de ce travail via Skype pour s'assurer qu'elle comprenait.

## Cahier d'encodage chercheur 2

J'indique ici les orientations établies à partir des hypothèses, des principes ou des raisonnements qui ont été utilisés pour diriger les propos vers les catégories valeur, attitude ou conviction. Il faut noter que ces listes se sont développées et élargies au cours de ce processus d'encodage, et donc le processus est devenu itératif. En d'autres termes, dans les derniers entretiens, j'ai toujours réfléchi sur les orientations qui facilitaient le processus d'encodage dans les entretiens précédents, et j'ai progressé à partir de cela.

Valeur: lorsque des propos reflètent un fort degré d'importance, comme «nous devons», «il faut que», «important».

- Lorsqu'ils reflètent un jugement fort ou de haute portée, par exemple «assurer», «inquiétant», «besoin», «espèce humaine».
- Lorsqu'ils contiennent des mots d'emphase ou grandiloquents, par exemple «assurer», «souffrance humaine», «inévitable».
- Lorsqu'ils reflètent une chose considérée d'importance ou ambitieuse, «besoins (exigences)», «réflexivité».
- Lorsqu'ils constituent une partie significative d'une histoire ou d'une discussion, ou illustre quelque chose d'important, par exemple «exigence», «moralité».
- Lorsqu'ils illustrent une chose jugée digne d'intérêt, par exemple «justifier», «droit humanitaire».

Attitude: Lorsque les propos dénotent des tendances ou une direction, par exemple «non seulement», «trop».

- Lorsqu'il s'agit d'opinion ou de préférence personnelles, par exemple «vague», «pas possible», «vous voulez».
- Lorsqu'il s'agit d'une position relative, par exemple «n'a pas d'importance», «vraiment difficile», «vague».
- Lorsqu'il s'agit d'une chose qui pourrait affecter le jugement, par exemple «pas pour», «plus que», «nous voulons».
- Lorsqu'on emploie des mots illustrant des points de vue ou qui sont répétés, comme «contre», «vague», «terrible».
- Lorsque les mots reflètent une certaine prise de position, comme «fait délibérément», «devrait être», «terriblement», «séduit».

Conviction: lorsque les propos sont basés sur des hypothèses, ou concernent une chose considérée comme vraie

- Lorsqu'ils sont basés sur des attentes ou des hypothèses, comme «nous pourrions», «cela sera».
- Lorsqu'on discute de données inconnues ou des perspectives supposées d'autres personnes.
- Lorsque les propos reflètent des connaissances / expériences personnelles, comme «survient / arrive» ou «ne sais pas».
- Lorsqu'ils concernent des questions, spéculations ou des hypothèses, comme «ils n'ont pas», «sera».
- Lorsqu'ils concernent des perceptions interprétatives ou futures, comme «il n'y a que», «sera».

APPENDICE H

# Résultats processus d'encodage

| Entretiens | Valeurs | | |
|---|---|---|---|
| | *Chercheur* | *Second réviseur* | *Valeurs similaires* |
| Entretien 1 (Personne E) | Intervenir n'importe quand<br>Contrôler<br>Les rappeler<br>Ne peut être contrôlé<br>Niveau de contrôle | Nous devons réfléchir à la manière de les utiliser ainsi avec prudence.<br>Le fait qu'il doit être possible d'intervenir n'importe quand est très important pour la Marine.<br>Mais nous voulons toujours contrôler. Nous voulons toujours pouvoir les rappeler lorsque la situation a changé. Ou pouvoir redéfinir leur mission si nécessaire.<br>On doit pouvoir les rappeler.<br>On doit intégrer davantage de ces systèmes de sécurité.<br>Nous devons progresser. | Intervenir n'importe quand.<br>Contrôler.<br>Les rappeler. |
| *Entretien 2 (Personne D)* | Echappe au contrôle.<br>Contrôle.<br>Contrôle.<br>Risque.<br>Contrôlé.<br>Risque.<br>Faire bien.<br>Risque. | S'assurer que nous les utilisons de manière aussi responsable que possible.<br>Minimiser le risque de voir la situation échapper à tout contrôle.<br>Dans le contexte militaire la promptitude de la prise de décision est importante. | Minimiser le risque.<br>Contrôler.<br>Bonne volonté.<br>Fixer des limites. |

 | DOI:10.1163/9789004449084_015

| Entretiens | Valeurs | | |
|---|---|---|---|
| | *Chercheur* | *Second réviseur* | *Valeurs similaires* |
| | Risque incontrôlable.<br>Echappe au contrôle non intentionnellement.<br>Limites.<br>Bonne volonté<br>Fixer des limites.<br>Missions avec limitations.<br>Limites.<br>Sans limites.<br>Bonne volonté.<br>Risque.<br>Risqué. | Ce n'est pas une situation que nous souhaitons pour l'espèce humaine.<br>S'assurer que nous ne tombons pas de la falaise.<br>Empêcher que ces technologies créent leur propre vision.<br>Les technologies à barrières peu élevées, mais potentiellement risquées et incontrôlables sont inquiétantes pour l'humanité.<br>Considérer les conséquences possibles lorsque l'on se lance dans ce type de technologie, et ce qui pourrait mal tourner si tel n'était pas le cas.<br>Nous aurons besoin d'une intelligence du contexte développée ainsi qu'une sorte de « bonne volonté ».<br>Mission accomplie mais pas de la manière dont nous l'avions voulu.<br>Devrons coopérer et la détermination des objectifs deviendra extrêmement importante.<br>Les systèmes doivent connaître nos intentions. | |

| Entretiens | Valeurs | | |
|---|---|---|---|
| | *Chercheur* | *Second réviseur* | *Valeurs similaires* |
| | | Nous devons former nos personnels à travailler avec ces systèmes, et à communiquer les objectifs à ces systèmes.<br>Il faut former nos personnels à assigner des missions avec des limitations.<br>Bonne volonté.<br>Restreindre le risque.<br>Construire l'intelligence artificielle avec le bon état d'esprit mais détruire le monde quand même. | |
| *Entretien 3* (*Personne C*) | Décisions.<br>Décider<br>Forme fonctionnelle<br>Décisions<br>Décision<br>Décision<br>Décisions<br>Décision<br>On a le droit d'être tué par une personne<br>Souffrance humaine | Comment être sûr que c'est là la limite<br>On a le droit d'être tué [par une personne]<br>L'intégration de l'autonomie dans les armes est inévitable<br>Réduire la souffrance humaine<br>Ces stratégies sont très intéressantes<br>Vous donne l'avantage sur le champ de bataille ?<br>On ne peut pas être assuré que de l'autre côté on ne construit pas des armes autonomes. | Droit à être tué par une personne<br>Souffrance humaine |

| Entretiens | Valeurs | | |
|---|---|---|---|
| | *Chercheur* | *Second réviseur* | *Valeurs similaires* |
| *Entretien 4* (*Personne A*) | Cadre éthique.<br>On ne peut pas laisser le contrôle nous échapper.<br>Contrôle humain.<br>Distance.<br>Contrôle humain significatif.<br>Responsabilité humaine.<br>Contrôle humain significatif.<br>Distance.<br>Distance.<br>Distance.<br>Distance.<br>Absence de responsabilité [et] imprévisibilité.<br>On ne s'en soucie pas.<br>Réflexivité.<br>Distance.<br>Invincibilité.<br>Déterminisme.<br>Inévitabilité.<br>Besoin humain d'être impliqué dans la décision de supprimer des vies.<br>Besoin qu'ont les humains de contrôler une machine. | Une arme sans contrôle humain dans les fonctions critiques, comme le ciblage, doit être interdite.<br>Contrôle humain significatif.<br>Responsabilité humaine.<br>Le contrôle humain doit être appliqué au moment du ciblage.<br>Les systèmes d'armes actuels doivent être révisés concernant l'intégration du contrôle humain.<br>Délibérément conçu pour quelque chose.<br>Créer une distance ou réduire le tort causé.<br>Peur, absence de responsabilité et imprévisibilité.<br>Réflexivité.<br>Distance.<br>Invincibilité.<br>Déterminisme.<br>Inévitabilité.<br>Besoin humain d'être impliqué dans la décision de supprimer des vies.<br>Besoin qu'ont les humains de contrôler une machine à l'intérieur de certains paramètres.<br>Comment le contrôle humain est garanti. | Contrôle humain significatif.<br>Responsabilité humaine.<br>Prévisibilité.<br>Réflexivité.<br>Distance.<br>Invincibilité.<br>Déterminisme.<br>Inévitabilité.<br>Besoin humain d'être impliqué dans la décision de supprimer des vies. |

| Entretiens | Valeurs | | |
|---|---|---|---|
| | *Chercheur* | *Second réviseur* | *Valeurs similaires* |
| *Entretien 5*<br>(*Personne F*) | Contrôle humain significatif.<br>Contrôle humain significatif.<br>Contrôle significatif.<br>Intention.<br>Significatif.<br>Significatif.<br>Significatif.<br>Contrôle humain significatif.<br>Contrôle humain significatif.<br>Contrôle humain significatif.<br>Contrôle humain significatif.<br>Souffrance inutile.<br>Souffrance non nécessaire.<br>Blessures superflues.<br>Contrôle significatif.<br>Subjectivité automate.<br>Contrôle humain significatif.<br>Contrôle.<br>Dignité humaine. | Contrôle humain effectif.<br>Intelligence ou compréhension contextuelle ou situationnelle.<br>Vérification de la légalité ou moralité.<br>Intention ou objectif militaire dans le choix d'une cible.<br>Contrôle humain significatif<br>Souffrance non nécessaire et blessures superflues.<br>Valide, moralement et légalement.<br>Systèmes intelligents sophistiqués.<br>Perte de contrôle.<br>Légalité et contrôle opérationnel.<br>Dignité humaine.<br>Action de tuer légale et morale.<br>Tuer dans la dignité.<br>Besoin militaire.<br>Jugement moral ou légal.<br>Assurer un processus moral, légal et significatif.<br>Prévisibilité.<br>Contrôle.<br>Responsable. | Contrôle humain significatif.<br>Souffrance non nécessaire.<br>Blessures superflues.<br>Dignité humaine.<br>Suppression de la vie dans la dignité.<br>Prévisibilité.<br>Contrôle.<br>Imputabilité.<br>Responsabilité. |

| Entretiens | Valeurs | | |
|---|---|---|---|
| | *Chercheur* | *Second réviseur* | *Valeurs similaires* |
| | Degré du significatif. | Imputabilité et responsabilité. | |
| | Digne. | Lois et normes. | |
| | Dignité. | Chaîne de commandement. | |
| | Intention. | Autorité. | |
| | Sens. | Transparent. | |
| | Digne. | Conserver ces données. | |
| | Significatif. | | |
| | Contrôle humain. | | |
| | Subjectivité. | | |
| | Subjectivités. | | |
| | Prévisibilité. | | |
| | Non prévisibilité. | | |
| | Contrôle. | | |
| | Contrôle. | | |
| | Responsable. | | |
| | Imputabilité. | | |
| | Responsabilité. | | |
| | Imputable. | | |
| | Responsable. | | |
| | Contrôle humain significatif. | | |
| | Responsable. | | |
| | Contrôle humain significatif | | |
| | Contrôle humain significatif | | |
| | Contrôle humain significatif. | | |

| Entretiens | Valeurs | | |
|---|---|---|---|
| | *Chercheur* | *Second réviseur* | *Valeurs similaires* |
| *Entretien 6 Personne B*) | Défense.<br>Qui défend.<br>Défend.<br>Défend.<br>Irresponsable.<br>Avantages.<br>Sans défense.<br>Tort causé. | Justifier.<br>Droit humanitaire.<br>Efficacité.<br>Barrières.<br>Lois et régulations.<br>Défense contre ceci.<br>Avantages militaires.<br>En vous défendant.<br>Précision.<br>Vérifier.<br>Se comporter.<br>Se comporter d'une certaine façon.<br>Audit et vérification.<br>Environnement.<br>Comportement attendu.<br>Gouvernance éthique.<br>Enveloppes bien définies.<br>Sauvegardes éthiques.<br>Tort causé.<br>Se comporter comme on le voudrait.<br>Ne peut être piraté. | Défense.<br>Tort causé. |

Basée sur les entretiens et la comparaison de l'encodage des valeurs du chercheur et du deuxième réviseur, la liste suivante de 23 valeurs uniques peut être établie:

1. Intervenir à tout moment
2. Contrôler
3. Les rappeler
4. Minimiser le risque
5. Contrôler
6. Bonne volonté
7. Fixer les limites
8. Droit à être tué par une personne
9. Cadre éthique
10. Contrôle humain significatif
11. Prévisibilité
12. Réflexivité
13. Distance
14. Invincibilité
15. Déterminisme
16. Inévitabilité
17. Souffrance non nécessaire
18. Blessures superflues
19. Dignité humaine
20. Imputabilité
21. Responsabilité
22. Défense
23. Tort causé

*TRAD JFP*

# Bibliographie


AIV, & CAVV. (2016). *Autonomous weapon systems: the need for meaningful human control*. (No. 97, No. 26). Extrait de http://aiv-advice.nl/8gr.

Aldewereld, H., Dignum, V., & Tan, Y.-h. (2015). Design for Values Information and communication technologies in Software Development Software development. *Handbook of Ethics, Values, and Technological Design: Sources, Theory, Values and Application Domains*, 831–845.

Alfano, M., & Loeb, D. (2014). Experimental Moral Philosophy. Extrait de https://plato.stanford.edu/entries/experimental-moral/

Altmann, J., Asaro, P., Sharkey, N., & Sparrow, R. (2013). Armed military robots: editorial. *Ethics and Information Technology, 15*(2), 73.

Asaro, P. (2012). On banning autonomous weapon systems: human rights, automation, and the dehumanization of lethal decision-making. *International Review of the Red Cross, 94*(886), 687–709.

Asaro, P. (2016, 14-10-2017). Talk on Autonomous Weapon Systems. Extrait de https://livestream.com/nyu-tv/ethicsofAI/videos/138822041

Awad, E. (2017). *MORAL MACHINE: Perception of Moral Judgment Made by Machines*. (Master), Massachusetts Institute of Technology, Boston.

Bandura, A. (2001). Social cognitive theory: An agentic perspective. *Annual review of psychology, 52*(1), 1–26.

Beauchamp, T. L., & Walters, L. R. (1999). *Contemporary Issues in Bioethics*: Wadsworth Pub.

Bonnefon, J.-F., Shariff, A., & Rahwan, I. (2016). The social dilemma of autonomous vehicles. *Science, 352*(6293), 1573–1576.

Borning, A., & Muller, M. (2012). *Next steps for value sensitive design*. Etude présentée au cours des Travaux de la Conférence SIGCHI sur les facteurs humains dans les systèmes informatiques.

Bradley, B. (2006). Two concepts of intrinsic value. *Ethical Theory and Moral Practice, 9*(2), 111–130.

Bratman, M. E., Israel, D. J., & Pollack, M. E. (1988). Plans and resource-bounded practical reasoning. *Computational intelligence, 4*(3), 349–355.

Bryson, J. J., Kime, P. P., & Zürich, C. (2011). *Just an artifact: why machines are perceived as moral agents*. Etude présentée au cours des travaux de la Conférence Internationale sur l'Intelligence Artificielle.

Campaign to Stop Killer Robots. (2015). Campagne organisée contre les "robots tueurs". Extrait de https://www.stopkillerrobots.org/

Campaign to Stop Killer Robots. (2017). The Problem. Extrait de http://www.stopkiller-robots.org/the-problem/

Carpenter, J. (2016). *Culture and Human-Robot Interaction in Militarized Spaces: A War Story*: Taylor & Francis.

Chappelle, W., Goodman, T., Reardon, L., & Thompson, W. (2014). An analysis of post-traumatic stress symptoms in United States Air Force drone operators. *Journal of anxiety disorders, 28*(5), 480–487.

Cheng, A. S., & Fleischmann, K. R. (2010). Developing a meta-inventory of human values. *Proceedings of the American Society for Information Science and Technology, 47*(1), 1–10.

Coeckelbergh, M. (2013). Drones, information technology, and distance: mapping the moral epistemology of remote fighting. *Ethics and Information Technology, 15*(2), 87–98.

Cointe, N., Bonnet, G., & Boissier, O. (2016). *Ethical Judgment of Agents' Behaviors in Multi-Agent Systems.* Etude présentée au cours des travaux de la Conférence Internationale (2016) sur les systèmes d'agents et multi-agents autonomes.

Cummings, M. L. (2006). Integrating ethics in design through the value-sensitive design approach. *Science and Engineering Ethics, 12*(4), 701–715.

Cummings, M. L., Mastracchio, C., Thornburg, K. M., & Mkrtchyan, A. (2013). Boredom and distraction in multiple unmanned vehicle supervisory control. *Interacting with Computers, 25*(1), 34–47.

Cushman, F., & Young, L. (2011). Patterns of moral judgment derive from nonmoral psychological representations. *Cognitive Science, 35*(6), 1052–1075.

Davis, J., & Nathan, L. P. (2015). Value Sensitive Design: Applications, Adaptations, and Critiques. *Handbook of Ethics, Values, and Technological Design: Sources, Theory, Values and Application Domains*, 11–40.

Dietvorst, B. J. (2016). People Reject (Superior) Algorithms Because They Compare Them to Counter-Normative Reference Points.

Dignum, F., Kinny, D., & Sonenberg, L. (2002). From desires, obligations and norms to goals. *Cognitive Science Quarterly, 2*(3–4), 407–430.

Dignum, V. (2016). Introduction to AI. Extrait de https://rai2016.tbm.tudelft.nl/contents/

Docherty, B. (2012). *Losing humanity: The case against killer robots.*

Docherty, B. (2015). *Mind the gap: The lack of accountability for killer robots*: Human Rights Watch.

Etzioni, A., & Etzioni, O. (2017). Pros and Cons of Autonomous Weapons Systems. *Military Review*(May-June 2017).

Floridi, L., & Sanders, J. W. (2004). On the morality of artificial agents. *Minds and Machines, 14*(3), 349–379.

Friedman, B., & Kahn Jr, P. H. (2003). Human values, ethics, and design. *The human-computer interaction handbook*, 1177–1201.

Friedman, B., Kahn Jr, P. H., Borning, A., & Huldtgren, A. (2013). Value sensitive design and information systems. In *Early engagement and new technologies: Opening up the laboratory* (pp. 55–95): Springer.

Future of Life Institute. (2016). AI Open Letter. In.

Galliott, J. (2015). *Military robots: Mapping the moral landscape*: Ashgate Publishing, Ltd.

Assemblée Générale des Nations--Unies. (2016). (A/HRC/31/66). Rapport conjoint du rapporteur spécial sur les droits à la liberte d'assemblée pacifique et d'association et du rapporteur spécial sur les exécutions sommaires, arbitraires et sans jugement sur une bonne gestion des assemblées.

Graduation Portal. (2017). Graduation Portal. Extrait de https://teams.connect.tudelft.nl/sites/tbm/graduate/SitePages/Home.aspx

Graham, J., Haidt, J., Koleva, S., Motyl, M., Iyer, R., Wojcik, S. P., & Ditto, P. H. (2012). Moral foundations theory: The pragmatic validity of moral pluralism.

Gray, H. M., Gray, K., & Wegner, D. M. (2007). Dimensions of mind perception. *Science, 315*(5812), 619–619.

Gray, K., & Wegner, D. M. (2012). Feeling robots and human zombies: Mind perception and the uncanny valley. *Cognition, 125*(1), 125–130.

Haidt, J., & Joseph, C. (2004). Intuitive ethics: How innately prepared intuitions generate culturally variable virtues. *Daedalus, 133*(4), 55–66.

Harper, D. (2002). Talking about pictures: A case for photo elicitation. *Visual studies, 17*(1), 13–26.

Himma, K. E. (2009). Artificial agency, consciousness, and the criteria for moral agency: what properties must an artificial agent have to be a moral agent? *Ethics and Information Technology, 11*(1), 19–29.

Horowitz, M. C. (2016). The Ethics & Morality of Robotic Warfare: Assessing the Debate over Autonomous Weapons. *Daedalus, 145*(4), 25–36.

Hristova, E., & Grinberg, M. (2015). *Should Robots Kill? Moral Judgments for Actions of Artificial Cognitive Agents.* Etude présentée à l' EAPCogSci.

Hurka, T. (2005). Proportionality in the Morality of War. *Philosophy & Public Affairs, 33*(1), 34–66.

IA program. (2017). Information Architecture. Extrait de https://www.tudelft.nl/onderwijs/opleidingen/masters/cs/msc-computer-science/special-programmes/information-architecture/

ICRC. (2010, 29-10-2010). War and international humanitarian law. Extrait de https://www.icrc.org/eng/war-and-law/overview-war-and-law.htm

Johnson, A. M., & Axinn, S. (2013). The morality of autonomous robots. *Journal of Military Ethics, 12*(2), 129–141.

Kaag, J., & Kaufman, W. (2009). Military frameworks: Technological know-how and the legitimization of warfare. *Cambridge Review of International Affairs, 22*(4), 585–606.

Kasher, A. (2016). 6 The Threshold of Killing Drones. *Drones and Responsibility: Legal, Philosophical and Socio-Technical Perspectives on Remotely Controlled Weapons*, 119.

Kominsky, J. F., Phillips, J., Gerstenberg, T., Lagnado, D., & Knobe, J. (2015). Causal superseding. *Cognition, 137*, 196–209.

Kreps, S. (2014). Flying under the radar: A study of public attitudes towards unmanned aerial vehicles. *Research & Politics, 1*(1), 2053168014536533.

Kuptel, A., & Williams, A. (2014). Policy Guidance: Autonomy in Defence Systems.

Le Dantec, C. A., Poole, E. S., & Wyche, S. P. (2009). *Values as lived experience: evolving value sensitive design in support of value discovery.* Etude présentée au cours des travaux de la Conférence SIGCHI sur les facteurs humains dans les systèmes informatiques.

Lin, J., & Singer, P. W. (2014). China's New Military Robots Pack More Robots Inside (Starcraft-Style).

Logg, J. M. (2017). Theory of Machine: When Do People Rely on Algorithms?

Malle, B. F. (2015). Integrating robot ethics and machine morality: the study and design of moral competence in robots. *Ethics and Information Technology*, 1–14.

Malle, B. F., & Thapa Magar, S. (2017). *What Kind of Mind Do I Want in My Robot?: Developing a Measure of Desired Mental Capacities in Social Robots.* Etude présentée au cours des travaux de la Conférence Internationale ACM/IEEE sur les interactions Homme/Robots.

Maslow, A. H. (1943). A theory of human motivation. *Psychological review, 50*(4), 370.

Nass, C., Moon, Y., Fogg, B., Reeves, B., & Dryer, D. C. (1995). Can computer personalities be human personalities? *International Journal of Human-Computer Studies, 43*(2), 223–239.

Neapolitan, R. E., & Jiang, X. (2012). *Contemporary artificial intelligence*: CRC Press.

Noone, G. P., & Noone, D. C. (2015). The debate over autonomous weapons systems. *Case W. Res. J. Int'l L., 47*, 25.

NOS. (2016). Video: Vliegtuig redt piloot. Extrait de http://nos.nl/artikel/2132527-video-vliegtuig-redt-piloot.html

Oehlert, G. W. (2010). *A first course in design and analysis of experiments.*

Ohlin, J. D. (2016). The Combatant's Stance: Autonomous Weapons on the Battlefield. *92 International Law Studies, Cornell Legal Studies Research Paper No. 16–12*, 1–30.

Open Roboethics initiative. (2015, 5-11-2015). The Ethics and Governance of Lethal Autonomous Weapons Systems: An International Public Opinion Poll. Extrait de http://www.openroboethics.org/laws_survey_released/

Perugini, M., & Bagozzi, R. P. (2004). The distinction between desires and intentions. *European Journal of Social Psychology, 34*(1), 69–84.

Pommeranz, A., Detweiler, C., Wiggers, P., & Jonker, C. M. (2011). *Self-reflection on personal values to support value-sensitive design.* Etude présentée au cours des Travaux de la 25ème Conférence BCS sur l'interaction homme-ordinateur.

Reips, U.-D. (2002). Standards for Internet-based experimenting. *Experimental Psychology, 49*(4), 243.

Roff, H. M. (2016). Weapons autonomy is rocketing. Extrait de http://foreignpolicy.com/2016/09/28/weapons-autonomy-is-rocketing/

Rønnow-Rasmussen, T. (2002). Instrumental values–Strong and weak. *Ethical Theory and Moral Practice, 5*(1), 23–43.

Rosenberg, M., & Markoff, J. (2016). The Pentagon's 'Terminator Conundrum': Robots That Could Kill on Their Own. *The New York Times*. Extrait de http://www.nytimes.com/2016/10/26/us/pentagon-artificial-intelligence-terminator.html?_r=0

Royakkers, L., & Orbons, S. (2015). Design for Values in the Armed Forces: Nonlethal Weapons Weapons and Military Military Robots Robot. *Handbook of Ethics, Values, and Technological Design: Sources, Theory, Values and Application Domains*, 613–638.

RT. (2017, 5-07-2017). Kalashnikov develops fully automated neural network-based combat module Extrait de https://www.rt.com/news/395375-kalashnikov-automated-neural-network-gun/

Russell, S., Norvig, P., & Intelligence, A. (1995). A modern approach. *Artificial Intelligence. Prentice-Hall, Egnlewood Cliffs, 25.*

Saldaña, J. (2015). *The coding manual for qualitative researchers*: Sage.

Scalable Cooperation Group. (2016). Moral Machine. Extrait de http://moralmachine.mit.edu/

Schroeder, M. (2016). Value Theory. Extrait de https://plato.stanford.edu/entries/value-theory/

Schwartz, S. H. (1994). Are there universal aspects in the structure and contents of human values? *Journal of social issues, 50*(4), 19–45.

Schwartz, S. H. (2012). An overview of the Schwartz theory of basic values. *Online readings in Psychology and Culture, 2*(1), 11.

Searle, J. R. (1980). Minds, brains, and programs. *Behavioral and brain sciences, 3*(03), 417–424.

Searle, J. R. (1995). The Construction of Social Reality.

Sharkey, N., & Suchman, L. (2013). *Wishful mnemonics and autonomous killing machines.* Paper presented at the Proceedings of the AISB.

Shelley, C. (2013). Fairness and regulation of violence in technological design. *Moral, Ethical, and Social Dilemmas in the Age of Technology: Theories and Practice: Theories and Practice*, 182.

Stewart, P. (2016). U.S. military christens self-driving 'Sea Hunter' warship. Extrait de http://www.reuters.com/article/us-usa-military-robot-ship-idUSKCN0X42I4

Stewart, W. (2015). Russia has turned its T-90 tank into a robot – and plans to hire gamers to fight future wars. Extrait de http://www.dailymail.co.uk/news/article-3271094/Russia-turned-T-90-tank-robot-plans-hire-gamers-fight-future-wars.html

Strawser, B. J. (2010). Moral predators: The duty to employ uninhabited aerial vehicles. In *Handbook of Unmanned Aerial Vehicles* (pp. 2943–2964): Springer.

Sullins, J. P. (2006). When is a robot a moral agent. *Machine Ethics*, 151–160.

Thompson, W. T., Lopez, N., Hickey, P., DaLuz, C., Caldwell, J. L., & Tvaryanas, A. P. (2006). *Effects of shift work and sustained operations: Operator performance in remotely piloted aircraft (OP-REPAIR)*. Retrieved from

UNDIR. (2014). *Framing Discussions on the Weaponization of Increasingly Autonomous Technologies*. Extrait de http://www.unidir.org/files/publications/pdfs/framing-discussions-on-the-weaponization-of-increasingly-autonomous-technologies-en-606.pdf.

UNDIR. (2015). *The Weaponization of Increasingly Autonomous Technologies: Considering Ethics and Social Values*. Extrait de http://www.unidir.org/files/publications/pdfs/considering-ethics-and-social-values-en-624.pdf.

Van de Poel, I. (2013). Translating values into design requirements. In *Philosophy and engineering: Reflections on practice, principles and process* (pp. 253–266): Springer.

van Wynsberghe, A., & Robbins, S. (2014). Ethicist as Designer: a pragmatic approach to ethics in the lab. *Science and Engineering Ethics, 20*(4), 947–961.

Walsh, J. I. (2015). Precision weapons, civilian casualties, and support for the use of force. *Political Psychology, 36*(5), 507–523.

Walsh, J. I., & Schulzke, M. (2015). *The Ethics of Drone Strikes: Does Reducing the Cost of Conflict Encourage War?* Extrait de (?)

Waytz, A., Gray, K., Epley, N., & Wegner, D. M. (2010). Causes and consequences of mind perception. *Trends in cognitive sciences, 14*(8), 383–388.

Williams, A. P., Scharre, P. D., & Mayer, C. (2015). Developing Autonomous Systems in an Ethical Manner. In *Autonomous Systems: Issues for Defence Policymakers*: NATO Allied Command Transformation (Capability Engineering and Innovation).

www.ingramcontent.com/pod-product-compliance
Lightning Source LLC
LaVergne TN
LVHW021002300926
849160LV00007B/223
*9789004449077*